A COMPENDIUM OF GEOCHEMISTRY

A COMPENDIUM OF GEOCHEMISTRY

FROM SOLAR NEBULA TO THE HUMAN BRAIN

YUAN-HUI LI

PRINCETON UNIVERSITY PRESS

PRINCETON AND OXFORD

Published by Princeton University Press, 41 William Street,
Princeton, New Jersey 08540
In the United Kingdom: Princeton University Press, 3 Market Place,
Woodstock, Oxfordshire OX20 1SY

Library of Congress Cataloging-in-Publication Data

Li, Yuan-Hui, 1936-
A compendium of geochemistry : from solar nebula to the human brain/Yuan-Hui Li.
p. cm.
Includes bibliographical references and index.
ISBN 0-691-00938-4 (alk. paper)
1. Geochemistry. I. Title.

QE515 .L385 2000
551.9--dc21 99-089467

This book has been composed in Times Roman

The paper used in this publication meets the minimum requirements
of ANSI/NISO Z39.48-1992 (R1997) (*Permanence of Paper*)

http://pup.princeton.edu

Printed in the United States of America

10 9 8 7 6 5 4 3 2 1

TO THE INSPIRING TEACHERS

AND FRIENDS

Wallace S. Broecker and the late Werner Stumm

CONTENTS

PREFACE xi

CHAPTER I
Atoms, Nuclei, and Energy 3

Introduction 3
I-1. Periodic Table of the Elements and Electron Configurations of Atoms 3
I-2. Atomic Nuclei and Nuclear Binding Energies 13
I-3. Cohesive Energies among the Atoms of Pure Metals and Nonmetals 19
I-4. Ionization Energies and Electron Affinities of Gaseous Atoms and Ions 23
I-5. Ionic Radii and Ionic Potentials 28
I-6. Electric Polarizability 35
I-7. Electronegativity 39
I-8. Crystal Lattice Energies 44
I-9. Hydrolysis of Cations and Dissociation of Oxyacids 48
I-10. Solubility Products and Affinity of Aqueous Cations to Oxides 52
I-11. Concluding Remarks 53

CHAPTER II
The Solar Nebula and Nucleosynthesis 55

Introduction 55
II-1. Elemental and Isotopic Compositions of the Solar Nebula 55
II-2. Cosmological Nucleosynthesis 63
II-3. Stellar Nucleosynthesis 68
II-4. Concluding Remarks 82

CHAPTER III
Structure and Chemistry of the Solar System 83

Introduction 83
III-1. Motion of Interplanetary Objects 83
III-2. Asteroids and Comets 86
III-3. Planets, Their Satellites, and Their Rings 99
III-4. Condensation of Solid Dusts from the Solar Nebular Gas 107

CHAPTER IV
Distribution of Elements in Meteorites 119
Introduction 119
IV-1. Classification of Meteorites 119
IV-2. Oxygen Isotopes and Possible Genetic Relationships among Subclasses of Meteorites 129
IV-3. Bulk Compositions of Chondrites and Factor Analysis 133
IV-4. Cosmochemical Classification of Elements 144
IV-5. Matrices and Chondrules of Chondrites 154
IV-6. Ca-Al-Rich Inclusions in Chondrites 167
IV-7. Igneous Differentiation in Achondrites and Iron Meteorites 179
IV-8. Concluding Remarks 188

CHAPTER V
Igneous Rocks and the Composition of the Earth 189
Introduction 189
V-1. Classification Scheme for Igneous Rocks 189
V-2. Earth's Structure and Mineral Composition 193
V-3. Partial Melting and Fractional Crystallization Models of Igneous Rocks 197
V-4. Deduction of the Primitive Upper Mantle Composition 209
V-5. Partition of the Elements between Mantle and Core 215
V-6. Continental and Oceanic Crusts 221
V-7. Relationship between the Compositions of Mantle and Crust 231
V-8. Isotopic Heterogeneity of the Mantle 234
V-9. Case Studies of Elemental Association in Igneous Rocks 242
V-10. Concluding Remarks 249

CHAPTER VI
Weathering and Sedimentary Rocks 253
Introduction 253
VI-1. Weathering of Igneous Rocks 254
VI-2. Dissolved Products of Chemical Weathering 260
VI-3. Major Classes of Sedimentary Rocks 264
VI-4. Relative Abundances of Major Sedimentary Rock Types and Mass Balance 268
VI-5. Shales and Related Materials 273
VI-6. Trace Elements in Sandstone and Limestone 283
VI-7. Iron Formations 288
VI-8. Partition of Elements between River-Suspended Particles and River Water, and the Adsorption Model 296
VI-9. Concluding Remarks 302

CHAPTER VII
Distribution of Elements in the Ocean 303

Introduction 303
VII-1. Concentrations of Elements in the Oceans 303
VII-2. Chemical Speciation of Elements in the Ocean 312
VII-3. Marine Algae and Plankton 317
VII-4. Zooplankton Fecal Pellets and Sediment Trap Material 325
VII-5. Marine Sediments 329
VII-6. Marine Manganese Nodules and Seamount Manganese Crusts 341
VII-7. Marine Phosphorite 347
VII-8. Hydrothermal Vents of the Mid-Ocean Ridges 348
VII-9. Concluding Remarks 352

CHAPTER VIII
Biosphere and Homo Sapiens 355

Introduction 355
VIII-1. Are all Creatures Created Equal? 355
VIII-2. Human Body 366
VIII-3. Coals, Crude Oils, and Organic-Rich Shales 376
VIII-4. Relative Volatility of Elements and Compositions of Aerosol Particles 387
VIII-5. Effects of Fossil Fuel Burning on the Chemistry of Rain and River Waters: A Case Study 405
VIII-6. Concluding Remarks 411

APPENDIX
Rayleigh Condensation and Evaporation Models 413

APPENDIX TABLE A-1
Ionization Energies 415

APPENDIX TABLE A-2
Abundance of the Nuclides 418

APPENDIX TABLE A-3
Minerals in Meteorites 424

APPENDIX TABLE A-4
Minerals in Igneous Rocks 426

REFERENCES 429

INDEX 465

PREFACE

SINCE the publication of the classic works *Data of Geochemistry* (Clark, 1924; sixth edition, edited by Fleischer, 1963–1979), *Geochemistry* (Goldschmidt, 1954), *Geochemistry* (Rankama and Sahama, 1950), and *Handbook of Geochemistry* (edited by Wedepohl, 1969–1978), the chemical data for a wide variety of extraterrestrial and terrestrial materials have been increasing exponentially and are published in widely different scientific journals, books, and reports. This growth derives from the introduction of many faster and more accurate modern analytical techniques (Potts, 1993), and the ever-increasing concern over environmental degradation and pollution problems.

As illustrated by Goldschmidt (1954), the composition of natural materials does not change randomly, and their variations can be explained through cosmological, geological, and biological processes and by a common, unifying set of underlying physicochemical principles. A general understanding of these processes and principles is essential for us to better decipher the formative environment of any given sample from chemical data and to predict transport pathways and final sinks of anthropogenic pollutants in our environment.

This book summarizes many up-to-date compositional data and related references for various natural substances. The first objective is to serve as a handbook of geochemistry; the second is to illustrate the physicochemical principles and various natural processes that can explain the observed compositional changes of these natural substances. Chapter I summarizes atomic properties of elements that are critical for explaining and predicting geochemical behaviors of the elements. Chapters II through VIII illustrate consecutively (1) the synthesis processes of the elements and their isotopes immediately following the "Big Bang" event and inside numerous stars; (2) the processes of condensation and aggregation of dust particles from the gaseous solar nebula during the formation of our solar system; (3) complex formative processes of meteorites as well as of our solar system; (4) magmatic differentiation processes of the Earth into core, mantle, and crust; (5) weathering, transportation, and sedimentation processes for the formation of soils and major sedimentary rock types on the Earth's surface; (6) the role of biological uptake-release and inorganic adsorption-desorption processes in controlling elemental cycles in the oceans; and finally (7) causes for the similarity of composition among marine and terrestrial organisms including human beings, and effects of human activities on the chemistry of aerosols and surficial waters.

In the geochemical literature, much effort has been expended on obtaining the so-called average composition of many broadly defined geological and biological units (e.g., Vinogradov, 1953, 1956, 1959; Turekian and Wedepohl, 1961; Bowen, 1966, 1979; Mason, 1979; Taylor and McLennan, 1985; Condie, 1993; and Wedepohl, 1995). Whether these average estimates truly offer meaningful and representative values has never been fully resolved. In the process of averaging, important information on the variability of data, interrelationships among measured variables (i.e., the concentrations of various elements), and chemical kinship or uniqueness of individual samples is lost. Therefore, this book emphasizes the use of original raw data as much as possible, and applies the statistical technique of factor analysis to elucidate any underlying interrelationships among elements and among given samples. Whenever applicable, simple chemical thermodynamic models are also introduced to explain the observed partitioning of elements among different phases, e.g., metallic iron–iron sulfide–silicate phases in meteorites; molten magma and crystallized minerals; and dissolved and adsorbed species in aquatic environments. Readers unfamiliar with thermodynamics may skip these sections and accept the conclusions at face value.

Isotope geochemistry is an important and rapidly evolving branch of geochemistry, but is beyond the scope of this book. One may refer to *Principles of Isotope Geology* by Faure (1986) for an introductory exposure. However, some pertinent isotopic data are introduced to illustrate the classification and origin of meteorites (Chapter IV) and the heterogeneity of Earth's mantle (Chapter V). The chemistry and geochemistry of Earth's atmosphere are treated extensively by others (e.g., Holland, 1978, 1984; Ozime and Podosck, 1983; Warneck, 1988), and thus are not covered here. However, the effects of fossil fuel burning and volcanic activities on the chemistry of aerosols are discussed in Chapter VIII.

Acknowledgments are due to many colleagues: Fred Mackenzie and Rolf Arvidson patiently read through the whole manuscript; Dick Holland, Klaus Keil, John Lewis, Steven Smith, Mark Van Baalen, the late Keith Chave, and others read various chapters. Their comments and suggestions have improved the book greatly. Diane Sakamoto, Kathy Kozuma, and Diane Henderson have typed and edited several versions of the original manuscript. Many thanks are also due to my wife, Gertraude Roth Li. Without her constant encouragement and editorial help, this book would never have been published. Editorial assistance from Jack Repcheck, Kristin Gager, Linda Chang, Jennifer Slater, Karen Fortgang, and Molan Chun Goldstein at Princeton University Press was indispensable.

Finally, I dedicate this book to my teachers and friends, Wallace S. Broecker and the late Werner Stumm, who greatly influenced my scientific thinking during the formative years of my career. Their dedication to science has always been an inspiration to their students.

Honolulu, Hawaii

A COMPENDIUM OF GEOCHEMISTRY

Chapter I

ATOMS, NUCLEI, AND ENERGY

INTRODUCTION

IN PRINCIPLE, the distribution of the elements among coexisting phases in equilibrium can be predicted, if one knows the Gibbs free energies of formation and activity coefficients of all chemical species in all phases as functions of temperature, pressure, and compositions of phases. Because this is not always possible, many geochemists attempt to explain and predict the geochemical behavior of various elements by their atomic properties, such as electron configuration, ionization energies, ionic radii, ionic potentials, electric polarizability, crystal lattice energies, hydrolysis constants, solubility products, and electronegativity. This kind of approach has proved to be useful but has also caused confusion when the exact physical meaning of those parameters is ignored.

In order to provide a framework for the discussion in the following chapters, these atomic properties and their closely interrelated nature are reviewed in the following sections. For ease of explaining the observed relative abundance of nuclides in the solar nebula in the next chapter, some pertinent nuclear properties are also reviewed here.

I-1. PERIODIC TABLE OF THE ELEMENTS AND ELECTRON CONFIGURATIONS OF ATOMS

An atom consists of a nucleus containing protons and neutrons, with electrons orbiting around the nucleus. In an electrically neutral atom, the number of electrons is equal to the number of protons. The number of protons in the nucleus of an atom is called the **atomic number** (Z). There are 112 known chemical elements so far, as shown in the periodic table (Table I-1a). Elements with atomic numbers between 104 and 112 are artificial radionuclides with short half-lives. The names and symbols for these elements are still contested (Table I-1b). The periodic table consists of seven horizontal rows or periods. The vertical columns under the arabic numerals with A and B are called groups of chemical elements. Elements in the same group are called **congeners** and have similar physicochemical properties. Therefore, the periodic law states that the physicochemical properties of the elements change in periodic or repeated cycles as their atomic numbers increase. For example,

Table I-1a
Periodic table

Period	Group																	
	1	2	3	4	5	6	7	8	9	10	11	12	13	14	15	16	17	18
	1A	2A	3B	4B	5B	6B	7B	8B			1B	2B	3A	4A	5A	6A	7A	8A
	s-block		d-block										p-block					
1	1 H																	2 He
2	3 Li	4 Be											5 B	6 C	7 N	8 O	9 F	10 Ne
3	11 Na	12 Mg											13 Al	14 Si	15 P	16 S	17 Cl	18 Ar
4	19 K	20 Ca	21 Sc	22 Ti	23 V	24 Cr	25 Mn	26 Fe	27 Co	28 Ni	29 Cu	30 Zn	31 Ga	32 Ge	33 As	34 Se	35 Br	36 Kr
5	37 Rb	38 Sr	39 Y	40 Zr	41 Nb	42 Mo	43 Tc	44 Ru	45 Rh	46 Pd	47 Ag	48 Cd	49 In	50 Sn	51 Sb	52 Te	53 I	54 Xe
6	55 Cs	56 Ba	*57 La	72 Hf	73 Ta	74 W	75 Re	76 Os	77 Ir	78 Pt	79 Au	80 Hg	81 Tl	82 Pb	83 Bi	84 Po	85 At	86 Rn
7	87 Fr	88 Ra	#89 Ac	104 Rf	105 Db	106 Sg	107 Bh	108 Hs	109 Mt	110	111	112						
Lanthanides			*57 La	58 Ce	59 Pr	60 Nd	61 Pm	62 Sm	63 Eu	64 Gd	65 Tb	66 Dy	67 Ho	68 Er	69 Tm	70 Yb	71 Lu	
Actinides			#89 Ac	90 Th	91 Pa	92 U	93 Np	94 Pu	95 Am	96 Cm	97 Bk	98 Cf	99 Es	100 Fm	101 Md	102 No	103 Lr	
				f-block														

TABLE I-1B

Some atomic properties of the elements

Z	*Symbol*	*Name*	*Atomic weight*	*Electron configuration*	*Oxidation states*	*Density* (g/cm^3)	*Melting point (K)*	*Boiling point (K)*
1	H	Hydrogen	1.00794	$1s^1$	1	(0.0899)	14.0	20.28
2	He	Helium	4.00260	$1s^2$		(0.1785)	0.95	4.216
3	Li	Lithium	6.941	$1s^2 2s^1$	1	0.53	453.7	1615
4	Be	Beryllium	9.01218	$1s^2 2s^2$	2	1.85	1560	2757
5	B	Boron	10.81	$1s^2 2s^2 p^1$	3	2.34	2365	4275
6	C	Carbon	12.011	$1s^2 2s^2 p^2$	2, ±**4**	2.26	3825	5100
7	N	Nitrogen	14.0067	$1s^2 2s^2 p^3$	2, ±**3**, 4, 5	(1.251)	63.15	77.35
8	O	Oxygen	15.9994	$1s^2 2s^2 p^4$	−2	(1.429)	54.8	90.18
9	F	Flourine	18.9984	$1s^2 2s^2 p^5$	−1	(1.696)	53.5	85
10	Ne	Neon	20.1797	$1s^2 2s^2 p^6$		(0.90)	24.55	27.1
11	Na	Sodium	22.98977	$[Ne]3s^1$	1	0.97	371.0	1156
12	Mg	Magnesium	24.305	$[Ne]3s^2$	2	1.74	922	1380
13	Al	Aluminum	26.98154	$[Ne]3s^2 p^1$	3	2.70	933.5	2740
14	Si	Silicon	28.0855	$[Ne]3s^2 p^2$	4	2.33	1683	3522
15	P	Phosphorus	30.97376	$[Ne]3s^2 p^3$	±3, 4, **5**	1.82	317.3	553
16	S	Sulfur	32.066	$[Ne]3s^2 p^4$	±2, 4, **6**	2.70	392.2	717.8
17	Cl	Chlorine	35.4527	$[Ne]3s^2 p^5$	±**1**, 3, 5, 7	(3.214)	172.2	239.1
18	Ar	Argon	39.948	$[Ne]3s^2 p^6$		(1.784)	83.9	87.4
19	K	Potassium	39.0983	$[Ar]4s^1$	1	0.86	336.8	1033
20	Ca	Calcium	40.078	$[Ar]4s^2$	2	1.55	1112	1757
21	Sc	Scandium	44.9559	$[Ar]3d^1 4s^2$	3	2.99	1814	3109
22	Ti	Titanium	47.88	$[Ar]3d^2 4s^2$	3,**4**	4.54	1945	3560
23	V	Vanadium	50.9415	$[Ar]3d^3 4s^2$	2,3,4,**5**	6.11	2163	3650
24	Cr	Chromium	51.996	$[Ar]3d^5 4s^1$	2, **3**, 6	7.19	2130	2945

Table I-1b *Continued*

Z	*Symbol*	*Name*	*Atomic weight*	*Electron configuration*	*Oxidation states*	*Density* (g/cm^3)	*Melting point (K)*	*Boiling point (K)*
25	Mn	Manganese	54.9380	$[Ar]3d^5 4s^2$	**2**, 3, 4, 6, 7	7.44	1518	2335
26	Fe	Iron	55.847	$[Ar]3d^6 4s^2$	2, **3**	7.874	1808	3135
27	Co	Cobalt	58.9332	$[Ar]3d^7 4s^2$	**2**, 3	8.90	1768	3143
28	Ni	Nickel	58.6934	$[Ar]3d^8 4s^2$	**2**, 3	8.90	1726	3187
29	Cu	Copper	63.546	$[Ar]3d^{10} 4s^1$	1, **2**	8.96	1357	2840
30	Zn	Zinc	65.39	$[Ar]3d^{10} 4s^2$	2	7.13	692.7	1180
31	Ga	Gallium	69.723	$[Ar]3d^{10} 4s^2 p^1$	3	5.91	302.9	2478
32	Ge	Germanium	72.61	$[Ar]3d^{10} 4s^2 p^2$	4	5.32	1211	3107
33	As	Arsenic	74.9216	$[Ar]3d^{10} 4s^2 p^3$	±**3**, 5	5.78	1090	876
34	Se	Selenium	78.96	$[Ar]3d^{10} \mathbf{4}s^2 p^4$	−2, **4**, 6	4.79	494	958
35	Br	Bromine	74.904	$[Ar]3d^{10} 4s^2 p^5$	±**1**, 5	3.12	265.9	332
36	Kr	Krypton	83.80	$[Ar]3d^{10} 4s^2 p^6$		(3.75)	116	120
37	Rb	Rubidium	85.4678	$[Kr]5s^1$	1	1.532	312.6	961
38	Sr	Strontium	87.62	$[Kr]5s^2$	2	2.54	1042	1655
39	Y	Yttrium	88.9059	$[Kr]4d^1 5s^2$	3	4.47	1795	3611
40	Zr	Zirconium	91.224	$[Kr]4d^2 5s^2$	4	6.51	2128	4682
41	Nb	Niobium	92.9064	$[Kr]4d^4 5s^1$	3, **5**	8.57	2742	5015
42	Mo	Molybdenum	95.94	$[Kr]4d^5 5s^1$	2, 3, 4, 5, **6**	10.22	2896	4912
43	Tc*	Technetium	98	$[Kr]4d^5 5s^2$	7	11.5	2477	4538
44	Ru	Ruthenium	101.07	$[Kr]4d^7 5s^1$	2, **3**, **4**, 6, 8	12.37	2523	4425
45	Rh	Rhodium	102.9055	$[Kr]4d^8 5s^1$	2, **3**, 4	12.41	2236	3970
46	Pd	Palladium	106.42	$[Kr]4d^{10} 5s^0$	**2**, 4	12.0	1825	3240
47	Ag	Silver	107.868	$[Kr]4d^{10} 5s^1$	1	10.5	1235	2436
48	Cd	Cadmium	112.41	$[Kr]4d^{10} 5s^2$	2	8.65	594.2	1040

49	In	Indium	114.82	$[Kr]4d^{10}5s^2p^1$	3	7.31	429.7	2350
50	Sn	Tin	118.71	$[Kr]4d^{10}5s^2p^2$	2, **4**	7.31	505.1	2876
51	Sb	Antimony	121.757	$[Kr]4d^{10}5s^2p^3$	±**3**, 5	6.69	904	1860
52	Te	Tellurium	127.60	$[Kr]4d^{10}5s^2p^4$	−2, **4**, 6	6.24	722.7	1261
53	I	Iodine	126.9045	$[Kr]4d^{10}5s^2p^5$	±**1**, 5, 7	4.93	386.7	457.5
54	Xe	Xenon	131.29	$[Kr]4d^{10}5s^2p^6$		(5.90)	161.4	165.1
55	Cs	Cesium	132.9054	$[Xe]6s^1$	1	1.87	301.5	944
56	Ba	Barium	137.33	$[Xe]6s^2$	2	3.59	1002	2078
57	La	Lanthanum	138.9055	$[Xe]5d^16s^2$	3	6.15	1193	3737
58	Ce	Cerium	140.12	$[Xe]4f^15d^16s^2$	**3**, 4	6.77	1071	3715
59	Pr	Praseodymium	140.9077	$[Xe]4f^36s^2$	**3**, 4	6.77	1204	3785
60	Nd	Neodymium	144.24	$[Xe]4f^46s^2$	3	7.01	1294	3347
61	Pm*	Promethium	145	$[Xe]4f^56s^2$	3	7.22	1315	3273
62	Sm	Samarium	150.36	$[Xe]4f^66s^2$	2, **3**	7.52	1345	2064
63	Eu	Europium	151.965	$[Xe]4f^76s^2$	2, **3**	5.24	1090	1870
64	Gd	Gadolinium	157.25	$[Xe]4f^86s^2$	3	7.9	1585	3539
65	Tb	Terbium	158.9253	$[Xe]4f^96s^2$	**3**, 4	8.23	1630	3496
66	Dy	Dysprosium	162.50	$[Xe]4f^{10}6s^2$	3	8.55	1682	2835
67	Ho	Holmium	164.9303	$[Xe]4f^{11}6s^2$	3	8.80	1743	2968
68	Er	Erbium	167.26	$[Xe]4f^{12}6s^2$	3	9.07	1795	3136
69	Tm	Thulium	168.9342	$[Xe]4f^{13}6s^2$	2, **3**	9.32	1818	2220
70	Yb	Ytterbium	173.04	$[Xe]4f^{14}6s^2$	2, **3**	6.97	1097	1467
71	Lu	Lutetium	174.967	$[Xe]4f^{14}5d^16s^2$	3	9.84	1936	3668
72	Hf	Hafnium	178.49	$[Xe]4f^{14}5d^26s^2$	4	13.31	2504	4875

TABLE I-1B *Continued*

Z	Symbol	Name	Atomic weight	Electron configuration	Oxidation states	Density (g/cm^3)	Melting point (K)	Boiling point (K)
73	Ta	Tantalum	180.9479	$[Xe]4f^{14}5d^3 6s^2$	5	16.65	3293	5730
74	W	Tungsten	183.85	$[Xe]4f^{14}5d^4 6s^2$	2, 3, 4, 5, **6**	19.3	3680	5828
75	Re	Rhenium	186.207	$[Xe]4f^{14}5d^5 6s^2$	−1, 2, 4, 6, **7**	21.0	3455	5870
76	Os	Osmium	190.2	$[Xe]4f^{14}5d^6 6s^2$	2, 3, **4**, 6, 8	22.6	3300	5300
77	Ir	Iridium	192.22	$[Xe]4f^{14}5d^7 6s^2$	2, 3, **4**, 6	22.6	2720	4700
78	Pt	Platinum	195.08	$[Xe]4f^{14}5d^9 6s^1$	2, **4**	21.45	2042	4100
79	Au	Gold	196.9665	$[Xe]4f^{14}5d^{10} 6s^1$	1, **3**	18.88	1337	3130
80	Hg	Mercury	200.59	$[Xe]4f^{14}5d^{10} 6s^2$	1, **2**	13.55	234.3	630
81	Tl	Thallium	204.383	$[Xe]4f^{14}5d^{10} 6s^2 p^1$	**1**, 3	11.85	577	1746
82	Pb	Lead	207.2	$[Xe]4f^{14}5d^{10} 6s^2 p^2$	**2**, 4	11.35	600.6	2023
83	Bi	Bismuth	208.9804	$[Xe]4f^{14}5d^{10} 6s^2 p^3$	**3**, 5	9.75	544.5	1837
84	Po	Polonium	209	$[Xe]4f^{14}5d^{10} 6s^2 p^4$	2, **4**	9.3	527	1235
85	At	Astatine	210	$[Xe]4f^{14}5d^{10} 6s^2 p^5$	±**1**, 3, 5, 7		575	610
86	Rn	Radon	222	$[Xe]4f^{14}5d^{10} 6s^2 p^6$		(9.73)	202	211
87	Fr	Francium	223	$[Rn]7s^1$	1		300	950
88	Ra	Radium	226.0254	$[Rn]7s^2$	2	5	973	1900
89	Ac	Actinium	227.0278	$[Rn]6d^1 7s^2$	3	10.07	1324	3470
90	Th	Thorium	232.0381	$[Rn]6d^2 7s^2$	4	11.72	2028	5061
91	Pa	Protactinium	231.0359	$[Rn]5f^2 6d^1 7s^2$	4, **5**	15.4	1845	4300
92	U	Uranium	238.029	$[Rn]5f^3 6d^1 7s^2$	3, 4, 5, **6**	18.95	1408	4407
93	Np*	Neptunium	237.0482	$[Rn]5f^4 6d^1 7s^2$	3, 4, **5**, 6	20.2	912	4175
94	Pu*	Plutonium	224	$[Rn]5f^6 7s^2$	3, **4**, 5, 6	19.84	913	3505
95	Am*	Americium	243	$[Rn]5f^7 7s^2$	**3**, 4, 5, 6	13.7	1449	2880
96	Cm*	Curium	247	$[Rn]5f^7 6d^1 7s^2$	3	13.5	1620	

97	Bk*	Berkelium	247	$[Rn]5f^{9}7s^{2}$	**3**, 4		
98	Cf*	Californium	251	$[Rn]5f^{10}7s^{2}$	3		1170
99	Es*	Einsteinium	252	$[Rn]5f^{11}7s^{2}$			1130
100	Fm*	Fermium	257	$[Rn]5f^{12}7s^{2}$			1800
101	Md*	Mendelevium	258	$[Rn]5f^{13}7s^{2}$			1100
102	No*	Nobelium	259	$[Rn]5f^{14}7s^{2}$			1100
103	Lr*	Lawrencium	260	$[Rn]5f^{14}6d^{1}7s^{2}$			1900
104	Rf*	Rutherfordium	261	$[Rn]5f^{14}6d^{2}7s^{2}$			
105	Db*	Dubnium	262	$[Rn]5f^{14}6d^{3}7s^{2}$			
106	Sg*	Seaborgium	263	$[Rn]5f^{14}6d^{4}7s^{2}$			
107	Bh*	Bohrium	262	$[Rn]5f^{14}6d^{5}7s^{2}$			
108	Hs*	Hassium	265	$[Rn]5f^{14}6d^{6}7s^{2}$			
109	Mt*	Meitnerium	266	$[Rn]5f^{14}6d^{7}7s^{2}$			
110	Uun*	Ununnilium	269	$[Rn]5f^{14}6d^{9}7s^{1}$			
111	Uun*	Unununium	272	$[Rn]5f^{14}6d^{10}7s^{1}$			
112	Uub*	Ununbium	277	$[Rn]5f^{14}6d^{10}7s^{2}$			

Notes: Density is at 300 K and one atmosphere, except that those in parentheses, which are in the gaseous state, are in units of g/ℓ. The bold numbers are the most common oxidation states for a given element. Elements with asterisks are artificial radionuclides.

the molar atomic volume of the elements (= atomic weight/density) at 25°C and 1 atmosphere changes periodically as a function of the atomic number (Figure I-1). For some elements that are in the gaseous state at 25°C and 1 atmosphere pressure, the density in the liquid state at the melting point is used.

The similarity in physicochemical properties among certain elements is often a reflection of their similar **electron configurations**. According to quantum mechanics, the energy levels (or states) and most probable spatial distribution of electrons in the orbital cloud around an atomic nucleus can be described by four **quantum numbers**, i.e., n (principal), l (angular momentum), m_l (magnetic), and m_s (spin). For the physical meaning of the four quantum numbers, one should refer to any standard physical chemistry textbook. According to the Pauli exclusion principle, no two electrons in an atom can have the same four quantum numbers.

The **principal quantum number**, n, is a positive integer that ranges from 1 to 7 for the known atoms in the ground (i.e., lowest-energy) state. The principal energy levels or states with $n = 1$ to 7 are often denoted as the K, L, M, N, O, P, and Q shells, respectively. For a given n, the **angular momentum quantum numbers**, l, are positive integers from 0 to $n - 1$. For example, if $n = 4$ then $l = 0, 1, 2, 3$. The energy levels with $l = 0, 1, 2, 3, 4$ are designated as the s, p, d, f, and g subshells, respectively.

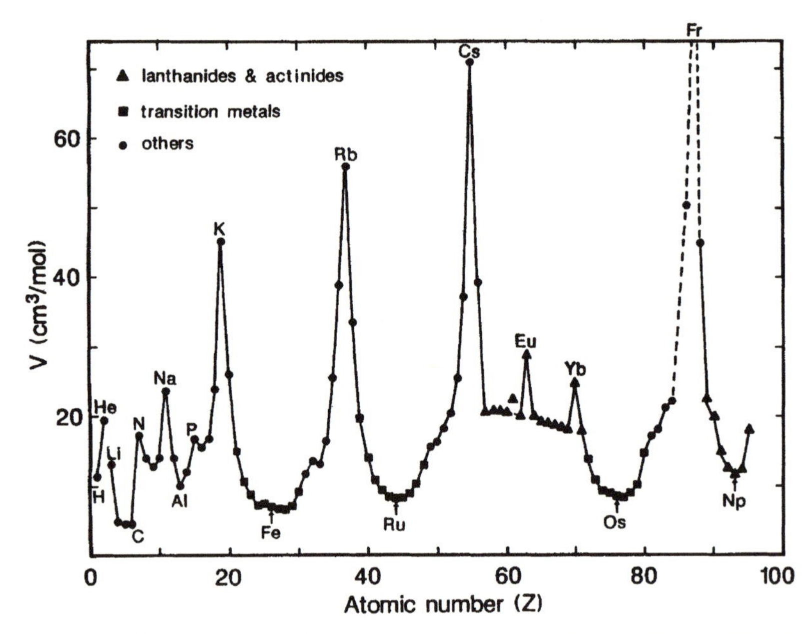

FIGURE I-1. Periodicity of molar volumes of elements at 25°C (except those for H, N, O, F, Cl, and noble gases are at or below their boiling points) and 1 atmosphere.

The combinations of n and l can define all the major subshells in atoms. For example, $nl = 1s$; $2s, 2p$; $3s, 3p, 3d$; $4s, 4p, 4d, 4f$; $5s, 5p, 5d, 5f, 5g$, etc. The relative energy levels of these subshells for neutral atoms in the ground state are schematically shown in Figure I-2 (from the lowest $1s$ to the highest $6d$; notice that the g subshell is unnecessary for accommodating electrons of atoms in the ground state). One should be aware that the relative energy levels of subshells are slightly different from those of Figure I-2 when the atoms are positively charged (i.e., cations). This point will be discussed later.

For a given l value, the **magnetic quantum number**, m_l, can have all integers between $-l$ and l including 0 (e.g., $l = 3$, $m_l = -3, -2, -1, 0, 1, 2, 3$). Thus, there are $(2l + 1)$ values of m_l for a given l. In addition, **the spin quantum number**, m_s, can have only two values (i.e., $+1/2$ and $-1/2$). Therefore, the combination of m_l and m_s produces $2(2l + 1)$ numbers of distinctive

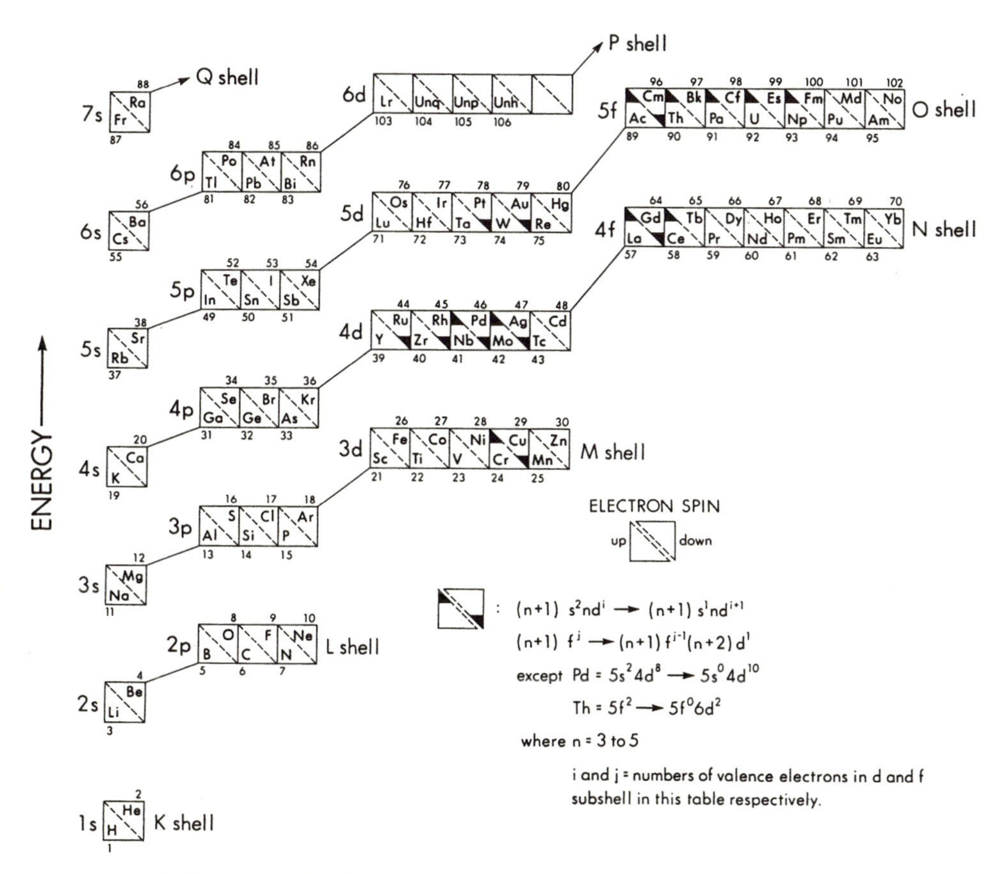

FIGURE I-2. Energy levels of subshells and the electron configurations of neutral elements. The exceptions to the given configuration are indicated by solid triangles.

quantum states of electrons with the same energy level, i.e., $2(2l+1)$-fold energy degeneracy. In other words, the s subshell ($l = 0$) consists of one orbital; the p subshell ($l = 1$) has 3 orbitals; the d subshell ($l = 2$) has 5 orbitals; and the f subshell ($l = 3$) has 7 orbitals with the same energy levels. Each orbital can accommodate one pair of electrons spinning in opposite directions, schematically shown in Figure I-2 as cubic boxes with up and down triangles. The electron configuration of any atom with atomic number Z is constructed by feeding electrons one by one into the lowest empty orbitals of subshells until a number of electrons equal to the atomic number Z are all accommodated. The orbitals in the p, d, and f subshells should each be first filled by a single electron with parallel spin before electrons with opposite spins can enter any of those orbitals. For example, the electron configuration of Mn ($Z = 25$) is $[Ar]3d^5 4s^2$, where the superscripts represent the numbers of electrons in subshells and [Ar] the electron configuration equivalent to the Ar atom. $3d^5$ represents 5 electrons in the $3d$ subshell, and these 5 electrons are spinning in the same direction (parallel spins), each in one orbital. There are, however, many exceptions in Figure I-2: the energy levels of $(n+1)s$ and $(n-1)f$ are usually slightly lower than those of nd for most of the elements in Figure I-2, but the reverse is the case for some atoms (indicated by solid triangles). In this case, one or two electrons from the $(n+1)s$ and $(n-1)f$ subshells should move into the nd subshell. For example, the electron configurations of Cu ($Z = 29$) and Ce ($Z = 58$) are $[Ar]3d^9 4s^2$ and $[Xe]4f^2 6s^2$ according to Figure I-2 but in fact should be $[Ar]3d^{10} 4s^1$ and $[Xe]4f^1 5d^1 6s^2$, respectively.

The outermost two or three subshells are usually called the subshells of valence electrons. By convention, the subshells of the valence electrons are listed in increasing order of the principal quantum number as shown in Table I-1b (not in order of the subshell energy levels for the neutral atoms as shown in Figure I-2). The advantage of this convention is that the sequence of subshells of valence electrons is listed exactly in the increasing order of energy levels for cations. Therefore, one can easily obtain the electron configuration for any cation, M^{+z}, by simply taking away z electrons from the outermost subshell(s) of the neutral atom as given in Table I-1b.

The heavy zigzag line in Table I-1a separates the elements into **metals** on the left and **nonmetals** on the right. The elements bordering the zigzag line, such as B, Si, Ge, As, Sb, Te, and Po, are called **metalloids**. The characteristic properties of all metals are high electric and thermal conductivity, high reflectivity (metallic luster), mechanical ductility, and malleability, and their power to replace hydrogen in acids. The elements can also be classified into four blocks according to which subshell is filled by the last valence electron. They are the *s*-**block** of the alkali (group 1A) and alkaline earth metals (group 2A), the *d*-**block** of the transition metals (group 1B to 8B), the f-**block** of the lanthanides and actinides, and the p-**block** of group 3A to 8A as shown in Table I-1a and Figure I-2.

I-2. Atomic Nuclei and Nuclear Binding Energies

The total number of protons Z and neutrons N in an atomic nucleus is called the **mass number** $A(= Z + N)$ of the nucleus. Nuclei with identical Z but different N are called **isotopes**, whereas nuclei with identical N but different Z are called **isotones**. Nuclei with identical A but variable N and Z are **isobars**. By convention, an atomic nucleus is represented by a chemical symbol with its mass number A as a left-hand side superscript and its atomic number Z as a left-hand side subscript, e.g., ${}^{16}_{8}O$ and ${}^{12}_{6}C$. Since the elemental symbol already implies the atomic number, the subscript of the atomic number is often omitted, e.g., ${}^{16}O$ and ${}^{12}C$.

There are about 268 stable nuclei in nature: 159 nuclei with even Z and even N; 55 with even Z and odd N; 49 with odd Z and even N; and only 5 with odd Z and odd N combinations (i.e., ${}^{2}_{1}H$, ${}^{6}_{3}Li$, ${}^{10}_{5}B$, ${}^{14}_{7}N$, and ${}^{50}_{23}V$). It is apparent that nuclei with paired protons and/or paired neutrons are favored among stable nuclei. Figure I-3 also shows that nuclei with even Z (or even N) usually have more than three stable isotopes (or isotones), and nuclei with odd Z (or odd N) usually have only one isotope (or isotone), or occasionally two. Nuclei with odd A (i.e., even Z–odd N or odd Z–even N) have only one stable isobar, while nuclei with even A (i.e., even Z–even N, except for the five odd Z–odd N nuclei mentioned earlier) have one to three stable isobars. The neutron to proton ratio (N/Z) is about one for stable nuclei with Z (or N) less than 20, but increases gradually with increasing Z and reaches about 1.5 for the heaviest stable nucleus ${}^{209}_{83}Bi$ (Figure I-3). Any nucleus with Z and N values falling outside the stable nuclei field in Figure I-3 is unstable, and will undergo various radioactive decay steps (such as α and β decays and spontaneous fission) so that the N/Z ratio will eventually approach one of those stable nuclei.

According to international convention, the mass of one mole of ${}^{12}C$ atoms ($= 6.022045 \times 10^{23}$ atoms = Avogadro's number) equals 12 grams, and the mass of one ${}^{12}C$ atom is defined to equal 12 **atomic mass units** (amu). Therefore one atomic mass unit (amu) is equivalent to $1/(6.022045 \times 10^{23}) = 1.66056 \times 10^{-24}$ g. Its energy equivalent is 1.492442×10^{-3} erg = 931.5023 MeV (1 MeV = 10^6 electronvolts). Accordingly, the rest mass of one hydrogen atom M_H (one proton plus one electron) is 1.007825037 amu = 938.79134 MeV; the rest mass of a neutron M_n = 1.008665012 amu = 939.57378 MeV; and the rest mass of an electron M_e = 0.00054858026 amu = 0.5110034 MeV. The masses of various neutral atoms are summarized by Bievre et al. (1984).

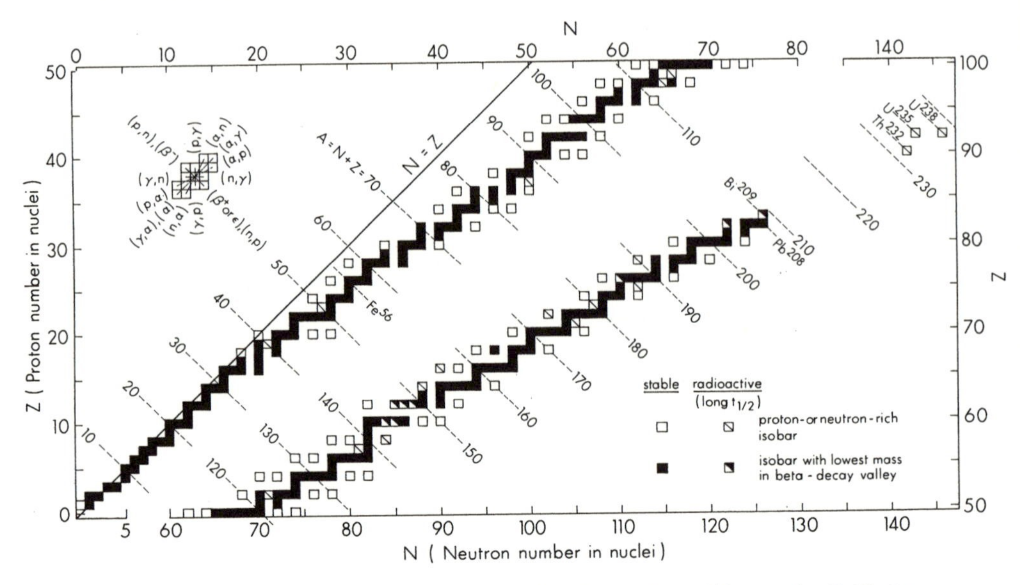

FIGURE I-3. Known stable nuclei and long-half-life radionuclides on the Z-N plane. Notice that the diagram is split into two at $Z = 50$ and $N = 60$.

The energy released in the formation of a nucleus from its component nucleons (i.e., hydrogen atoms and neutrons) is called the **binding energy of the nucleus**, Q_b, i.e.,

$$Z \cdot M_{\mathrm{H}} + N \cdot M_n \rightarrow {}_Z^AM + Q_b, \text{ i.e., } Q_b = Z \cdot M_{\mathrm{H}} + N \cdot M_n - {}_Z^AM. \quad \text{(I-1)}$$

For example, the binding energy of the ${}_2^4\mathrm{He}$ nucleus is

$$Q_b = 2M_{\mathrm{H}} + 2M_n - {}^4\mathrm{He} = 2 \times 1.0078250 + 2 \times 1.0086650 - 4.0026033 \text{ amu}$$
$$= 0.030376 \text{ amu} = 28.2960 \text{ MeV}.$$

The relative stability of nuclei can be represented by the binding energy per nucleon (Q_b/A). Figure I-4 shows the Q_b/A values calculated from equation I-1 as a function of A for all stable nuclei and for some heavy radioactive nuclei ($A > 226$). The complete listing of Q_b/A values is given by Yoshihara et al. (1985). The Q_b/A values generally increase with increasing A up to a broad maximum around $A = 56$, which corresponds to the ${}^{56}\mathrm{Fe}$ nucleus, then decrease gradually. Therefore ${}^{56}\mathrm{Fe}$ is the most stable of all nuclei. Energetically it is possible to release nuclear energies (exothermic process) by fusing more than two lighter nuclei ($A < 56$) to form a new nucleus with a higher Q_b/A value (fusion process). For example,

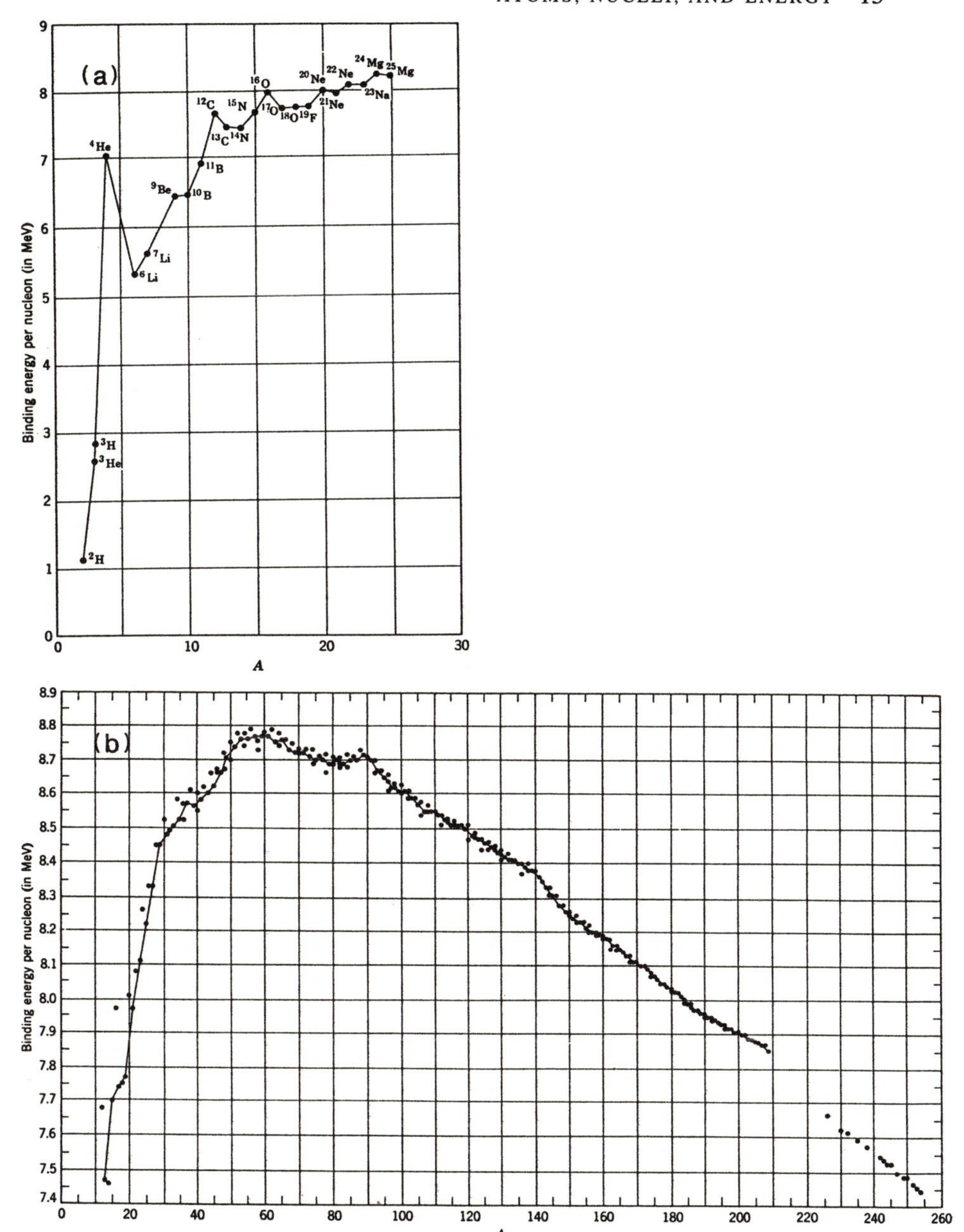

FIGURE I-4. Binding energy per nucleon as a function of mass number A for nuclei between (a) $2 \leq A \leq 25$ and (b) $12 \leq A \leq 250$ (the solid line connects the odd-A nuclei). The highest point in the broad maximum is ^{56}Fe. Reprinted by permission of John Wiley & Sons Inc.

$^4_2\text{He} + ^{12}_6\text{C} \rightarrow ^{16}_8\text{O} + Q$(energy released);

$$Q = ^4_2\text{He} + ^{12}_6\text{C} - ^{16}_8\text{O} \tag{I-2a}$$
$$= 4.00260 + 12 - 15.99491$$
$$= 0.00769\,\text{amu} = 7.16\,\text{MeV}.$$

For nuclei with $A > 56$, the only way to release the nuclear energy is to split a nucleus into lighter nuclei with higher Q_b/A values (fission process). For example,

$$^{235}_{92}\text{U} + n \rightarrow ^{140}_{56}\text{Ba} + ^{93}_{36}\text{Kr} + 3n + Q;$$
$$Q = 174\,\text{MeV}. \tag{I-2b}$$

In Figure I-4, the Q_b/A values for ^4_2He, $^{12}_6\text{C}$, $^{16}_8\text{O}$, $^{20}_{10}\text{Ne}$, and $^{24}_{12}\text{Mg}$ are higher than those of their immediate neighbors. In other words, the nuclei that are multiples of the helium nucleus have extra stability. This extra stability can be seen clearly in the theoretical minimum energy required to remove one neutron,

$$Q_n\left(= {}^{A-1}_{Z}M + M_n - {}^{A}_{Z}M\right), \tag{I-2c}$$

or one proton,

$$Q_p\left(= {}^{A-1}_{Z-1}M + M_{\text{H}} - {}^{A}_{Z}M\right), \tag{I-2d}$$

from a nucleus as shown in Figure I-5 for Q_n.

Analogous to the ions with noble gas electron configuration are the nuclei with neutron or proton numbers of 2, 8, 20, 28, 50, 82, and 126, which also have extra stability. Those numbers are called **magic numbers**. Nuclei with N and/or $Z = 8$, 50, 82, and 126 also have very low capture cross sections for thermal neutrons (Friedlander et al., 1981).

For a set of isobars, the term $(Z \cdot M_{\text{H}} + N \cdot M_n)$ in equation I-1 is near constant; therefore the lower the mass of the isobar, ${}^A_ZM$, the higher is Q_b, i.e., the isobar with minimum nuclear mass is the most stable. For example, the mass of a set of isobars with odd mass number ($A = 135$) can be represented by a parabolic function of Z (Figure I-6 left). The stable nucleus ^{135}Ba lies at the bottom of the parabola. The isobars on the neutron-rich side of the parabola (left arm) are unstable and will be transformed into ^{135}Ba by successive negative β decays (${}^A_ZM \rightarrow {}_{Z+1}^{\;\;A}M^+ + e^- + \bar{\nu}$, where $\bar{\nu}$ is the antineutrino and e^- is the electron). The isobars on the proton-rich side of the parabola (right arm) will again be transformed into ^{135}Ba by successive positive β decays (${}^A_ZM \rightarrow {}_{Z-1}^{\;\;A}M^- + e^+ + \nu$, where ν is the neutrino and e^+ is

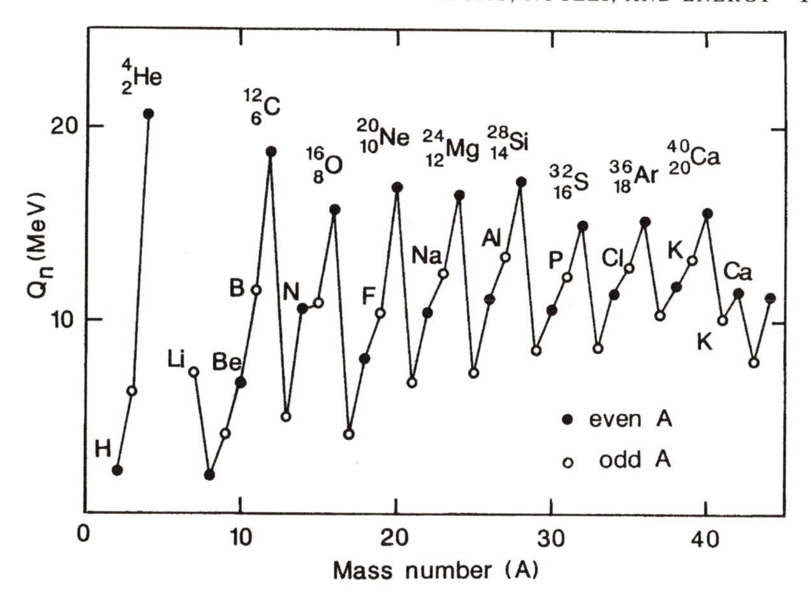

FIGURE I-5. Separation energy of a neutron (Q_n) from nuclei ($1 \leq A \leq 44$). Notice the peaks for the multiple-α-particle nuclei.

the positron) or by electron capture ($^A_ZM + e^- \rightarrow {}_{Z-1}^{\ \ A}M + \nu$), i.e., a nucleus captures an electron from the inner electron orbitals of the atom. Therefore this kind of parabola is often called the **β-stability valley**.

In the case of isobars with even A (= 136, Figure I-6 right), the mass of the isobars can be represented by two parabolas (the upper one for odd Z–odd N and the lower one for even Z–even N). There are two metastable nuclei, ^{136}Xe and ^{136}Ce, in addition to the most stable ^{136}Ba. In principle, ^{136}Xe and ^{136}Ce, can decay into ^{136}Ba by a so-called double β decay process (i.e., simultaneous emission of two $e^{\pm}$ particles or capture of two electrons). However, the probability of a double β decay is extremely low. Metastable nuclei such as ^{136}Xe and ^{136}Ce are also often called the **shielding** nuclei (open squares in Figure I-3), whereas the stable ^{136}Ba is called the **shielded** nucleus. The shielding nuclei preclude the formation of the shielded nucleus by way of successive β decay of neutron- or proton-rich unstable isobars.

Nuclear reactions such as that in equation I-2a can be expressed by the short-hand notation ^{12}C(α, γ)^{16}O; i.e., the lighter reacting particle and the lighter resulting particle are written within parentheses between the heavier initial and final nuclei and separated by a comma. For a spontaneous decay reaction, there is often no light reacting particle, thus it is omitted in the notation. For example, $^3\text{H} \rightarrow {}^3\text{He} + e^- + \bar{\nu}$ is written as ^{3}H($e^-\bar{\nu}$)^{3}He. A

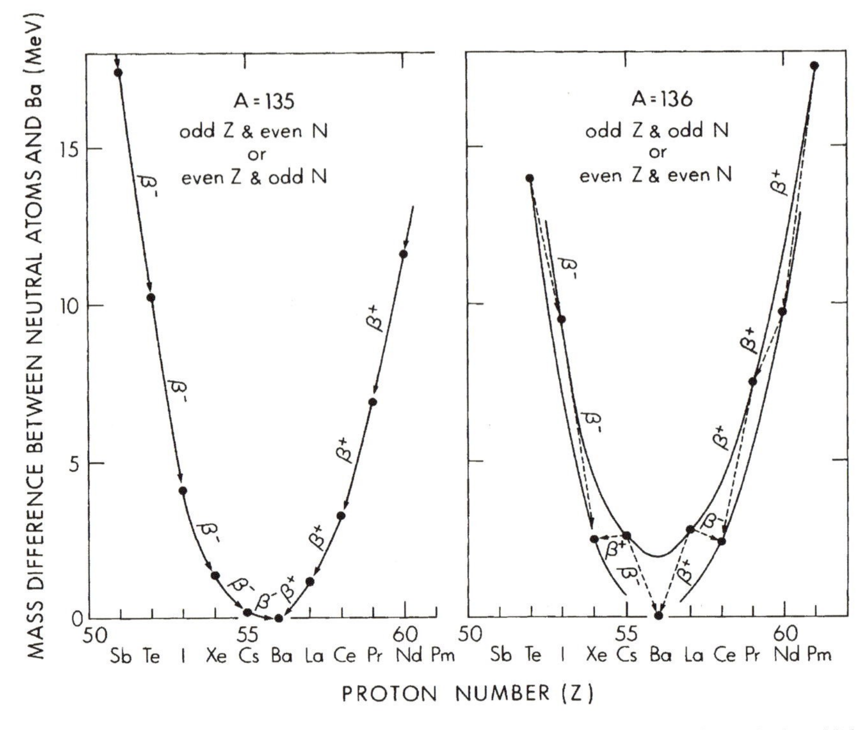

FIGURE I-6. Beta-stability-valley diagrams for isobars $A = 135$ (left) and $A = 136$ (right). The β symbols and arrows indicate consecutive beta decays.

spontaneous electron capture reaction such as $^{23}\text{Mg} + e^- \rightarrow {}^{23}\text{Na} + \nu$ can be expressed as $^{23}\text{Mg}(e^-, \nu)^{23}\text{Na}$. A consecutive reaction

$$^{12}\text{C} + {}^{4}\text{He} \rightarrow {}^{16}\text{O} + \gamma$$

$$^{16}\text{O} + {}^{4}\text{He} \rightarrow {}^{20}\text{Ne} + \gamma$$

can be simplified as

$$^{12}\text{C}(\alpha, \gamma)^{16}\text{O}(\alpha, \gamma)^{20}\text{Ne}.$$

The isotopic ratios of an element in natural samples are usually compared to those of a chosen standard material. For example, **standard mean ocean water**, abbreviated as SMOW (Craig, 1961), is the standard material for oxygen and hydrogen isotopes. The deviation of oxygen and hydrogen isotopic ratios of any sample from those of SMOW is given by the following convenient notations:

$$\delta^{18}O = \left[\frac{(^{18}O/^{16}O)_{sample}}{(^{18}O/^{16}O)_{SMOW}} - 1\right] \times 1000\,‰,$$

$$\delta^{17}O = \left[\frac{(^{17}O/^{16}O)_{sample}}{(^{17}O/^{16}O)_{SMOW}} - 1\right] \times 1000\,‰,$$

$$\delta D = \left[\frac{(D/H)_{sample}}{(D/H)_{SMOW}} - 1\right] \times 1000\,‰,$$

where D is deuterium (^2H). A $\delta^{18}O$ value of +8 ‰ for basaltic rock means that the sample is 8 per mil (‰) enriched in ^{18}O relative to SMOW. The absolute oxygen and hydrogen isotopic ratios of SMOW are

$$^{18}O/^{16}O = 200520 \pm 45 \times 10^{-8} \qquad \text{(Baertschi, 1976)},$$

$$^{17}O/^{16}O = 38309 \pm 34 \times 10^{-8} \qquad \text{(McKeegan, 1987)},$$

$$D/H = 15576 \pm 5 \times 10^{-8} \qquad \text{(Hagemann et al., 1970)}.$$

If two coexisting phases A and B are isotopically in equilibrium, the ratio of $^{18}O/^{16}O$ in phase A to the ratio in phase B is called the **fractionation factor** ($\alpha_{A/B}$), i.e.,

$$\alpha_{A/B} = \frac{(^{18}O/^{16}O)_A}{(^{18}O/^{16}O)_B}.$$

For example, $\alpha_{water/vapor} = 1.0092$ at 25°C. This means that the water phase is enriched in ^{18}O relative to the vapor phase. The $\alpha_{A/B}$ can be related to δ_A and δ_B by

$$\alpha_{A/B} = \frac{\delta_A/1000 + 1}{\delta_B/1000 + 1}$$

or approximated by $10^3 \ln \alpha_{A/B} \approx \delta_A - \delta_B$.

I-3. Cohesive Energies among the Atoms of Pure Metals and Nonmetals

The atoms of a pure solid metal are held together mainly by the so-called **metallic bonds**, which can be visualized as resulting from the attraction between a network of metal cations and the "sea" of freely mobile electrons (mainly valence electrons). The atoms or polyatomic molecules of nonmetals in solid state are held together mainly by the **weak Van der Waals force**.

The weak Van der Waals force results from the interaction of the instantaneous electric dipole moments of nonmetal atoms. The electrons of an atom may have their center of charge coincident with that of a nucleus at one instant (thus no electric dipole moment), but at another instant it may lie beside the nucleus, thus forming an instantaneous electric dipole moment (two opposite but equal charges multiplied by distance). Rather than pointing in random fashion, the vectors of the instantaneous electric dipole moments of the neighboring atoms are more often synchronized (due to dipole-induced dipoles), thus resulting in a weak net attraction (London, 1930).

The magnitude of the cohesive energy holding the atoms or polyatomic molecules of the pure solid elements together can be best represented by the **enthalpy of sublimation**, ΔH^0_{sbl}, which is the energy required to convert one mole of atoms or polyatomic molecules from solid to gas at the standard state (25°C and 1 atmosphere). For elements that are gas or liquid at the standard state, the ΔH^0_{sbl} values at their melting points are adopted. The periodicity of ΔH^0_{sbl} as a function of atomic number is shown in Figure I-7 and listed

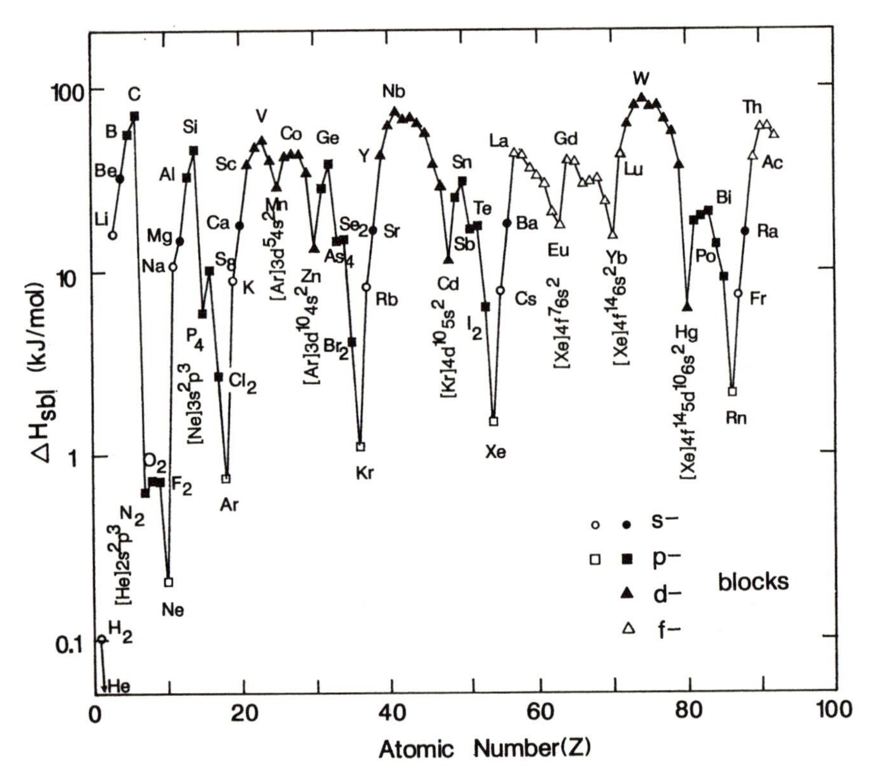

FIGURE I-7. Heats of sublimation $(\Delta H^0_{\text{sbl}})$ of solid elements as a function of Z. The minima are always associated with full or half-full outermost subshells.

in Table I-2. The minima of ΔH^0_{sbl} are always associated with atoms that have full or half-full outermost subshells. The polyatomic molecules shown in Table I-2 and Figure I-7 represent the most stable gaseous species for those elements at the standard state.

TABLE I-2
Heats of sublimation (ΔH^0_{sbl}) and boiling points (T_b) of the elements

Z	ΔH^0_{sbl} (kJ/mol)	T_b (K)	Z	ΔH^0_{sbl} (kJ/mol)	T_b (K)	Z	ΔH^0_{sbl} (kJ/mol)	T_b (K)
H_2	*1.01	20.3	Ge	379.6	3107	Pm	293	3273
He	*0.13	4.2	As	302.5	876	Sm	206.7	2064
			As_4	144	876	Eu	175.3	1870
Li	159.4	1615	Se	227.1	958	Gd	397.5	3539
Be	324.3	2757	Se_2	146	958	Tb	388.7	3496
B	562.7	4275	Br_2	*41.1	332	Dy	290.4	2835
C	716.7	5100	Kr	*10.7	120	Ho	300.8	2968
N_2	*6.25	77.4				Er	317.1	3136
O_2	*7.26	90.2	Rb	80.9	961	Tm	232.2	2220
F_2	7.2	85	Sr	164.4	1655	Yb	152.3	1467
Ne	2.04	27.1	Y	421.3	3611	Lu	427.6	3668
			Zr	608.8	4682	Hf	619.2	4876
Na	107.3	1156	Nb	725.9	5015	Ta	782	5730
Mg	147.7	1380	Mo	658.1	4912	W	849.4	5828
Al	326.4	2740	Tc	678	4538	Re	769.9	5870
Si	455.6	3522	Ru	642.7	4425	Os	791	5300
P	314.6	553	Rh	556.9	3970	Ir	665.3	4701
P_4	58.9	553	Pd	378.2	3240	Pt	565.3	4100
S	278.8	717.8	Ag	284.6	2436	Au	366.1	3130
S_8	102.3	717.8	Cd	112	1040	Hg	61.3	630
Cl_2	*26.8	239.1	In	243.3	2346	Tl	182.2	1746
Ar	*7.5	87.3	Sn	302.1	2876	Pb	195	2023
			Sb	262.3	1860	Bi	207.1	1837
K	89.2	1033	Sb_2	167	1860	Po	146	1235
Ca	178.2	1757	Te	196.7	1261	At	90.4	610
Sc	377.8	3109	Te_2	172	1261	Rn	*21.3	211
Ti	469.9	3560	I	107.8	458			
V	514.2	3650	I_2	62.4	458	Fr	73	950
Cr	396.6	2945	Xe	*14.9	165	Ra	159	1900
Mn	280.7	2335				Ac	406	3470
Fe	416.3	3135	Cs	76.1	944	Th	598.3	5061
Co	424.7	3143	Ba	180	2078	Pa	607	4300
Ni	429.7	3187	La	431	3737	U	535.6	4407
Cu	338.3	2840	Ce	423	3715	Np		4175
Zn	130.7	1180	Pr	355.6	3785	Pu	352	3505
Ga	277	2478	Nd	327.6	3347	Am		2880

Notes: ΔH^0_{sbl} of elements with asterisks is estimated at the melting point, otherwise at 25°C. ΔH^0_{sbl} are mostly from Wagman et al. (1982) with some additional data from Dean (1985). The boiling point data are copied from Table I-1b.

Interestingly, the **boiling points** of the elements (Table I-2) are closely related to ΔH^0_{sbl} as shown in Figure I-8 (notice the different correlation lines for metals and nonmetals). Thus, the stronger the cohesive energy among the atoms of a pure element in the solid state, the higher is the boiling point of the element. Also, the boiling points and ΔH^0_{sbl} values for the *p*- and *s*-block elements are generally lower than those for the *d*- and *f*-block elements in the same period (Figure I-7).

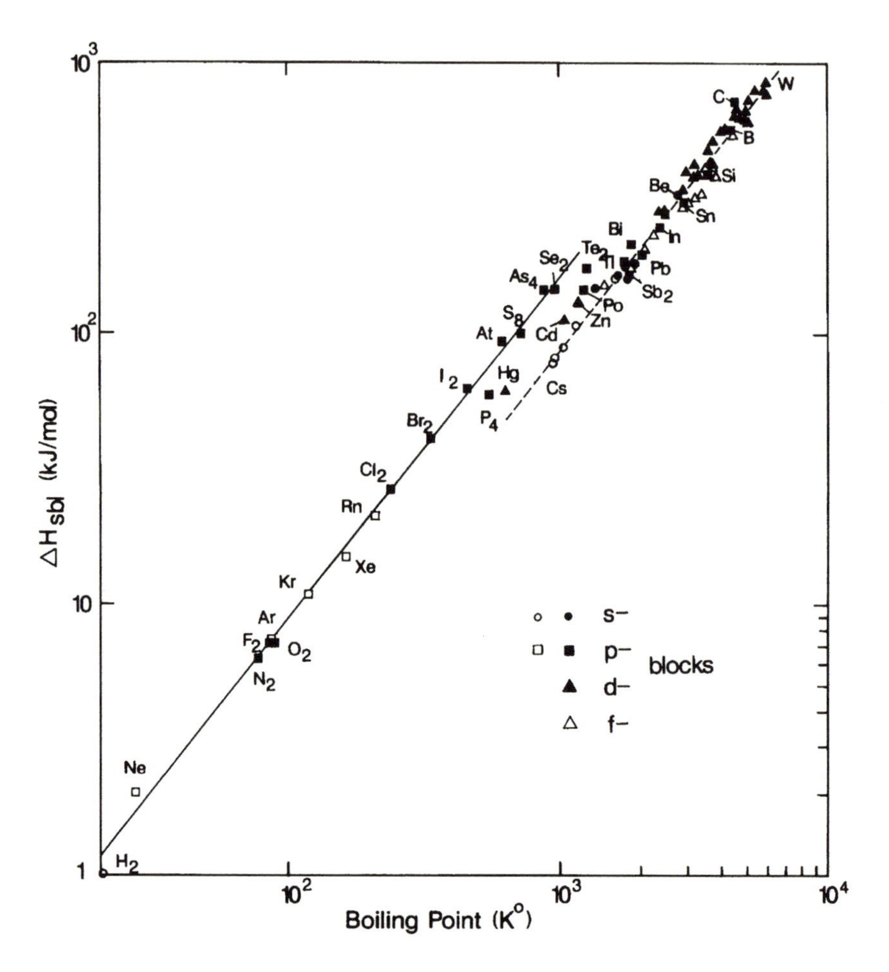

FIGURE I-8. The good correlation between the boiling point of elements and the heats of sublimation (ΔH^0_{sbl}) of atoms or polyatomic molecules of solid elements. Noble gases and polyatomic molecules (except Sb_2) fall on the same correlation line.

I-4. Ionization Energies and Electron Affinities of Gaseous Atoms and Ions

The energy required to remove the least strongly bound electron from a gaseous atom M(g), to infinite distance is called the first **ionization energy** (or **ionization potential**) of the atom, I_1. The energies required to remove subsequent electrons are called the second (I_2), third (I_3), and up to zth (I_z) ionization energies, i.e.,

$$M(\mathrm{g}) + I_1 \rightarrow M^{+}(\mathrm{g}) + e^{-},$$

$$M(\mathrm{g}) + I_2 \rightarrow M^{+2}(\mathrm{g}) + e^{-},$$

$$\vdots$$

$$M^{z-1}(\mathrm{g}) + I_z \rightarrow M^{+z}(\mathrm{g}) + e^{-}.$$

The zth ionization energy, I_z, of M^{z-1}(g) is equivalent to the binding energy of an electron to a gaseous cation M^{+z}(g) in the reverse reaction above. Adding the above equations, one obtains

$$M(\mathrm{g}) + \Sigma I_z \rightarrow M^{+z}(\mathrm{g}) + ze^{-},$$

where ΣI_z is the **cumulative ionization energy** ($= I_1 + I_2 + \cdots + I_z$). Some gaseous neutral atoms can react spontaneously with an electron to form a monovalent gaseous anion, i.e.,

$$M(\mathrm{g}) + e^{-} \rightarrow M^{-1}(\mathrm{g}) + I_{-1}.$$

The energy released in the above equation (I_{-1}) is called the (first) **electron affinity** to a gaseous atom M(g). The electron affinity is equivalent to the ionization energy of a monovalent anion. Similarly, we can define the cumulative electron affinity, ΣI_{-z}, as equal to the energy released in the spontaneous reaction

$$M(\mathrm{g}) + ze^{-} \rightarrow M^{-z}(\mathrm{g}) + \Sigma I_{-z}.$$

However, the data for ΣI_{-z} are scarce. By losing or gaining electrons, the neutral gaseous atoms can be transformed into various gaseous ions; but the actual numbers of stable ions in chemical compounds are quite limited. According to their electron configuration, the most common stable ions can be classified into four main types: (1) **A-type cations and anions**, which have the electron configurations of noble gases; (2) **B-type cations** with the outermost subshell configuration of nd^{10}, $nd^{10}(n+1)s^2$, as well as [He]$2s^2$ and [Ne]$3s^2$; (3) the transition metal cations with the outermost d shells partially filled; and (4) the lanthanide and actinide cations with the outermost f subshells partially or totally filled.

The successive ionization energies I_1 up to I_7 (Dean, 1985) and I_{-1} (Bratsch, 1983; Hotop and Lineberger, 1985), for each element at 0 K are given in the Appendix Table A-1. The I_z and the **average binding energy**, $\Sigma I_z/z$, for common stable cations are summarized in Table I-3. The successive ionization energies at 298 K (25°C) can be obtained by adding 0.064 eV to each value in Appendix Table A-1 as explained in the footnote of the table. The ionization energy is also often expressed in terms of the ionization of one mole of gaseous atoms instead of a single atom. For example, 1 eV/atom = 96.5 kJ/mol = 23.1 kcal/mol.

The data for I_1 to I_4 are plotted as a function of atomic number (Z) in Figure I-9. The maxima are again always associated with the cations that have full or half-full outermost subshells. The minima are the cations with the outermost subshell configurations of ns^1 and np^1, i.e., the single electrons in the outermost s and p subshells are loosely held by the atoms. Figure I-9 also indicates that the energy levels of ns subshells are higher than those of $(n-1)d$ subshells for cations with identical valency, i.e., the $(n-1)d$ subshell is filled in by electrons prior to the ns subshell. The converse is the case for neutral atoms, as already shown in Figure I-2.

The successive ionization energies as a function of charge z for Ca (s-block); Cl, As, and Br (p-block); and Zn and Mn (d-block) are plotted in Figure I-10 (see page 27). Whenever an electron is taken away from a full or half-full subshell, there is always a break in the slope. For clarity and convenience hereafter, the ionization energy I_z will be considered only as the **electron binding energy** to gaseous cation M^{+z} unless noted otherwise. The plot of I_z against $\Sigma I_z/z$ for most common cations indicates a good linear relationship (Figure I-11, see page 27). Thus both parameters are interchangeable.

The I_z divided by the charge z is called the **polarizing power** of cation M^{+z} by Goldschmidt (1954). A nice periodicity of I_z/z as a function of atomic number Z is apparent in Figure I-12 (see page 28). The close relationship between I_z/z and the electronegativity will be discussed in Section I-7.

Application of the concept of the electron binding energy or ionization energy to many geochemical problems has been amply demonstrated by Ahrens (1952, 1953, 1954, 1983). In the following sections, its applications will be demonstrated wherever relevant.

TABLE I-3
Electron binding energy to cation $M^{+z}(I_z)$ and average electron binding energy $(\Sigma I_z/z)$

Z	z	I_z	$\Sigma I_z/z$	Z	z	I_z	$\Sigma I_z/z$	Z	z	I_z	$\Sigma I_z/z$
Ac-89	3	(19)	(12)	Hf-72	4	33.3	19.5		4	(41)	(24)
Ag-47	1	7.58	7.58	Hg-80	1	10.4	10.4	Ra-88	2	10.1	7.7
Al-13	3	28.5	17.8		2	18.8	14.6	Rb-37	1	4.18	4.18
As-33	3	28.4	18.9	Ho-67	3	22.8	13.6	Re-75	4	37.7	21.2
	5	62.6	33.9	1-53	5	(71)	(35)		7	(79)	(41)
At-85	5	(51)	(30)		7	(104)	(52)	Rh-45	3	31.1	18.9
	7	(91)	(46)	In-49	3	28	17.6		4	(46)	(26)
Au-79	1	9.23	9.23	Ir-77	3	(27)	(18)	Ru-44	3	28.5	17.5
	3	(30)	(20)		4	(39)	(23)		4	(46)	(25)
B-5	3	37.9	23.8	K-19	1	4.34	4.34		8	(119)	(58)
Ba-56	2	10	7.61	La-57	3	19.2	11.9	S-16	4	47.3	29
Be-4	2	18.2	13.8	Li-3	1	5.39	5.39		6	88	46.1
Bi-83	3	25.6	18.1	Lu-71	3	21	13.4	Sb-51	3	25.3	16.8
	5	56	33.1	Mg-12	2	15	11.3		5	56	30.1
Br-35	5	59.7	35.3	Mn-25	2	15.6	11.5	Sc-21	3	24.8	14.7
	7	103	52.6		3	33.7	18.9	Se-34	4	42.9	26.2
C-6	4	64.5	37		4	51.2	27		6	81.7	42.5
Ca-20	2	11.9	8.99		7	119	56.4	Si-14	4	45.1	25.8
Cd-48	2	16.9	13	Mo-42	3	27.2	16.8	Sm-62	3	23.4	13.4
Ce-58	3	20.2	12.2		4	46.4	24.2	Sn-50	2	14.6	11
	4	36.8	18.3		5	61.2	31.6		4	40.7	23.3
Cl-17	5	67.8	39.5		6	68	37.7	Sr-38	2	11	8.4
	7	114	58.4	N-7	3	47.4	30.5	Ta-73	5	(45)	(25)
Co-27	2	17.1	12.5		5	97.9	53.4	Tb-65	3	21.9	13.1
	3	33.5	19.5	Na-11	1	5.14	5.14	Tc-43	7	(94)	(46)
Cr-24	2	16.5	11.6	Nb-41	3	25	15.4	Te-52	4	37.4	23.2
	3	31	18.1		5	50.6	27		6	70.7	37.1
	6	90.5	43.9	Nd-60	3	22.1	12.8	Th-90	4	28.8	16.6
Cs-55	1	3.89	3.89	Ni-28	2	18.2	12.9	Ti-22	3	27.5	16
Cu-29	1	7.73	7.73		3	35.2	20.3		4	43.3	22.8
	2	20.3	14	O-8	6	138	72.2	Tl-81	1	6.11	6.11
Dy-66	3	22.8	13.5	Os-76	3	(25)	(17)		3	29.8	18.8
Er-68	3	22.7	13.6		4	(40)	(23)	Tm-69	3	23.7	14
Eu-63	2	11.2	8.45		8	(99)	(50)	U-92	6	(57)	
	3	24.9	13.9	P-15	3	30.2	20.1	V-23	2	14.7	10.7
F-9	7	185	94.1		5	65	35.4		3	29.3	16.9
Fe-26	2	16.2	12	Pb-82	2	15	11.2		4	46.7	24.4
	3	30.7	18.2		4	42.3	24.2		5	65.2	32.5
Fr-87	1	3.98	3.98	Pd-46	2	19.4	13.9	W-74	6	(61)	(32)
Ga-31	3	30.7	19.1		3	32.9	20.2	Y-39	3	20.5	13
Gd-64	3	20.6	12.9	Pm-61	3	22.3	12.9	Yb-70	3	25	14.5
Ge-32	2	15.9	11.9	Po-84	4	(38)	(23)	Zn-30	2	18	13.7
	4	45.7	25.9	Pr-59	3	21.6	12.5	Zr-40	4	34.3	19.3
H-1	1	13.6	13.6	Pt-78	2	18.6	13.8				

Source: Dean (1985); Samsonov (1973) in parentheses.
Note: Energies in units of eV/electron.

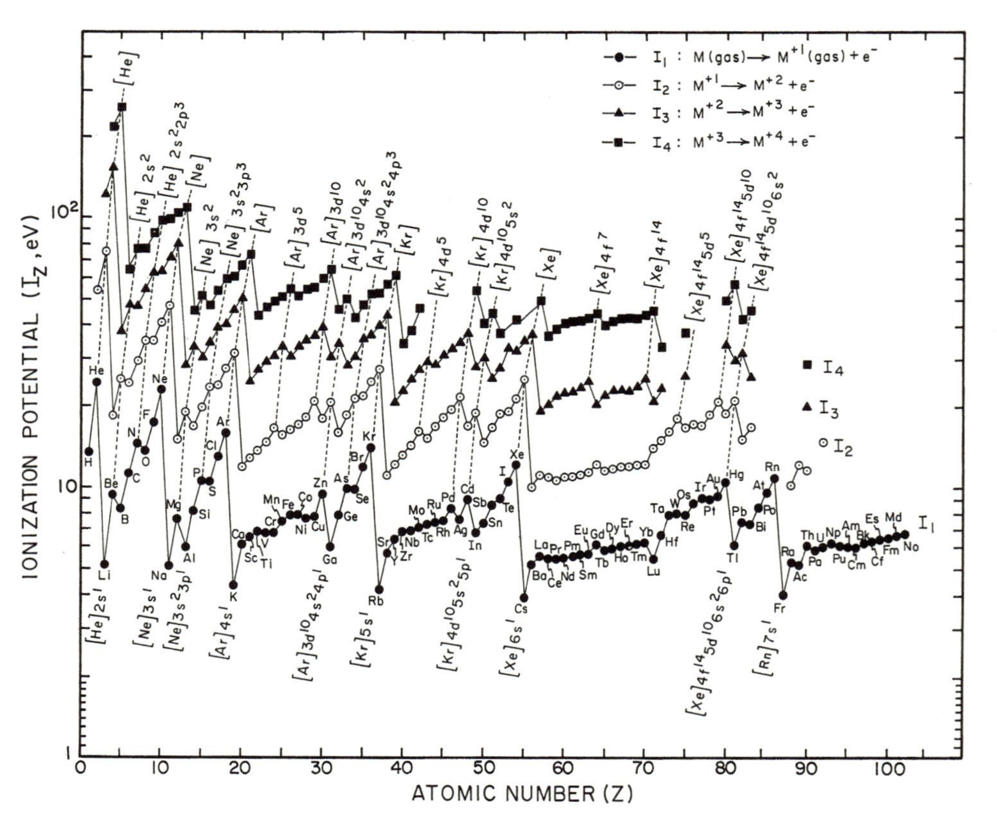

FIGURE I-9. First to fourth ionization energies (I_1 to I_4) as a function of Z. The maxima are always associated with full or half-full electron configurations.

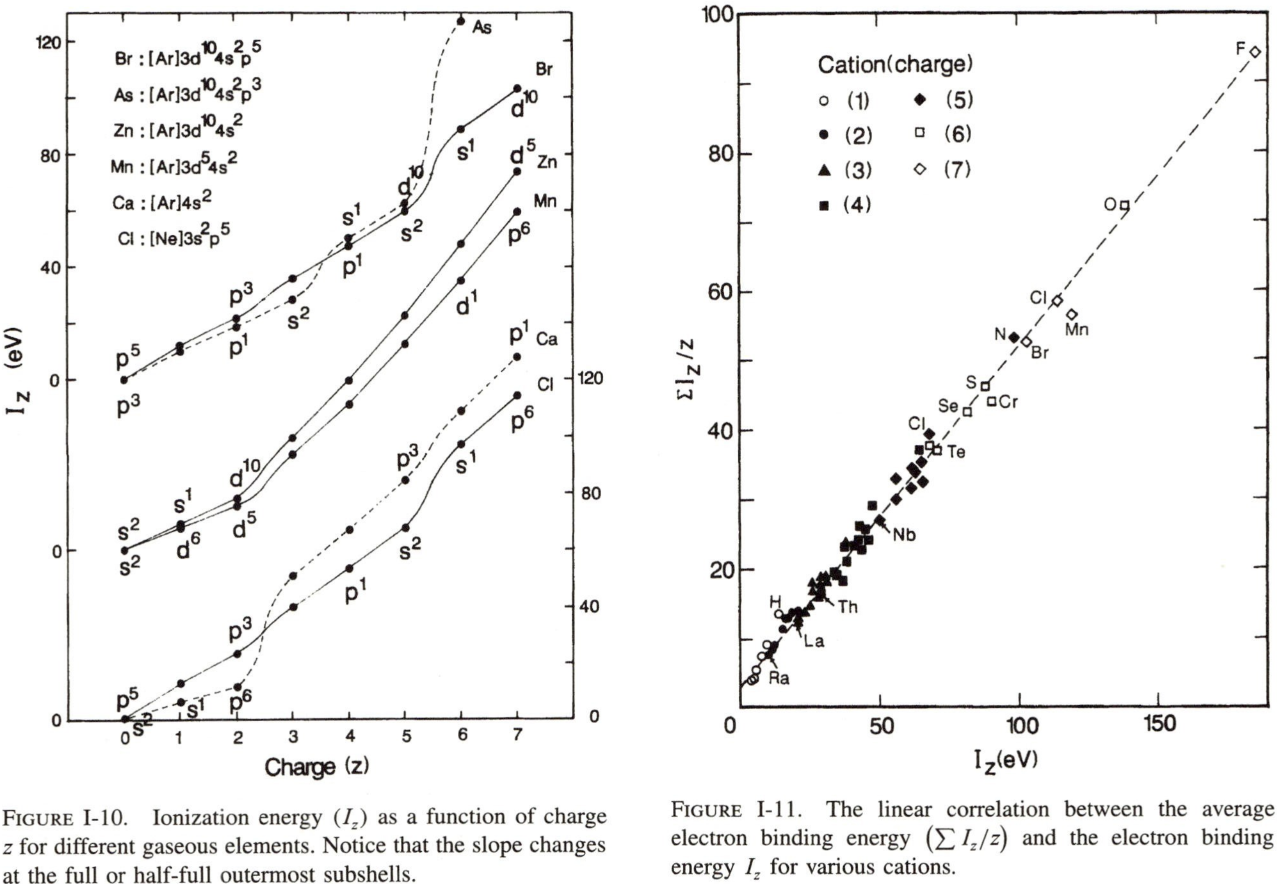

FIGURE I-10. Ionization energy (I_z) as a function of charge z for different gaseous elements. Notice that the slope changes at the full or half-full outermost subshells.

FIGURE I-11. The linear correlation between the average electron binding energy $\left(\sum I_z/z\right)$ and the electron binding energy I_z for various cations.

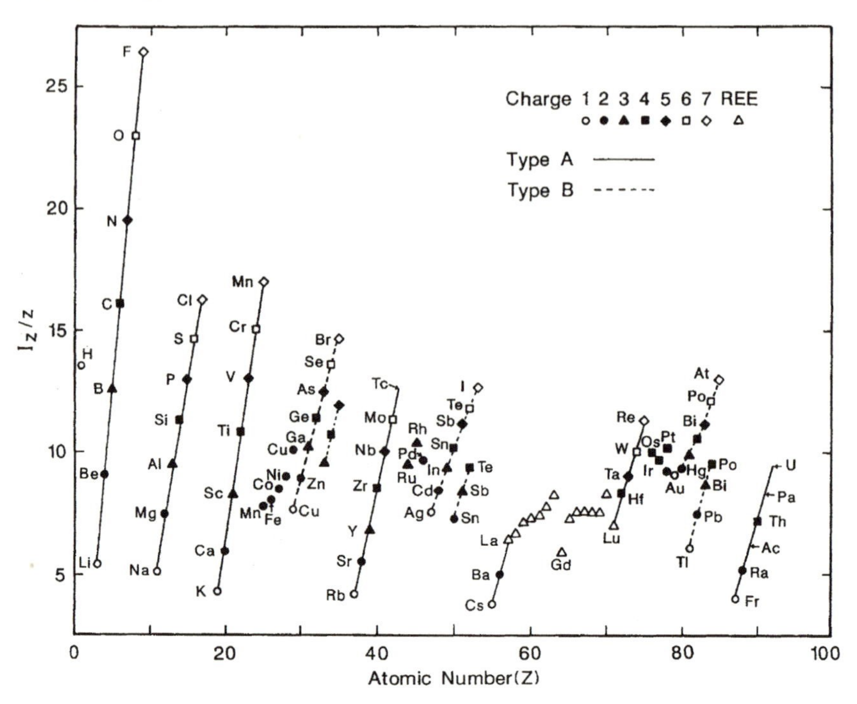

FIGURE I-12. Periodicity of polarizing power (I_z/z) as a function of atomic number Z.

I-5. IONIC RADII AND IONIC POTENTIALS

Ionic radii are derived from interatomic distances in oxides and fluorides (i.e., so-called ionic crystals), which consist of frameworks of cations and anions. One sets the ionic radius of O^{-2} with a coordination number of 6 (octahedral coordination) equal to 1.40 Å and assumes an additivity of ionic radii. Table I-4 summarizes the ionic radii (in Å) for sixfold coordination given by Shannon (1976), who revised and updated the earlier compilation by Shannon and Prewitt (1969) and adopted some values from Ahrens (1952). Though the ionic radii change with the coordination number (Shannon and Prewitt, 1969), it is still useful for comparison purposes to consider the ionic radii for the sixfold coordination as an intrinsic property of ions.

According to crystal field theory, when a transition metal cation is surrounded by six negatively charged ligands (e.g., O^{-2}) in octahedral coordination in a crystal, the energy level of the outermost *d* subshell is split in two with two orbitals at the higher energy level (e_g) and three orbitals at the lower energy level (t_{2g}). The energy separation between e_g and t_{2g} is called the **crystal field splitting parameter** Δ . The magnitude of Δ depends on the

TABLE I-4
Ionic radii (Å) for sixfold coordination

Z	*z*	*r*	*Z*	*z*	*r*	*Z*	*z*	*r*	*Z*	*z*	*r*
Ac-89	3	1.12		3L	0.54	Ni-28	2	0.69A		4	0.37A
Ag-47	1	1.15	Dy-66	2	1.07		3L	0.56		6	0.29A
	2	0.94		3	0.91A		3H	0.6	Sb-51	−3	2.45P
	3	0.75	Er-68	3	0.89A		4L	0.48		3	0.76A
Al-13	3	0.54	Eu-63	2	1.17	Np-93	2	1.1		5	0.6
Am-95	3	0.98		3	0.95		3	1.01	Sc-21	3	0.75
	4	0.85	F-9	−1	1.33A		4	0.87	Se-34	−2	1.98P
As-33	−3	2.22P		7	0.08A		5	0.75		4	0.50A
	3	0.58A	Fe-26	2L	0.61		6	0.72		6	0.42A
	5	0.46A		2H	0.78	O-8	−2	1.40A	Si-14	−4	2.71P
At-85	7	0.62A		3L	0.55		6	0.10A		4	0.4
Au-79	1	1.2a		3H	0.65A	OH	−1	1.37	Sm-62	3	0.96
	3	0.7a	Fr-87	1	1.80A	Os-76	4	0.63	Sn-50	−4	2.94P
	5	0.57	Ga-31	3	0.62A		5	0.58		2	0.93A
B-5	3	0.27	Gd-64	3	0.94		6	0.55		4	0.69
Ba-56	2	1.35A	Ge-32	−4	2.72P		7	0.53	Sr-38	2	1.18
Be-4	2	0.45		2	0.73A	P-15	−3	2.12P	Ta-73	3	0.72
Bi-83	3	1.03		4	0.53A		3	0.44A		4	0.68
	5	0.72a	Hf-72	4	0.71		5	0.34A		5	0.64
Bk-97	3	0.96	Hg-80	1	1.19	Pa-91	3	1.04	Tb-65	3	0.92A
	4	0.83		2	1.02		4	0.9		4	0.76
Br-35	−1	1.96P	Ho-67	3	0.90A		5	0.78	Tc-43	4	0.63a
	5	0.47A	I-53	−1	2.2	Pb-82	2	1.19A		5	0.6
	7	0.39A		5	0.62A		4	0.78		7	0.56A
C-6	−4	2.60P		7	0.53	Pd-46	2	0.86	Te-52	−2	2.21P
	4	0.16A	In-49	3	0.80A		3	0.71a		4	0.67a
Ca-20	2	1.00A	Ir-77	3	0.68A		4	0.62		6	0.56A
Cd-48	2	0.95		4	0.63	Pm-61	3	0.97	Th-90	4	0.94
Ce-58	3	1.01		5	0.57	Po-84	4	0.94	Ti-22	2	0.86
	4	0.87	K-19	1	1.38		6	0.67A		3	0.67
Cf-98	3	0.95	La-57	3	1.03	Pr-59	3	0.99		4	0.61
	4	0.82	Li-3	1	0.76		4	0.85	Tl-81	1	1.5
Cl-17	−1	1.81P	Lu-71	3	0.86A	Pt-78	2	0.80A		3	0.89
	5	0.34A	Mg-12	2	0.72		4	0.63	Tm-69	2	1.03
	7	0.27A	Mn-25	2L	0.67		5	0.57		3	0.88A
Cm-96	3	0.97		2H	0.83	Pu-94	3	1	U-92	3	1.03
	4	0.85		3L	0.58		4	0.86		4	0.89
Co-27	2L	0.65		3H	0.70a		5	0.74		5	0.76
	2H	0.75		4	0.53		6	0.71		6	0.73
	3L	0.55		7	0.46A	Ra-88	2	1.43A	V-23	2	0.79
	3H	0.61	Mo-42	3	0.69	Rb-37	1	1.52		3	0.64
	4L	0.49a		4	0.65	Re-75	4	0.63		4	0.58
	4H	0.53		5	0.61		5	0.58		5	0.54
Cr-24	2L	0.73		6	0.59		6	0.55	W-74	4	0.66
	2H	0.8	N-7	−3	1.71P		7	0.53		5	0.62
	3	0.62A		3	0.16A	Rh-45	3	0.67A		6	0.58a
	4	0.55		5	0.13A		4	0.6	Y-39	3	0.9
	5	0.51a	Na-11	1	1.02		5	0.55	Yb-70	2	1.02
	6	0.49a	Nb-41	3	0.72	Ru-44	3	0.68		3	0.87A
Cs-55	1	1.67A		4	0.68		4	0.62	Zn-30	2	0.74A
Cu-29	1	0.96A		5	0.64		5	0.57	Zr-40	4	0.72
	2	0.73A	Nd-60	3	0.98	S-16	−2	1.84P			

Notes: A: Ahrens' (1952) values adopted by Shannon (1976) or in agreement with Shannon's values within ±0.01; P: Pauling's Table 13-3 (1960); a: preferred interpolated values from Figure I-13. L: low spin; H: high spin. Some more ionic radii data are given by Dean (1985) though the data sources are not indicated.

nature of the ligands coordinated to the transition metal cation. The effect of crystal field splitting is to cause d subshell electrons to fill in first the lower energy level t_{2g} and then e_g, if Δ is large. On the other hand, the electric repulsion among electrons tends to cause electrons to be distributed first in all orbitals of d subshells with parallel spins, if the Δ is small. These two possibilities lead to **high-** and **low-spin electron configurations** for transition metal cations as shown in Table I-5. In oxides and silicates, those transition metal cations are in high-spin configuration. Burns (1970) discusses crystal field splitting under different coordination numbers of ligands and the application of crystal field theory to many geochemical problems.

As shown in Figure I-13, the radii of cations (for sixfold coordination; Shannon, 1976) with identical electron configuration decrease systematically with increasing charge of the cation. This can be explained by the contraction of the electron sheath by the increasing nuclear charges. The radii of A- and B-type cations with the same charge increase with the principal quantum number n, as one would have expected from the addition of new subshells. The radii of transition metal cations with identical charge and low-spin configurations decrease initially with increasing number of electrons in the outermost d subshell and then increase steadily when the d subshell electrons increase to more than 5 or 6. Also, the high-spin cations (open circles) are always larger than the low-spin cations (solid circles). In contrast, the radii of the lanthanide and actinide cations with identical charge decrease steadily when the number of electrons in the outermost f subshell increases. These phenomena are often called the **lanthanide** and **actinide contractions** (Goldschmidt, 1954). Apparently, the net effect of adding one electron to the electron sheath and one proton to the nucleus of lanthanides and actinides is a slight contraction of the electron sheath. The important consequence

TABLE I-5
Electron configurations of transition metal cations in octahedral coordination

No. of 3d electron	*Cation*			*High spin*		*Low spin*	
	$z = 2$	$z = 3$	$z = 4$	t_{2g}	e_g	t_{2g}	e_g
1		Ti	V	↑		↑	
2	Ti	V	Cr	↑ ↑		↑ ↑	
3	V	Cr	Mn	↑ ↑ ↑	↑	↑ ↑ ↑	
4	Cr	Mn		↑ ↑ ↑	↑ ↑	↑↓↑ ↑	
5	Mn	Fe		↑ ↑ ↑	↑ ↑	↑↓↑↓↑	
6	Fe	Co	Ni	↑↓↑ ↑	↑ ↑	↑↓↑↓↑↓	
7	Co	Ni		↑↓↑↓↑	↑ ↑	↑↓↑↓↑↓	↑
8	Ni			↑↓↑↓↑↓	↑ ↑	↑↓↑↓↑↓	↑ ↑
9	Cu			↑↓↑↓↑↓	↑↓↑	↑↓↑↓↑↓	↑↓↑
10	Zn	Ga	Ge	↑↓↑↓↑↓	↑↓↑↓	↑↓↑↓↑↓	↑↓↑↓

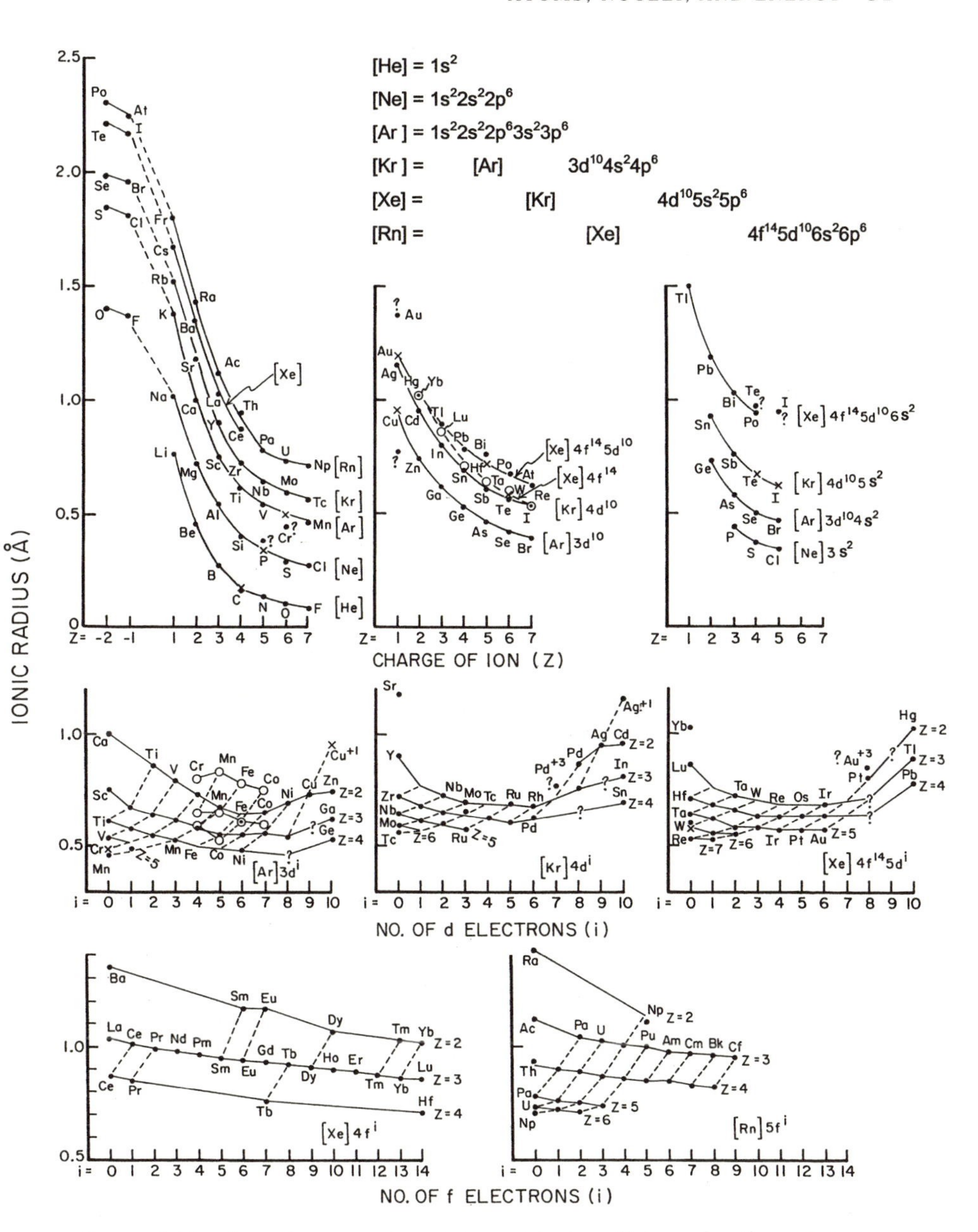

FIGURE I-13. Smooth systematic variations of ionic radii (r) as functions of electron configurations and charges. The question marks indicate doubtful data from Shannon (1976) and the crosses are the preferred values.

of the lanthanide contraction is that the ionic radii of the second and third series of transition metals become similar. Therefore, the Y-Lu, Zr-Hf, Nb-Ta, Mo-W, Ru-Os, Rh-Ir, Pd-Pt, and Ag-Au pairs often exhibit strikingly similar geochemical behavior.

Some data points given by Shannon (1976) do not fall on the smooth curves in Figure I-13 (indicated by question marks beside the data points), probably indicating unreliable data. The preferred ionic radii for those points are indicated by the crosses in the figure and by letter a in Table I-4.

The cationic charge z divided by its ionic radius r is called the **ionic potential** of the cation (Cartledge, 1928, 1930). Physically, the ionic potential, z/r, can be visualized as a quantity related to the energy required to separate an electron with charge of one from a positive point charge z at distance r to infinity, since this energy is equal to

$$\int_r^\infty ze^2/r^2\,dr = e^2\,z/r.$$

As discussed earlier, the zth ionization energy, I_z, is equivalent to the binding energy of an electron to a cation M^{+z}. Therefore, one would expect that z/r and I_z are somehow closely related, as shown in Figure I-14 (r is from Table I-4). For cations with identical electron configuration (connected by solid and dashed lines), I_z is equal to $a(z/r)^b$, where a is the intercept on the I_z axis at $z/r = 1$, and b is the slope of the straight lines in Figure I-14. One exception is the line connecting cations with [Ne] configuration (Li^+ to F^{+7}), where the slope exhibits a distinct break at Be^{+2}. Whether this break is real or not is uncertain. However, if one adopts the ionic radii of Be^{+2} (0.35 Å) and B^{+3} (0.23 Å) given by Ahrens (1952), the break disappears. Also, using only the r given by Ahrens (1952), the relationship $I_z = a(z/r)^b$ always holds. The slope, b, is not exactly equal to one, but not far from it (0.78 to 1.11). Comparing cations with the same charge (up to charge 4), I_z for B-type, transition and lanthanide metal cations is always higher than for A-type cations at similar z/r values. In short, the ionic potential alone cannot fully represent the interaction energy between cations and an electron. After all, a cation M^{+z} is not a simple sphere with a radius of r and a point charge of z at the center.

For comparison, Figure I-15 summarizes the ionic radii as a function of ionic charge and the types of electron configurations (A-, B-types, etc.). In hard rock geochemistry, we mostly deal with silicates and oxides. Therefore the high-spin radii of divalent and trivalent Mn, Fe, Co, and Ni are adopted in Figure I-15. Ions with similar ionic radius and charge can often substitute for one another in a mineral structure, thus exhibiting similar geochemical behavior. For example, in olivine ($[Mg, Fe]_2SiO_4$), the Mg and Fe can easily be substituted by Ni, Co, Mn, and Zn, since they have very similar ionic radii and identical charge. In chromite ($FeCr_2O_4$), the noble metals are enriched

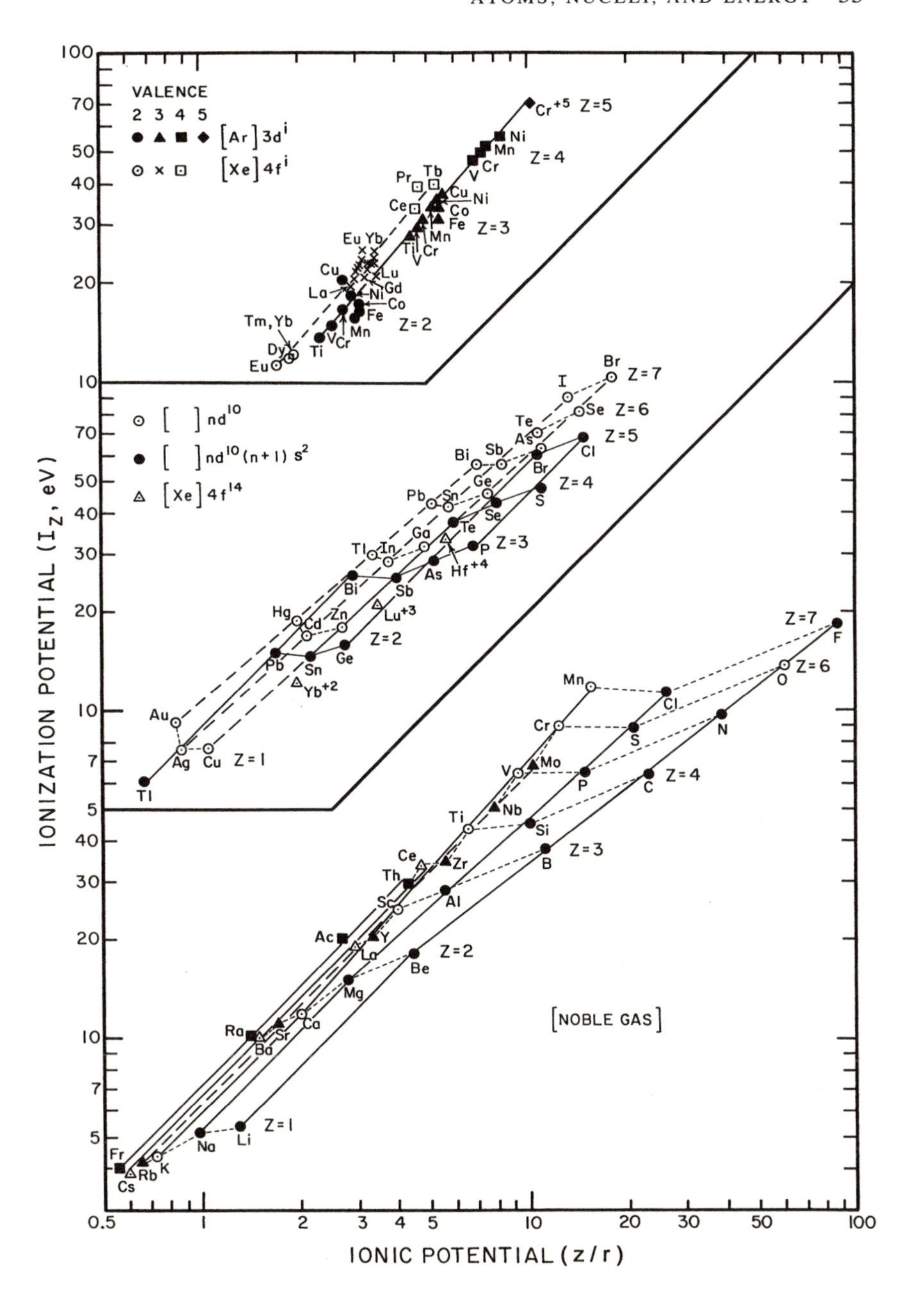

FIGURE I-14. Plots of ionic potentials (z/r) versus the electron binding energies (I_z) for cations with different electron configurations. Data are from Tables I-3 and I-4.

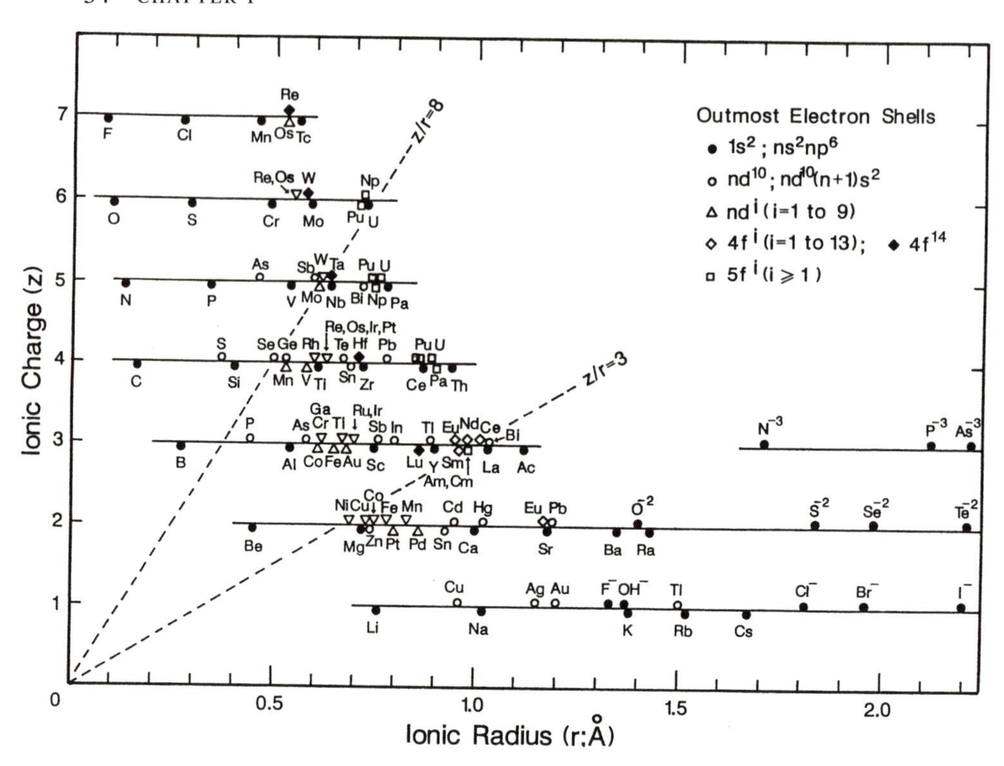

FIGURE I-15. Ionic radii (r) as functions of charges and electron configurations of outermost subshells.

because the ionic radii of Cr^{+3} and tri- and tetravalent noble metal cations are similar. Similarly, ilmenite ($FeTiO_3$) is often enriched in V, Hf-Zr, and Nb-Ta pairs. Also Y-REE (rare earth elements) and W-Mo pairs are closely associated in nature. Nonetheless, one should not predict the geochemical behavior of elements by ionic radius and charge alone. For example, according to ionic radius and charge, the B-type Cu^{+}, Ag^{+}, Au^{+}, Sn^{+2}, Cd^{+2}, and Hg^{+2} cations can easily substitute for K^{+}, Na^{+}, and Ca^{+2} in alkali and plagioclase feldspars of igneous rock. However, the concentrations of those B-type cations in these minerals are usually very low, because they may already have been segregated from the magma as sulfide minerals.

In general, cations with an ionic potential (z/r) of less than 3 exist as dissolved cations, with $8 > z/r > 3$ as least soluble hydroxide precipitates, and with $z/r > 8$ as dissolved oxyanions in oxygenated aqueous solutions (Mason, 1966a).

I-6. Electric Polarizability

Under the influence of an external electric field, E, the electron sheath around an atom is deformed to form an electric dipole moment, μ, which is proportional to E. The proportionality constant, k_α, is called the **electric polarizability**, i.e., $\mu = k_\alpha \cdot E$. The electric polarizability has units of volume (10^{-24} cm^3/atom or ion). The k_α values for some A- and B-type ions are summarized in Table I-6 (Pauling, 1927). For ions with the same electron configuration, k_α decreases systematically from the highly charged anions to the highly charged cations (Table I-6). Comparing cations with the same charge, the k_α for B-type cations (and probably transition metal cations) is always much higher than for A-type cations with similar ionic radii, as exemplified in Figure I-16. In other words, the electron sheaths of A-type cations are spherically symmetric and hard to deform under the electric field (low polarizability), whereas the electron sheaths of B-type and transition metal cations are highly polarizable, especially in the d subshell electrons. Since the electric polarizability of the S^{-2} anion is much higher than that of the O^{-2} anion (10.3 vs. 3.92×10^{-24} cm^3/ion in Table I-6), the B-type and some transition metal cations tend to form sulfides with strong covalent characteristics (i.e., electron pair sharing). A-type cations form oxides with strong ionic bond characteristics (i.e., electrostatic attraction between cation and anion). Also, the complexation constants of a given B-type cation with various ligands are in the order of $P^{-3} > N^{-3}$; $S^{-2} > O^{-2}$; $I^- > Br^- > Cl^- > F^-$ (in increasing order of k_α), and $NO_3^- > CO_3^{-2}$; $ClO_4^- > SO_4^{-2} > PO_4^{-3}$ (in decreasing order of k_α). The order is reversed for A-type cations (Stumm and Morgan, 1981).

One may consider a reaction

$$\frac{1}{z}M_2O_z(s) + S(s) \leftrightarrow \frac{1}{z}M_2S_z(s) + \frac{1}{2}O_2(g).$$

Its equilibrium constant is $K = P_{O_2}^{1/2} = \exp(-\Delta\tilde{G}/RT)$, or

$$-RT\ln K = -\frac{RT}{2}\ln P_{O_2} = \Delta\tilde{G} = (\Delta G_f^0 M_2S_z - \Delta G_f^0 M_2O_z)/z,$$

where $\Delta\tilde{G}$ is the Gibbs free energy of the reaction, $\Delta G_f^0 M_2S_z$ and $\Delta G_f^0 M_2O_z$ are the Gibbs free energies of formation for M_2S_z and M_2O_z solids, and z is the valence of M; for $z = 2$ and 4, $\Delta G_f^0 M_2O_z$ (or $\Delta G_f^0 M_2S_z$) is equivalent to 2 $\Delta G_f^0 MO$ and 2 $\Delta G_f^0 MO_2$, respectively. The calculated $(\Delta G_f^0 M_2S_z - \Delta G_f^0 M_2O_z)/z$ values for B-type and transition metal cations are summarized in Table I-7 (see page 38). The term $(\Delta G_f^0 M_2S_z - \Delta G_f^0 M_2O_z)/z$ is a measure of the relative affinity of cations for sulfides over oxides, or a measure of the **relative polarizability** of cations. The more negative the $(\Delta G_f^0 M_2S_z -$

TABLE I-6
Electric polarizability K_α

z												Electron
−4	−3	−2	−1	0	1	2	3	4	5	6	7	configuration
			H	He	Li	Be	B	C	N			[He]
			10.17	0.203	0.029	0.0079	0.003	0.0013	0.0007			
C	N	O	F	Ne	Na	Mg	Al	Si	P	S	Cl	[Ne]
2140	28.8	3.92	1.05	0.394	0.181	0.094	0.054	0.033	0.021	0.014	0.01	
Si	P	S	Cl	Ar	K	Ca	Sc	Ti	V	Cr	Mn	[Ar]
377	41.6	10.3	3.69	1.64	0.84	0.472	0.29	0.19	0.12	0.087	0.063	
					Cu	Zn	Ga	Ge	As	Se	Br	[Ar]$3d^{10}$
					0.43	0.29	0.2	0.14	0.1	0.075	0.059	
Ge	As	Se	Br	Kr	Rb	Sr	Y	Zr	Nb	Mo		[Kr]
109	28.8	10.6	4.81	2.48	1.42	0.86	0.56	0.38	0.26	0.19		
					Ag	Cd	In	Sn	Sb	Te	I	[Kr]$4d^{10}$
					1.72	1.09	0.73	0.5	0.36	0.26	0.19	
Sn	Sb	Te	I	Xe	Cs	Ba	La	Ce				[Xe]
90.4	31.9	14.1	7.16	4.03	2.44	1.56	1.05	0.74				
					Au	Hg	Ti	Pb	Bi			[Xe]$4f^{14}\,5d^{10}$
					1.88	1.24	0.87	0.62	0.46			

Note: Units are 10^{-24} cm^3/atom or ion.

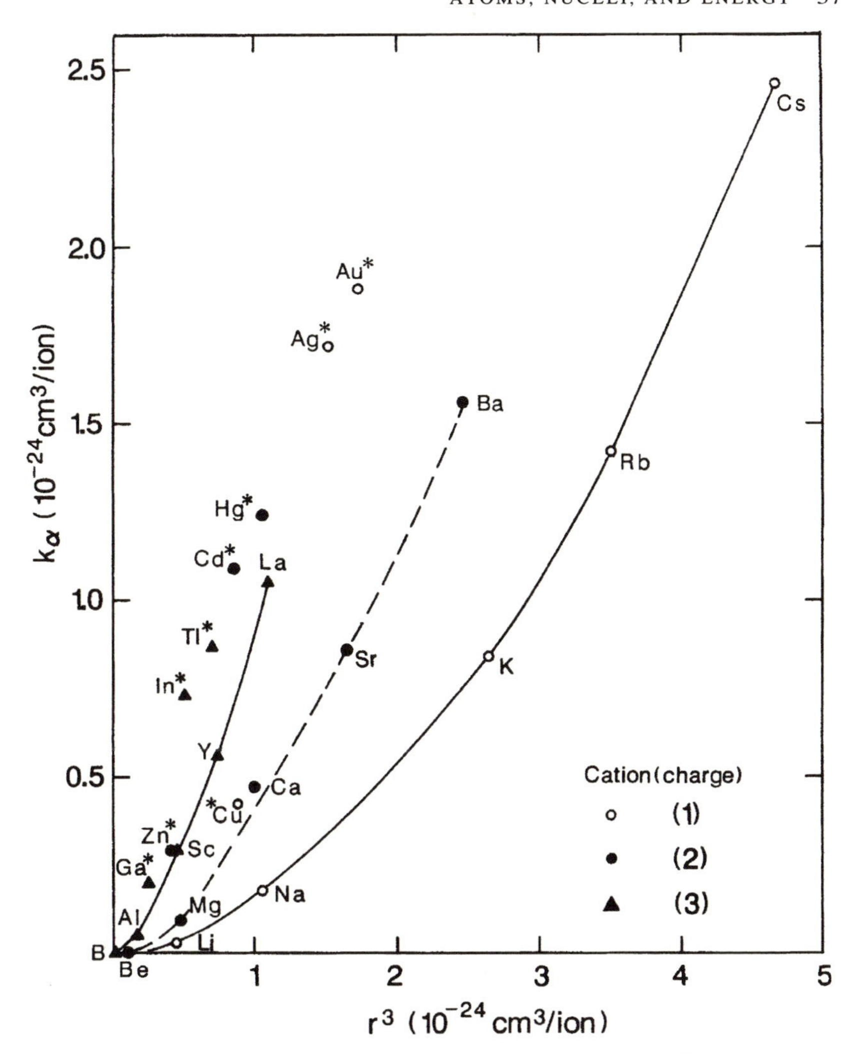

FIGURE I-16. Electric polarizability of cations (k_α) as a function of the cube of the ionic radius and the charge. Notice that the B-type cations (with asterisk) always fall above the lines connecting the A-type isovalent cations.

$\Delta G^0_f M_2O_z)/z$ values, the higher the equilibrium partial pressure of oxygen, (P_{O_2}), the higher the affinity to sulfides, and the higher the relative polarizability of cations. The relative polarizabilities of Cu^+, Cu^{+2}, Ag^+, Au^+, Au^{+3}, Hg^+, Hg^{+2}, Pd^{+2}, Pt^{+2}, Pt^{+4}, Tl^{+3}, Ir^{+3}, Ir^{+4}, Rh^{+3}, Os^{+4}, and Ru^{+4} are high as shown in Table I-7. These cations also form strong complexes with organic ligands and are highly toxic to living organisms, in part due to their blockage of vital functional groups in biomolecules (Ochiai, 1987).

TABLE I-7

$(\Delta G^0_f M_2S_z - \Delta G^0_f M_2O_z)/z$ values for B-type and transition metal cations

	$n = 3$				$n = 4$				$n = 5$			
	$z = 1$	2	3	4	$z = 1$	2	3	4	$z = 1$	2	3	4
nd^1		Sc	Ti									
			55									
nd^2		Ti	V					Mo				W
		53	51					37				41
nd^3		V	Cr					Tc				Re
		48	58					20				24
nd^4								Ru				Os
								8				8
nd^5		Mn	Fe									Ir
		37	39									−1
nd^6		Fe					Rh				Ir	Pt
		34					−4				−3	−3
nd^7		Co										
		31										
nd^8		Ni				Pd				Pt		
		32				−1				−8		
nd^9		Cu										
		17										
nd^{10}	Cu	Zn	Ga	Ge	Ag	Cd	In	Sn	Au	Hg		
	11	28	38	43	−7	18	33	42		2		
$nd^{10} s^1$									Hg			
									7			
$nd^{10} s^2$		Ge	As			Sn	Sb		Tl	Pb	Bi	
		33	35			31	34		13	22	28	

Source: Thermodynamic data are mainly from Dean (1985), Karapet'yants and Karapet'yants (1970), Mills (1974), and Wagman et al. (1982).

Note: Units are kcal/mole.

I-7. Electronegativity

Pauling (1932, 1960) described qualitatively the **electronegativity**, χ, as "the power of an atom in a molecule to attract electrons to itself." Therefore, we would again expect a close relationship between the electronegativity and the electron binding energy. According to Pauling (1932, 1960), the electronegativity difference between two unlike atoms of nonmetals A_i and A_j is related by the following equation:

$$\alpha \cdot (\chi_{A_i} - \chi_{A_j})^2 = D(A_i \leftrightarrow A_j) - \{D(A_i \leftrightarrow A_i) + D(A_j \leftrightarrow A_j)\}/2, \qquad \text{(I-3a)}$$

where $D(A_i \leftrightarrow A_j)$ is the single bond energy between two unlike atoms in a gaseous molecule, $D(A_i \leftrightarrow A_i)$ or $D(A_j \leftrightarrow A_j)$ is the single covalent bond energy between two identical atoms in a diatomic or polyatomic gaseous molecule, and α is a proportionality constant with the dimension of energy. Therefore, according to equation I-3a, the electronegativities of atoms are dimensionless.

The $D(A_i \leftrightarrow A_j)$ values, including $i = j$, for many gaseous compounds containing nonmetals and hydrogen at the standard state can be obtained by thermochemical and spectroscopic methods and are listed by Pauling (1960) and Pauling and Pauling (1975). By assigning 2.5 and 4.0 as the electronegativities of carbon and fluorine, respectively, Pauling obtained an α value of 23 kcal/mole and the χ_A values for other nonmetals and hydrogen, as summarized in Table I-8. Even though he mentioned that his electronegativity values for the elements refer to the most common oxidation states of the elements, he did not explicitly assign the oxidation states. One can, however, infer the oxidation states of elements for each χ_A from the gaseous compound by which the $D(A_i \leftrightarrow A_j)$ value is estimated, assuming the fluorine atom in the gaseous compound is always a monovalent anion. The inferred oxidation states of the nonmetals are also given in Table I-8. As discussed by Pauling (1960), the larger the electronegativity difference between two unlike nonmetal atoms, the stronger the ionic bond or the weaker the covalent bond characteristics (independent of the bond strength itself).

The metals usually cannot exist as diatomic gases (except for alkali metals), and metal-nonmetal compounds are mostly solid or liquid at the standard state. Therefore, one cannot apply equation I-3a to obtain the electronegativity of metal atoms (χ_M). Pauling (1960) postulated, however, that

$$\alpha \cdot (\chi_M - \chi_A)^2 = \frac{-(\Delta H_f M_m A_n)}{m \cdot n}, \qquad \text{(I-3b)}$$

TABLE I-8
The electronegativities of metal cations ($\chi_{M^{+z}}$) and nonmetal ions (χ_{A^z})

Z	z	$\chi_{M^{+z}}$	Z	z	$\chi_{M^{+z}}$	Z	z	$\chi_{M^{+z}}$	Z	z	$\chi_{M^{+z}}$
Ac-89	3	1	Ga-31	3	1.6	Pd-46	2	2	Tl-81	1	1.5
Ag-47	1	1.8	Gd-64	3	*1.1		3	*2.1		3	1.9
Al-13	3	1.5	Ge-32	4	1.8	Pm-61	3	*1.1	Tm-69	3	*1.1
Am-95	3	*1.1	Hf-72	4	1.4	Po-84	4	2	U-92	4	1.4
As-33	3	2	Hg-80	1	*1.7	Pr-59	3	*1.1		6	*1.6
	5	2.1		2	1.8	Pt-78	1–4	*2.2	V-23	3	*1.55
At-85	5	2.2	Ho-67	3	*1.1	Pu-94	3,4	1.1		4	1.7
Au-79	3	2.3	In-49	1	*1.5	Ra-88	2	*0.85		5	1.9
B-5	3	2		3	*1.6	Rb-37	1	0.8	W-74	2	*1.8
Ba-56	2	0.9	Ir-77	3,4	2.2	Re-75	3	*2.0		4	*1.9
Be-4	2	1.5	K-19	1	0.8		5	*2.1		6	2.0
Bi-83	3	1.8	La-57	3	1.1		7	2.2	Y-39	3	1.2
C-6	4	2.5	Li-3	1	0.95	Rh-45	2,3	2.1	Yb-70	2	*1.0
Ca-20	2	1	Lu-71	3	1.15	Ru-44	3	*2.1		3	*1.2
Cd-48	2	1.5	Mg-12	2	1.2	S-16	2,4	*2.5	Zn-30	2	1.5
Ce-58	3	1.1	Mn-25	2	1.5		6	(2.3)	Zr-40	4	1.4
	4	*1.4		3	*1.7	Sb-51	3	1.9			
Cl-17	7	(2.5)		7	2.5		5	2.1			
Cm96	3	*1.2	Mo-42	2	*1.8	Sc-21	3	1.3			
Co-27	2	1.8		4	*1.9	Se-34	4	*2.4	*Nonmetal ions*		
	3	1.9		6	*2.0	Si-14	4	1.8	Z	z	χ_{A^z}
Cr-24	2	1.5	N-7	5	(3)	Sm-62	2	*1.0			
	3	1.6	Na-11	1	0.9		3	*1.1	As-33	3	2
	6	2.2	Nb-41	3	*1.5	Sn-50	2	1.7	Br-35	−1	2.8
Cs-55	1	0.75		5	*1.7		4	1.8	C-6	4	2.5
Cu-29	1	1.8	Nd-60	3	1.1	Sr-38	2	1	Cl-17	−1	3
	2	2	Ni-28	2	1.8	Ta-73	3	*1.6	F-9	−1	4
Dy-66	3	*1.1		3	*1.9		5	1.7	H-1	1	2.1
Er-68	3	*1.1	O-8	6	(3.5)	Tb-65	3	*1.1	I-53	−1	2.5
Eu-63	2	*0.9	Os-76	3,4	2.2	Tc-43	7	2.3	N-7	3	3
	3	*1.2	P-15	3,5	2.1	Te-52	4	*2.1	O-8	−2	3.5
F-9	7	(4)	Pa-91	4	*1.35	Th-90	2	*1.1	P-15	3	2.1
Fe-26	2	1.7		5	*1.45		4	*1.2	S-16	−2	2.5
	3	1.8	Pb-82	2	1.6	Ti-22	2	1.4	Se-34	4	2.4
Fr-87	1	0.7		4	*2.0		4	1.6	Si-14	4	1.8

Source: Data are mainly from Gordy and Thomas (1956) and Pauling (1960) except for those with an asterisk, which are newly calculated. The $\chi_{M^{+z}}$ values in parentheses are extrapolated from Figure I-17.

Note: Z and z are the atomic number and the charge of the ions, respectively.

if A is a nonmetal other than oxygen and nitrogen, or

$$\alpha \cdot (\chi_M - \chi_A)^2 = \frac{-(\Delta H_f M_m A_n + a \cdot n)}{m \cdot n}, \qquad \text{(I-3c)}$$

if A is oxygen or nitrogen, where $\Delta H_f M_m A_n$ is the enthalpy of formation for a metal-nonmetal solid compound $M_m A_n$ at the standard state; n and m are the absolute values of the charge of metal M and nonmetal A, respectively, in the solid compound $M_m A_n$; α is 23 kcal/mole; and a is 26.0 kcal/mole for oxygen and 55.4 kcal/mole for nitrogen.

Because χ_A and α are already known, χ_M can be calculated from equations I-3b and I-3c if ΔH_f for a solid metal-nonmetal compound is known. The χ_M values given by Pauling (1960) and Gordy and Thomas (1956) are summarized in Table I-8 with some revisions and additions, using more recent ΔH_f data for chloride, bromide, and iodide compounds (Wagman et al., 1982; Dean, 1985; Karapet'yants and Karapet'yants, 1970). As exemplified in Table I-9, the χ_M values calculated from fluoride and oxide data for mono- to trivalent cations are very often much larger than those from other compound data. Therefore, strictly speaking, the χ_M values in Table I-8 are applicable only to chloride, bromide, iodide, sulfide, and selenide compounds, and not to every kind of solid compound. The corresponding oxidation states of metals for χ_M, as inferred from the solid compound $M_m A_n$, are also indicated in Table I-8. As suggested by Haissinsky (1946) and Gordy and Thomas (1956), the electronegativity for multivalent metals varies with valence (see Table I-8). Therefore, it will be more convenient to talk about the electronegativity of a cation, $\chi_{M^{+z}}$, than that of a neutral element hereafter. The fact that $\chi_{M^{+z}}$ are equal to $\chi_{A^{+z}}$ for metalloid elements (As, C, P, Se, and Si)

TABLE I-9
Electronegativities of cations calculated by equations I-3b and I-3c for halides, oxides, sulfieds, and selenides

	F^-	Cl^-	Br^-	I^-	O^{-2}	S^{-2}	Se^{-2}
Na^+	1.55	0.93	0.86	0.77	1.85	1.12	1.07
Mg^{+2}	1.58	1.17	1.15	1.12	1.58	1.16	1.21
Al^{+3}	1.71	1.44	1.47	1.46	1.64	1.38	1.41
Si^{+4}		1.66	1.7	1.8	1.79	1.76	
P^{+5}		2.04	2.05				
Cu^+	2.59	1.81	1.75	1.66	2.3	1.86	1.84
Zn^{+2}	1.99	1.53	1.49	1.46	1.54	1.5	1.48
Ga^{+3}	1.99	1.65	1.64	1.59	1.93	1.5	1.53
Ge^{+4}		1.82	1.85	1.89	1.8	1.86	
As^{+5}					2.26	2.11	

in Table I-8 proves the validity of equation I-3b as postulated by Pauling (1960).

As would be expected, the **cationic electronegativities** are linearly correlated with I_z/z values for the A-type cations, while B-type cations and transition metal cations with a more than half-full d subshell tend to have higher $\chi_{M^{+z}}$ values than those for A-type cations at a given I_z/z, value as shown in Figure I-17. Furthermore, the extrapolated $\chi_{M^{+z}}$ values for the A-type cations such as F^{+6}, O^{+6}, and N^{+5} in Figure I-17 are identical to χ_A values for F^-, O^{-2}, and N^{-3}. Therefore, the electronegativities for these elements are independent of charge. The periodicity of cationic electronegativities as a function

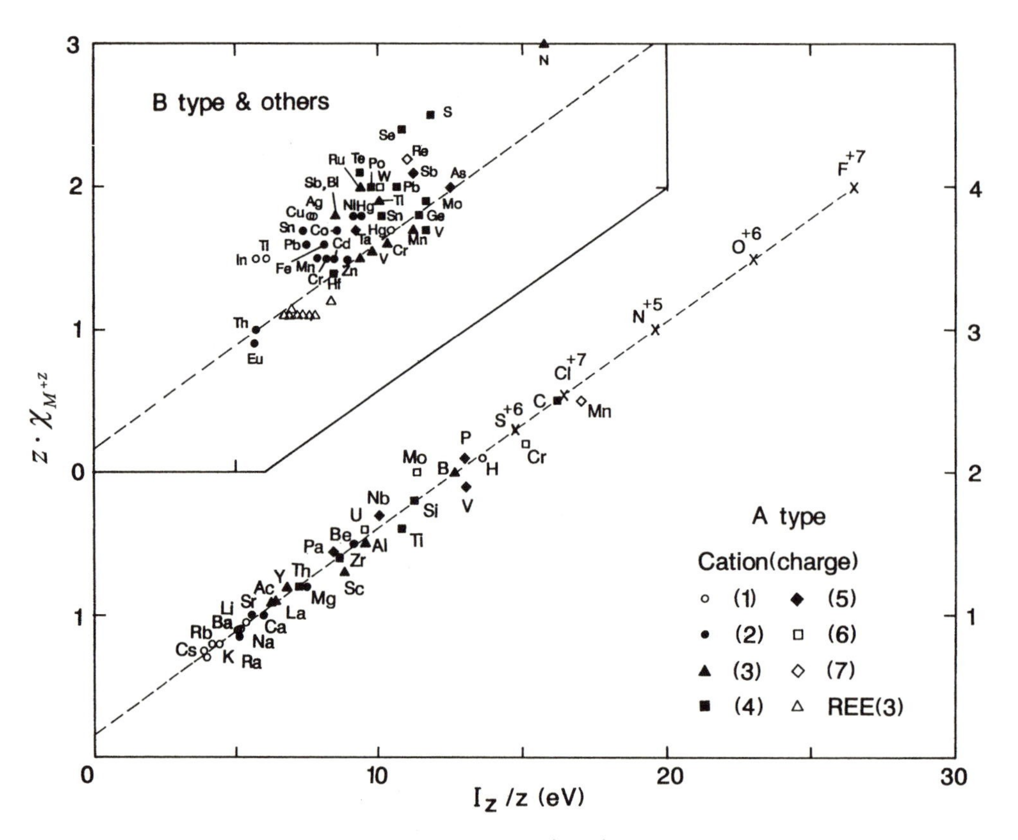

FIGURE I-17. Electronegativities of cations $(\chi_{M^{+z}})$ versus their polarizing powers (I_z/z). The crosses are extrapolated values. Data are from Tables I-3 and I-8.

of atomic number is shown in Figure I-18 and is very similar to the periodicity of I_z/z (Figure I-12).

Mulliken (1934) also defined the electronegativity of neutral elements as

$$\chi_M = \frac{I_1 + I_{-1}}{2}.$$

The original electronegativity scale by Pauling (his Table 3-8, 1960) is often assumed to be for the neutral elements and is compared with that by Mulliken (1934) as shown in Figure I-19. Though the general correlation between them is good, the scatter of O, H, B, Al, Ga, In, Tl, Pb, Bi, Cs, and many other transition metals is large, and those elements do not possess any common physicochemical property that explains the scatter.

In conclusion, Pauling's cationic electronegativity multiplied by cationic charge ($z \cdot \chi_{M^{+z}}$) is a quantity closely related to I_z for A-type cations. For

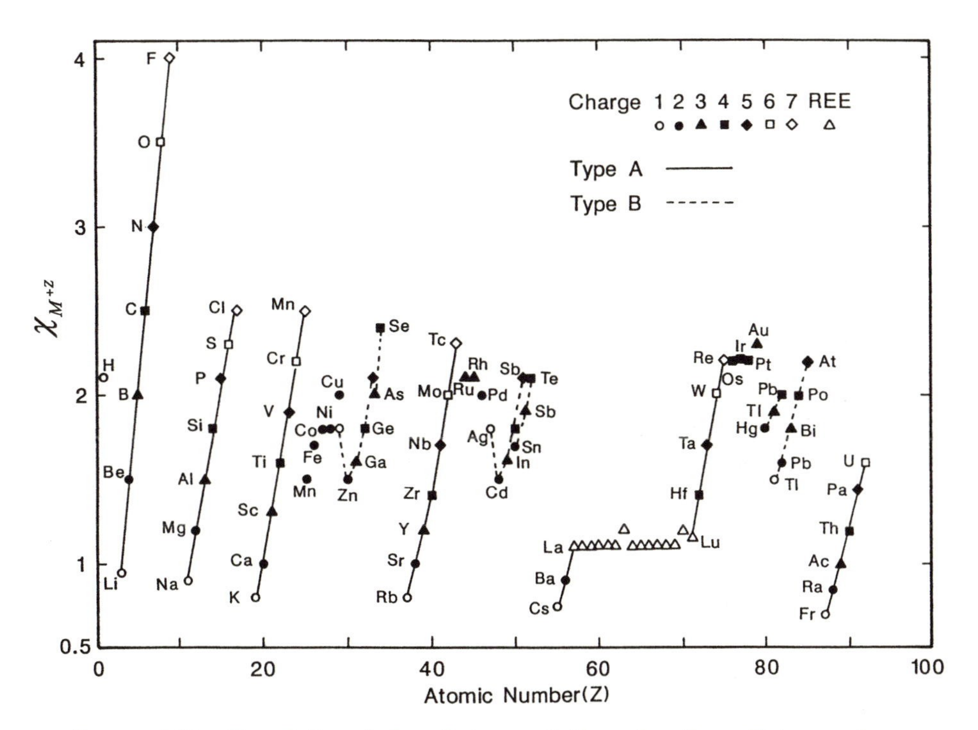

FIGURE I-18. Periodicity of the electronegativities of cations. Data are from Table I-8.

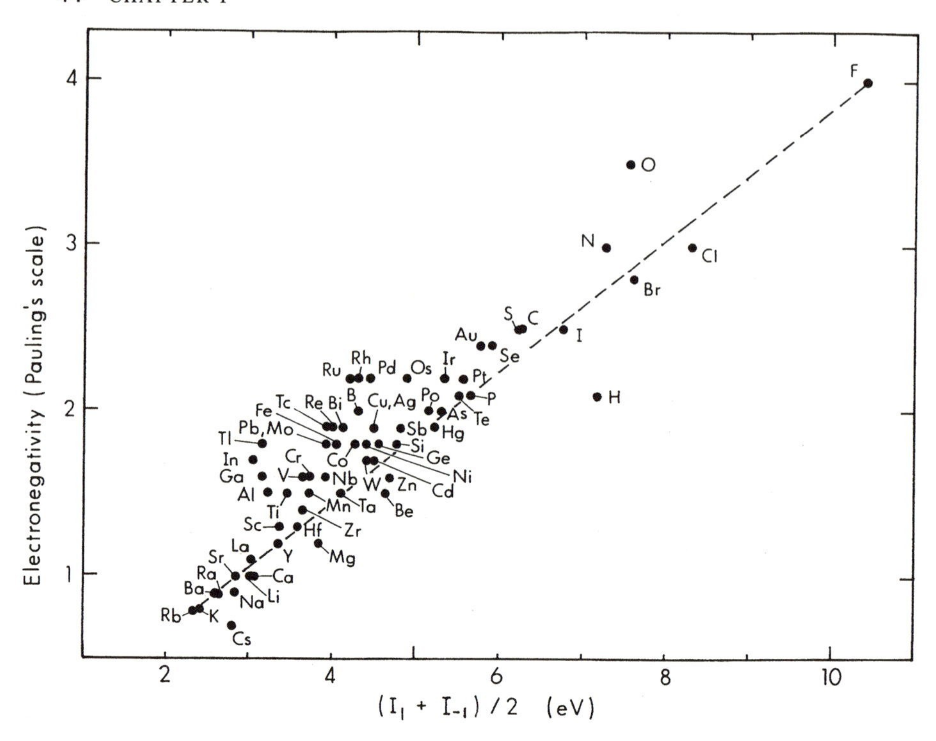

FIGURE I-19. Pauling's electronegativity scale versus Mulliken's $(I_1 + I_{-1})/2$.

B-type cations and some transitional metal cations, the I_z may underestimate the electron attraction power as compared to the corresponding $z \cdot \chi_{M^{+z}}$. We may call $z \cdot \chi_{M^{+z}}$ the **total cationic electronegativity**. As mentioned earlier, Pauling's electronegativity scale cannot apply to fluorides and oxides nor to silicates (Saxena, 1977). For those types of compounds, one has to adopt a new electronegativity scale specific to each type of compound, as Saxena (1977) did for silicates.

I-8. CRYSTAL LATTICE ENERGIES

The crystal lattice energy of a mineral, ΔH_L^0, is defined as the energy required to break up one mole of molecules in a crystalline form into gaseous cations and anions at 25°C and one bar. The anions can be monoatomic (e.g., O^{-2}, S^{-2}) or complex ions such as oxyanions (CO_3^{-2}, SiO_4^{-4}).

The ΔH_L^0 of an oxide $M_m O_n$ can be estimated by using the Born-Haber cycle (Born and Lande, 1918; Born, 1919; Haber, 1919), i.e.,

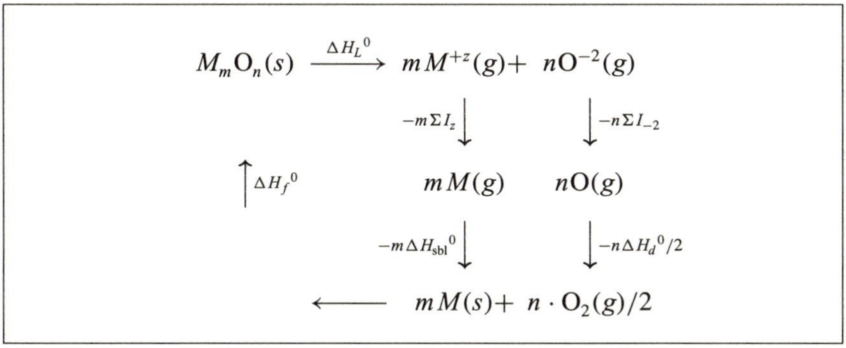

$$
\begin{array}{ccccc}
M_m O_n(s) & \xrightarrow{\Delta H_L{}^0} & mM^{+z}(g) + & nO^{-2}(g) \\
 & & \downarrow -m\Sigma I_z & \downarrow -n\Sigma I_{-2} \\
\uparrow \Delta H_f{}^0 & & mM(g) & nO(g) \\
 & & \downarrow -m\Delta H_{sbl}{}^0 & \downarrow -n\Delta H_d{}^0/2 \\
 & \longleftarrow & mM(s) + & n \cdot O_2(g)/2
\end{array}
$$

where m and n are the number of cations and oxygen atoms in the oxide formula, respectively. If the charge of the cation is z, then $m = 2n/z$; ΣI_z is the cumulative ionization energy of the metal M(g); ΔH_{sbl}^0 is the enthalpy of sublimation of the metal M(s); ΣI_{-2} is the affinity of two electrons to the gaseous oxygen atom = 601 kJ/mol; ΔH_d^0 is the enthalpy of dissociation of O_2 gas = 498 kJ/mol (Wagman et al., 1982); and ΔH_f^0 is the enthalpy of formation of the oxide $M_m O_n$.

Since the total enthalpy change for the cycle is equal to zero, i.e.,

$$\Delta H_L^0 - m(\Sigma I_z + \Delta H_{sbl}^0) - n(\Sigma I_{-2} + \Delta H_d^0/2) + \Delta H_f^0 = 0,$$

then

$$\Delta H_L^0 = m(\Sigma I_z + \Delta H_{sbl}^0) + n(\Sigma I_{-2} + \Delta H_d^0/2) - \Delta H_f^0, \qquad \text{(I-4a)}$$

or

$$\Delta H_L^0/n = 2/z \cdot (\Sigma I_z + \Delta H_{sbl}^0) + (\Sigma I_{-2} + \Delta H_d^0/2) - \Delta H_f^0/n. \qquad \text{(I-4b)}$$

$\Delta H_L^0/n$ can be considered as a measure of relative chemical bond strength between oxygen and various metals (O-M).

In equation I-4b, the $(\Sigma I_{-2} + \Delta H_d^0/2)$ term for oxygen is constant (= 850 kJ/mol), and the $2\Sigma I_z/z$ term is always much larger than the $2\Delta H_{\text{sbl}}^0/z$ and $-\Delta H_f^0/n$ terms. Therefore, a near-linear correlation for oxides between $\Delta H_L^0/n$ (summarized in Table I-10) and $\Sigma I_z/z$ is expected, and is proven in Figure I-20. As shown before, $\Sigma I_z/z$ and I_z are linearly related (Figure I-11), therefore $\Delta H_L^0/n$ should also correlate nicely with I_z. In short, the O-*M* bond strength as represented by the crystal lattice energy, $\Delta H_L^0/n$, is closely related to the electron binding energy, I_z, which is in turn closely related to the total cationic electronegativity $z \cdot \chi_{M^{+z}}$.

TABLE I-10
Crystal lattice energies of oxides (M_mO_n) per one oxygen, $\Delta H_L^0/n$, as calculated by the Born-Haber cycle

Oxide	$\Delta H_L^0/n$	*Oxide*	$\Delta H_L^0/n$	*Oxide*	$\Delta H_L^0/n$	*Oxide*	$\Delta H_L^0/n$
Ag_2O	2930	Eu_2O_3	4220	MoO_2	6160	SnO	3570
Al_2O_3	5060	FeO	3860	MoO_3	8600	SnO_2	5800
As_2O_3	4830	Fe_2O_3	4920	Na_2O	2490	SrO	3230
B_2O_3	6260	Fr_2O	2120	NbO	4040	Tb_2O_3	4270
BaO	3060	Ga_2O_3	5090	NbO_2	5700	ThO_2	4980
BeO	4450	GeO	3750	Nd_2O_3	4150	TiO	3840
Bi_2O_3	4690	GeO_2	6330	NiO	4020	Ti_2O_3	4760
CaO	3410	HfO_2	5510	Ni_2O_3	5240	TiO_2	5970
CdO	3730	HgO	3830	P_2O_3	5075	Tl_2O	2580
Ce_2O_3	4090	Ho_2O_3	4310	P_2O_5	8010	Tl_2O_3	4710
CeO_2	5160	In_2O_3	4720	PbO	3440	Tm_2O_3	4340
CoO	3930	K_2O	2240	PbO_2	5760	VO	3870
Cr_2O_3	5000	La_2O_3	4050	PdO	4000	V_2O_3	4870
Cs_2O	2120	Li_2O	2820	Rb_2O	2180	VO_2	6180
Cu_2O	3210	Lu_2O_3	4370	ReO_2	5550	V_2O_5	7660
CuO	4070	MgO	3800	Rh_2O_3	4970	Y_2O_3	4300
Dy_2O_3	4280	MnO	3760	Sc_2O_3	4590	ZnO	3980
Er_2O_3	4330	Mn_2O_3	5020	SiO_2	6520	ZrO_2	5450
EuO	3260	MnO_2	6470	Sm_2O_3	4190		

Source: Thermodynamic data sources are the same as in Table I-7.
Note: All are in units of kJ/mol oxygen.

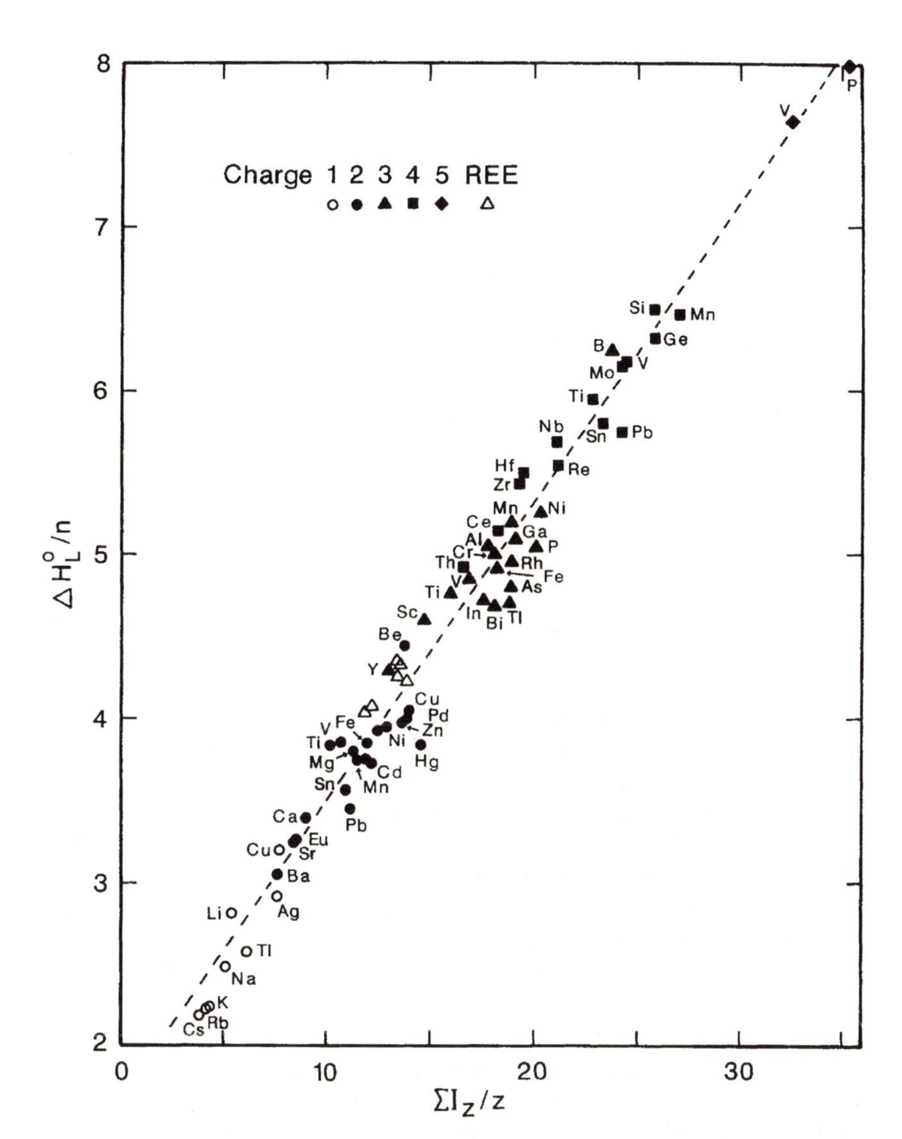

FIGURE I-20. Plot of crystal lattice energy for metal oxides divided by the number of oxygen atoms in the molar crystal formula $(\Delta H_L^0/n)$ versus the average electron binding energy $(\sum I_z/z)$ to metal cations.

I-9. Hydrolysis of Cations and Dissociation of Oxyacids

The hydrolysis of a cation M^{+z} in water can be represented by

$$M^{+z} + H_2O \rightarrow M(OH)^{z-1} + H^+,$$

$$M^{+z} + iH_2O \rightarrow M(OH)_i^{z-i} + iH^+,$$

where the first **hydrolysis constant** $^*K_1$ is defined by

$$^*K_1 = \frac{[M(OH)^{z-1}] \cdot [H^+]}{[M^{+z}]}, \qquad \text{(I-5a)}$$

and the ith cumulative hydrolysis constant $^*k_i$ by

$$^*k_i = \frac{[M(OH)_i^{z-i}] \cdot [H^+]^i}{[M^{+z}]}. \qquad \text{(I-5b)}$$

The $^*K_1$ values are summarized in Table I-11. The $\log {}^*K_1$ can be considered as a measure of relative bond strength between hydroxyl and metal ions (M-OH). The larger the $\log {}^*K_1$ value, the higher the tendency for the cation to form a hydroxide complex, or the higher the bond strength between cation and hydroxyl anion. For a natural water with pH = 8, $[M^{+z}]/[M(OH)^{z-1}]$ ratios are greater than one, if the $\log {}^*K_1$ values of the cations are less than −8 according to equation I-5a (mostly mono- and divalent cations in Table I-11). Cations with $\log {}^*K_1 > -8$ are strongly to fully hydrolyzed (mostly tri- and tetravalent cations). Cations with valence equal to or greater than four often form oxyanions.

As one would have expected, $\log {}^*K_1$ and $\log {}^*k_i$ are linearly correlated with I_z and $z \cdot \chi_M^{+z}$ for cations, as shown in Figure I-21B ($\log {}^*k_i$ values are not shown here) and Figure I-21A, respectively. The obvious exceptions are Be^{+2} and again some of the heavy B-type cations, such as Tl^+, Ag^+, Hg^{+2}, Pb^{+2}, Tl^{+3}, and Bi^{+3}. The $\log {}^*K_1$ and $\log {}^*k_i$ values are also linearly correlated with the logarithm of the complexation constants of cations with a given organic ligand (Balistrieri et al., 1981). Figure I-22 (see page 51) shows the example of the EDTA ligand for a reaction of $M^{+z} + L^{-4} \rightarrow ML^{z-4}$. In other words, the larger the $\log {}^*K_1$, the higher the tendency for the cation in solution to form hydroxyl and metal–organic ligand complexes and the higher the bond strengths between the cation and a given anion or ligand.

TABLE I-11
First hydrolysis constants of cations at 25°C and $I = 0$

Z	z	$-\log{}^*K_1$	Z	z	$-\log{}^*K_1$	Z	z	$-\log{}^*K_1$
Ac-89	3	10.4	Hf-72	4	0.3S	PuO_2^{+2}	6	5.6
Ag-47	1	12	Hg-80	2	3.4	Rb-45	3	3.3C
Al-13	3	5	Ho-67	3	8	Ru-44	3	2.2R
Am-95	3	5.8S	In-49	3	4	RuO_4^0	8	11.9
Ba-56	2	13.5	Ir-77	3	4.4C	Sc-21	3	4.3
Be-4	2	5.4	K-19	1	14.5	Sm-62	3	7.9
Bi-83	3	1.1	La-57	3	8.5	Sn-50	4	−1.5
Bk-97	3	5.5S	Li-3	1	13.6	Sr-38	2	13.3
Ca-20	2	12.9	Lu-71	3	7.6	Tb-65	3	7.9
Cd-48	2	10.1	Mg-12	2	11.4	Th-90	4	3.2
Ce-58	3	8.3	Mn-25	2	10.6	Ti-22	3	2.2
	4	−0.7		3	−0.3	TiO^{+2}	4	2.3
Cf-98	3	5.5S	Na-11	1	14.2	Tl-81	1	13.2
Cm-96	3	5.8S	Nd-60	3	8		3	0.62
Co-27	2	9.7	Ni-28	2	9.9	Tm-69	3	7.7
	3	1.3	Np-93	4	1.5	U-92	4	0.65
Cr-24	3	4	NpO_2^+	5	8.9	UO_2^{+2}	6	5.8
Cu-29	2	7.9	OsO_4^0	8	12.1	V-23	3	2.3
Dy-66	3	8	Pa-91	4	−0.84	VO^{+2}	4	5.4
Er-68	3	7.9	PaO_2^+	5	4.5	Y-39	3	7.7
Eu-63	3	7.8	Pb-82	2	7.7	Yb-70	3	7.7
Fe-26	2	9.5	Pd-46	2	(2.3)A	Zn-30	2	9
	3	2.2	Pr-59	3	8.1	Zr-40	4	−0.1S
Ga-31	3	2.6	Pu-94	3	7			
Gd-64	3	8		4	0.5			

Source: Data are mainly from Baes and Mesmer (1981). Other data are from (C) Cotton and Wilkinson (1988), (R) Rard (1985), and (S) Smith and Martell (1976).
Note: A: Pd value is at 17°C and in 0.1 M $NaClO_4$.

The dissociation of the hydrogen ion from an oxyacid (H_mA) can be represented by

$$H_mA \rightarrow H_{m-1}A^- + H^+ \qquad (K_1),$$

$$* \; H_{m-1}A^- \rightarrow H_{m-2}A^{-2} + H^+ \qquad (K_2),$$

where A is an oxyanion; m the number of hydrogen atoms in the oxyacid; and K_1 and K_2 the first and second **acid dissociation constants**.

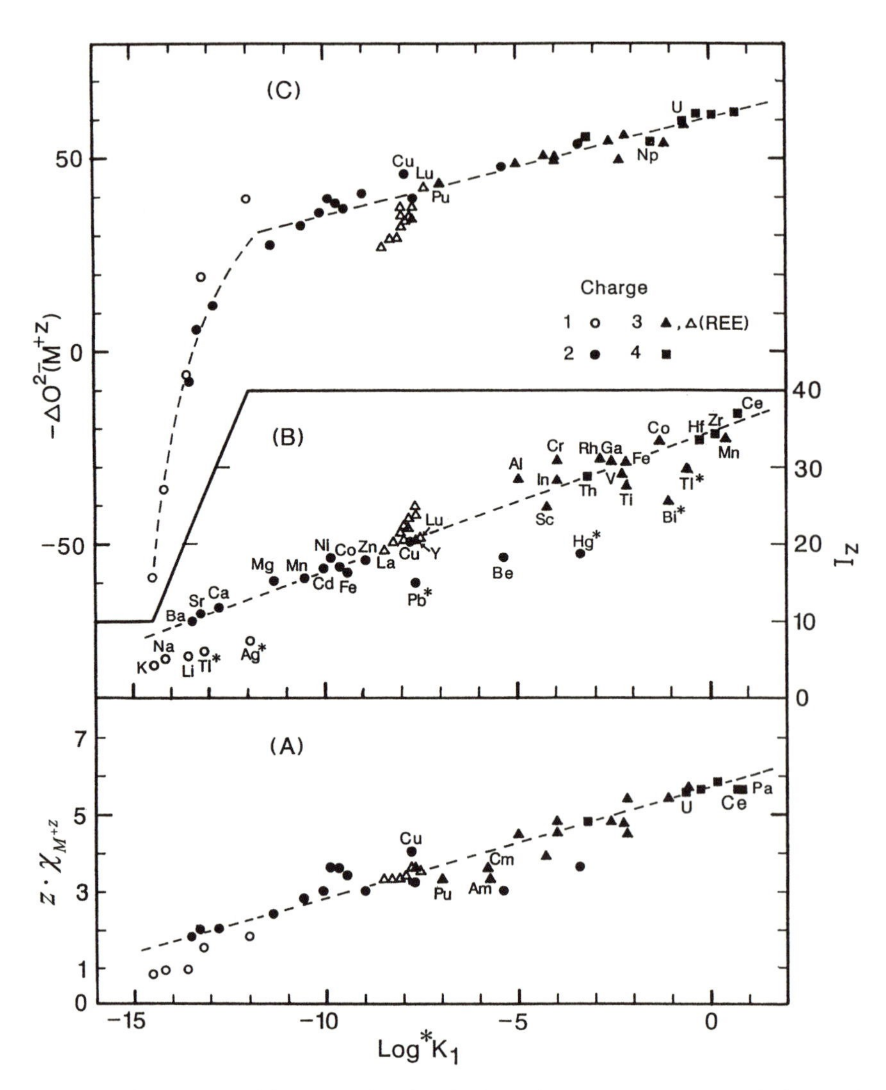

FIGURE I-21. Plots of $\log{}^*K_1$ (the first hydrolysis constant) versus (A) the total cationic electronegativity ($z \cdot \chi_{M^{+z}}$), (B) the electron binding energy I_z, and (C) the affinity of an aqueous cation to its oxide crystal ($-\Delta O^{2-}$). The ΔO^{2-} data are from Tardy and Garrels (1976).

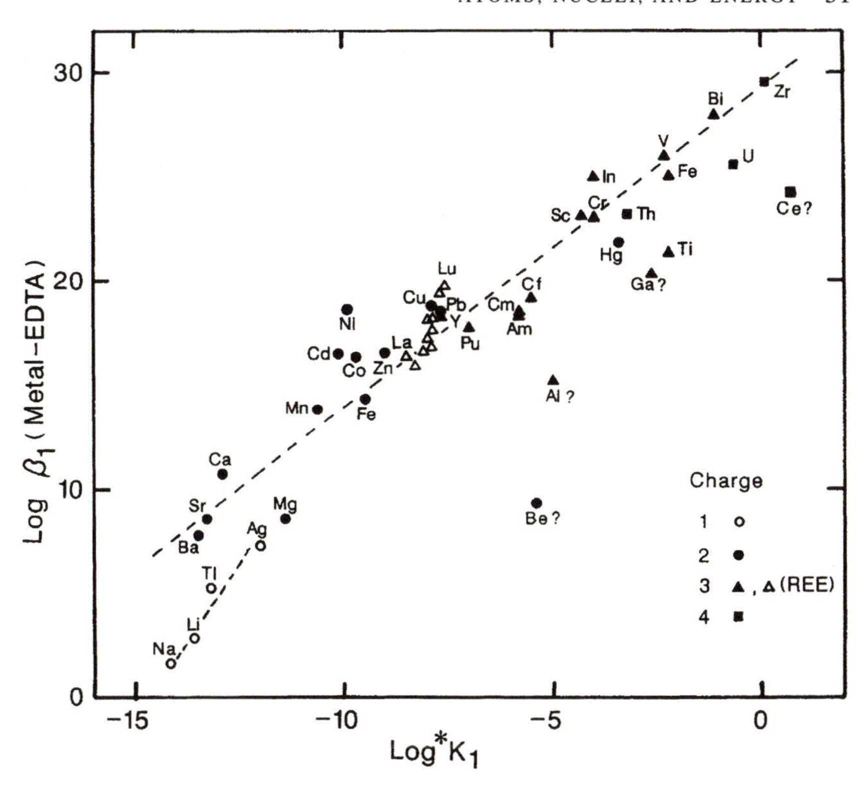

FIGURE I-22. Plot of log $^*K_1$ (the first hydrolysis constant) versus log β_1^{EDTA} (the complexation constant of the cation to EDTA for the reaction $M^{+z} + L^{-4} \rightarrow ML^{z-4}$). The log β_1 data are from Dean (1985) and Schwarzenback (1957).

The log K_1 and log K_2 values of various oxyacids are generally positively related to the I_z (or $z \cdot \chi_{M^{+z}}$) values of the central metals in the oxyacid as shown in Figure I-23 (only log K_1 vs. I_z is shown). In other words, the higher the I_z (or $z \cdot \chi_{M^{+z}}$) value of the central metal in the oxyacid, the stronger the chemical bond strength between the central metal and the oxygen of the oxyanion, and the weaker the bond strength between the hydrogen ion and the oxygen of the oxyanion (i.e., strong acids). In general, the higher the log K_1 and log K_2 values are, the lower the tendency for the oxyanion to be adsorbed onto hydrous oxide particles. In short, the first hydrolysis constants of cations and dissociation constants of oxyacids are useful parameters to predict the affinity of cations and oxyanions to the surface of a hydrous oxide, aluminosilicate, or organic particles in aquatic environments. A detailed discussion is given in Chapter VI.

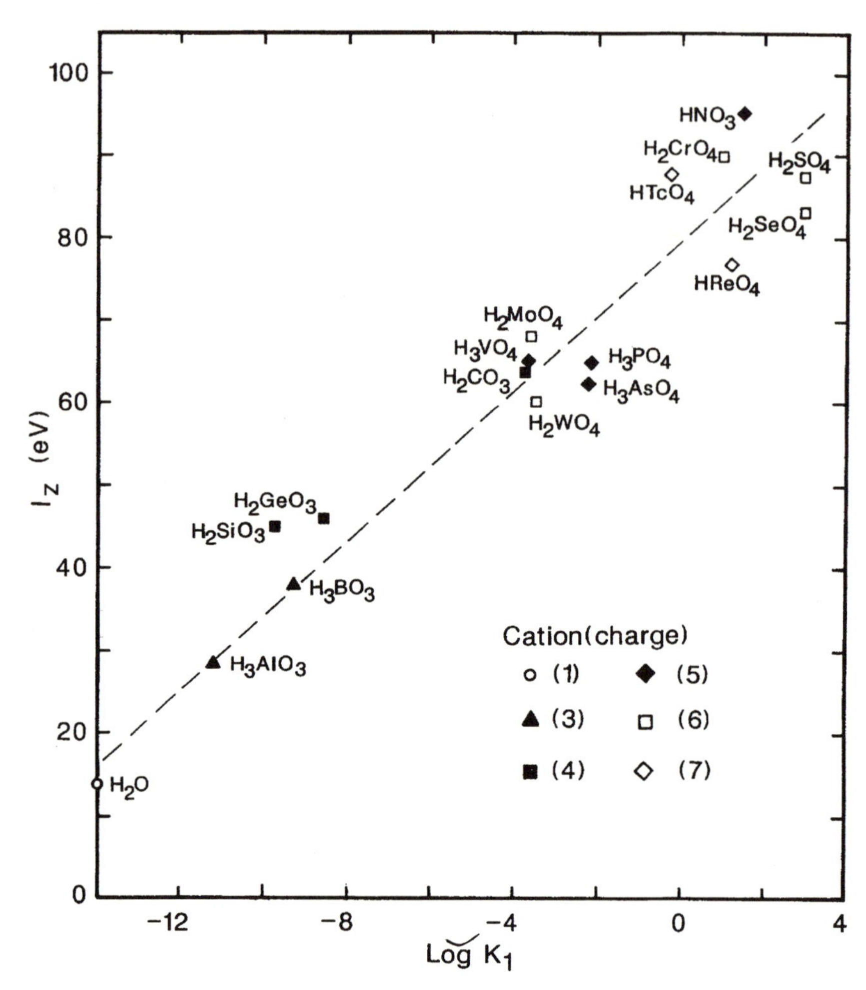

FIGURE I-23. log K_1 (the first dissociation constant of an oxyacid) versus the I_z of the central metal in those oxyacids. The log K_1 data are from Dean (1985).

I-10. SOLUBILITY PRODUCTS AND AFFINITY OF AQUEOUS CATIONS TO OXIDES

The precipitation or dissolution of metal oxides can be expressed by

$$\frac{1}{z}M^{+z} + \mathrm{OH}^- \leftrightarrow \frac{1}{2z} \cdot M_2\mathrm{O}_z + \frac{1}{2}\mathrm{H_2O},$$

where K_{pt} is the forward reaction (precipitation) constant,

$$K_{pt} = \frac{1}{[M^{+z}]^{1/z}[OH^-]} = 10^{-\Delta\tilde{G}/(2.3RT)},$$

$\Delta\tilde{G}$ is the Gibbs free energy of the reaction, and K_{sp} is the solubility product,

$$\begin{aligned} K_{sp} &= [M^{+z}][OH^-]^z \\ &= 1/(K_{pt})^z. \end{aligned}$$

Accordingly, $\log K_{pt}$ is equal to $-\log K_{sp}/z$ and proportional to $-\Delta\tilde{G} = \Delta G_f M^{+z}/z + \Delta G_f OH^- - \Delta G_f M_2O_z/2z - \Delta G_f H_2O/2$, or proportional to $(2\Delta G_f M^{+z} - \Delta G_f M_2O_z)/z$; thus these parameters are interchangeable. The higher these parameter values, the higher the tendency for the aqueous cation to form oxide precipitates, or the higher the average bond strength between cation and oxygen in an oxide crystal. We may call the parameter $(2\Delta G_f M^{+z} - \Delta G_f M_2O_z)/z$ the relative affinity of an aqueous cation to its oxide crystal, and it was designated as $-\Delta O^{2-}$ by Tardy and Garrels (1976). Therefore, $-\Delta O^{2-}$ is only another expression proportional to $-\log K_{sp}/z$.

The plot of $-\Delta O^{2-}$ (as well as $\log K_{pt}$ and $-\log K_{sp}/z$) versus $\log {}^*K_1$ shows again a roughly linear relationship (Figure I-21C). The obvious exceptions are alkali, alkaline earth, and some rare earth cations. One possible explanation is that the $\Delta\tilde{G}$ term contains not only the bond energy between cation and oxygen but also the cohesive energy for one mole of individual oxide molecules to aggregate into a solid crystal. As shown in Figures I-7 and I-8, ΔH^0_{sbl} values for alkali and alkaline earth metals are much lower than for other metals. Therefore, the oxide crystals of alkali and alkaline earth elements may also have much lower cohesive energies than other metal oxide crystals.

I-11. Concluding Remarks

The geochemical parameters discussed in this chapter are all closely interrelated quantities and all are in one way or another related to the concept of relative chemical bond strengths between cations and given ligands. The ionization energy data for various cations and anions are quite extensive and most are accurately determined. Therefore, the ionization energy or the electron binding energy, I_z, is one of the most useful geochemical parameters. I_z along with $\log {}^*K_1$ (first hydrolysis constant) and $\log K_1$ (first dissociation constant of oxyacids) are the most convenient overall parameters to relate the affinity of both cations and oxyanions to the surfaces of hydrous oxides, aluminosilicates, and organic particles in aquatic environments.

The total cationic electronegativities, $z \cdot \chi_{M^{+z}}$, are also linearly related to $\log {}^*K_1$ (Figure I-21) and to I_z (Figure I-17) except for some B-type and transition metal cations. Therefore the parameter $z \cdot \chi_{M^{+z}}$ should be as useful as the parameter I_z. The main drawbacks of the former are the larger uncertainty in determined values and somewhat limited data numbers.

The crystal lattice energies of oxides per oxygen ($\Delta H_L^0/n$) are essentially a linear function of both $\Sigma I_z/z$ and I_z (Figure I-20), and thus would not be more useful than the electron binding energies. The ionic potentials (z/r) of cations with a similar electron configuration are positively correlated to their I_z values (Figure I-14), but the ionic potential ignores the shielding effect of the electron sheath around the positively charged nucleus.

The classification of ions according to their electron configuration (A-type, B-type, etc.) and related parameters such as the electric polarizability (Table I-6) and the relative polarizability (Table I-7) is also a very useful tool to predict their chemical and geochemical behavior.

Chapter II

THE SOLAR NEBULA AND NUCLEOSYNTHESIS

INTRODUCTION

THIS CHAPTER discusses how the average composition of the solar nebula is estimated and highlights the main features of its composition. The "Big Bang" theory and stellar nucleosynthesis theory are introduced to explain the observed abundance patterns of elements and isotopes in various stars, including our solar system.

II-1. ELEMENTAL AND ISOTOPIC COMPOSITIONS OF THE SOLAR NEBULA

Our solar system presently consists of one burning star (the Sun), nine planets and their satellites, asteroids, and comets. Except for the Sun, all solar system bodies are supposed to have accreted from the condensates of solar nebular gases about 4.5 billion years (Ga) ago. The abundances of chemical elements in the original solar nebula have been deduced mainly from those of primitive meteorites called the **type 1 carbonaceous chondrites** (C1), which are composed mainly of fine-grained hydrous silicates and complex organic compounds, and will be discussed in chapter IV; and those of the present **solar photosphere** (Goldschmidt, 1937; Brown, 1949; Suess and Urey, 1956; Cameron 1968, 1973, 1982; Mason, 1979; Anders and Ebihara, 1982; Wasson, 1985; Anders and Grevesse, 1989). Terrestrial and lunar rock samples are chemically too severely fractionated to be useful for this purpose. Terrestrial materials along with meteorites, however, often serve as our primary sources for obtaining the relative isotopic abundances of most chemical elements in the solar nebula. The assumption is that the isotopic ratios for most elements are quite uniform throughout our solar system. Rare inclusions from some carbonaceous chondrites, however, show some heterogeneity for isotopes of C, N, O, Mg, Al, Si, Ca, Ti, Cr, Sr, Ba, Nd, Sm, and noble gases (Wasserburg et al., 1980; Lee, 1979, 1987; Anders, 1991; Ott, 1993; Amari et al., 1993; Nittler et al., 1994).

The average compositions of C1 chondrites and the solar atmosphere, as compiled by Anders and Grevesse (1989), are given in Table II-1. These compilations improved upon many earlier ones (e.g., Anders and Ebihara, 1982; Cameron, 1982; Allen, 1987; and references therein). The general

TABLE II-1

Chemical compositions of the primitive carbonaceous chondrites (C1) and the solar atmosphere

Z		C1 (ppm)	C1 (atoms/10^6 Si)	Sun (atoms/10^6Si)	Z		C1 (ppm)	C1 (atoms/10^6 Si)	Sun (atoms/10^6Si)
1	H	20,200	5.29E + 06	2.82E + 10	44	Ru	0.712	1.86	1.9
2	He	0.01	0.66	2.75E + 09	45	Rh	0.134	0.344	0.37
3	Li	1.5	57.1	0.41	46	Pd	0.56	1.39	1.4
4	Be	0.025	0.73	0.4	47	Ag	0.199	0.486	(0.25)
5	B	0.87	21	(11)	48	Cd	0.686	1.61	2
6	C	34,500	758,000	1.00E + 07	49	In	0.08	0.184	1.3
7	N	3,180	60,000	3.20E + 06	50	Sn	1.72	3.82	2.8
8	O	464,000	7.66E + 06	2.40E + 07	51	Sb	0.142	0.31	0.28
9	F	60.7	843	1,000	52	Te	2.32	4.81	
10	Ne	1.80E – 04	2.40E – 03	3.50E + 06	53	I	0.433	0.9	
11	Na	5,000	57,000	60,000	54	Xe	5.00E – 05	1.00E – 04	4.35
12	Mg	98,990	1.08E + 06	1.10E + 06	55	Cs	0.187	0.372	
13	Al	8,680	84,900	83,000	56	Ba	2.34	4.49	3.8
14	Si	106,400	1.00E + 06	1.00E + 06	57	La	0.235	0.446	0.47
15	P	1,220	10,400	7,900	58	Ce	0.603	1.14	1
16	S	62,500	515,000	460,000	59	Pr	0.089	0.167	0.14
17	Cl	704	5240	18,900	60	Nd	0.452	0.828	0.89
18	Ar	1.34E – 03	8.80E – 03	102,000	62	Sm	0.147	0.258	0.28
19	K	558	3,770	3,700	63	Eu	0.056	0.097	0.091
20	Ca	9,280	61,100	65,000	64	Gd	0.197	0.33	0.37
21	Sc	5.82	34.2	35	65	Tb	0.0363	0.06	(0.022)
22	Ti	436	2,400	2,800	66	Dy	0.243	0.394	0.35

23	V	56.5	293	280	67	Ho	0.0556	0.089	(0.051)
24	Cr	2,660	13,500	13,000	68	Er	0.159	0.251	0.24
25	Mn	1,990	9,550	6,900	69	Tm	0.0242	0.0378	(0.028)
26	Fe	190,400	900,000	1.30E + 06	70	Yb	0.163	0.248	0.34
27	Co	502	2,250	2,300	71	Lu	0.0243	0.0367	(0.16)
28	Ni	11,000	49,300	50,000	72	Hf	0.104	0.154	0.21
29	Cu	126	522	460	73	Ta	0.014	0.0207	
30	Zn	312	1,260	1,100	74	W	0.093	0.133	(0.36)
31	Ga	10	37.8	21	75	Re	0.0365	0.0517	
32	Ge	32.7	119	72	76	Os	0.486	0.675	0.79
33	As	1.86	6.56		77	Ir	0.481	0.661	0.63
34	Se	18.6	62.1		78	Pt	0.99	1.34	1.8
35	Br	3.57	11.8		79	Au	0.14	0.187	0.29
36	Kr	3.30E − 05	1.00E − 04	45.3	80	Hg	0.258	0.34	
37	Rb	2.3	7.09	11	81	Tl	0.142	0.184	(0.22)
38	Sr	7.8	23.5	22	82	Pb	2.47	3.15	1.9
39	Y	1.56	4.64	4.9	83	Bi	0.114	0.144	
40	Zr	3.94	11.4	11	90	Th	0.0294	0.0335	0.037
41	Nb	0.246	0.7	0.74	92	U	0.0081	0.009	(0.01)
42	Mo	0.928	2.55	2.9					

Source: Data are mainly from Anders and Grevesse (1989), except that H, C, N, O, and noble gas data for Cl are adopted from Orgueil carbonaceous chondrite (Anders and Ebihara, 1982). For the solar nebula composition (in units of atoms/10^6 Si), Cl data can be used except that values for H, C, N, O, and noble gases should be substituted by the solar atmosphere data.

Notes: E ± 06 is equivalent to $10^{\pm 6}$. Values in parentheses are less reliable.

agreement between the C1 and solar atmosphere compositions is surprisingly good (within a factor of two or better), as shown in Figure II-1. The apparent exceptions are H, He, C, N, O, and other noble gases, which are greatly depleted in meteorites compared to the solar atmosphere. Those elements are called **atmophiles** by Goldschmidt (1954). Li is also greatly depleted in the solar atmosphere, probably due to the destruction of Li by nuclear reactions involving thermal or suprathermal protons in the Sun,

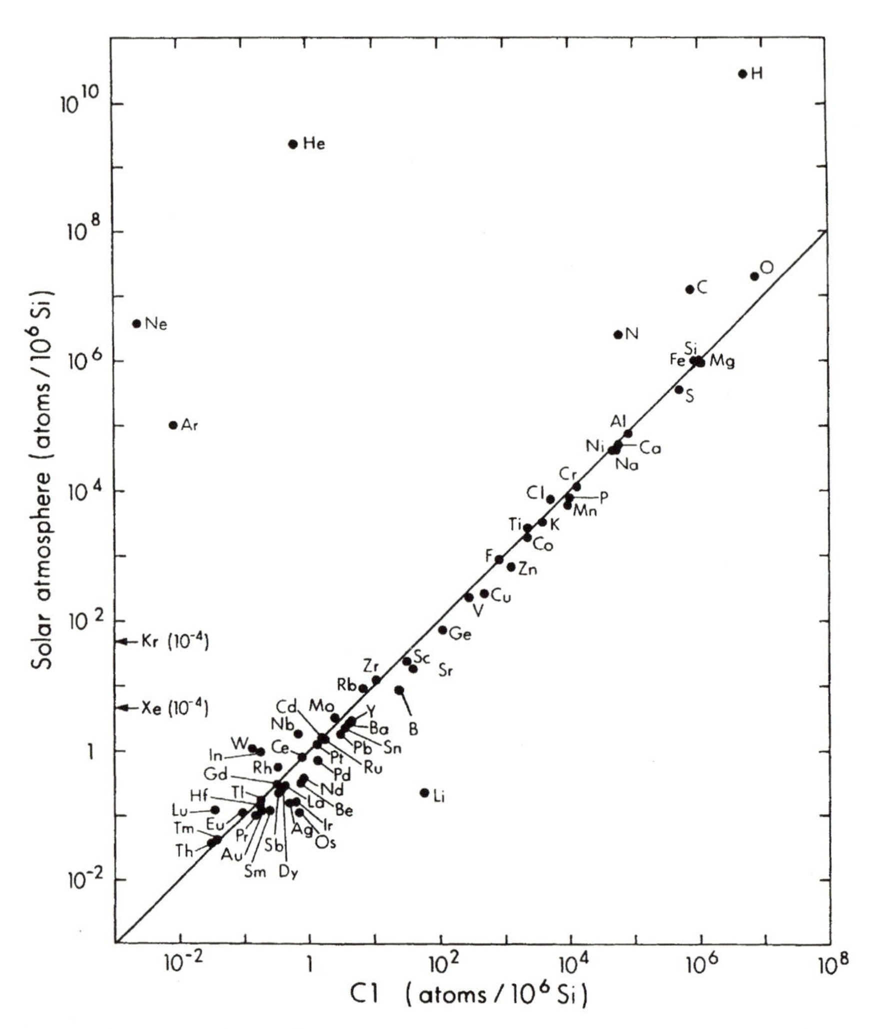

FIGURE II-1. Correlation plot of the abundances of elements in the type 1 carbonaceous chondrites against those in the solar atmosphere (Table II-1). Perfect agreement is indicated by the solid line.

for example, $^7Li + p \rightarrow 2 \times {}^4He$ (Ross and Allen, 1976; Holweger, 1979). Whether the higher Fe content in the solar atmosphere than in C1 is real (Anders and Grevesse, 1989) or not (Anders and Ebihara,1982; Allen, 1987) needs further verification.

The C1 data can be adapted to represent the solar nebula composition (in units of atoms per 10^6 Si atoms), except for H, He, C, N, O, and other noble gases, which should be obtained from the solar atmospheric data. A slight change in the solar nebula composition as a function of heliocentric distance will be discussed in Chapter III. In general, the abundances of the elements in the solar nebula decrease exponentially with increasing atomic number Z (Figure II-2). Also, except for He and Be, the abundances of even-Z elements are always higher than those of the neighboring odd-Z elements. The causes of the broad maxima around Fe, Sn to Ba, and Os to Pb will become clearer in the following sections. Margolis and Black (1985) demonstrated a surprisingly good agreement between the composition data of the solar nebula and the galactic cosmic ray sources by applying suitable corrections

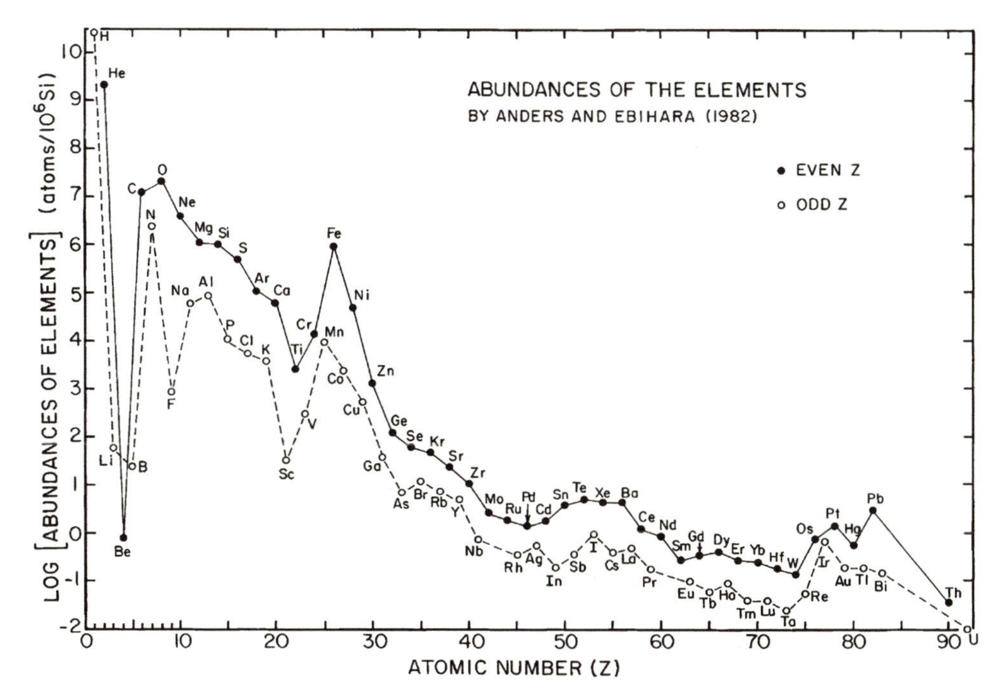

FIGURE II-2. Abundances of elements in the solar nebula as a function of atomic number Z. The even-Z elements are always more abundant than neighboring odd-Z elements except for He and Be.

to the observed galactic cosmic ray composition data (Binns et al., 1984; Engelmann et al., 1981). The corrections include the changes caused by nuclear reactions between cosmic ray nuclei and interstellar matter during the galactic propagation, and the fractionation caused by the acceleration of cosmic ray nuclei in the source region (this fractionation is shown to be related to the first ionization energy of the gaseous elements).

In astronomical literature, the capital letters X, Y, and Z often refer, respectively, to the weight fractions of H, He, and all other elements heavier than He. Cameron (1982) gave $X = 0.77 \pm 0.04$, $Y = 0.21 \pm 0.04$, and $Z = 0.02$; whereas Anders and Grevesse (1989) reported $X = 0.707$, $Y = 0.274$, and $Z = 0.019$ for the solar nebula. Other useful parameters are the weight fractions of Fe and Ni ($f_{Fe} = 0.0015$), of metal oxides of Na, K, Mg, Ca, Mn, Fe, Co, Ni, Al, Si, etc. ($f_{oxides} = 0.005$), and of Ne, Ar, H_2O, CH_4, and NH_3($f_{hydrides} = 0.017$) in the solar nebula (based on Table II-1).

The abundances of isotopes of all elements in the solar nebula as compiled by Anders and Grevesse (1989) are given in Appendix Table A-2. Because this new compilation differs only slightly from the earlier one by Anders and Ebihara (1982), the earlier reference is the source of data for the plots of abundances of isotopes in Figures II-3 and II-4. In Figure II-3, only the most abundant isobar is plotted for each mass number. For those nuclei with even A having two isobars with near equal abundances, the sum is plotted. Nuclei with odd A have only one stable isobar. Five exceptions are actually unstable nuclei having very long decay half-lives (^{87}Rb, $t_{1/2} = 4.8 \times 10^{10}$y; ^{113}Cd, 9×10^{15}y; ^{115}In, 4.4×10^{14}y; ^{123}Te, 1.3×10^{13}y; and ^{187}Re, 4.5×10^{10}y; their stable counterparts are ^{87}Sr, ^{113}In, ^{115}Sn, ^{123}Sb, and ^{187}Os). In Figure II-4, all stable isobars with even A and $A > 70$ are plotted. Those isobars are further separated, according to Figure I-3, into neutron-rich (solid circles) and proton-rich isobars (triangles), and isobars shielded (open circles) or unshielded (small dots) by neutron-rich isobars.

The main features that stand out in Figures II-3 and II-4 follow:

- there is a roughly exponential decrease of the abundances from hydrogen to $A \approx 100$, then a much slower decrease for $A > 100$
- except for ^{2}H, ^{6}Li, and ^{10}B, the abundances of even-A nuclei are mostly higher than, but in some cases equal to, those of neighboring odd-A nuclei
- Li, Be, and B isotopes are rare as compared to their neighboring nuclei ^{4}He and ^{12}C, and nuclei with $A = 5$ and 8 are absent
- there are high abundances of nuclei with A a multiple of 4, such as ^{4}He, ^{12}C, ^{16}O, ^{20}Ne, ^{24}Mg, ^{28}Si, ^{32}S, ^{36}Ar, ^{40}Ca (all are multiples of the helium nucleus), ^{44}Ca, and ^{48}Ti
- there are pronounced peaks around ^{56}Fe for even-A nuclei and around ^{57}Fe for odd-A nuclei

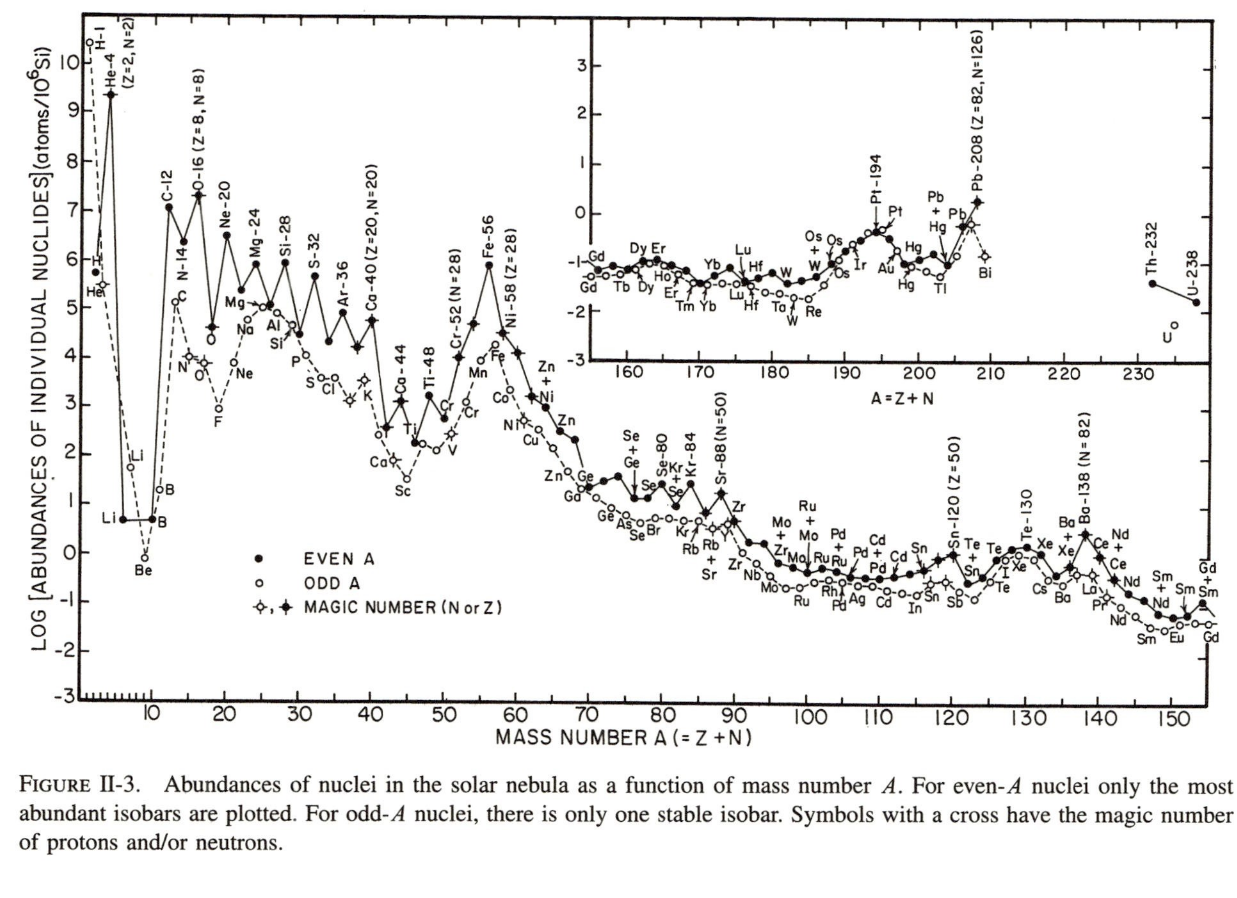

FIGURE II-3. Abundances of nuclei in the solar nebula as a function of mass number A. For even-A nuclei only the most abundant isobars are plotted. For odd-A nuclei, there is only one stable isobar. Symbols with a cross have the magic number of protons and/or neutrons.

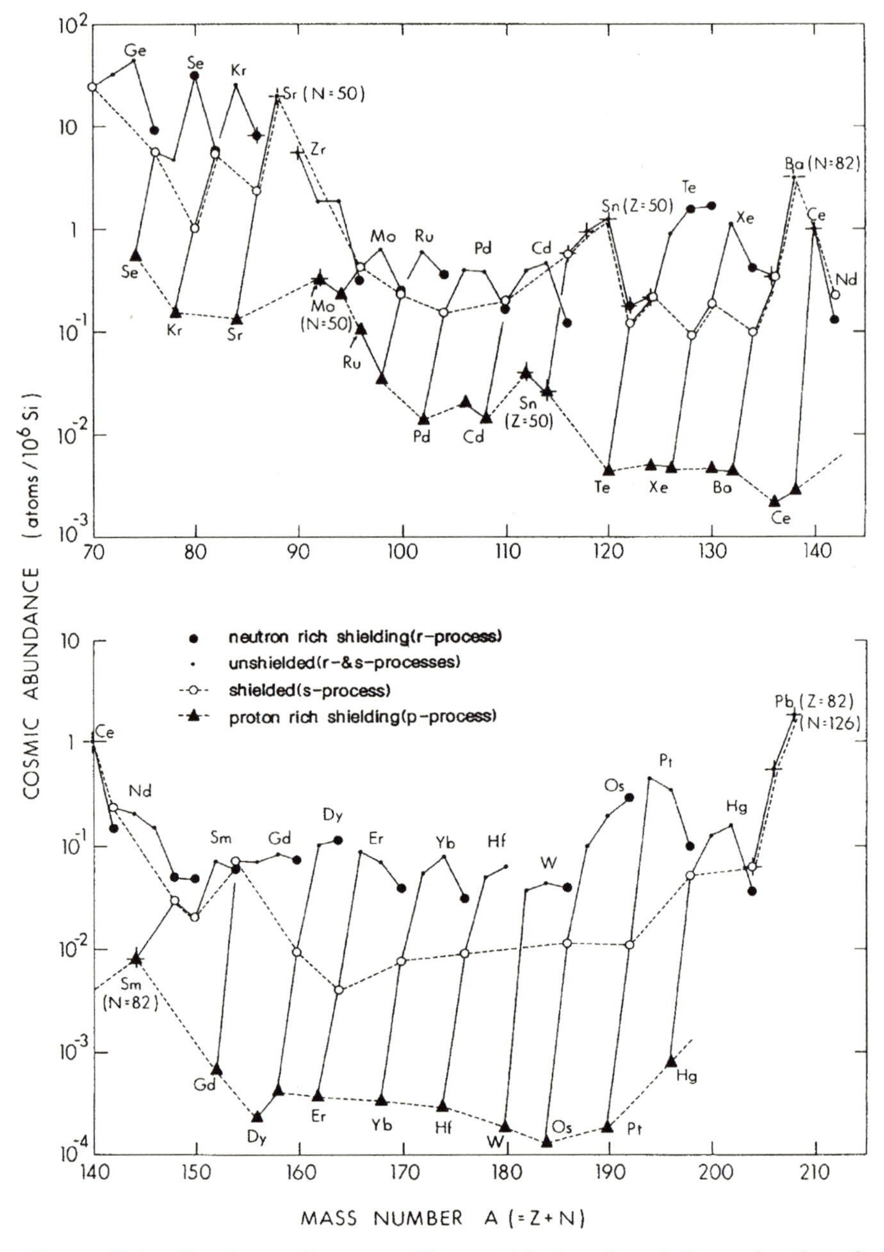

FIGURE II-4. Abundance of isotopes with even A in the solar nebula as a function of mass number. The nuclei are further separated into neutron-rich shielding, unshielded, shielded, and proton-rich shielding isobars according to Figure I-3. The major nucleosynthesis processes responsible for the formation of these isobars are also indicated. Symbols with a cross have the magic number of protons and/or neutrons.

- there are peaks around the nuclei with magic neutron and/or proton numbers, such as ^{88}Sr(N = 50), ^{120}Sn(Z = 50), ^{138}Ba(N = 82), and ^{208}Pb(Z = 82, N = 126)
- peaks around A = 80, 130, and 194 are broader for both even-A and odd-A nuclei
- there is little or no difference in abundance between even-A and odd-A nuclei around A = 130, 164, 194
- proton-rich heavy nuclei are rare but show peaks at nuclei with magic numbers again, e.g., ^{92}Mo(N=50), ^{112}Sn(Z = 50), and ^{144}Sm(N = 82) in Figure II-4

Some of these features are directly related to the stability of nuclei. As discussed in Section I-2, nuclei with even A (i.e., even Z and even N, except five odd-Z and odd-N nuclei) are more stable than their neighboring nuclei with odd A (i.e., even Z and odd N or odd Z and even N). Nuclei with magic proton and/or neutron numbers as well as multiples of α particles have extra stability (Figure I-5). Furthermore, the nuclei around ^{56}Fe have the highest bonding energy per nucleon (Figure I-4). However, in order to explain all the features in the solar abundance curves, Hoyle et al. (1956) and Cameron (1957) proposed various nucleosynthesis processes in stars. Burbidge et al. (1957) treated the nucleosynthesis processes in much greater detail. Their contributions became the basis for almost all later modeling works on nucleosynthesis in stars. In addition, there are many reviews and books on this important and evolving subject (e.g., Clayton, 1968; Barnes et al., 1982; Trimble, 1975, 1982, 1983; Truran, 1984; Fowler, 1984; Arnett and Truran, 1985; Audouze and Mathieu, 1985; Woosley and Weaver, 1986a, 1989; Arnett, 1996; and Arnett and Bazan, 1997).

II-2. Cosmological Nucleosynthesis

The most abundant elements in the Universe are H and He. Heavier elements must form from H and He via intrastellar reactions. This begs the question of how and where H and He were formed to begin with. The most probable explanation is the **Big Bang theory**, nicely summarized by Riordan and Schramm (1991), Silk (1989), and Kaufman (1985).

About 15 billion years ago, the entire mass and energy of the present Universe was concentrated in one point called the **cosmic singularity**. The cosmic singularity had infinite density, where space and time did not exist as separate entities. For unknown reasons, the cosmic singularity suddenly expanded with violence in the event called the Big Bang.

Our understanding of what happened to the Universe between the Big Bang (time zero) and 10^{-12} seconds after is highly speculative. According to the **Theory of Everything**, the four fundamental forces of nature, i.e., the forces of gravitational, electromagnetic, strong, and weak interactions (Table II-2a), are believed to be totally indistinguishable between time zero and 10^{-43} seconds (the so-called **Planck time**) when the temperature of the Universe was extremely high ($\geq 10^{32}$ K) and the average particle energy was $\geq 10^{22}$ MeV (Figure II-5). The temperature (T) and average particle energy (E) are related by the equation $E = kT$, where k is Boltzmann's constant $= 8.6 \times 10^{-5}$ eV/K. At Planck time, the gravitational force was differentiated from the other three, still unified forces. According to the **Grand Unified Theories** (GUTs), the strong, weak, and electromagnetic forces acted as a single unified force between 10^{-43} and 10^{-34} seconds after the Big Bang (the so-called GUT epoch in Figure II-5). At 10^{-34} seconds when the temperature was 10^{28} K and the average particle energy was 10^{18} MeV, the strong force was separated from the still unified electromagnetic and weak forces, and the Universe expanded greatly (the so-called epoch of inflation in Figure II-5). At 10^{-12} seconds ($T = 10^{16}$ K, $E = 10^{12}$ MeV), the electromagnetic and weak forces also split off. Between 10^{-34} and 10^{-12} seconds, the Universe was filled with leptons, quarks, and their antiparticles (Table II-2b) as well as free $W^{\pm}$ and Z^0 particles (weak-force carriers), free gluons (strong-force carriers), and photons (carriers of electromagnetic force) (Table II-2a). However, when the temperature fell below 10^{15} K at 10^{-10} seconds, $W^{\pm}$ and Z^0 particles disappeared all together as free particles.

At 10^{-6} seconds when the temperature fell to about 10^{13} K and the average particle energy to about 10^3 MeV, quarks and antiquarks combined to form baryons (protons, neutrons, etc.), mesons (pions, kaons, etc.) (Table II-2b), and their corresponding antiparticles. A baryon consists of three quarks, and a meson of one quark and one antiquark. All are held together by strong-force-carrier gluons.

Because the rest masses of the neutrons (n) and antineutrons ($\bar{n}$) are both about 939.6 MeV, the energy of a photon in equilibrium ($n + \bar{n} \leftrightarrow 2\gamma$) should

TABLE II-2A
Four fundamental forces and their carrier particles

Type	Relative strength	Range (meters)	Carrier of force	Rest mass (MeV)	Electric charge	Spin of carrier
Strong	1	$<10^{-15}$	gluons	0	0	1
			mesons	>100	0, ±1	0
Electromagnetic	10^{-2}	infinite	photons	0	0	1
Weak	10^{-13}	$<10^{-17}$	weakons $W^{\pm}$	81,000	±1	1
			weakon Z^0	93,000	0	1
Gravitational	10^{-38}	infinite	graviton	0	0	2

Source: Modified from Ostdiek and Bord, (1991).

Note: Gluons hold quarks together in baryons and mesons, whereas mesons hold baryons together in nuclei.

TABLE II-2B
Some physical properties of quarks, baryons, mesons, and leptons

Quarks (one-half spin; interacting via strong force)						
Name	*Symbol*	*Anti-particle*	*Rest mass (MeV)*	*Electric charge*		
down	d	$\bar{d}$	310	−1/3		
up	u	$\bar{u}$	310	+2/3		
strange	s	$\bar{s}$	505	−1/3		
charm	c	$\bar{c}$	1500	+2/3		
bottom	b	$\bar{b}$	5000	−1/3		
top	t	$\bar{t}$	>22500	+2/3		
Baryons (one-half spin; interacting via strong force)						
Name	*Symbol*	*Anti-particle*	*Rest mass (MeV)*	*Electric charge*	*Quark content*	*Threshold temperature (10^{12} K)*
proton	p	$\bar{p}$	938.3	+1	uud	10.91
neutron	n	$\bar{n}$	939.6	0	udd	10.93
lambda	Λ^0	$\bar{\Lambda}^0$	1115.6	0	uds	12.97
sigma	Σ^+	$\bar{\Sigma}^-$	1189.4	+1	uus	13.83
	Σ^-	$\bar{\Sigma}^+$	1197.3	−1	dds	13.92
Mesons (integer spin; interacting via strong force)						
pion	Π^0	self	135.0	0	$u\bar{u}, d\bar{d}$	1.57
	Π^+	Π^-	139.6	+1	$u\bar{d}$	1.62
kaon	K^+	K^-	493.7	+1	$u\bar{s}$	5.74
	K^0	$\bar{K}^0$	497.7	0	$d\bar{s}$	5.79
eta	η^0	self	548.8	0	$u\bar{u}, d\bar{d}, s\bar{s}$	6.38
Leptons (one-half spin; interacting via weak force)						
eletron	e^-	e^+	0.511	−1		0.006
e-neutrino	ν	$\bar{\nu}$	$0(<5 \times 10^{-5})$	0		
muon	μ^-	μ^+	105.7	−1		1.23
μ-neutrino	ν_μ	$\bar{\nu}_\mu$	$0(<0.52)$	0		
tauon	τ^-	τ^+	1784	−1		20.7

Notes: (1) Quarks, baryons, and leptons all have one-half spin, and are collectively called fermions, which follow Pauli's exclusion principle, i.e., no two fermion particles can have the exact same quantum states. Mesons and force carrier particles all have integer spin, and are called bosons, which do not follow Pauli's principle. Baryons also include other particles like Xi and Omega.

(2) For antiparticles, electric charges change sign.

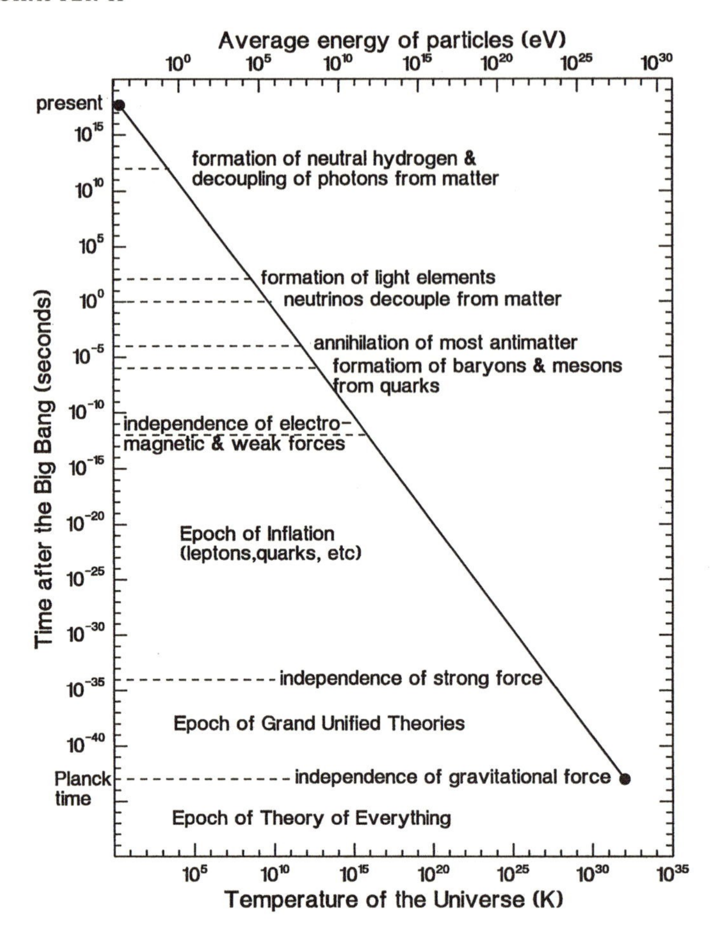

FIGURE II-5. Major events in the expanding Universe from the Big Bang to the present. The temperature of the Universe or the average energy of particles and antiparticles as a function of time is shown schematically as a solid diagonal line.

be 939.6 MeV also. The photon energy of 939.6 MeV for black body radiation corresponds to a temperature of 10.9×10^{12} (according to $T = E/k$, where k is the Boltzmann constant, E the average photon energy). This temperature is called the **threshold temperature** for neutrons and antineutrons. The threshold temperature for protons and antiprotons is only slightly lower. When the temperature of the Universe fell below the threshold temperature of a given particle-antiparticle pair, this pair would be annihilated into photons.

About 10^{-4} seconds after the Big Bang, the temperature of the Universe fell below 10^{13} K because of its continuous expansion; thus neutron-antineutron and proton-antiproton pairs were annihilated into photons. When the universe was about one second old, the temperature fell below 5.9×10^9 K and the annihilation of electron-positron pairs began. The end result was an increase of photon content and a decrease of the matter-antimatter content in the Universe. However, the presently observable Universe seems to be dominated by matter over antimatter. Therefore slightly more quark particles than antiquarks must have been created at the end of the GUT epoch in order to produce surplus protons and neutrons. This imbalance is called **symmetry breaking**.

At a temperature slightly below 5.9×10^9 K, the numbers of protons and neutrons were kept roughly equal by the reversible reactions

$$p + e^- \leftrightarrow n + \nu,$$
$$p + \bar{\nu} \leftrightarrow n + e^+.$$

Two seconds after the Big Bang, the matter in the Universe was so thinned out that neutrinos (ν) and antineutrinos ($\bar{\nu}$) no longer interacted significantly with protons and neutrons to maintain the above reversible reactions. In other words, the Universe became transparent to neutrinos and antineutrinos. Meanwhile, the forward neutron-producing reactions slowed down relative to the backward proton-producing reactions, because neutrons are slightly heavier than protons. Therefore the neutron-to-proton ratio fell to about 1/3. Also, neutrons decayed spontaneously by the following reaction with a decay half-life of 12 minutes:

$$n \rightarrow p + e^- + \bar{\nu}.$$

At this point, the photons were still energetic enough to prevent the formation of ^{2}H nuclei (a combination of a proton and a neutron).

About two to three minutes after the Big Bang, the temperature fell to about 10^9 K, and the following nuclear reactions became possible:

$$^1\mathrm{H}(n, \gamma)\,^2\mathrm{H}(p, \gamma)\,^3\mathrm{He},$$
$$^2\mathrm{H}(^2\mathrm{H}, n)\,^3\mathrm{He}\,(^3\mathrm{He}, 2p)\,^4\mathrm{He},$$
$$^2\mathrm{H}(^2\mathrm{H}, p)\,^3\mathrm{H}\,(^2\mathrm{H}, n)\,^4\mathrm{He}(^3\mathrm{H}, \gamma)\,^7\mathrm{Li}.$$

The main end result was the creation for every 16 protons of one ^{4}He nucleus, which is observed in the Universe today; ^{2}H, ^{3}He, and ^{7}Li were minor but important products. The protons and neutrons in nucleons are held together by the strong-force-carrier pions.

When the Universe was about 10^6 years old, the temperature decreased to about 3000 K. At this temperature, the photons were no longer energetic enough to prevent the formation of neutral hydrogen atoms ($p + e^- \rightarrow H$). Once uncharged hydrogen atoms were formed throughout the Universe, the scattering of photons by free electrons ceased. Subsequently, photons were able to move freely from matter. We see these same photons (decoupled from matter) today as the 2.7 K cosmic microwave background (Penzias and Wilson, 1965). This period is called the **era of recombination or decoupling**.

Whether or not the present expansion of the Universe will continue forever is very much dependent on the total mass of the Universe. According to theoretical calculation, if the present average density of the Universe is equal to or less than 5×10^{-30} g/cm^3 (the so-called critical density), the Universe will expand forever and the temperature of the Universe will eventually approach absolute zero (an event called the **Big Chill**). Otherwise, the Universe will recontract some day and come back to the cosmic singularity (the **Big Crunch**).

The best estimate of the total mass of the detectable Universe, including visible galaxies, invisible but detectable clouds of gas, brown dwarfs, dark stars, and even galactic black holes, can provide only 10 to 15% of the critical density. In order to have a contracting Universe, we need to look for yet undetected "hidden matter" (or "dark matter"). Possible candidates for the hidden matter are various neutrinos. If the average rest mass of these neutrinos in space is about 20 to 30 eV, the Universe will have enough mass to contract. However, if the recent discovery that the expansion of the Universe is accelerating is true (Perlmutter et al., 1998; Glanz, 1998), the Big Chill will be the fate of the Universe. Certainly further studies are needed.

II-3. Stellar Nucleosynthesis

In the early Universe (after the Big Bang event), the H and He gases were probably more or less evenly distributed but with some local density fluctuations. Density fluctuations would have caused gravitational instability. Eventually the gases collapsed into numerous superclusters and clusters of galaxies. The recent discovery of anisotropy in cosmic microwave background radiation by NASA's cosmic background explorer (COBE) does indeed suggest that density fluctuations existed in the early Universe (Levi, 1992). Within each galaxy, smaller fragments of gas clouds contracted gravitationally and gradually coalesced into lumps called protostars. The gravitational potential energy of a protostar is proportional to the square of its mass (M^2); therefore, the more massive a protostar is, the more gravitational potential energy per unit mass is available to be converted into heat as it contracts, and the lower is its central density when it reaches a given central temperature. If the mass of a protostar is greater than 0.08 solar mass ($M_\odot = 2 \times 10^{30}$ kg), the central temperature can eventually become high enough (1 to 4 × 10^7 K) to induce a series of nuclear reactions called **core hydrogen burning**. At this instant, the protostar becomes a full-fledged star on the **main sequence** of the **Hertzsprung-Russell** (luminosity versus temperature) **diagram** (Figure II-6a). In Figure II-6a the luminosity (L) is a measure of total energy output by a star (usually expressed relative to that of the sun), and the temperature (T) is the surface temperature of the star. According to the Stefan-Boltzmann law, L is equal to $4\pi r^2 \sigma T^4$, where r is the radius of the main sequence star and σ is the Stefan-Boltzmann constant ($= 5.67 \times 10^{-5}$ erg cm^{-2}K^{-4} sec^{-1}). Therefore, the high luminosity of a star can result from high surface temperature and/or large radius of the star. For example, the **red giant** stars and **red supergiant** stars in Figure II-6a have low surface temperatures but have high luminosity due to their large size. Similarly, the **white dwarf** stars have high surface temperatures but low luminosity due to their small size.

Furthermore, it was observed that the luminosity of the main sequence star is roughly proportional to the third to fourth power of its mass (M^3 to M^4) when M is greater than $0.5M_\odot$ (Figure II-6a). Therefore, the more massive a star, the faster it burns hydrogen fuel, and the shorter is its life span as a main sequence star. In short, the evolutionary path of a main sequence star very much depends on its initial mass, as shown in Figure II-6b for stars with 1, 5, and 15 solar masses. For ease of illustration, a possible scenario of nucleosynthesis processes in the stars with masses of 10 to 70 $M_\odot$ is given

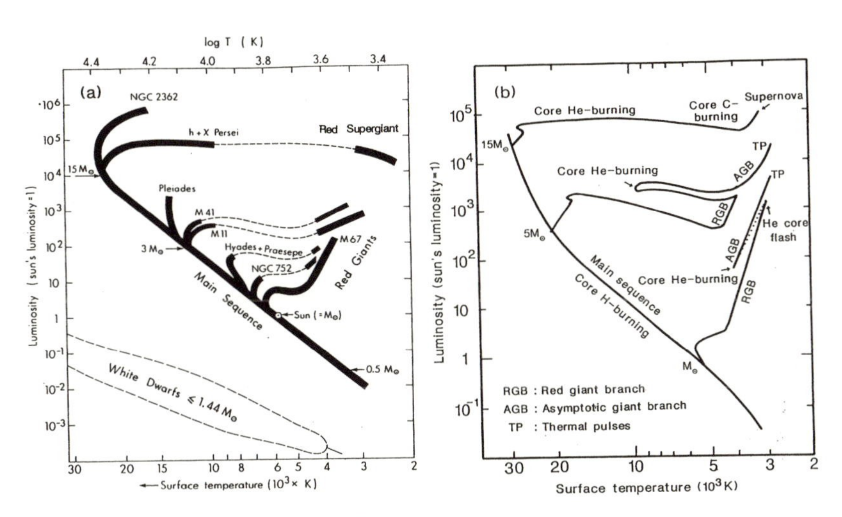

FIGURE II-6. (a) The Hertzsprung-Russell diagram for galactic clusters in the plane of our Milky Way galaxy (modified from Sandage, 1957). Thick solid bars represent numerous stars overlapping one another in each given galactic cluster. (b) Theoretical evolutionary tracks of stars with 1, 5, and 15 solar masses in the Hertzsprung-Russell diagram (modified from Ott, 1993).

first. It must be kept in mind that this scenario is highly dependent on model parameterization and the choice of values for given parameters, and thus does not represent the only possible reaction sequence.

Hydrogen Burning

The main thermonuclear reaction chains are

$$^{1}\mathrm{H}(p, e^{+}\nu)\,^{2}\mathrm{H}(p, \gamma)\,^{3}\mathrm{He}\,(^{3}\mathrm{He}, 2p)\,^{4}\mathrm{He}.$$

At a higher temperature (3×10^{7} K), the following chain reactions also become important:

$$^{4}\mathrm{He}\,(^{3}\mathrm{He}, \gamma)\,^{7}\mathrm{Be}(e^{-}, \nu)\,^{7}\mathrm{Li}(p, {}^{4}\mathrm{He})\,^{4}\mathrm{He},$$
$$^{7}\mathrm{Be}(p, \gamma)\,^{8}\mathrm{B}(e^{+}\nu)\,^{8}\mathrm{Be}(\gamma)2 \times {}^{4}\mathrm{He}.$$

The main results are a fusion of four hydrogens into ^{4}He, a large release of nuclear energy, and the formation of minor isotopes such as ^{2}H, ^{3}He, ^{7}Be, and ^{7}Li. One should be aware that the gaseous atoms are completely dissociated into bare nuclei and electrons at the high temperature deep inside the star. If the main sequence star is one of the so-called **population I stars**, which by definition are enriched with elements heavier than He (i.e., Z elements), then the preexisting C, N, and O nuclei (products of earlier supernova explosions of very massive stars) can act as catalysts to facilitate hydrogen burning by the following so-called **CNO cycle**:

$$^{12}\mathrm{C}(p, \gamma)\,^{13}\mathrm{N}(e^{+}\nu)\,^{13}\mathrm{C}(p, \gamma)\,^{14}\mathrm{N}(p, \gamma)\,^{15}\mathrm{O}(e^{+}\nu)\,^{15}\mathrm{N}(p, {}^{4}\mathrm{He})\,^{12}\mathrm{C},$$
$$^{15}\mathrm{N}(p, \gamma)\,^{16}\mathrm{O}(p, \gamma)^{17}\mathrm{F}(e^{+}\nu)\,^{17}\mathrm{O}(p, {}^{4}\mathrm{He})\,^{14}\mathrm{N},$$
$$^{17}\mathrm{O}(p, \gamma)\,^{18}\mathrm{F}(e^{+}\nu)\,^{18}\mathrm{O}(p, {}^{4}\mathrm{He})\,^{15}\mathrm{N}.$$

The end result of the CNO cycle, besides helium production, is a conversion of most C and O isotopes in the center of the star into ^{14}N, because of the relatively slow reaction rate of $^{14}\mathrm{N}(p, \gamma)^{15}\mathrm{O}$ in the cycle.

Similarly, at even higher temperatures, hydrogen burning can be facilitated by preexisting Na and Ne nuclei through the so-called **Ne-Na cycle**, i.e.,

$$^{20}\mathrm{Ne}(p, \gamma)\,^{21}\mathrm{Na}(e^{+}\nu)\,^{21}\mathrm{Ne}(p, \gamma)\,^{22}\mathrm{Na}(e^{+}\nu)\,^{22}\mathrm{Ne}(p, \gamma)\,^{23}\mathrm{Na}(p, {}^{4}\mathrm{He})\,^{20}\mathrm{Ne}.$$

The metal-rich population I stars are often members of star clusters called **galactic clusters** in the plane or disk of a galaxy. These stars were accreted from the interstellar gases that were steadily enriched by heavy elements from the ashes of exploded dead stars, as will be discussed later. Another type of star cluster is the **globular cluster**, which is generally located outside the galaxy's disk and is relatively metal poor. These stars, called **population II** stars, were formed relatively early, when the interstellar gases were not yet greatly enriched with heavy elements. So far, stars consisting of only H and He have not been detected.

When all the hydrogen fuel in the core of an aging main sequence star is converted into helium, the core begins to contract. The gravitational energy released by the contraction heats the hydrogen layer around the He core. Therefore hydrogen burning can continue in a thin shell surrounding the helium core. This is called **shell hydrogen burning**. The mass of the helium core continues to increase during shell hydrogen burning and the core slowly contracts, thus causing an increase in the core temperature. The increased energy output of the core by gravitational contraction and shell hydrogen burning causes the star's outer layer to expand. Thus the star evolves to the right of the main sequence star curve (Figure II-6b for 15 solar masses). When the core temperature reaches about 0.3×10^8 K to 0.8×10^8 K, ^{26}Al becomes an important product (Arnett, 1996). The Compton Gamma Ray Observatory Satellite (CGRO) did detect 1.809 MeV gamma ray emission from the radioactive decay of $^{26}\mathrm{Al}(t_{1/2} = 7.2 \times 10^5$ years) from the disk of the Milky Way, where massive stars ($M > 10M_{\odot}$) are abundant. The important implication is that the ^{26}Al produced must be advected to the surface and ejected into space by the mass loss process of massive stars (Arnett and Bazan, 1997). When the central temperature of the star reaches about 2×10^8 K, ^{4}He in the core center is ignited into ^{12}C by the so-called **core helium burning**.

Helium Burning

The main reaction chain is

$$^{4}\mathrm{He}(2\alpha, \gamma)\,^{12}\mathrm{C}(\alpha, \gamma)\,^{16}\mathrm{O}(\alpha, \gamma)\,^{20}\mathrm{Ne}(\alpha, \gamma)\,^{24}\mathrm{Mg}(\alpha, \gamma)\,^{28}\mathrm{Si}.$$

The major products are ^{16}O, followed by ^{12}C, ^{20}Ne, and ^{24}Mg. A new C-O core will grow steadily.

If ^{14}N is present from the earlier CNO cycle, additional α-capture reactions are possible, i.e.,

$$^{14}\mathrm{N}(\alpha, \gamma)\,^{18}\mathrm{F}(e^{+}\nu)\,^{18}\mathrm{O}(\alpha, \gamma)\,^{22}\mathrm{Ne}(\alpha, n)\,^{25}\mathrm{Mg},$$

$$^{22}\mathrm{Ne}(\alpha, \gamma)\,^{26}\mathrm{Mg}.$$

The last reaction, $^{22}\mathrm{Ne}(\alpha, n)^{25}\mathrm{Mg}$, is an important source of neutrons. An effective production of neutrons at high temperature initiates the so-called

s (slow) process, i.e., successive neutron captures by seed nuclei, such as preexisting Ne and Fe peak nuclei, at a slow rate compared to the intervening beta decays; for example,

$$^{20}\mathrm{Ne}(n,\gamma)\,^{21}\mathrm{Ne}(n,\gamma)\,^{22}\mathrm{Ne}(n,\gamma)\,^{23}\mathrm{Ne}(e\bar{\nu})\,^{23}\mathrm{Na}(n,\gamma)^{24}\mathrm{Na},$$

$$^{24}\mathrm{Na}(e\bar{\nu})\,^{24}\mathrm{Mg}(n,\gamma)\,^{25}\mathrm{Mg},$$

$$^{25}\mathrm{Mg}(n,\gamma)\,^{26}\mathrm{Mg}(n,\gamma)\,^{27}\mathrm{Mg}(e\bar{\nu})\,^{27}\mathrm{Al}(n,\gamma)\,^{28}\mathrm{Al},$$

and further s-processing along the valley of beta stability up to ^{209}Bi as shown in Figure II-7 and Figure I-3. The so-called **r (rapid) process** will be discussed later.

When helium fuel at the center of the core is used up, the core again begins to contract. The energy released by the contraction heats the helium-rich layer around the C-O core and initiates the so-called **shell helium burning**. The renewed outpouring of energy causes a further expansion of the outer layer

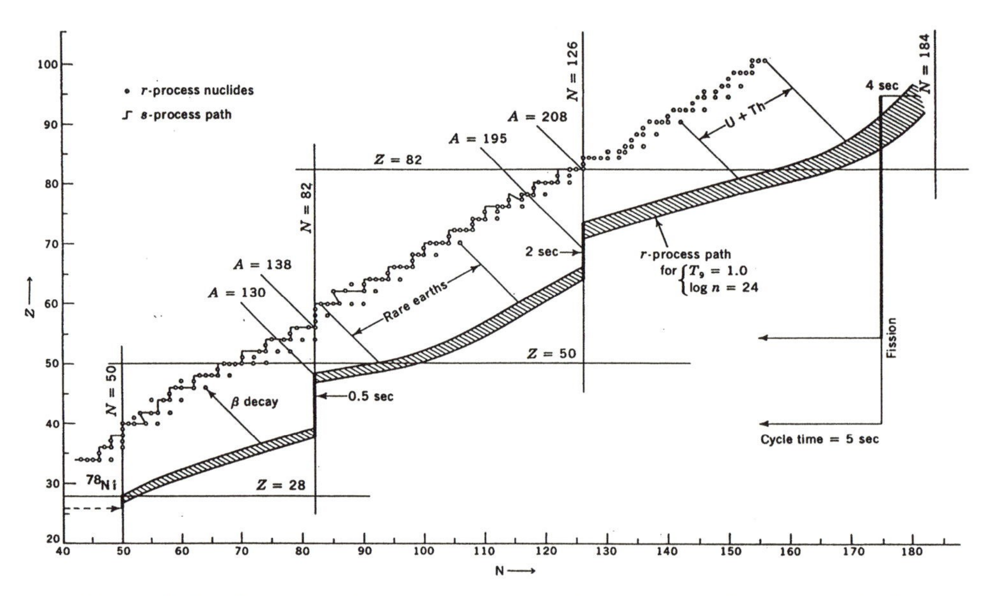

FIGURE II-7. The successive neutron capture paths for the s- and r-processes in the *N-Z* plane. The s-process follows the path along the bottom of the beta-stability valley (solid zigzag line). The r-process starts upward from nuclei around ^{78}Ni along the shaded band until neutron-induced fission occurs around $A = 270$. This r-process path is for the case of $T = 10^9$ K and neutron density $= 10^{24}/\mathrm{cm}^3$. After successive beta decays, the nuclei in the r-process path become stable neutron-rich shielding isobars (unconnected open circles) or unshielded isobars on the s-process path (after Seeger et al., 1965; with permission from Clayton, D.D., and the University of Chicago Press).

of the star, and the star becomes a **red supergiant** (Figures II-6a and II-6b). The internal structure of the star now consists of a growing C-O core, a thin helium burning shell, a helium-rich layer, a thin hydrogen burning shell, and a massive hydrogen outer layer. Further gravitational contraction can heat up the central core to about 6 to 8 $\times$ 10^8 K and initiate **carbon burning**.

Carbon Burning

The major reactions are

$$
\begin{aligned}
&{}^{12}\mathrm{C}({}^{12}\mathrm{C}, \alpha)\,{}^{20}\mathrm{Ne}(\alpha, \gamma)\,{}^{24}\mathrm{Mg}(\alpha, \gamma)\,{}^{28}\mathrm{Si},\\
&{}^{12}\mathrm{C}({}^{12}\mathrm{C}, 2\alpha)\,{}^{16}\mathrm{O}(\alpha, \gamma)\,{}^{20}\mathrm{Ne},\\
&{}^{12}\mathrm{C}({}^{12}\mathrm{C}, p)\,{}^{23}\mathrm{Na}(p, \alpha)\,{}^{20}\mathrm{Ne},\\
&\qquad\qquad\quad {}^{23}\mathrm{Na}(p, \gamma)\,{}^{24}\mathrm{Mg}.
\end{aligned}
$$

The major products are ^{20}Ne followed by ^{24}Mg, ^{28}Si, ^{23}Na, and energetic α and p particles. Of course, the core still retains most of the original ^{16}O intact. The following p- and α-capture reactions represent an important neutron source during the carbon burning stage:

$$
{}^{12}\mathrm{C}(p, \gamma)\,{}^{13}\mathrm{N}(e^{+}\nu)\,{}^{13}\mathrm{C}(\alpha, n)\,{}^{16}\mathrm{O}.
$$

If there are enough p particles, ^{20}Ne can be converted to ^{22}Ne in the central core (as shown in the Ne-Na cycle), and the reaction ${}^{22}\mathrm{Ne}(\alpha, n){}^{25}\mathrm{Mg}$ can be a neutron source also. Again, the neutron-capturing s-process proceeds along the valley of beta stability up to ^{209}Bi.

At the temperature of carbon burning ($\approx 8 \times 10^8$ K) and above, most of the energy generated inside the star escapes mainly as $\nu\bar{\nu}$ pairs (instead of photons), which are produced inside the highly excited plasma (γ + plasma $\rightarrow \nu + \bar{\nu} +$ plasma). Neutrino pairs will freely escape without much interaction with stellar gas. At the end of the core carbon burning stage, the new inner core consists mainly of ^{16}O, ^{20}Ne, and ^{24}Mg. Between the new O-Ne-Mg inner core and the C-O layer is a thin carbon burning shell. The further gravitational contraction of the O-Ne-Mg core raises the central temperature to about 10^9 K and starts a new process called **neon burning**.

Neon Burning

At 10^9 K, the photons are energetic enough to photodisintegrate part of ^{20}Ne to ^{16}O by the reaction ${}^{20}\mathrm{Ne}(\gamma, \alpha){}^{16}\mathrm{O}$. The newly produced α particles, in turn, are reabsorbed by ^{20}Ne to form ^{24}Mg and other heavier α-particle multiples through the following reactions:

$$^{20}\mathrm{Ne}(\alpha,\gamma)\,^{24}\mathrm{Mg}(\alpha,\gamma)\,^{28}\mathrm{Si}(\alpha,\gamma)\,^{32}\mathrm{S}(\alpha,\gamma)\,^{36}\mathrm{Ar}(\alpha,\gamma)^{40}\mathrm{Ca}(\alpha,\gamma)\,^{44}\mathrm{Ti},$$

$$^{44}\mathrm{Ti}(e,\nu)^{44}\mathrm{Sc}(e^+\nu)\,^{44}\mathrm{Ca}(\alpha,\gamma)\,^{48}\mathrm{Ti}.$$

The production of α-particle multiples heavier than ^{36}Ar is, however, still minor. When all of ^{20}Ne is converted into α-particle multiples, the new inner core consists mainly of ^{28}Si, ^{32}S, ^{16}O, ^{36}Ar, ^{40}Ca, and ^{24}Mg and is enveloped by the thin neon burning shell, a O-Ne-Mg layer, and so on. When the central temperature reaches about 1.5×10^9 K, **oxygen burning** begins.

Oxygen Burning

The major reactions are

$$^{16}\mathrm{O}(^{16}\mathrm{O},\gamma)\,^{32}\mathrm{S},$$

$$^{16}\mathrm{O}(^{16}\mathrm{O},n)\,^{31}\mathrm{S}(e^+\nu)\,^{31}\mathrm{P},$$

$$^{16}\mathrm{O}(^{16}\mathrm{O},p)\,^{31}\mathrm{P}(p,\alpha)\,^{28}\mathrm{Si},$$

$$^{16}\mathrm{O}(^{16}\mathrm{O},\alpha)\,^{28}\mathrm{Si},$$

$$^{16}\mathrm{O}(^{16}\mathrm{O},2p)\,^{30}\mathrm{Si};$$

and also

$$^{24}\mathrm{Mg}(\alpha,\gamma)\,^{28}\mathrm{Si},$$

$$^{24}\mathrm{Mg}(\alpha,p)\,^{27}\mathrm{Al}(p,\gamma)\,^{28}\mathrm{Si}.$$

The major products are ^{28}Si, ^{32}S, and some ^{30}Si. The new energetic α particles interact with ^{24}Mg, ^{28}Si, and ^{32}S to produce heavier α-particle multiples such as ^{36}Ar, ^{40}Ca, and others. The α particles can also react with ^{30}Si to produce

$$^{30}\mathrm{Si}(\alpha,\gamma)\,^{34}\mathrm{S}(\alpha,\gamma)^{38}\mathrm{Ar}(\alpha,\gamma)\,^{42}\mathrm{Ca}(\alpha,p)\,^{45}\mathrm{Sc}(p,\gamma)\,^{46}\mathrm{Ti}.$$

The neutrons produced from reactions such as $^{16}\mathrm{O}(^{16}\mathrm{O},n)^{31}\mathrm{S}$ can result in some s-processes but cannot go beyond the Fe peak nuclei. The reason is that the photons are so energetic at the temperature of oxygen burning that nuclei heavier than Fe, which were formed by the s-process during the earlier burning stages, are photodisintegrated back into Fe peak nuclei. When the ^{16}O fuel is used up and the central temperature rises to about 3×10^9 K, **silicon burning** begins.

Silicon Burning and Nuclear Statistical Equilibrium

The first step of silicon burning is photodisintegration of ^{28}Si and lighter nuclei to release protons, α particles, and neutrons, e.g.,

$$^{28}\text{Si}(\gamma, p)\,^{27}\text{Al}(\gamma, p)\,^{26}\text{Mg}(\gamma, n)\,^{25}\text{Mg}(\gamma, n)\,^{24}\text{Mg},$$

$$^{28}\text{Si}(\gamma, \alpha)\,^{24}\text{Mg}(\gamma, \alpha)\,^{20}\text{Ne}(\gamma, \alpha)\,^{16}\text{O}(\gamma, \alpha)\,^{12}\text{C}(\gamma, 2\alpha)\,^{4}\text{He},$$

$$^{24}\text{Mg}(\alpha, p)\,^{27}\text{Al}(\alpha, p)\,^{30}\text{Si};$$

then n, p, and α particles are recaptured by the remaining ^{28}Si and other heavy nuclei (such as ^{30}Si, ^{32}S, ^{24}Mg, ^{27}Al) to build up Fe peak nuclei. The nuclear reactions include hundreds of forward and backward reactions of nuclei ($12 \leq A \leq 76$) involving n, p, α particles, photodisintegration, and beta decays. Figure II-8 shows how the composition of major nuclei in the silicon burning core may change with the extent of silicon burning. Eventually, the forward and backward reaction rates of all the reactions become equal. This condition is called the **nuclear statistical equilibrium** (NSE) and the most abundant nuclei in the core are Fe peak nuclei around ^{56}Fe. Because these nuclei have maximum nuclear binding energy per nucleon (Figure I-4), the iron core is incapable of further exothermic nuclear reactions. So far, nucleosynthesis has proceeded in a star that maintains hydrostatic equilibrium, i.e., the gravitational weight of the overlying layer at any depth in the star is balanced by the pressure exerted by the gas. Therefore the hydrogen to silicon burning stages are also called hydrostatic nucleosynthesis processes, and the onion-shell-like structure of the model star is often called a spherically symmetric simulation. However, thermal instability may cause convective mixing of material among different shells and produce more complicated nucleosynthesis reactions (Arnett and Bazan, 1997).

The size and density of the iron core increases steadily as the shell silicon burning continues. For a star smaller than about 20 $M_{\odot}$, however, the iron core density ceases to increase when the free electrons inside the iron core are tightly packed into a so-called **degeneracy** state. According to the Pauli exclusion principle, no two identical fermion elementary particles can occupy the same quantum state simultaneously; therefore, at an extremely high pressure, all electrons will fill in all possible lowest quantum states and occupy a finite minimum volume. This condition is called electron degeneracy. When electrons are in a state of degeneracy, a further increase in external pressure cannot squeeze the electrons any closer. In other words, the degenerate electrons can exert counterpressure against an extremely high external gravitational weight.

However, when the mass of the iron core exceeds the so-called **Chandrasekhar limit** (about 1.4 solar mass), even the degenerate electron

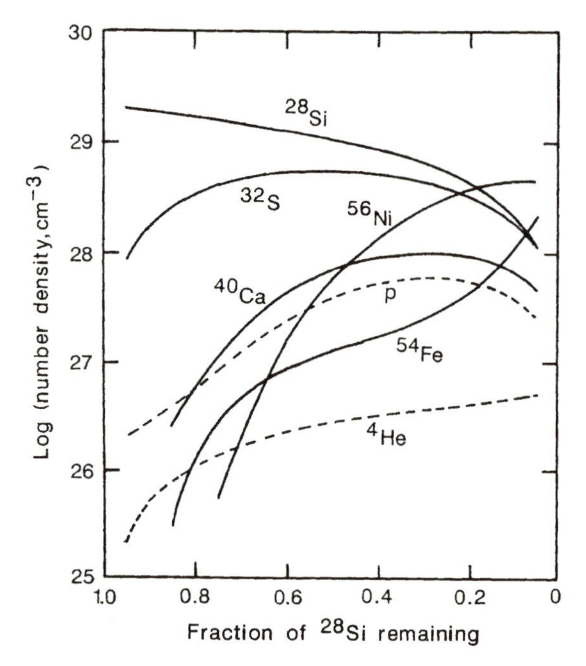

FIGURE II-8. Evolution of nuclear abundances during silicon burning at $T = 4 \times 10^9$ K and $\rho = 10^7$ g/cm^3 (after Bodansky et al., 1968; with permission from Clayton, D.D., and the University of Chicago Press).

pressure can no longer support the enormous weight of the star and the free electrons will suddenly collapse into Fe peak nuclei (i.e., there is an electron capture process that invariably emits neutrinos). The sudden implosion of the core causes the central temperature to soar to about 5×10^9 K and initiates the photodisintegration of now-neutron-rich nuclei into α, p, and n particles (e.g., $^{56}\mathrm{Fe} + \gamma \rightarrow 13\alpha + 4n$; $\alpha + \gamma \rightarrow 2p + 2n$). These photodisintegration reactions are endothermic and thus further facilitate the collapse of the iron core. For stars greater than $20M_{\odot}$, the central temperature can reach about 5×10^9 K before the onset of electron capture. Therefore, the photodisintegration may first trigger core collapse.

During core collapse, the central density continues to climb sharply and force the fusion of electrons and protons into neutrons, releasing a flood of neutrinos ($e^- + p \rightarrow n + \nu$). The implosive pressure will soon squeeze neutrons of the new core into a state of degeneracy (neutron degeneracy), i.e., the neutron core becomes virtually incompressible. As the remaining infalling materials crash onto the now-rigid neutron core, they bounce back outward with an explosive force, generating a shock wave capable of ejecting the entire envelope. Also, the flood of neutrinos (released during the formation of the neutron core) may transfer enough energy to the base of the envelope to facilitate the generation of a shock wave. The star explodes into a so-called **type II supernova** (by definition, a type II supernova is a star explosion with energy of $\geq 10^{51}$ ergs, and its spectra near maximum light has hydrogen lines) and leaves behind a degenerate **neutron star**. The predicted outburst of neutrinos during a type II supernova explosion was verified by the supernova event of early 1987 (Hirata et al., 1987; Bionta et al., 1987). This supernova event is well summarized by Woosley and Weaver (1989), and Arnett et al. (1989).

Figure II-9 shows one possible distribution of various nuclei in a $25M_\odot$ star just at the onset of core collapse (Weaver et al., 1983). Table II-3 also summarizes the major nucleosynthesis stages, corresponding central temperatures and densities, and duration of each stage for a $25M_\odot$ star. A higher temperature is required at each advanced burning stage, because the burning

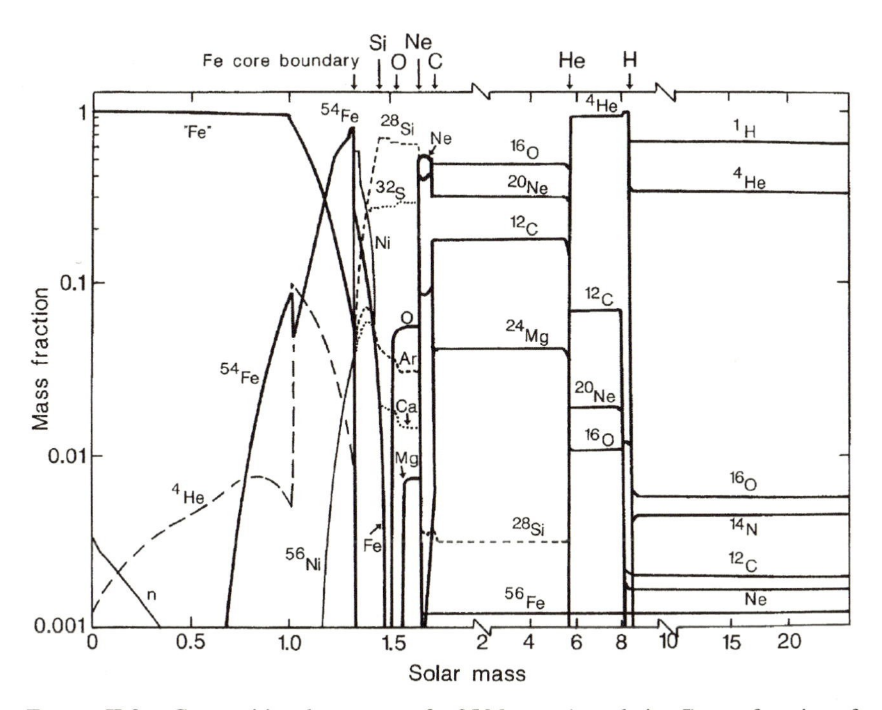

FIGURE II-9. Compositional structure of a $25M_\odot$ star (population I) as a function of interior mass (in units of solar mass) at the onset of core collapse. The positions of the shell hydrogen burning to shell silicon burning and the Fe core boundary are shown as arrowheads on the upper abscissa. Horizontal abundance lines for all isotopes in any regime may indicate effective mixing by convection. "Fe" includes all iron-group isotopes more neutron rich than ^{54}Fe (modified from Weaver et al., 1983).

TABLE II-3
Nucleosynthesis of a $25M_{\odot}$ population I star (after Kaufmann, 1985)

Stage	*Temperature (K)*	*Density (g/cm³)*	*Duration time*
Hydrogen burning	4×10^7	5	7×10^6 yrs
Helium burning	2×10^8	700	5×10^5 yrs
Carbon burning	6×10^8	2×10^5	600 yrs
Neon burning	1.2×10^9	4×10^6	1 yr
Oxygen burning	1.5×10^9	10^7	6 months
Silicon burning	2.7×10^9	3×10^7	1 day
Core collapse	5.4×10^9	3×10^9	1.4 sec
Core bounce	2.3×10^{10}	4×10^{14}	0.001 sec
Star explosion	10^9 to 5×10^9	Varies	0.1 to 10 sec

of hydrogen to silicon fuels is characterized by increasing Coulomb potential barriers ($= Z_1 Z_2 e^2/r$, where Z_1 and Z_2 are the charges of the projectile and target nuclei, and r is the sum of the projectile and target radii). The duration of each advanced burning stage also becomes shorter, because the energy lost by neutrino pairs becomes extremely large as temperature increases with each advanced stage. One should notice that Figure II-9 is only a model result, which is very much dependent on the parameterization and the chosen values of parameters. For example, if one adopts the newly measured nuclear reaction cross section for $^{12}C(\alpha, \gamma)^{16}O$, which is about three times the old value, the production of ^{12}C during the helium burning, and thus of ^{20}Ne during the carbon burning, become so low that the distinct carbon and neon burning stages may disappear altogether (Woosley and Weaver, 1986b).

The r (Rapid) Process of Successive Neutron Captures and Explosive Nucleosynthesis

At the instant of core bouncing, the innermost ejected layer can have a temperature of 10^{10} to 10^{11} K, a density of about 10^{11} g/cm³, and a total neutron-to-proton ratio of 1.5 to 8. When the temperature falls below 10^{10} K, thermonuclear reactions proceed rapidly through the recombination of free n and p to ^2H and ^4He, then $3\,^4He \rightarrow {}^{12}C$, followed by the production of heavier nuclei all the way up to extremely neutron-rich (or neutronized) nuclei around ^{78}Ni. These nuclei are in nuclear statistical equilibrium under the high-neutron-concentration environment. As soon as the temperature falls below about 3 to 4×10^9 K, most nuclear reaction rates diminish greatly except for neutron capture, beta decay, and some photoemission of neutrons. The high neutron flux results in successive neutron capture by the neutronized "seed" nuclei along the path shown in Figure II-7 (shaded area) at a rapid rate compared to the intervening beta decays. This is called the **r-process.**

Whenever the number of neutrons in the neutronized nucleus along the path adds up to the magic number (e.g., $N = 82, 126, 184$ in Figure II-7), the neutron binding energy (or neutron capture cross section) becomes so low that no more neutrons are captured by the nucleus until a beta decay increases its charge. Then the nucleus captures one more neutron and becomes neutron-magic again. Only after repeating several beta decay and neutron capture reactions does the neutron-magic nucleus have a Z/N ratio high enough to let it break away from the trap of the magic number and become an even heavier neutronized nucleus. Finally, any addition of a neutron will simply break up the nucleus by the fission process. The time scale of the r-process is on the order of 10 to 10^2 seconds. Because the neutron-magic nuclei have relatively low neutron capture cross sections and small beta decay rate constants, they reach a greater abundance than other neutronized nuclei along the r-process path. After cessation of the r-process (decrease of neutron flux), all neutronized nuclei undergo a series of beta decays until they become either stable neutron-rich shielding isobars or unshielded isobars at the bottom of the beta-stability valley. Thus, the relatively abundant neutron-magic nuclei end up as the observed high peaks around $A = 80, 130$, and 194 in Figure II-3. In short, the stable neutron-rich shielding nuclei can be produced only by the r-process, and the stable shielded nuclei only by the s-process. The unshielded nuclei can be produced by both r- and s-processes. It is possible, however, to make a rough estimate of the relative contribution of r- and s-processes to the formation of unshielded nuclei with $A > 70$ (Cameron, 1982; Binns et al., 1984; and Appendix Table 2). For example, in Figure II-4, the unshielded nuclei of some rare earth isotopes, Hf, W, Os, and Pt, are obviously dominated by the r-process, because these isotopes are much more abundant than those of neighboring shielded nuclei.

While the r-process is proceeding vigorously, the shock wave propagates outward through the envelope, and compresses and temporarily heats up the unburned nuclear fuels to initiate additional nuclear reactions. For example, if the shock peak temperature is $\geq 4.5 \times 10^9$ K when the shock wave passes through the Si-rich layer, it triggers the **explosive silicon burning** phase and produces mostly the iron peak nuclei around ^{56}Ni (compare Figure II-10 and Figure II-9). ^{56}Ni($t_{1/2} = 6.1$ days) decays into ^{56}Co($t_{1/2} = 77.7$ days), which then decays into stable ^{56}Fe. Indeed, the gamma lines from the decay of ^{56}Co have been detected in the 1987A supernova, whose brightness also has been decaying with a half-time similar to the half-life of ^{56}Co (Arnett et al., 1989; Arnett and Bazan, 1997). The shock peak temperature will decrease rapidly as the shock wave propagates outward. If the shock peak temperature is about 3–4 $\times\ 10^9$ K in the oxygen (neon)-rich layer, it causes the **explosive oxygen (neon) burning** phase and produces mainly nuclei from

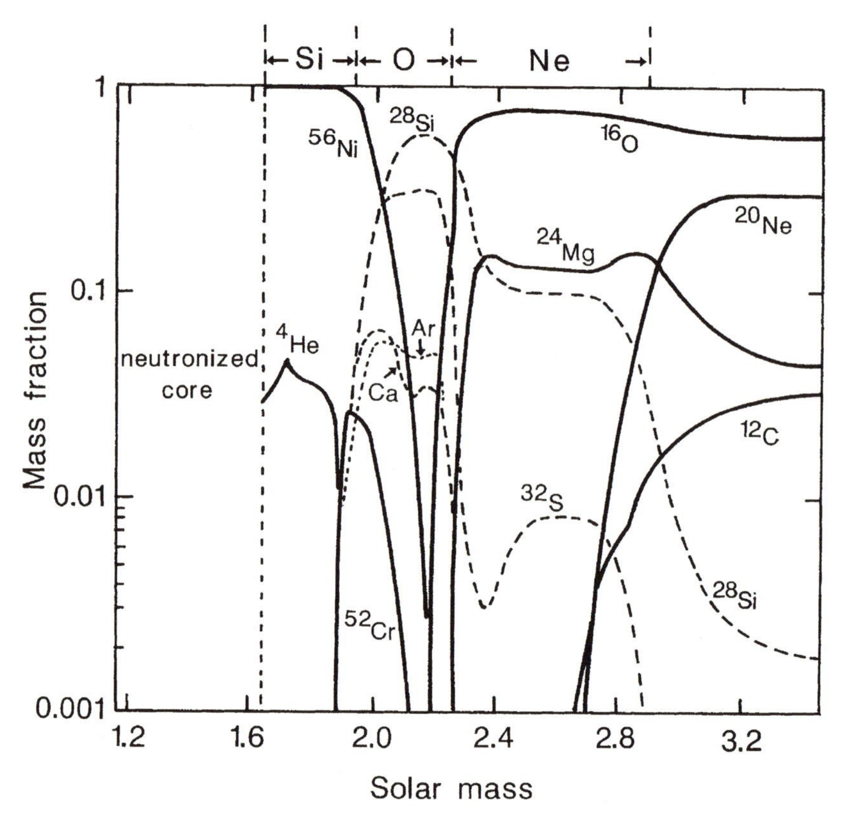

FIGURE II-10. Composition of the inner region of a $25M_\odot$ supernova soon after shock wave passage. The "neutronized core" region may contain extremely neutron-rich nuclei around ^{78}Ni and be an important site for r-process. The ^{56}Ni peak region is caused by explosive Si burning; the ^{28}Si to ^{40}Ca peak region by explosive O burning; and the ^{16}O to ^{32}S peak region by explosive Ne burning (modified from Weaver and Woosley, 1980).

^{28}Si to ^{40}Ca(^{16}O to ^{32}S) (Figure II-10). In principle, a shock peak temperature of about 2×10^9 K in the carbon-rich layer will cause an **explosive carbon burning** phase and produce nuclei from ^{20}Ne to ^{32}S. However, for the present example of a $25M_\odot$ star (Figure II-10), the shock peak temperature in the carbon-rich layer is already below 10^9 K, and thus no explosive carbon burning occurs. If the shock peak temperature and density conditions in the hydrogen-rich outer layer are about $T = 2 \times 10^9$ K and $\rho = 10^4$ g/cm^3, reactions such as (p, γ), (p, n), and (γ, n) on the preexisting nuclei produce proton-rich nuclei as shown in Figures II-4 and I-3. This is called the p **(proton) process**. The (γ, n) reaction can also occur at temperatures higher than 2–3 $\times$ 10^9 K. Thus it is not necessarily confined to the hydrogen-rich layer. In summary, a supernova explosion modifies the isotopic composition produced by the series of hydrostatic nucleosyntheses and ejects the modified products back into outer space as interstellar gas and dust. Reaggregation of interstellar gas and dust into a new generation of protostars and stars repeats the nucleosynthesis cycle, and the Universe progressively becomes enriched in the elements heavier than He (i.e., the Z fraction).

For stars more massive than about $70M_\odot$, oxygen burning occurs at a relatively high central temperature, which favors the endothermic electron-positron pair production ($2\gamma \rightarrow e^- + e^+$). Hence endothermic $e^\pm$ pair production can trigger a core implosion that, in turn, triggers silicon burning, iron core formation, photodisintegration, formation of a neutron core greater than $2.3M_\odot$ (Chandrasekhar limit for degenerate neutrons), and the final collapse of the neutron core into a **black hole**, where space and time fold together (a singularity) again. The gravitational pull of a black hole is so extreme that even photons cannot escape. Massive black holes are thought to exist at the centers of many galaxies (van der Marel et al., 1997).

For a low-mass star (0.5 to $2M_\odot$), the star's outer hydrogen layer expands greatly during the gravitational contraction of the helium core and during the shell hydrogen burning stage, and the star evolves into the **red giant branch** (Fig II-6b, 1 solar mass track). Due to its low mass, the helium core must contract greatly in order to attain the helium burning temperature. Therefore, more often than not, the electrons in the core are already in a state of degeneracy when the helium burning starts. When the core is supported by degenerate electron pressure, it does not expand to cool (as is the case for a star in hydrostatic equilibrium) when it is heated by the newly released energy from helium burning. Therefore, the rising temperature of the core causes helium to burn at an even greater rate, producing a thermonuclear runaway which results in an explosion, called a **helium core flash**, and the star jumps backward into the red giant branch again. When shell helium burning starts after core helium burning, the star expands again along the so-called **asymptotic giant branch** (Figure II-6b). In the end, the helium burning shell becomes unstable and develops a rapid series of brief bursts called **thermal pulses**. The thermal pulses eject the star's envelope and its hot core becomes a **white dwarf star**, as shown in Fig II-6a.

Protostars smaller than $\sim 0.08M_\odot$ eventually become planet-like objects called **brown dwarfs** (Kulkarni, 1997). Main sequence stars less massive than $0.5M_\odot$ never evolve into a helium burning stage and end up as helium white dwarfs after shedding their hydrogen outer layer. However, we cannot observe any helium white dwarfs today because these small stars are still in the hydrogen burning stage, even if they were formed at the beginning of the universe.

Finally, there is one more nucleosynthesis process called the x-**process**, which is responsible for the synthesis of rare isotopes such as ^2H, ^3He, ^6Li,

^{9}Be, ^{10}B, and ^{11}B. These light isotopes are all unstable in stellar interiors at temperatures even below hydrogen burning. It is presently accepted that ^{6}Li, ^{9}Be, ^{10}B, and ^{11}B are mostly the spallation products of interaction between galactic cosmic rays (mostly high-energy p and α particles) and interstellar gases such as ^{4}He, ^{12}C, ^{14}N, ^{16}O, and ^{20}Ne. ^{2}H and ^{3}He are mainly the products of cosmological nucleosynthesis as discussed earlier.

In summary, the chemical and isotopic compositions of interstellar gas and dust represent a mixture of nucleosynthesis products from large stars of various generations. The major nucleosynthesis processes which are responsible for the formation of each nucleus in the Universe are summarized in Appendix Table A-2. One can also expect isotopic compositions of various celestial entities to be quite different. For example, the isotopic compositions of C, O, Ne, Mg, Si, and Fe in galactic cosmic ray sources have been proven to be different from those in the solar nebula (Simpson, 1983). As mentioned earlier, rare fine inclusions in some meteorites also show isotopic heterogeneity. Those rare inclusions were formed in the circumstellar atmospheres of mass-losing red giant stars, massive supergiant stars, etc. (Ott, 1993; Arnett and Bazan, 1997).

II-4. Concluding Remarks

The general agreement of the relative abundance of elements among type 1 carbonaceous chondrites, the solar atmosphere, and galactic cosmic ray sources strongly suggests a first-order uniformity of the chemical composition of our Universe, but not of isotopic composition.

Cosmological and stellar nucleosynthesis theories provide the framework for relating the evolutionary paths of stars to their compositional changes. Nucleosynthesis theories will certainly evolve with new astronomical observations and advances in knowledge of the elementary particles.

Chapter III

STRUCTURE AND CHEMISTRY OF THE SOLAR SYSTEM

INTRODUCTION

THIS CHAPTER summarizes the distribution and chemical compositions of planets, their satellites, asteroids, and comets in the solar system. Basic terminologies related to the motion of planetary and interplanetary objects are briefly reviewed first, because these parameters describe the distribution and classification of asteroids and comets.

Compositions of asteroids and comets are mainly deduced from meteorites, albedo and reflectance spectra of asteroids and comets, interplanetary dust particles collected from the Earth's stratosphere, and the coma of the comet Halley. Compositions of bulk planets and satellites are deduced from estimated bulk density of these bodies at 1 bar pressure, and albedo and reflectance spectra of satellites.

The observed changes in chemical composition of planetary and interplanetary objects can be related as a function of distance from the Sun using condensation models of solar nebular gas. These are discussed at the end of the chapter. Additional information can be found in the book *Physics and Chemistry of the Solar System* by Lewis (1997).

III-1. MOTION OF INTERPLANETARY OBJECTS

The planets, asteroids, and comets all revolve around the Sun in elliptic orbits with the Sun as one of the elliptic foci. Similarly, satellites also circle around their central planets in elliptic orbits. The heliocentric elliptic orbit (Figure III-1) can be represented by

$$d = \frac{a(1-e^2)}{1+e\cos\theta}, \tag{III-1}$$

where d = distance between the Sun and the revolving body; a and b = semimajor and semiminor axes of the ellipse, respectively; e = eccentricity = $(1-b^2/a^2)^{1/2}$; θ = angle between the semimajor axis and the line connecting two bodies.

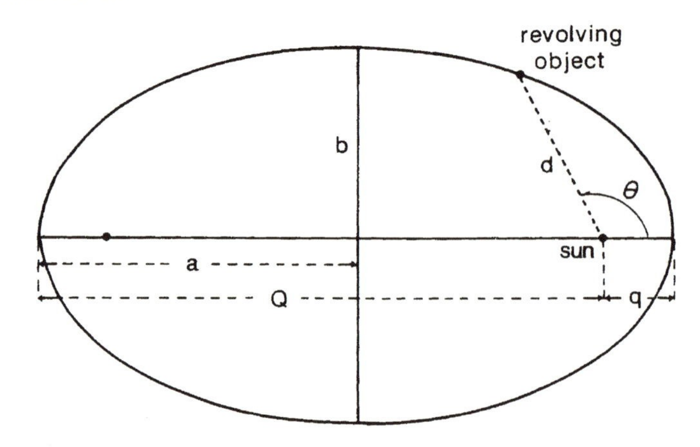

FIGURE III-1. Elliptical orbit showing the useful parameters: semimajor axis a, semiminor axis b, aphelion Q, perihelion q, polar coordinates of revolving object (d = radius vector, θ = polar angle) with the Sun as the origin.

When $\theta = 0°$, d is equal to $a(1 - e)$ and is called the **perihelion**, q, i.e., the closest distance between the two bodies. When $\theta = 180°$, d is equal to $a(1 + e)$ and is called the **aphelion**, Q, i.e., the longest distance between the two bodies. It is obvious that $Q + q = 2a$. If e approaches zero, the ellipse approaches a perfect circle and d equals a. The semimajor axis of the Earth's orbit (= 150×10^6 km) is defined as one **astronomical unit** (AU).

If an object is in a circular orbit with radius of d, the gravitational force between the Sun and the object (= $GmM_\odot/d^2$) should be balanced by the imaginary centripetal force (= mv^2/d), i.e.,

$$\frac{GmM_\odot}{d^2} = \frac{mv^2}{d}, \qquad \text{(III-2)}$$

where G is Newton's gravitational constant = 6.67×10^{-8} dyne cm^2/g^2; $M_\odot$ and m are the mass of the Sun and the object; and v is the velocity of the object. Also, the time for the object to make one complete revolution around the Sun (i.e., the orbital period, p) is

$$p = 2\pi d/v. \qquad \text{(III-3)}$$

Solving equations III-2 and III-3, one gets

$$p^2 = 4\pi^2 d^3/(GM_\odot). \qquad \text{(III-4)}$$

The above equation also applies to an object with an elliptic orbit, except that d is replaced by the semimajor axis a and is a mathematical expression of Kepler's famous third law. If one can measure p and d or a of any planet accurately, one can estimate the mass of the Sun from equation III-4.

Rearranging equation III-2, one obtains the angular momentum (mvd) of the object in a circular orbit as

$$mvd = m(dGM_{\odot})^{1/2}. \qquad \text{(III-5a)}$$

The angular momentum (mvd) of the object in an elliptic orbit is also a constant with the value of

$$mvd = m\left[a(1-e^2)GM_{\odot}\right]^{1/2}, \qquad \text{(III-5b)}$$

where v is the velocity of the object perpendicular to the d vector. Standard textbooks, such as *Introductory Astronomy and Astrophysics* by Smith and Jacobs (1973), provide rigorous derivations of equations III-4 and III-5b for an object with an elliptic orbit.

The north pole vector of the orbital plane of any heliocentric object is a vector perpendicular to the orbital plane and pointing in the direction of the thumb of the right hand when the other fingers bend around in the direction of the object's motion in the orbit. The orbital plane of the Earth is called the **ecliptic plane**. The angle between the north pole vectors of the ecliptic plane and the orbital plane of an object is called the **orbital inclination** i. If an object has an orbital inclination of greater than 90° and up to 180°, its north pole vector and its direction of motion are opposite to those of the Earth. If the orbital inclination is less than 90°, the angle between the ecliptic plane and the orbital plane of the object is identical to the orbital inclination. The spin vectors of the spinning Sun and planets are also determined by the direction of the thumb of the right hand when the other fingers move round in the direction of the spinning movement of those objects. The **spin inclination** i_s is the angle between the spin vector and the north pole vector of a planet's orbit. If i_s is less than 90°, the angle between the orbital plane and the equatorial plane of the spin is identical to its i_s value.

Equations III-1 to III-5 and the definitions given above are all applicable to the satellite system of any planet with appropriate substitution of satellite parameters for those of the planet, and the planet for the Sun.

The work needed to remove a particle having mass m_1 from the surface of an object having mass m to an infinite distance is

$$\int_r^{\infty} \frac{Gmm_1}{x^2}\,dx = \frac{Gmm_1}{r},$$

where r is the radius of the object and x is the distance of the particle from the surface of the object. Therefore, if the kinetic energy of the particle ($= m_1 v_1^2/2$) is equal to or greater than Gmm_1/r, i.e., $v \geq (2Gm/r)^{1/2}$, the particle will escape the influence of the gravitational pull of the object. The quantity $(2Gm/r)^{1/2}$ is called the **escape velocity** (v_e). For example, the v_e

from the Sun is about 616 km/sec; thus, for a hydrogen atom to escape from the gravitational pull of the Sun, it should have a kinetic energy of about 2 keV; and for He, 8 keV (1 erg = 1 dyne cm = 0.624×10^9 keV).

Finally, as mentioned in Section II-3, the total energy output of the Sun is $L_\odot = 4\pi r_\odot^2 \sigma T_\odot^4$, where $r_\odot$ and $T_\odot$ are the radius and surface temperature of the Sun, and σ is the Stefan-Boltzmann constant. An object with radius r and heliocentric distance d intercepts only a small fraction of $L_\odot$, i.e., $L_\odot \pi r^2/(4\pi d^2)$. If the dissipation of heat by the object through **black body radiation** (i.e., $4\pi r^2 \sigma T^4$, where T is the effective surface temperature of the object) is balanced by the solar heat input, then

$$4\pi r^2 \sigma T^4 = 4\pi r_\odot^2 \sigma T_\odot^4 (\pi r^2/4\pi d^2),$$

so

$$T = \frac{T_\odot r_\odot^{1/2}}{(2d)^{1/2}}, \qquad \text{(III-6)}$$

i.e., T is proportional to $d^{-1/2}$.

III-2. Asteroids and Comets

Asteroids and comets are small heliocentric objects. The main operational distinction between the two is that a comet is capable of producing a coma, an envelope of gases and dust around the cometary nucleus in its present orbit, especially when it passes near the Sun. In contrast, an asteroid does not show any coma activity. A comet can become an asteroid-like object if it has lost all its volatiles, or if its volatiles are sealed inside its interior.

Asteroids

The majority of asteroids occupy the region between the orbits of Mars (a = 1.5 AU) and Jupiter (a = 5.2 AU). There are only three asteroids with radii greater than 250 km (Table III-1), about 35 greater than 100 km, 250 greater than 50 km, 1000 greater than 25 km, and probably more than 100,000 greater than 2.5 km. The total mass of all asteroids is estimated to be about 3×10^{24} g or only 0.0005 Earth mass (m_E; Kresak, 1987). The three largest asteroids (Ceres, Vesta, and Pallas) actually account for more than one-half of the total mass of asteroids (Table III-1). The orbital parameters (a, e, and i) of the asteroids are shown in the frequency diagrams of Figure III-2 and are plotted in the a-e plane in Figure III-3 (see page 89).

TABLE III-1
Properties of the three largest asteroids

	Semimajor axis a (AU)	*Eccentricity e*	*Inclination i (degrees)*	*Radius r (km)*	*Mass m (10^{24} g)*	*Density ρ (g/cm^3)*	*Albedo*[a]
4 Vesta	2.36	0.090	7.1	275	0.28	3.1 ± 0.5	0.22
1 Ceres	2.76	0.077	10.6	510	1.20	2.1 ± 0.3	0.06
2 Pallas	2.77	0.233	34.8	270	0.22	2.6 ± 0.5	0.07

[a] albedo = fraction of incident visible light (from the Sun) reflected from the surface of the asteroid.

In Figure III-3, the individual data points are shown as small solid circles and the areas enclosed by irregular solid and dashed lines represent, respectively, the high and moderately high data density zones. Most inclinations are less than 30°, and the eccentricity shows a broad maximum around 0.14 in Figure III-2. The minima in Figure III-2 (top), or the gaps between the high-data-density zones in Figure III-3, are called **Kirkwood gaps**. One should be aware of the fact that the Kirkwood gaps are not real spatial gaps in the distribution of asteriods in the solar system.

From equation III-4, one obtains

$$p/p_J = (a/a_J)^{3/2}, \qquad \text{(III-7)}$$

where p_J and a_J are the period and semimajor axis for Jupiter, and p and a for an asteroid. Equation III-7 is another expression of Kepler's third law. Interestingly, if one puts any semimajor axis value at the Kirkwood gaps into equation III-7, p/p_J is always a quotient of two integers with absolute values of less than one. Thus the existence of Kirkwood gaps can be explained in the following way: whenever a faster-moving asteroid passes by massive Jupiter ($p_J > p$), its orbit is changed slightly by the gravitational tug of Jupiter. If p/p_J is not a quotient of two integers, these close passes occur at random points along the asteroid's orbit; thus, the effects of the tugs of Jupiter cancel out over time. If p/p_J is 1/2, the asteroid will pass Jupiter every second trip and always at the same location. Therefore the gravitational perturbations add up and cause a marked increase in orbital eccentricity, resulting in either the ejection of the asteroid from the solar system or transformation of its orbit into a new path, which passes into the inner solar system. If any asteroid or fragment of a postcollision asteroid is thrown into any Kirkwood gap, it may either be ejected from the solar system or eventually evolve into the so-called **Mars- and Earth-crossing orbits**.

Provided two orbits are in the same plane, the conditions for the orbit of an asteroid with known a and e to intersect with that of a given planet x (with low eccentricity) are approximately

$$Q_x \leq Q = a(1 + e),$$

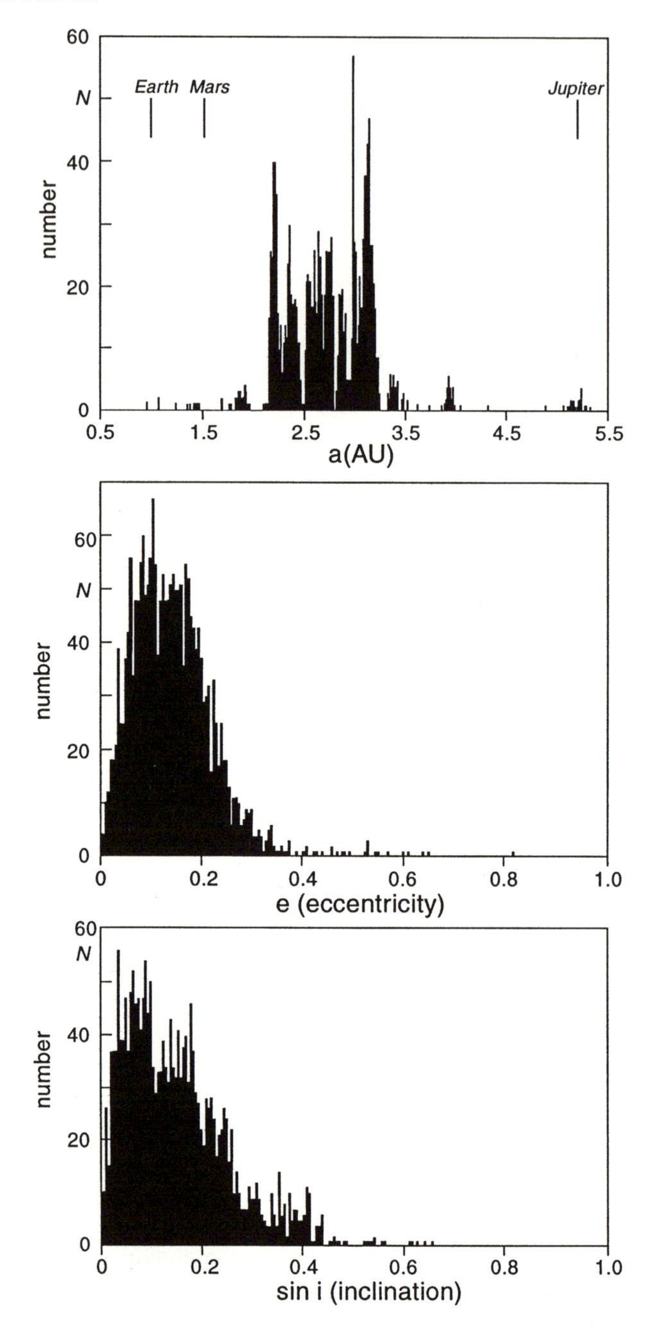

FIGURE III-2. Frequency distribution of (top) asteroidal semimajor axis a; (center) eccentricities e; (bottom) orbital inclination as $\sin i$ (Scholl, 1987).

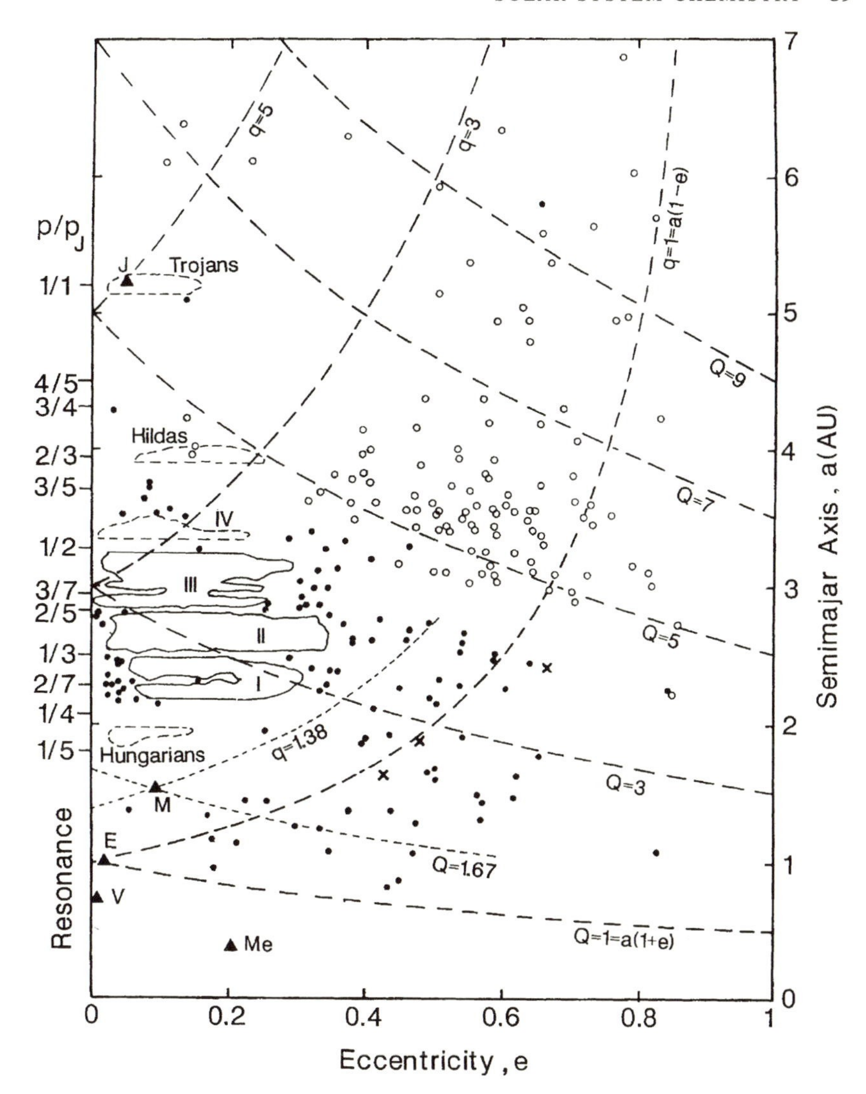

FIGURE III-3. Distributions of asteroids (dots and areas enclosed by irregular solid and dashed lines) and Jupiter family comets (open circles) in an eccentricity–semimajor axis plane. The positions of Mercury (Me), Venus (V), Earth (E), Mars (M), and Jupiter (J) are indicated by solid triangles. The semimajor axis corresponding to a given orbit resonance with Jupiter (p/p_J) is shown on the left abscissa. The Earth-crossing Apollo objects lie on the right-hand side of the contour lines $q = 1 = a(1 - e)$ and $Q = 1 = a(1 + e)$. The crosses in this field are the observed a-e values for three meteorite falls. The Mars-crossing Amor objects are enclosed by the contour lines $q = 1$ and $q = 1.38$ AU (data points are mainly from Kresak, 1967, 1979).

and

$$q_x \geq q = a(1 - e).$$

If two bodies happen to be at the orbital intersection at the same time, they will collide. If the orbits do not lie in the same plane, they may not necessarily intersect. Therefore the above equations may approximate only the conditions for the close encounter of two orbits and possible perturbation of the asteroid's orbit by the more massive planet when the asteroid approaches within the effective radius of the gravitational pull of the planet. For example, Q_x and q_x for the Earth are respectively 1.017 and 0.983 AU, so Earth-crossing asteroids should have values of a and e such that $0.983 \geq a(1 - e)$ and $1.017 \leq a(1 + e)$, i.e., the points within the area roughly enclosed by the $1 = a(1+e)$ and $1 = a(1-e)$ curves in Figure III-3. Those Earth-crossing asteroids are called **Apollo objects**. Similarly, asteroids with $q \leq 1.38$ and $Q \geq 1.67$ AU are Mars crossing. Of these asteroids, those with $1.017 \leq q \leq 1.30$ are called **Amor objects**. Amor objects can become Apollo objects by the gravitational perturbation of Mars. If an Earth-crossing asteroid encounters the Earth and its fragments survive the transit through the Earth's atmosphere, it becomes a meteorite. Accurate orbits are known for three photographed meteorite falls (three crosses in Figure III-3; i.e., Pribram-1959, $a = 2.42$, $e = 0.674$; Lost City–1969, $a = 1.66$, $e = 0.417$; Innisfree-1971, $a = 1.87$, $e = 0.473$). All belong to the ordinary chondrite class. The data (McCrosky et al., 1971; Halliday et al., 1978) clearly indicate that these three meteorites are Apollo objects. Various classes of meteorites provide first-hand knowledge of the density and mineralogical, and chemical compositions of asteroids. Furthermore, by comparing the characteristic visible to infrared light reflectance spectra of minerals and meteorites as observed in laboratories to those of asteroids observed through Earth-based telescopes and spacecraft, one may infer the spatial distribution of meteorite classes in the different asteroidal zones, as discussed in the following section.

Classes of Meteorites and Reflectance Spectrum Types of Asteroids

Chapter IV will discuss in detail the various classes of meteorites and their mineralogical, chemical, and isotopic compositions. Therefore it will suffice here to briefly summarize some classes pertinent to the discussion in this chapter. As mentioned earlier, the most primitive meteorite is the **type 1 carbonaceous chondrite** (C1). It consists mostly of extremely fine grained hydrous silicates and complex organic compounds. The average bulk density is about 2.2 g/cm^3. The **type 2** and **type 3 carbonaceous chondrites** (C2 and C3) consist of 30 to 50% fine-grained matrix materials. Embedded in these matrix materials are various proportions of the following three components: (1) Millimeter- to submillimeter-sized spheres and spheroids called **chondrules**. Many chondrules have glassy to microcrystalline textures, indicative of rapid cooling of once molten droplets. Chondrules mainly consist of olivine, pyroxene, and glass. (2) Millimeter-sized aggregates of loosely packed micron-sized crystals. One kind of aggregate consists mainly of olivine and is called **amoeboid olivine-rich inclusion** (AOI). Other common aggregates are **calcium-aluminum-rich inclusions** (CAI) containing micro- and submillimeter-sized Ca-Al-rich minerals such as hibonite ($CaAl_{12}O_{19}$), perovskite ($CaTiO_3$), melilite ($Ca_2Al_2SiO_7$), spinel ($MgAl_2O_4$), and anorthite ($CaAl_2Si_2O_8$). (3) Individual minerals and mineral fragments scattered in the matrix. These minerals are similar to those found in chondrules and regular aggregates but also include some secondary minerals grown within the matrix, e.g., calcite and gypsum veins. The densities of C2 and C3 are 2.5 to 2.9 g/cm^3 and about 3.4 g/cm^3, respectively.

The most abundant meteorite type is the so-called **ordinary chondrite** (OC), which is mainly composed of chondrules with various textures and a fine- to coarse-grained matrix of olivine, pyroxene, plagioclases, Fe-Ni alloy, and troilite (FeS) with or without minor carbonaceous materials. The **enstatite chondrites** (EC) are named for their major mineral constituent, enstatite ($MgSiO_3$). Other components are Fe-Ni alloy, troilite, and unusual minor sulfides, such as oldhamite (CaS), alabandite ([Mn,Fe]S), niningerite ([Mg,Fe]S), and djerfisherite ($K_3[Na,Cu][Fe,Ni]_{12}S_{14}$). This mineral assemblage is indicative of a very reducing environment during the formation of enstatite chondrites. The densities of ordinary and enstatite chondrites range from 3.5 to 3.8 g/cm^3, with an average of 3.7 g/cm^3.

It is generally accepted that the total or partial melting of chondritic parent bodies separated the Fe-Ni alloy plus troilite fraction from the silicate fraction. The silicate fraction became the **achondrites** which now totally lack chondrules and are low in Fe-Ni metal and troilite contents. The Fe-Ni plus troilite fraction became the **iron meteorites**. Incomplete separation of silicate and iron fractions resulted in **stony iron meteorites**. However, there are always some exceptions. For example, some stony iron meteorites are mixtures of achondrites and iron meteorites from totally different origins. The densities of achondrites, iron meteorites, and stony irons range, respectively, from 3.2 to 3.5 (average 3.3), 7.8 to 7.9, and 4.6 to 6.2 g/cm^3. Further subdivision of those meteorites according to their mineralogy and chemistry will be discussed in Chapter IV.

Typical reflectance spectra of some meteoritic minerals, bulk ordinary chondrites, and basaltic achondrites (also called eucrites, designated Aeu) are shown in Figure III-4 (all spectra are scaled to 1.0 at 0.56 microns wavelength). The minima in the spectra are mainly caused by the absorption of those wavelengths when electrons of Fe^{+2} in the crystal lattice move from

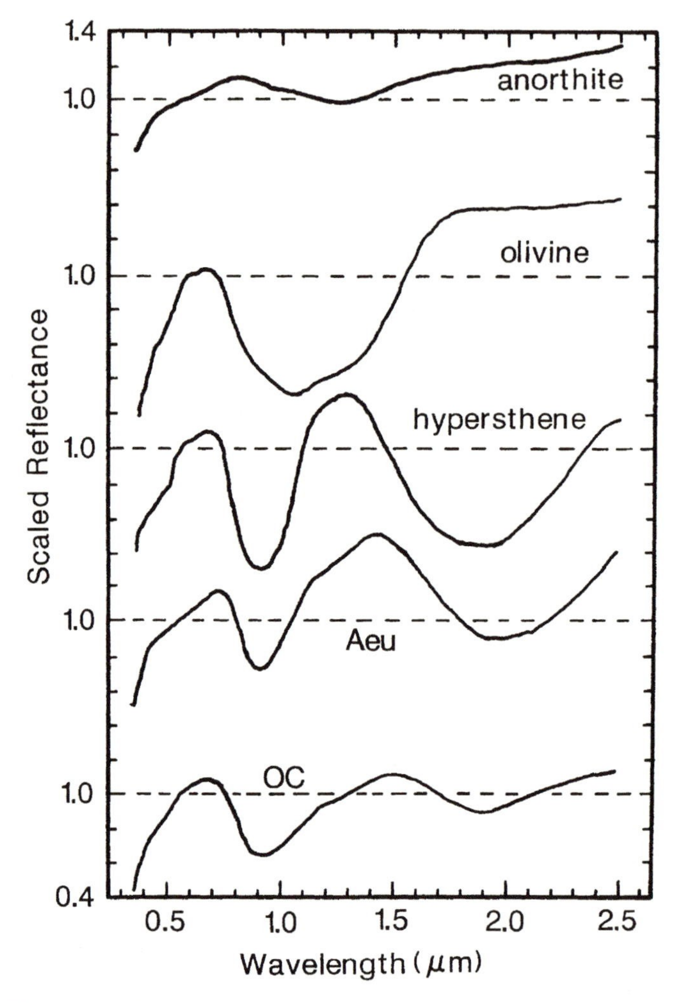

FIGURE III-4. Scaled reflectance spectra for anorthite, olivine, hypersthene, basaltic achondrite (or eucrite, Aeu), and ordinary chondrite (OC). Data curves are from Gaffey (1976).

lower (t_{2g}) to higher (e_g) energy levels, as already discussed in Section I-5. Detailed treatments of the reflectance spectra can be found elsewhere (e.g., minerals, Adams, 1975; meteorites, Gaffey, 1976).

Figure III-5 summarizes the major reflectance spectrum types of asteroids (Bell et al., 1987). Probable corresponding meteoritic classes are also indicated. Asteroids of types C, P, and D are very common far from the Sun. They are characterized by low **albedo** (the fraction of incident light reflected by the asteroid surface) of 0.02 to 0.07 and have nearly featureless but ever-reddening spectra (i.e., increasing reflectance in the infrared) from

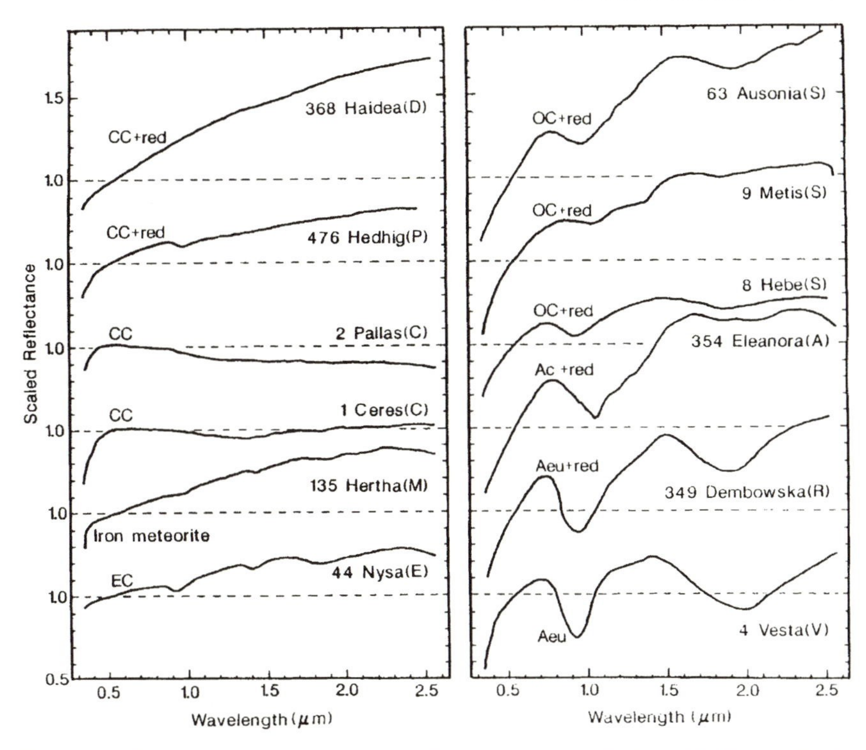

FIGURE III-5. Representative types of reflectance spectra (D, P, C, M, E, S, A, R, V) for various asteroids, according to the classification of Tholen (1984). The probable meteoritic classes whose spectra may match those asteroids are also indicated. The "+red" signs indicate the presence of "unknown" materials or structure (on the surface of asteroids) that reflect preferentially infrared light. Data curves are selected from Bell et al. (1987).

C to P to D. The spectra of C and P are similar to those of carbonaceous chondrites (Gaffey, 1976). Ceres and Pallas are type C asteroids, and their densities (Table III-1) are consistent with those of types 1 and 2 carbonaceous chondrites. There is no known carbonaceous chondrite that has the highly reddish spectrum of asteroid type D.

The asteroid types P, M, and E are almost identical spectrally, but differ greatly in their albedo (0.02 to 0.07; 0.07 to 0.23; and greater than 0.23, respectively). Type M is comparable to the iron meteorite spectra and albedo. Type E is similar to the enstatite chondrite spectra and albedo (Figure III-5).

Asteroid types A, R, and V are rare and all have medium albedo (0.07 to 0.23). The spectrum of the Vesta asteroid (type V) is almost identical to that of basaltic achondrites (Aeu). The density of Vesta (Table III-1) is also consistent with its achondritic composition. The type R spectrum is also similar to that of Aeu but is much more reddish. The type A spectrum is similar to that of olivine achondrites (also called chassignite, Ac) but again is much more reddish (compared with that of olivine in Figure III-4). There are no ready explanations for the reddish spectra of types R and A.

Type S asteroids have medium albedo and are most common. Most type S spectra lie between those of Hebe and Ausonia (Figure III-5) and are similar to those of ordinary chondrites (Figure III-4) but with varying degrees of reddishness. A few type S spectra are similar to the spectrum of the Metis asteroid, exhibiting olivine absorption characteristics (Figure III-5). The reddish spectra of types S, D, P, A, and R (Figure III-5) may result from so-called space weathering (including abrasion of the asteroid surface by solar wind, solar radiation, and micrometeorites etc.; Pieters and McFadden, 1994). One should keep in mind that the spectra of stony iron meteorites also resemble type S spectra (Gaffey, 1976, 1990; Gaffey et al., 1989; Lewis, 1997). However, stony iron meteorites account for only one percent of total meteorite falls. This is hard to reconcile with the fact that type S is one of the most abundant asteroid types. Pieters and McFadden (1994) suggested that the S type asteroids could be ordinary chondrite regolith (layer of unconsolidated rock debris), primitive achondrites, urelites, and pallasites (stony iron). Those meteorites are discussed in Chapter IV. Certainly, further studies are needed, especially into the causes of the enhanced reflection of infrared light at the asteroid surfaces.

The relative abundances of reflectance spectrum types in each asteroidal zone (in %) are summarized in Table III-2, based on the classification data given by Tholen (1984; his B, F, and G types are lumped together with C type; his T type with P type; and his R, V, and Q types with A type). The systematic changes from the innermost zone (high temperature) to the outermost zone (low temperature) are evident. For example, E type is most abundant in the Hungaria family; S type in Mainbelt I; C type in Mainbelt III; P type in the Hildas family; and D type in the Trojans family. The most interesting result in Table III-2 is that the relative abundances of meteorite classes in the meteorite falls match quite well with those of corresponding asteroid types in Apollo and Amor objects (Mars and Earth crossers), assuming S type asteroids as ordinary chondrites. Small numbers of C type and the absence of P and D types in Apollo and Amor objects may suggest that the loose materials of C, P, and D types did not survive well at the collision events of asteroids or during their transit from asteroid zones to Mars- and Earth-crossing orbits. The paucity of C, P, and D type meteorites may also be due to atmospheric breakup.

Comets

The heliocentric comets are widely distributed in the space between 4 AU and 10^5 AU. The size can range from about 1 km to 60 km in radius. The comets

TABLE III-2
Percentages of various meteoritic classes and of spectral types of asteroids in each given asteroid belt or family

Meteorites (674)			*CC*	*OC*	*EC*	*Achon.*	*Iron*	*Stony Iron*
			4.2	79.6	1.6	8.3	5.1	1.2
Asteroids	D	P	C	S	E	A	M	
Appollo + Amor (19)			5	79		11	5	
Hungaria (12)				42	58			
Mainbelt I (112)		1	25	68	2	1	3	
Mainbelt II (136)		3	41	43	2	1	10	
Mainbelt III (130)	3	4	49	32		2	10	
Mainbelt IV (24)	21	33	42	4				
Hildas (21)	38	57	5					
Trojans (16)	88	12						

Note: Number in parentheses represent the number of classified meteorite falls or classified asteroids in a given belt or family. Whether or not the S type asteroids really correspond to the ordinary chondrites (OC) needs further study.

are divided according to their period into the short-period ($p < 200$ yr) and the long-period ($p > 200$ yr) groups. The short period comets are further separated into the **Jupiter family** ($p < 20$ yr, or $a = 3$ to 9 AU) and the **Halley type** ($200 > p > 20$ yr, or $a = 9$ to 40 AU). The long-period comets with $p > 10^6$ years or $a = 10^4$ to 10^5 AU are called **Oort cloud comets** (Oort, 1950). The observed numbers of comets that have well-determined orbital parameters are plotted as a function of their revolution period p or semimajor axis a in Figure III-6a and of their inclination i in Figure III-6b. The "corrected" curve in Figure III-6a is the distribution of long-period comets corrected for the probability of observation, because long-period comets have a reduced probability of being found within observable distance during human history. The Halley type and long-period comets cover the whole range of orbital inclinations (0 to 180°) and most have eccentricities of greater than 0.99. This indicates that these comets have very stretched orbits and approach the inner solar system almost equally from every direction.

The Oort cloud is a hypothesized spherical cloud containing numerous comet-like bodies in the outer edge of the solar system (10^4 to 10^5 AU) as proposed first by Oort (1950). The total number of comet-like bodies in the Oort cloud is estimated to be about 2×10^{12} currently, and was probably 2×10^{13} at the beginning of the solar system (Remy and Mignard, 1985). Because the size distribution of comets is uncertain, the current total mass of the Oort cloud can only be roughly estimated; estimates range from 2 to 100 times Earth mass (m_E; Mendis and Marconi, 1986). For comparison, the total asteroidal mass of $0.0005 m_E$ is negligibly small. When the solar system as a whole rotates around the center of our Milky Way galaxy, passing stars

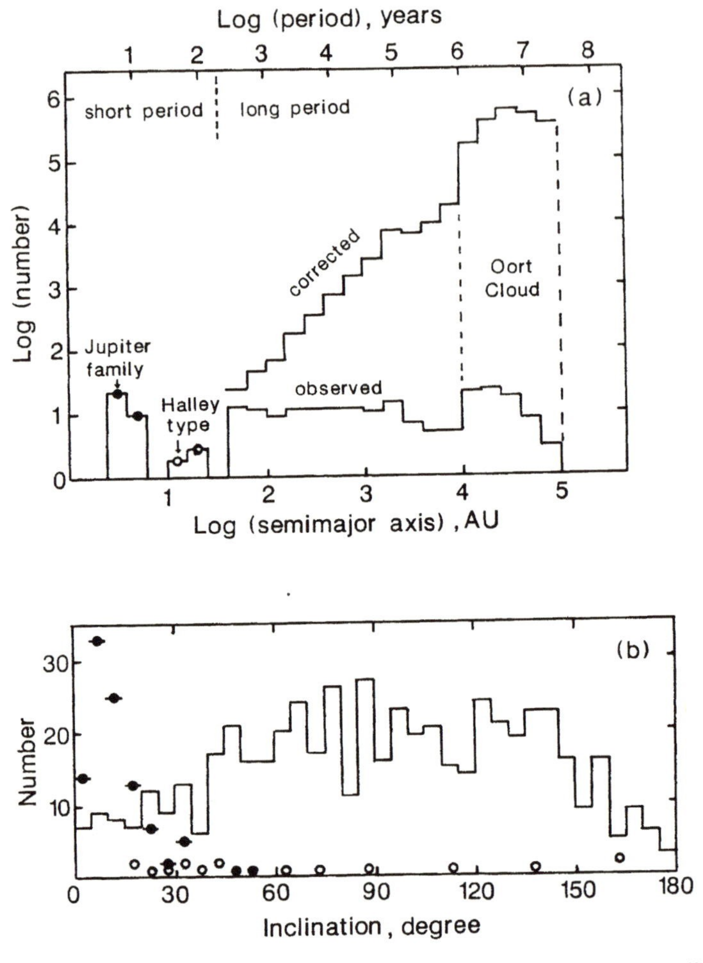

FIGURE III-6. Frequency distribution of Jupiter family (solid circles), Halley type (open circles), and long-period comets as a function of (a) their semimajor axis or period and (b) their inclination (modified after Kresak, 1982; reprinted by permission of the University of Arizona Press).

and galactic gas clouds can gravitationally jostle comet-like bodies out of the Oort cloud and into the orbits of long-period comets. Because of the loss of gas and dust and the gravitational perturbation by planets whenever the comet passes near the Sun, the orbit of the comet can be changed drastically into either a hyperbolic path (out of the solar system altogether) or an elliptic path with a shorter period (and thus finally into a Halley type orbit). The mean lifetime of Halley type comets is on the order of 10^4 years. It is still debatable as to where and how the comets in the Oort cloud were originally formed. Weissman (1985) gives a nice review on this matter.

The orbital inclinations of Jupiter family comets are mostly less than 30° and peak around 8° (Figure III-6b), similar to those of asteroids. Model simulations by Duncan et al. (1988) strongly suggest that the orbits of Jupiter comets cannot evolve from those of long-period comets and favor the hypothesis of the **Kuiper belt**, lying just outside the orbit of Neptune, as the source of Jupiter family comets. Jewitt and Luu (1993) first discovered the Kuiper belt objects. So far, about 40 Kuiper belt objects have been identified with semimajor axes of 30 to 50 AU, eccentricities less than 0.3, and inclinations less than 20°. Since only a small fraction of the sky has been searched, the total number of Kuiper belt objects with diameter larger than 100 km may be more than 3500 with a probable total mass of about $0.05m_E$ (Stewart, 1997). The recent discovery of two trans-Neptune objects ($1996TL_{66}$, with $a = 84$ AU, $e = 0.58$, and $i = 24°$; and $1996RQ_{20}$, with $a = 47$ AU, $e = 0.3$, and $i = 32°$) by Luu et al. (1997) and Helin et al. (1997) may also indicate the existence of a disk of scattered icy objects (SIO). The SIOs have a greater range of eccentricities and inclinations between the Kuiper belt and the Oort cloud as predicted by the simulation model of Duncan and Levison (1997). The scattered icy objects might originate in the region between Uranus and Neptune, or in the inner edge of the Kuiper belt and may have been ejected by the gravitational force of Neptune into eccentric orbits during the formation of the solar system. These scattered icy objects could also be the sources for both Jupiter family and Halley type comets (Duncon and Levison, 1997). Wandering of the Kuiper belt objects and SIOs into Earth-crossing orbits is a real possibility.

Figure III-3 (open circles) shows the distribution of Jupiter family comets in the a-e plane along with asteroid data. Most of them have eccentricities of greater than 0.3 and aphelion Q of greater than 4.5 AU. It is apparent from Figure III-3 that many Jupiter family comets are Jupiter-crossing. The spectacular collision of about 20 fragments of the comet Shoemaker-Levy 9 onto Jupiter from July 16 to 22, 1994 was the most recent example (Spencer and Mitton, 1995). A few asteroids within the Jupiter family field (e.g., Hidalgo, $a = 5.86$, $e = 0.66$; Chiron, $a = 13.7$, $e = 0.38$; 1983SA, $a = 4.2$, $e = 0.71$; 1984BC, $a = 3.45$, $e = 0.54$) might have been comets at one time but have lost all or part of their volatiles so that they are incapable of producing a coma. Interestingly, those "comet-origin" asteroids as well as comets P/Neujmin ($a = 6.85$, $e = 0.78$), P/Arend-Rigaux ($a = 3.6$, $e = 0.6$), and P/Schwassmann-Wachmann 1 ($a = 6.05$, $e = 0.04$) all have D type reflectance spectra (Hartman et al., 1987).

The results from the dust impact mass spectrometer on board the spacecraft Vega 1 indicate that most of the submicron-size dust particles in the coma of the comet Halley have a fluffy mineral core ($\rho^0 = 1$ to $2\,g/cm^3$), which is often covered by even more fluffy organic material ($\rho^0 = 0.3$ to $1\,g/cm^3$) and ices (Kissel and Krueger, 1987). As shown in Table III-3 (Jessberger et al., 1988), the mean composition of dust particles as a whole is similar to that of type 1 carbonaceous chondrites, roughly within a factor of two (which is also the estimated uncertainty of the comet data). The obvious exceptions are hydrogen, carbon, and probably nitrogen, which are enriched in the dust particles of the comet Halley. The gas phase of Halley's coma consists of more than 80% water vapor. The probable upper limits of volume mixing ratios of CO_2/H_2O, NH_3/H_2O, and CH_4/H_2O in the gas phase are, respectively, 3.5, 10, and 7% (Krankowsky et al., 1986). Additional information on organic compounds in comet Halley is summarized by Delsemme (1991). The estimated bulk density of comet Halley is about $0.6\,g/cm^3$ with uncertainty of $+0.9$ and $-0.4\,g/cm^3$ (Sagdeev et al., 1988).

Another interesting way to deduce the composition of comets is to study the **interplanetary dust particles** (IDP), which are extraterrestrial dust particles between 2 and 50 microns in size, collected from the Earth's stratosphere and probably of cometary origin (Brownlee, 1985; Mackinnon and Rietmeijer, 1987). The average composition of this interplanetary dust is again similar to C1 composition (Table III-3, Schramm et al., 1988). The interplanetary dust has a higher carbon content than C1, which is consis-

TABLE III-3

Compositions of comet Halley particles, interplanetary dust particles (IDP), and type 1 carbonaceous chondrite (solar nebula) normalized to A1 = 10 atoms

	Halley particles (1)	*IDP* (2)	*Cl (solar nebula)* (3)
H	3000		620 (3.3×10^6)
C	1200	230	90 (1200)
N	62		7.1(370)
O	1300	560	910 (2800)
Na	15	6.9	6.8
Mg	150	130	127
Al	10	10	10
Si	270	130	118
S	110	48	61
Ca	9.3	6.9	7.2
Ti	0.6	0.29	0.28
Cr	1.3	2	1.6
Mn	0.7	2	1.1
Fe	80	93	106
Co	0.4		0.27
Ni	6	3.5	5.8

Sources: (1) Jessberger et al. (1988); (2) Schramm et al. (1988), except Ti and Mn (Brownlee, 1978, 1985; Brownlee et al., 1977); (3) Anders and Grevesse (1989).

tent with a cometary origin. However, the tendency to have higher Na/Al, Mg/Al, Si/Al, C/Al, and N/Al ratios in Halley particles and IDP than in C1 may represent real compositional change of the solar nebula as a function of heliocentric distance (Rietmeijer, 1988; Ringwood, 1989).

Collision of large comets and asteroids with the Earth might have triggered mass extinctions of living species on the Earth (Alvarez et al., 1980). This subject has been studied extensively and discussed intensively for the last twenty years (Albritton, 1989; Clube, 1989; Kerr, 1997).

III-3. Planets, Their Satellites, and Their Rings

Tables III-4a and III-4b summarize some useful orbital parameters and physical properties of the Sun, its nine planets, and their satellites (radii larger than 200 km, except for Mars). The narrow range of orbital inclinations for the nine planets (0 to 17.2°) and the small spin inclination of the Sun relative to the ecliptic plane (7°) indicate that all solar system planets are orbiting around the Sun more or less in the same plane or disk and are moving in the same sense of direction as the Sun's rotation or spin. Furthermore, except for Venus, Uranus, and Pluto, the other six planets and the Sun are also spinning more or less in the same direction. The spin inclination of greater than 90° for Venus, Uranus, and Pluto might be caused by later collision events with some other interplanetary bodies. Except for a few satellites (e.g., Triton and Charon), most satellites also revolve and spin in the same sense of direction as their respective central planets. Because of the tidal effect between the central planet and its satellites, the orbital period and spin period of each satellite are already nearly identical, with only a few exceptions. About 99.9% of the present solar system mass is concentrated in the Sun, and only 0.1% in the planets. In contrast, the angular momentum (mvd) distribution is 99.5% in the planets and only 0.5% in the Sun (Table III-4a). Among nine planets, both the mass and angular momentum are mainly carried by the four **outer planets** (i.e., Jupiter, Saturn, Uranus, Neptune; excluding Pluto).

Planets

The average density, ρ, of planets or satellites is estimated by $\rho = 3m/(4\pi r^3)$. Although the masses of the **inner planets** (i.e., Mercury, Venus, Earth, and Mars) are much smaller than those of the outer planets, the former have much higher average densities than the latter, indicating fundamental differences in chemical composition between the two groups. The inner planets have an Earth-like composition, i.e., iron-nickel core and silicate mantle and crust. For ease of comparison, the estimated average densities for the inner planets

TABLE III-4A
Orbital and physical parameters of the Sun and nine planets

	Relative distance a/a_E *(AU)*	*Eccentricity* e	*Orbital period* p *(years)*	*Orbital inclination* i *(degrees)*	*Spin period* p_s *(days)*	*Spin inclination* i_s *(degrees)*
Sun	0				25 to 35[a]	~7[b]
Mercury	0.3871	0.206	0.24	7	58.6	~0
Venus	0.7233	0.007	0.062	3.4	243	177
Earth	1	0.017	1	0	1	23.4
Mars	1.524	0.093	1.88	1.9	1.03	25.2
Jupiter	5.203	0.048	11.9	1.3	0.41	3.1
Saturn	9.588	0.056	29.5	2.5	0.43	26.7
Uranus	19.19	0.046	84.1	0.77	0.65	97.9
Neptune	30.06	0.01	165	1.8	0.77	29
Pluto	39.53	0.248	249	17.2	6.4	~90

	Relative radius r/r_E	*Relative mass* m/m_E	*Angular momentum* $(10^{36}$ *g* $km^2/sec)$	*Escape velocity (km/sec)*	*Effective surface temperature* (K)	*Average density* $\rho(g/cm^3)$
Sun	109	3.3×10^5	190	616	5800	1.41
Mercury	0.38	0.055	0.92	4.3	448	5.43(5.2)[c]
Venus	0.95	0.82	18.5	10.4	328	5.24(4.0)[c]
Earth	1	1	26.6	11.2	279	5.52(4.0)[c]
Mars	0.53	0.107	3.5	5	226	3.94(3.7)[c]
Jupiter	11.2	318	19300	60	122	1.31
Saturn	9.41	94.3	7830	36	91	0.69
Uranus	3.98	14.5	1690	21	64	1.3
Neptune	3.81	17.2	2520	24	50	1.66
Pluto	0.18	~0.0026	~0.4	~1	45	2.1

Notes: a_E = semimajor axis of the Earth = 150×10^6 km = one astronomical unit (AU); r_E and m_E are the radius and mass of the Earth = 6380 km and 6×10^{27} g, respectively.

[a] 25 days on equator and 35 days near the pole.

[b] Relative to the ecliptic plane.

[c] Value in parentheses is the estimated density at 1 atmospheric pressure (Wasson, 1985).

at zero or one bar pressure, ρ^0, are also given in Table III-4 (numbers in parentheses). If one assumes zero pressure densities of silicates, ρ_1^0, equal to 3.3 g/cm^3 (similar to average achondritic meteorites) and of the Fe-Ni core, ρ_2^0, of 7.8 g/cm^3 (iron meteorites), one can estimate the mass fraction of the Fe-Ni core, f, to be about 0.63 for Mercury, 0.30 for Venus and Earth, and 0.19 for Mars, using the relationship

$$f/\rho_2^0 + (1 - f)/\rho_1^0 = 1/\rho_0^0. \tag{III-8}$$

TABLE III-4B
Orbital and physical parameters of satellites

	a (10^3 *km*)	*p* (*days*)	*r* (*km*)	*m* (10^{24} *g*)	$\rho(\rho^0)$ (g/cm^3)	*Albedo*	*Surface composition*[b]
Earth (1)							
Moon	384	27.3	1738	73.5	3.34	0.12	basalts
Mars (2)							
Phobos	9.38	0.319	~11[a]	1.3×10^{-3}	2.0 ± 0.5	0.06	cc
Deimos	23.5	1.26	~6[a]	1.8×10^{-4}	1.9 ± 0.5	0.06	cc
Jupiter (≥17)							
Io	422	1.77	1815	89.2	3.55 (3.4)	0.6	sulfur, SO_2
Europa	671	3.55	1570	48.7	3	0.6	ice
Ganymede	1070	7.16	2640	149	1.9 (~1.4)	0.4	dirty ice
Callisto	1880	16.7	2410	106	1.8 (~1.4)	0.2	dirty ice
Saturn (≥24)							
Mimas	186	0.942	195	0.038	1.4 ± 0.2	0.77	ice
Enceladus	238	1.37	250	0.074	1.2 ± 0.5	1	~ pure ice
Tethys	295	1.89	525	0.63	1.2 ± 0.1	0.8	ice
Dione	377	2.74	560	1.05	1.4 ± 0.1	0.55	ice
Rhea	527	4.52	765	2.3	1.2 ± 0.1	0.65	ice
Titan	1222	15.95	2560	136	1.9 (~1.4)	0.2	dirty ice; N_2 (g)
Iapetus	3561	79.3	720	1.93	1.2 ± 0.1	0.5/0.06	ice/cc

TABLE III-1B *Continued*

	a (10^3 km)	p (days)	r (km)	m (10^{24} g)	$\rho(\rho^0)$ (g/cm^3)	Albedo	Surface composition[b]
Uranus (≥15)							
Miranda	130	1.41	242	0.071	1.3 ± 0.4	0.22	dirty ice
Ariel	192	2.52	580	1.44	1.7 ± 0.3	0.4	dirty ice
Umbriel	267	4.14	595	1.18	1.4 ± 0.2	0.16	dirty ice
Titania	438	8.71	800	3.43	1.6 ± 0.1	0.23	dirty ice
Oberon	586	13.46	775	2.87	1.5 ± 0.1	0.2	dirty ice
Neptune (≥8)							
Triton	354	5.88	1360	~140	2.02	0.4	CH_4 + hydro-carbon + N_2 ice
Nereid	5520	360	~200			0.1	
Pluto (≥1)							
Charon	19.1	6.39	~595		~2		CH_4 + hydro-carbon + N_2 ice

Sources: Data are mostly from Farinelia (1987) and Burns (1986).

Notes: cc = carbonaceous chondrite; ice = water ice unless specified, and always mixed with various amount of carbonaceous chondrite materials. The distinction between ice and dirty ice is arbitrarily set at the surface albedo of 0.5. Number in parentheses after each planet indicates the number of satellites so far discovered. Density values in parentheses are at one atmosphere pressure.

[a] Phobos and Deimos are ellipsoidal with principal axes of 13.5 × 10.5 × 9 and 7.5 × 6 × 5, respectively.

[b] Surface compositions are deduced from the spectral reflectance curves of satellites.

The similar ρ^0 values for Mars (Table III-4a) and the ordinary chondrites (3.6 to 3.8 g/cm^3) may suggest their similarity in chemical composition. Because the mass fraction of Fe-Ni in the solar nebula is about 0.0015, the minimum masses of the solar nebular gas required to form Mercury, Venus, Earth, and Mars are, respectively, 23, 164, 200 and $14m_E(= mf/\{0.0015m_E\})$. We may call those minimum masses **normalized planetary masses**.

Figure III-7 shows the predicted mass-radius relationships for hypothetical celestial bodies of various composition (pure H; 0.75H + 0.25He; "ice" = $H_2O + NH_3 + CH_4$ in the relative cosmic abundance; and "rock" = metallic oxides of major cations in the relative cosmic abundance), assuming adiabatic accretion and certain equations of state for those materials (Stevenson, 1982). The superimposed positions of the outer planets in Figure III-7 suggest that Jupiter and Saturn are mainly composed of H and He with some ice + rock

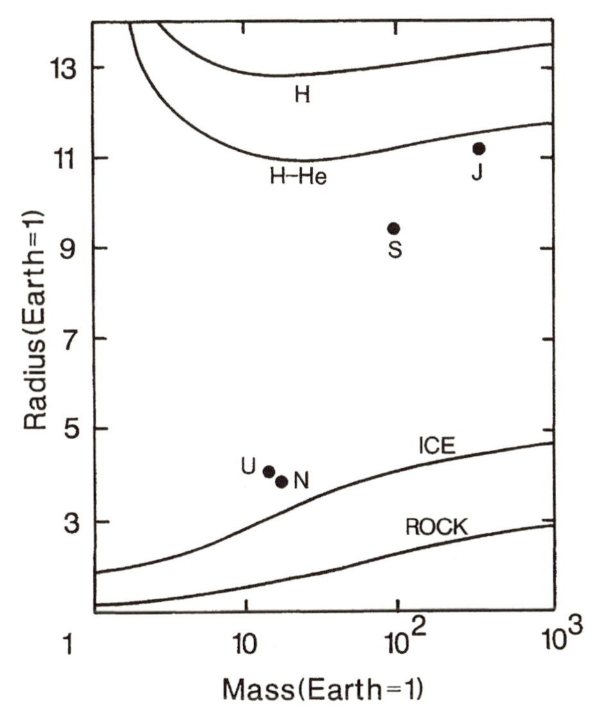

FIGURE III-7. Theoretical relationship between the radius and mass of celestial objects that are made of pure hydrogen (H), of a $0.75\,H_2 + 0.25\,He$ mixture (H-He), of the cosmic mixture of H_2O, NH_3, and CH_4 (ice), and of the cosmic mixture of metal oxides (rock) and formed under adiabatic conditions. Positions of Jupiter (J), Saturn (S), Uranus (U), and Neptune (N) in the figure indicate that J and S are mainly made of H_2-He, while U and N contain larger fractions of "ice" and "rock" components (modified after Stevenson, 1982.)

component; while Uranus and Neptune are mainly composed of ice and rock, with some H and He. An interesting result of model calculations is that the masses of ice+rock core for the four outer planets are all about $15m_E$, within a factor of two (Stevenson, 1982; Pollack, 1985). Because the mass fraction of ice + rock in the solar nebula is about 0.022, the normalized mass of each outer planet is about $15m_E/0.022 = 680m_E$. The total normalized mass of nine planets would be about $3120m_E$ or about 1% of the present solar mass.

The normalized mass of each planet can be dispersed back in a preplanetary solar nebular disk according to the assumptions that (1) the present orbit of each planet represents the orbit of the center of mass for its normalized planetary mass, and (2) the boundary between two adjacent normalized planetary masses in the disk is set to where the gravitational force on a unit mass by the two is equal and opposite. The outer boundary for Neptune is set equal to the distance between Neptune and the zero-force boundary with Uranus. The calculated **area loading** (g/cm^2) of the proto-solar nebular disk (the normalized planetary mass divided by the area of its disk zone) is shown in Figure III-8a. According to Figure III-8a, the area loading of the proto-solar nebular disk is roughly proportional to $d^{-3/2}$ except for the disk zones of Mercury and Mars. The low area loading for the Mercury disk zone is probably real, indicating a decrease in the nebular gas density toward the Sun. The low area loading for the Mars disk zone may be caused by the continuous loss of mass of asteroids and comets in the area between Mars and Jupiter since the formation of the solar system 4.5 billion years ago. According to the model calculation by Hayashi (1981), the thickness of the nebular disk (h) in hydrodynamic equilibrium is proportional to $d^{-5/4}$. Therefore, the density of the nebular disk, ρ, should be proportional to (area loading)$/h$ or $d^{-11/4}$, in fair agreement with the observed relationship $\rho \propto d^{-3}$ for the circumstellar disk around the main sequence star β-Pictoris (Smith and Terrile, 1984). If the proportionality of the area loading to $d^{-3/2}$ in the solar nebula can extend to 60 or 90 AU, then the mass of solar nebula between 30 (Neptune) and 60 or between 30 and 90 AU is estimated to be about $1600m_E$ or $2800m_E$, which in turn can condense into comets with total mass of 32 or $56m_E$ (using $Z = 0.02$). Therefore, it is feasible that the solar nebula beyond Neptune might provide the necessary materials for the comets in the Oort cloud and the Kuiper belt just outside the orbit of Neptune.

Satellites

For satellites less massive than the Moon, one can approximate $\rho = \rho^0$. For those more massive than the moon, the estimated ρ^0 values are also given in Table III-4b (Lewis and Prinn, 1984; their Figure 4.50). The density of

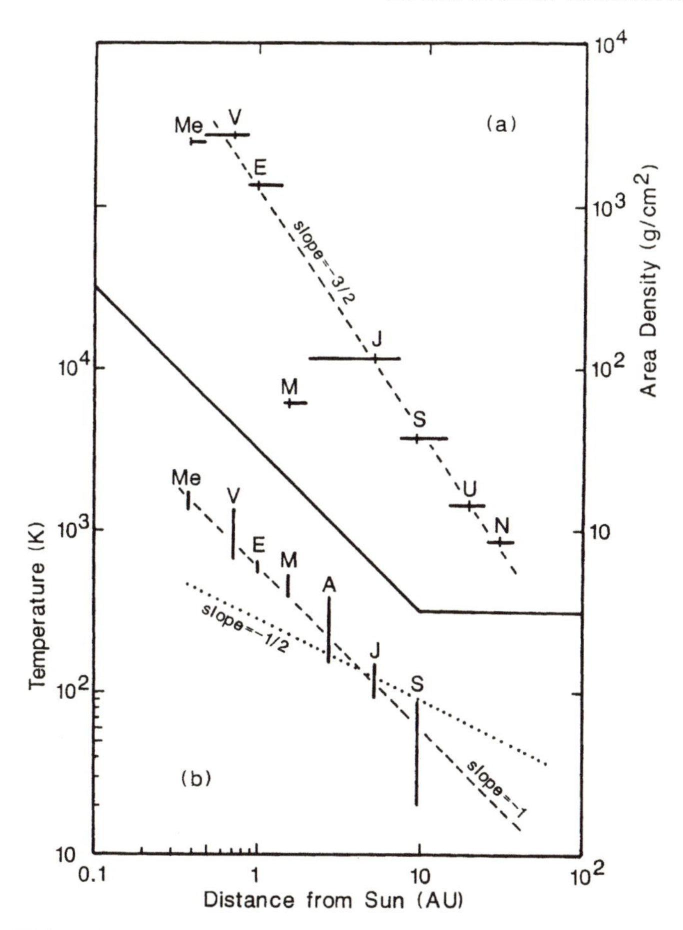

FIGURE III-8. Linear relationship between logarithms of the heliocentric distance (AU) and (a) the area loading (g/cm^2) of the solar nebular disk as estimated from the normalized planetary masses; (b) the formation temperatures of various planets. The dotted line with a slope of $-1/2$ shows the effective surface temperatures of planets today as calculated by equation III-6.

3.3 g/cm^3 at zero pressure for the Moon suggests that it may have a bulk composition similar to achondritic meteorites. Furthermore, the oxygen isotope composition of the Moon is almost identical to that of the enstatite achondrites, enstatite chondrites, and terrestrial basaltic rocks (Clayton et al., 1984). Therefore, one possible scenario for the origin of the Moon is that a Mars-sized body or larger with oxygen isotope composition similar to enstatite meteorites and the Earth first impacted on the proto-Earth. Then, the blasted materials formed a prelunar disk from which the Moon accreted later (Boss, 1986; Ida et al., 1997). The densities of Phobos and Deimos are similar to those of type 1 carbonaceous chondrites ($2.2\,g/cm^3$). The low albedo and flat spectral reflectance curves for those bodies (Clark et al., 1986) are also consistent with a carbonaceous chondrite composition. However, Phobos and Deimos have no detectable 3-micron water band. Evidently, their surface material has been shock-heated, ejected, and reaccreted. Phobos and Deimos were probably Mars-crossing asteroids captured by Mars.

The density of Io is similar to that of C3 carbonaceous chondrites and achondrites. According to the reflectance spectra, the surface of Io is covered by SO_2 frost and sulfur compounds, indicating intense magmatic differentiation sustained by some kind of tidal frictional heating by nearby Jupiter (Nash et al., 1986). Because achondrites are low in sulfur content, the bulk composition of Io may be closer to C3 or C2 before degassing (Lewis and Prinn, 1984). The reflectance spectrum of Europa indicates a layer of water ice on its surface. If one assumes Europa is composed of C3 and water ice, then the mass fraction of the ice crust should be about 5% (using an equation similar to III-8, $\rho_1^0 = 3.4\,g/cm^3$ for C3 and $\rho_2^0 = 0.9$ for water ice). Ganymede, Callisto, and satellites of Saturn and Uranus are all characterized by low bulk density (1.2 to $1.7\,g/cm^3$), high surface albedo, and a strong water ice signal in the reflectance spectra. Those satellites may be composed mainly of water ice mantles and cores of carbonaceous chondrite composition. Assuming densities of carbonaceous chondrites equal to $2.2\,g/cm^3$ (C1) and of water ice equal to $0.9\,g/cm^3$, the mass fraction of water ice in those satellites can range from 0.51 to 0.15 (using a relationship similar to equation III-8). Clark et al. (1986) suggest that the water ice of Saturn's satellites may contain 1 to 10% of $NH_3 \cdot H_2O$. The existence of N_2 and some CH_4 in the atmosphere of Titan also suggests the incorporation of $NH_3 \cdot H_2O$ and $CH_4 \cdot 7H_2O$ in Titan's water ice (Schubert et al., 1986). The encounter of Neptune by *Voyager 2* showed that Triton's surface is mainly N_2 ice mixed with some CH_4 and hydrocarbon ices, with no spectroscopic evidence of H_2O ice (Tsurutani, 1989). The high density of Triton (Tsurutani, 1989) and Pluto + Charon (Tholen, 1989) may suggest that Triton was captured by Neptune, and Pluto was deflected into its present orbit during a passage close to Neptune.

Rings

Jupiter, Saturn, and Uranus have so-called ring systems. Rings consist of numerous particles and blocks which orbit around the equatorial planes of these planets and inside the orbits of principal satellites (Greenberg and Brahic, 1984).

Jupiter has only a faint ring which lies between 1.73 and 1.81 Jupiter radii and is composed of dust and rock fragments. Saturn has seven bright major rings lying between 1.21 and 3.95 Saturn radii. Each ring consists of many narrow ringlets and is composed of numerous particles of ice and ice-coated rock fragments ranging from a few microns to about 10 meters in size. Nine rings are counted for Uranus between 1.60 and 1.95 Uranian radii. The low albedo of Uranian rings may indicate carbonaceous composition.

The existence of rings may indicate the failure of accretion of particles into satellites under the gravitational influence of the central planets or the breakup of celestial bodies that come too close to the planets. For example, we may consider a sphere consisting of dust particles which are held together by a gravitational cohesive force. In the gravitational field of a planet, the sphere is subjected to a stronger force on the side facing the planet than on the opposite side. This differential force is called the tidal force. The closer the sphere to the planet, the higher is the gravitational tidal effect. There is a distance at which the disruptive tidal effect equals the sphere's gravitational cohesive force. This distance is called the **Roche limit** (d_R), and can be expressed as $d_R = 2.5r(\rho_p/\rho)^{1/3}$ where r is the radius of a given central planet, ρ_p is the density of the planet, and ρ that of the sphere. Inside the Roche limit, the dust participles never accrete to form a large body. Moreover, if any large celestial body were to come inside the Roche limit of a planet, it might break apart due to the planet's gravitational tidal effect.

III-4. Condensation of Solid Dusts from the Solar Nebular Gas

It is generally believed that a slowly rotating cloud of interstellar gas and dust contracted and collapsed into a proto-Sun about 4.6×10^9 years ago. Meanwhile, continuous outflow of gas from the proto-Sun resulted in condensation of some gas into dust and the formation of a rotationally flattened disk of gas and dust, called the **solar nebula**, around the proto-Sun. The solar nebula might have a mass of about 0.02 to 0.05 solar mass (Safronov and Ruzmaikina, 1985; Hayashi et al., 1985) and carry most of the angular momentum in the proto-solar system. Possible mechanisms for transferring gas and angular momentum from the proto-Sun to the solar nebula include a strong magnetic field, outflow of gases as winds and collimated jets from the proto-Sun (O'Dell and Beckwith, 1997), and turbulent flow accompanying viscous stress within the nebula (Levey and Lunine, 1993).

When turbulent and convective motions in the solar nebula decayed gradually, dust grains sank gravitationally toward the mid-plane of the nebular disk and grew in size by coalescence. Consequently, a thin dust layer was formed. The thin dust layer eventually became too dense to be stable and fragmented into numerous clumps, which then coalesced into planetesimals through mutual collisions, and finally into proto–inner planets as well as cores of proto–outer planets. In principle, protoplanets with masses greater than 10^{26} g can gravitationally attract nebular gas and be surrounded by a primordial atmosphere. The mass of gas gathered by the growing proto–outer planets became so large that the gas (mainly H_2 and He) collapsed into liquid and/or solid phases on the surface of the planets. Because the area loading of the nebular disk decreases drastically from Venus outward to Uranus and Neptune (Figure III-8), the inner planets might form first, and Uranus and Neptune last.

Finally, during the so-called **T-Tauri** wind phase of the proto-Sun, the greatly enhanced solar wind and radiation blew away most of the uncondensed gas phase in the solar nebula. This mechanism, however, has been questioned (Levy and Lunine, 1993). For example, Trivedi (1984, 1987) argued that the solar nebula was formed during the T-Tauri phase of the proto-Sun instead. Cameron (1985, and references therein) favored a massive solar nebula as an accretionary disc with a mass of about one solar mass. In his model, the massive nebula fragmented into giant gaseous protoplanets with dust grains accumulated at the centers to form cores. The giant gaseous protoplanets evolved easily into the giant outer planets and into inner planets by dissipation of the thick gaseous envelope by yet unknown mechanisms. The main difficulties of Cameron's model are how to get rid of the large amount of unused nebular gas after planetary formation and how to form a proto-Sun with a massive nebula without evolving into a system of binary or multiple stars.

If the solar nebula were formed mainly from the ejected gas of the proto-Sun itself, the original interstellar dust should have been mostly destroyed by evaporation inside the proto-Sun. However, there is always a finite possibility that some interstellar dust might have been captured by the solar nebula directly and survived inside some primitive objects such as carbonaceous chondrites or comets. This possibility will be discussed in Chapter IV. In any case, one of the most important cosmochemical processes during the evolution of the solar nebula was the condensation of various mineral dusts from the nebular gas according to temperature and pressure conditions, which varied both spatially (Figure III-8) and temporally within the solar nebula (e.g., Urey, 1952; Larimer, 1967; Lewis, 1972, 1997; Grossman and Larimer, 1974, and references therein).

Equilibrium Condensation Models

If a hypothetical element M can exist only as a monoatomic gas and solid in the solar nebula, then the only possible reaction is $M(\mathrm{s}) \leftrightarrow M(\mathrm{g})$. The condition for equilibrium between the gas and solid phases is

$$K = \exp\left(-\frac{\Delta \widetilde{G}}{RT}\right) = \frac{f_M}{a_M} \approx \frac{P_M}{a_M} \qquad \text{(III-9)}$$

or

$$\log \frac{P_M}{a_M} \approx \log K = \frac{-\Delta\widetilde{G}}{2.3RT} \approx \frac{-\Delta\widetilde{H}^0_{298}}{2.3RT} + \frac{\Delta\widetilde{S}^0_{298}}{2.3R} \qquad \text{(III-10a)}$$

if the total pressure P of the system is one atmosphere or less and $-\Delta\widetilde{C}^0_{p,298}$ is negligibly small. Rearranging the above equation, one obtains the equilibrium temperature

$$T \approx \frac{\Delta\widetilde{H}^0_{298}}{\Delta\widetilde{S}^0_{298} - 2.3R\log P_M/a_M}, \qquad \text{(III-10b)}$$

where K = thermodynamic equilibrium constant for the reaction; $\Delta\widetilde{G}$ = Gibbs free energy of the reaction; $\Delta\widetilde{H}^0_{298}$, $\Delta\widetilde{S}^0_{298}$, and $\Delta\widetilde{C}^0_{p,298}$ are, respectively, the enthalpy, entropy, and heat capacity of the reaction at 298 K and total pressure of one atmosphere [$\Delta\widetilde{H}^0_{298}$ is essentially equal to the heat of sublimation of metal $M(\Delta\widetilde{H}^0_{\text{sbl}}$; see Section I-3)]; P_M = partial pressure of M(g) in equilibrium with M(s); f_M = fugacity of $M(\text{g}) \approx P_M$ at low total pressure P; and a_M = activity of M(s) = 1 for pure solid phase.

On the other hand, P_M in the solar nebula can be estimated by mass balance, i.e.,

$$\frac{P_M}{P_{H_2} + P_{He}} = \frac{(1-\alpha_M)A_M}{A_{H_2} + A_{He}},$$

or

$$P_M = \frac{P(1-\alpha_M)A_M}{A_{H_2} + A_{He}}, \qquad \text{(III-11)}$$

where P_{H_2} and P_{He} are partial pressures of hydrogen and helium gas, and $P_{H_2} + P_{He} \approx P$; A_M, A_{H_2}, and A_{He} are solar or cosmic abundances of M, H_2, and He, respectively (in atoms or molecules/10^6 atoms Si, Table II-1); and α_M is the fraction of M condensed to solid, i.e., $(1-\alpha_M)$ is the fraction of M remaining in the gas phase.

Therefore, at a given P, T can be obtained from equations III-10b and III-11 as a function of α_M, if M(s) is a pure phase. In the special case when $\alpha_M = 0$, i.e., the instant when M(g) starts to condense to M(s), T is called the **condensation temperature** ($T_{C,0}$) of M at a total pressure of P. When $\alpha_M = 0.5$, T is called the **50% condensation temperature** ($T_{C,50}$). For example, at $P = 10^{-4}$atm and $\alpha_{Fe} = 0$, P_{Fe} in the solar nebula is

$$P_{Fe} = \frac{10^{-4} \times 900{,}000}{(13.6 + 2.2) \times 10^9} = 5.7 \times 10^{-9}\ \text{atm}.$$

Also, $\Delta\widetilde{H}^0_{298}$ and $\Delta\widetilde{S}^0_{298}$ for the reaction Fe(s) $\leftrightarrow$ Fe(g) are, respectively, 416.3 kJ/mole and 0.155 kJ/mol/deg. Therefore, $T_{C,0} = 1332$ K and $T_{C,50} =$ 1307 K for the condensing pure Fe solid phase.

If a group of elements all exist as monoatomic gases and condense more or less in the same temperature range to form a metal alloy (or solid solution), then a_M is no longer 1 but $a_M = \gamma_M X_M$. In the general case $a_i = \gamma_i X_i$ for metal i in an alloy, where γ_i is the activity coefficient and X_i the atomic or molar fraction of metal i in the alloy. There are not many data on γ_i; therefore, γ_i is often assumed to be 1 (i.e., the ideal solution for all metals in solid solution); a_i or X_i can be related to α_i (fraction of metal i condensed) by

$$a_i \approx X_i = \frac{\alpha_i A_i}{\sum_{j=1}^{n} \alpha_j A_j}.$$

Meanwhile from equation III-11

$$P_i = \frac{P(1-\alpha_i)A_i}{A_{H_2} + A_{He}}.$$

Substituting the above two equations into an equation similar to III-9, one obtains

$$\frac{(1-\alpha_i)\sum_{1}^{n}\alpha_j A_j}{\alpha_i} = \frac{A_{H_2} + A_{He}}{P} \exp\left(\frac{-\Delta\widetilde{H}^0_{298,i} + T\Delta\widetilde{S}^0_{298,i}}{2.3RT}\right). \quad \text{(III-12)}$$

With an n-component alloy, one obtains n nonlinear equations like III-12 and can solve for α_i as a function of T for a given P using the Newton-Raphson iteration technique (Van Zeggeren and Storey, 1970). Palme and Wlotzka (1976) and Fegley and Palme (1985) calculated condensation temperatures as a function of α_i for the noble metals, Fe, Co, and Ni, which form a metal alloy (Figure III-9).

The above examples are relatively simple cases. In reality, most elements form various gas compounds and condense as pure phase or solid solutions of oxides, silicate, or sulfides. For example, aluminum can exist in the gas phase as Al(g), Al(OH)(g), AlH(g), and AlF(g) and first condenses as pure corundum (Al_2O_3), i.e.,

$$Al_2O_3(s) \leftrightarrow 2Al(g) + 3O(g), \quad \text{(III-13a)}$$

$$K = \frac{f_{Al}^2 \cdot f_O^3}{a_{Al_2O_3}} = \exp(-\Delta\widetilde{G}/RT) \approx P_{Al}^2 \cdot P_O^3. \quad \text{(III-13b)}$$

In order to solve for the equilibrium temperature T as a function of α_{Al} from equation III-13b, one should know P_{Al} and P_O as functions of T and

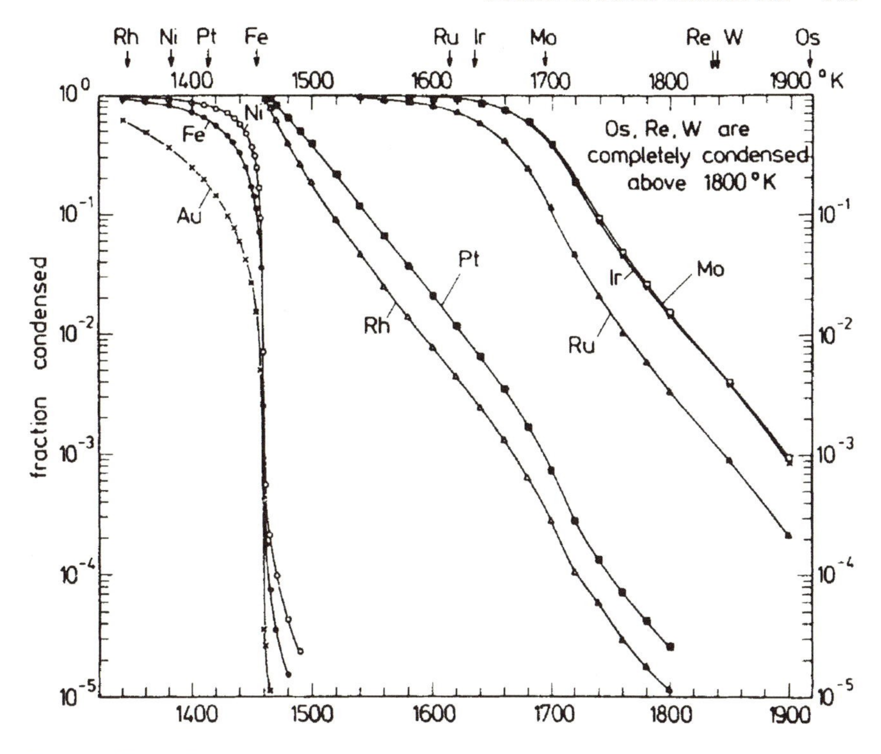

FIGURE III-9. Mole fractions of condensed metals (α_i) in an alloy of 13 metals (Os, Re, W, Mo, Ir, Ru, Rh, Pt, Ni, Fe, Co, Pd, and Au; ideal solid solution was assumed for all metals) are plotted as a function of temperature, under the total nebular gas pressure of 10^{-3} atmosphere (Palme and Wlotzka, 1976; with permission from Elsevier Science).

α_{Al} first. Grossman and Larimer (1974, and references therein) provide some examples for solving this kind of problem.

In the case of condensates forming solid solutions, the equations become more complicated but not impossible. Kornacki and Fegley (1986) provided an example of how to calculate the 50% condensation temperatures for oxides of 26 refractory trace elements in solid solution with perovskite ($CaTiO_3$).

Model Results and Implications

Table III-5 summarizes the 50% condensation temperatures for elements condensed as pure phase or as solid solution at a total pressure of 10^{-4} and 10^{-6} atmospheres. The $T_{C,50}$ at other pressures can easily be interpolated by the linear relationship of $1/T_{C,50}$ vs. log P. Figure III-10 (see page 114) shows

TABLE III-5

The 50% condensation temperature ($T_{C,50}$; K) of elements at two different total pressures, and their major condensed forms and host phase

Z	$T_{C,50}$ (10^{-4} atm)	$T_{C,50}$ (10^{-6} atm)	Condensed form (host phase)	Z	$T_{C,50}$ (10^{-4} atm)	$T_{C,50}$ (10^{-6} atm)	Condensed form (host phase)
Re-75	1818	1666	Re (Ia)	Ni-28	1354	1201	Ni (Ib)
Os-76	1812	1663	Os (Ia)	Mg-12[b]	1340	1203	Mg_2SiO_4
W-74	1794	1625	W (Ia)	Eu-63	1338	1232	Eu_2O_3 (II)
Zr-40	1717	1594	ZrO_2	Fe-26	1337	1185	Fe (Ib)
Hf-72	1690	1577	HfO_2	Pd-46	1321	1165	Pd (Ib)
Sc-21	1652	1524	Sc_2O_3	Si-14[b]	1311	1193	Mg_2SiO_4; $CaMgSi_2O_6$
Al-13[b]	1650	1531	Al_2O_3	Cr-24	1301	1154	Cr (Ib)
Y-39	1622	1499	Y_2O_3	Au-79[a]	1225	1074	Au (Ib)
Ir-77	1603	1465	Ir (Ia)	Li-3[a]	1225	1091	$Li_2SiO_3(Mg_2SiO_4)$
Ti-22	1598	1463	$CaTiO_3$	Sr-38	1217	1116	SrO (II)
Lu-71	1598	1463	Lu_2O_3 (II)	Mn-25[a]	1190	1078	$Mn_2SiO_4(Mg_2SiO_4)$
Er-68	1598	1463	Er_2O_3 (II)	Ba-56	1163	1052	BaO (II)
Th-90	1598	1463	ThO_2 (II)	P-15	1151	1070	Fe_3P
Ho-67	1598	1463	Ho_2O_3 (II)	As-33[a]	1135	1000	As (Ib)
Tb-65	1598	1463	Tb_2O_3 (II)	Cu-29[a]	1037	910	Cu (Ib)
Tm-69	1598	1463	Tm_2O_3 (II)	K-19[a]	1000	890	feldspar
Dy-66	1598	1463	Dy_2O_3 (II)	Rb-37[b]	~1000	~890	feldspar
Gd-64	1597	1462	Gd_2O_3 (II)	Na-11[b]	970	861	feldspar

Mo-42	1595	1456	Mo (Ia)	Ag-47[a]	952	843	Ag (Ib)
U-92	1580	1437	UO_2 (II)	Ga-31[a]	918	738	Ga (Ib)
Pu-94	1578	1445	PuO_2; Pu_2O_3 (II)	Sb-51[a]	912	767	Sb (Ib)
Ru-44	1565	1432	Ru (Ia)	Cl-17	863	775	$Na_4Al_3Si_3O_{12}Cl$
Nd-60	1563	1430	Nd_2O_3 (II)	Ge-32[a]	825	702	Ge (Ib)
Sm-62	1560	1433	Sm_2O_3 (II)	F-9	736	721	$Ca_5(PO_4)_3F$
Pr-59	1557	1420	Pr_2O_3 (II)	Sn-50[a]	720	625	Sn (Ib)
La-57	1544	1410	La_2O_3 (II)	Se-34[a]	684	684	FeSe (III)
Ta-73	1543	1418	Ta_2O_5 (II)	Te-52[a]	680	600	Te (III)
Ca-20[b]	1518	1382	$Ca_2Al_2Si_2O_8$	Zn-30[a]	660	605	ZnS(III)
Nb-41	1517	1362	NbO; NbO_2 (II)	S-16[a]	648	648	FeS (III)
Yb-70	1493	1392	Yb_2O_3 (II)	Pb-82[b]	521	472	Pb (IV)
V-23	1455	1308	VO; V_2O_3 (II)	Bi-83[b]	472	432	Bi (IV)
Ce-58	1440	1286	Ce_2O_3 (II)	In-49[b]	470	437	InS (IV)
Pt-78	1411	1275	Pt (Ia)	Tl-81[b]	448	409	Tl (IV)
Rh-45	1392	1250	Rh (Ia)	Cd-48[b]	429	392	CdS (IV)
Co-27	1356	1202	Co (Ib)	Br-35[b]	~350	~350	$Ca_5(PO_4)_3Br$

Sources: [a] Data from Wai and Wasson (1977, 1979), Wai et al. (1978). [b] Data from Larimer (1973) and Grossman and Larimer (1974). Others from Fegley and Lewis (1980), Fegley and Palme (1985), and Kornacki and Fegley (1986)

Notes: Host phases Ia and Ib: noble and iron metal group alloys with $T_{C,50}$ respectively greater and less than that of Fe-Co-Ni. Host phase II: solid solution with perovskite ($CaTiO_3$). Host phase III: solid solution with troilite (FeS). Host phase IV: in part solid solution with metal alloy and troilite and in part pure phase as indicated.

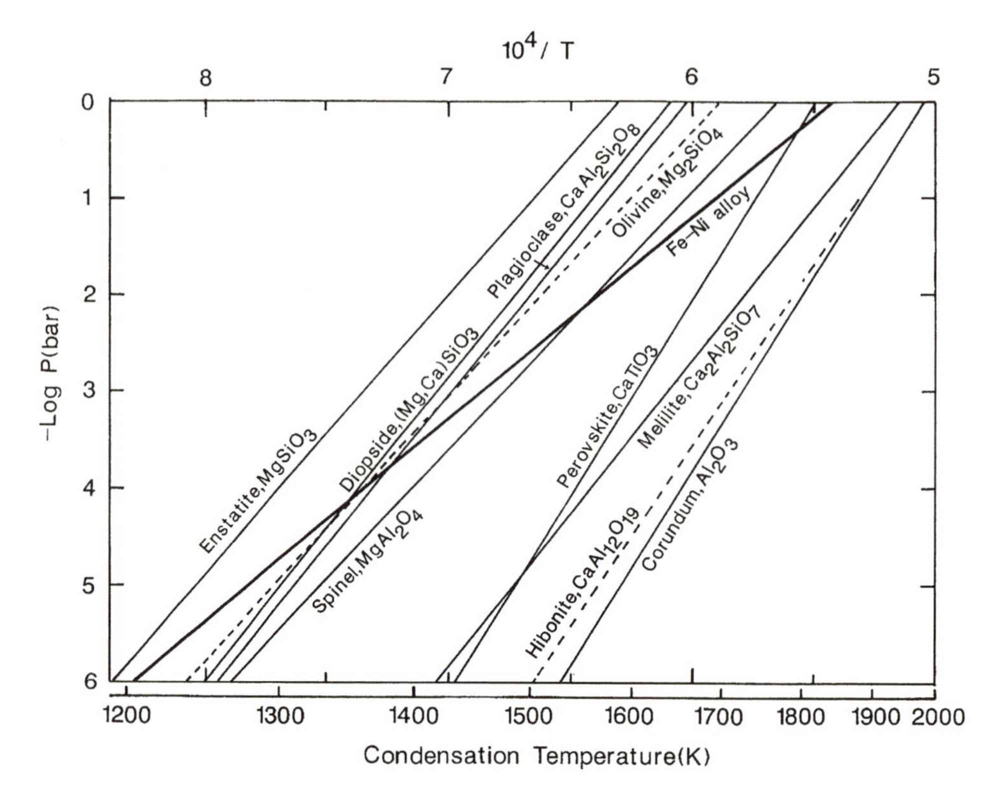

FIGURE III-10. Condensation temperatures ($T_{C,0}$) of oxide and silicate solids as a function of the total nebular gas pressure (modified after Saxena and Eriksson, 1983).

the condensation temperatures or the temperatures where oxide and silicate solids first appear as a function of the nebular gas pressure (based on Saxena and Eriksson, 1983). In general, the results agree with earlier calculations by Grossman and Larimer (1974). Minor differences are expected because different researchers may have used different sources of thermodynamic data and/or assumed different solid phase behavior (solid solution vs. pure phase; ideal vs. nonideal solid solution). One should also keep in mind that all model calculations always assume thermodynamic equilibrium among all the phases. In other words, all chemical reactions are reversible. In addition, the solar nebula is considered as a closed system. Actually, the solar nebular system might have violated from these assumptions in various degrees. For example, if the C/O ratio in the solar nebula changes locally by some mechanism, the condensed minerals and their condensation sequences can be quite different (Larimer and Bartholomay, 1979).

Figure III-11a shows another example of changes in the condensation temperatures for major compounds of the most abundant elements as the total pressure of the nebula varies (Lewis, 1974). The temperature and pressure at the surface of the present Sun are estimated to be about 5800 K and 0.134 atmosphere (Kaufmann, 1985). If a gas parcel from the surface of the Sun is ejected adiabatically outward until its temperature falls to 2000 K, then its pressure should be about 10^{-2} atm according to the adiabatic relationship for an ideal gas:

$$\frac{T_1}{T_2} = \left(\frac{P_1}{P_2}\right)^{\frac{\gamma-1}{\gamma}} = \left(\frac{V_2}{V_1}\right)^{\gamma-1} = \left(\frac{\rho_1}{\rho_2}\right)^{\gamma-1}, \quad \text{(III-14)}$$

where γ is the ratio of heat capacity at constant pressure C_P and constant volume C_V, and is equal to $5/3(\approx 1.67)$ for monoatomic hydrogen gas (at temperatures below 2000 K, hydrogen becomes a diatomic gas and γ should be 9/7), and ρ is the density ($= n/V$); therefore the ideal gas equation becomes $P = nRT/V = \rho RT$.

The two dotted lines in Figure III-11a represent the P-T relationship (equation III-14) for further adiabatic expansion of the solar gas with $\gamma = 5/3$ and 4/3 (= 1.33, which represents γ for a gas mixture of $0.9\,H_2 + 0.1\,He$). The condensation temperatures of major phases derived from these two dotted lines do not differ much. Tremolite and serpentine are proxy minerals to represent hydrous mineral condensates. The solid curve in Figure III-11b depicts how the zero pressure density ρ^0 of cumulative condensates from the solar nebula decreases stepwise with decreasing temperature when each new phase condenses (Lewis, 1972; Figure 4-46 in Lewis and Prinn, 1984). By comparing the ρ^0 of the cumulative condensates with the "observed" ρ^0 of the inner planets and three largest asteroids (Figure III-11b) along with Figure III-11a, one may infer possible variations of temperature and dust composition within the solar nebula. For example, Mercury was probably formed from dusts that were condensed within the solar nebula with a temperature range of about 1400 to 1700 K, and is enriched in Fe and depleted in silica, alkali metals, and other volatiles. Even though Venus and Earth have similar ρ^0, Venus might have formed at higher temperatures (650 to 1400 K) than the Earth (550 to 650 K) and is low in sulfur and water. The lower ρ^0 of Mars than Earth may be indicative of higher FeO and water contents in Mars than Earth. The asteroids with carbonaceous chondrite compositions were probably formed in a temperature range of 150 to 400 K (Lewis, 1974).

It is difficult to estimate the formation temperatures of the giant outer planets and their satellites with masses less than 10^{26} g. The problem is that giant planets accreted not only condensed dust but also huge amounts of gaseous species from the nebula, whereas small satellites might have lost some condensed components when their surface temperatures became higher

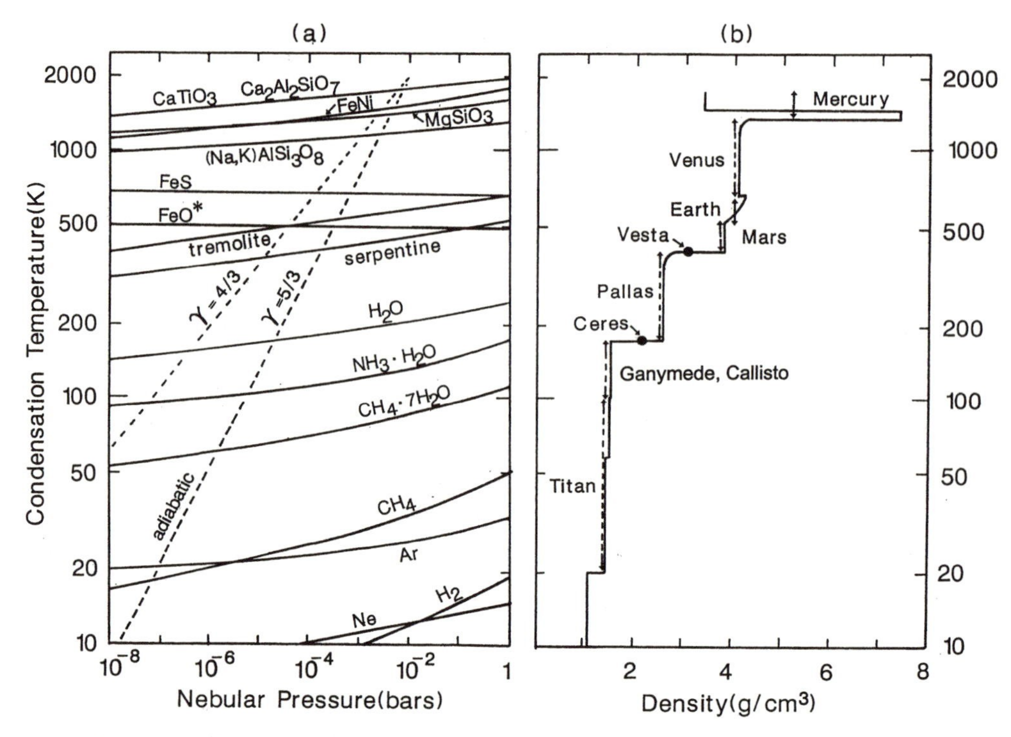

FIGURE III-11. (a) Condensation temperatures as a function of the solar nebular pressure for some compounds of the most abundant elements. FeO* represents the ferrous oxide component in olivine and pyroxene. Tremolite, $Ca_2Mg_5Si_8O_{22}(OH)_2$, and serpentine, $Mg_3Si_2O_5(OH)_4$, are important water-bearing silicates. (b) The solid line represents the predicted average density of all condensates formed above a given temperature during an adiabatic expansion of a gas parcel (with solar composition) along the dashed lines in Figure 11a. Superimposed are the "observed" zero-pressure densities of the inner planets, the three largest asteroids, and the large-size satellites. These figures are modified after Lewis (1972, 1974) and Lewis and Prinn (1984).

after their formation. Therefore, we are left with only four large satellites of masses slightly larger than 10^{26} g (Table III-4b). If Triton was captured by Neptune as mentioned earlier, we cannot use Triton's surface composition to deduce the formation temperature of Neptune. Ganymede, Callisto (satellites of Jupiter), and Titan (satellite of Saturn) all have a zero-pressure density of about 1.4 g/cm^3, but the surfaces of the former two satellites are composed mostly of water ice; therefore their formation temperature was in the range of about 90 to 150 K. In contrast, Titan's atmospheric composition suggests the existence of $NH_3 \cdot H_2O$ and $CH_4 \cdot 7H_2O$ in its water ice, as mentioned earlier. Therefore its formation temperature should be less than 90 K but higher than 20 K. Figure III-8b summarizes the formation temperatures of the inner planets, asteroids, and outer planets (deduced from their large satellites) as a function of distance from the Sun. They give a general relationship of

$$\frac{T_1}{T_2} = \frac{d_2}{d_1}, \quad \text{i.e., } T \text{ is proportional to } d^{-1}, \qquad \text{(III-15)}$$

as also shown earlier by Lewis (1974). Even though the uncertainty for Saturn is large in Figure III-8b, it does not affect the slope greatly. The dotted line with a slope of $-1/2$ in Figure III-8b represents the effective surface temperatures of the planets today (Table III-4a) for comparison. Substituting equation III-15 into equation III-14, one obtains

$$\frac{\rho_1}{\rho_2} = \left(\frac{d_2}{d_1}\right)^{1/(\gamma-1)}, \quad \text{or } \rho \text{ is proportional to } d^{-3} \text{ for } \gamma = 4/3,$$

again in good agreement with the prediction from the area-loading distribution of the solar nebula at hydrodynamic equilibrium (Hayashi, 1981), and with the observation for the circumstellar disk around β-Pictoris (Smith and Terrile, 1984) as discussed earlier. Similarly, by substituting equation III-15 into equation III-14, one also can show that P (pressure) is proportional to d^{-4} for $\gamma = 4/3$.

In summary, the temperature, pressure, and density of the solar nebula in its mid-plane decrease with roughly the -1st, -4th and -3rd power of the distance from the Sun. These distributions can be explained by the transport of gas mass from the proto-Sun outward under adiabatic conditions. According to the equilibrium condensation model of the solar nebula, the systematic change of zero-pressure density from the inner planets to the big satellites of the outer planets reflects the compositional change of condensates that were dependent on P-T conditions of the solar nebula, and thus the heliocentric distance.

Chapter IV

DISTRIBUTION OF ELEMENTS IN METEORITES

INTRODUCTION

THE PURPOSES OF THIS chapter are to present the extremely diverse nature of meteorites and their constituents (matrix, chondrules, calcium-aluminum rich inclusions, Fremdlinge, interstellar grains, etc.) with regard to mineralogy, petrology, chemistry, and isotopic compositions; and to demonstrate how this voluminous information has been digested into a systematic understanding of their origins and formative environments in the solar nebula. The field of meteorite study is evolving rapidly. Therefore, the approach of this chapter is not to review the extensive literature, but rather to provide some specific examples for the purpose of illustration.

This chapter summarizes the classification of various meteorites based on mineralogy, chemistry, and petrology and discusses a possible genetic relationship among those meteorites suggested by oxygen isotope data. The methods of enrichment factor calculation and factor analysis of multivariance are introduced to highlight the chemical differences among various classes of chondrites. Subsequently, the classification of elements into lithophile, siderophile, and chalcophite in ordinary chondrites by Goldschmidt (1954) is elaborated further, and the chemical thermodynamic basis for this classification is examined.

This chapter also presents variations of mineralogical, petrological, chemical, and isotopic compositions in major constituents of chondrites (e.g., matrix, chondrules, calcium-aluminum-rich inclusions) and postulates their possible formative environments in the solar nebula.

Finally, examples of igneous magmatic differentiation in parental bodies of achondrites and iron meteorites are briefly discussed.

IV-1. CLASSIFICATION OF METEORITES

As discussed in Section III-1, the major classes of meteorites are chondrites, achondrites, stony irons, and irons. The further division of major classes into subclasses is summarized in Tables IV-1 and IV-2 along with some pertinent information. In Table IV-1, olivine with $MgSiO_4$ equal to or greater than 90 mole % is called forsterite. For easy reference, Appendix Table A-3 also lists many minerals (including their formulas) found in meteorites.

TABLE IV-1

Classification of chondrites and achondrites

Classes and subclasses	C (wt. %)	S (wt. %)	Fe/Si (wt. ratio)	Principal minerals
Carbonaceous chondrites (CC)				
Type 1 (C1 or CI)	3.5	6.3	1.79	phy + org; mag; /fo + ol(10–20)
Type 2 (C2 or CM)	2.5	3.4	1.59	phy + org; fo > ol(15–50); /en + opx
Type 3 (C3 or CO	0.45	2	1.55	fo; ol(10–60); phy; m + t; CAI
and CV)	0.26	2.1	1.8	fo; ol(10–60); phy; m + t; CAI
Renazzo (CR)	1.5	1.3	1.58	fo > en; phy + org; m + t; mag
Kakangari (CK)	1	5.3	1.35	en > fo; phy + org; m + t
Ordinary chondrites (OC)				
Low-low iron (LL)	0.12	2.3	1.07	ol(26–33) > opx(22–25) > pl > m + t
Low iron (L)	0.09	2.2	1.17	ol(21–25) > opx(19–22) > pl > m + t
High iron (H)	0.11	2	1.56	ol(16–20) > opx(15–18) > pl > m + t
Netschaevo (HH)*	(0.21)	(1.4)	(2.25)	opx(15) > ol(14) > pl > m + t
Enstatite chondrites (EC)				
Low iron (EL)	0.36	3.3	1.11	en; m + t; /fo; Si polymorphs; rare non-oxides
High iron (EH)	0.39	5.9	1.73	en; m + t; /fo; Si polymorphs; rare non-oxides
Primitive achondrites (PA)				
Winonaite	0.46	1.5	0.85	en > fo; m + t
Acapulcoite		1.3	1.65	ol(11) ≈ opx(11) > pl; m + t
Lodranite*			2.14	ol(13) ≈ opx(13); m + t
Brachinaite	0.07	1.5	1.16	ol(20) ≫ aug ≈ pl; m + t
Carbonaceous achondrites				
Ureilites	2.6	0.5	0.76	ol(8–23) > pig(8–20); C polymorphs; m + t

AMP achondrite group				
Eucrite	0.06	0.2	0.66	pig(40–70) ≈ Ca-pl; Si polymorphs
Diogenite	0.04	0.38	0.56	opx(23–27); /Ca-pl; tridymite; ol(25–32)
Howardite	0.11	0.27	0.62	mixture of Eucrite and Diogenite
Mesosiderite*	0.08	1.1	4.2	ol(7–47); opx(17–40); Ca-pl; 17–80 wt% m + t
Pallasite*	0.08	0.19	5.2	ol(9–21) > m + t
SNC achondrite group				
Shergottite			0.64	pig(30–58); aug; maskelynite; /ol
Nakhlite		0.06	0.74	aug > ol(67–77) > pl + K – fp
Chassignite		0.14	1.18	ol(30) ≫ dio > pl + K – fp
Enstatite achondrites (EA)				
Aubrite	0.07	0.4	0.04	en; rare non-oxides
Bencubbin*			(0.32)	en > fo; m + t; pl
Other achondrites				
Angrite		0.45	0.36	dio ≫ Ca-ol(47) > spinel
Eagle Station*			(0.4)	ol(20); m + t

Sources: Bild and Wasson (1976, 1977); Boynton et al. (1976); Buseck (1977); Dodd (1981); Floran et al. (1978); Fukuoka et al. (1977); Graham et al. (1977); Janssens et al. (1987); Johnson et al. (1977); Ma et al. (1977); Manson and Wilk (1962, 1966); Mason and Jarosewich (1967); Mason (1979); Palme et al. (1981); Prinz et al. (1985); Sears and Dodd (1988); Smith et al. (1984); Watters and Prinz (1979).

Notes: Mineral abbreviations: aug = augite; CAI = Ca-Al–rich inclusions; dio =diopside; en = enstatite; fo = fosterite ($Mg_2SiO_4 \geq 90\%$); fp = feldspar; mag = magnetite; m + t = metal + troilite; ol = olivine ($Mg_2SiO_4 < 90\%$); opx = orthopyroxene; org = complex organic matter; phy = phyllosilicates; pig = pigeonite; pl = plagioclase; rare non-oxides = mainly sulfides and nitrides (see Appendix Table 3); numbers in parantheses after olivine, orthopyroxene, and pigeonite are the mole % of Fe_2SiO_4 or $FeSiO_3$; minerals after the forward slash have only trace amounts. Asterisks indicate silicate fraction in stony irons. C, S, and Fe/Si values in parentheses are for the silicate fraction only.

TABLE IV-2
Classification of iron meteorites and metal phase in stony irons

Iron meteorites	*Ni* (*mg/g*)	*Ga* ($\mu g/g$)	*Ge* ($\mu g/g$)
IAB	64–600	11–100	25–520
IC	61–68	49–55	212–247
IIAB	53–64	46–62	107–185
IIC	93–115	37–39	88–114
IID	96–113	70–83	82–98
IIE	75–97	21–28	62–75
IIF	106–104	9–12	99–193
IIIAB	71–105	16–23	24–47
IIICD	62–230	1.5–92	1.4–380
IIIE	82–90	17–19	34–37
IIIF	68–85	6.3–7.2	0.7–1.1
IVA	74–94	1.6–2.4	0.09–0.14
IVB	160–180	0.17–0.27	0.03–0.07
Eagle Station*	13–16	4–7	70–110
Bencubbin*	61	2	1

Note: Asterisks indicate metal phase in stony irons.

The subclasses of carbonaceous chondrites C1 and C2 correspond to Wasson's (1974) CI (Ivuna) and CM (Mighei), respectively. The C3 subclass is separated into CO (Ornans) and CV (Vigarano) by Wasson (1974). CR (Renazzo) and CK (Kakangari) are additional subclasses of carbonaceous chondrites (CC). According to the total iron content, the ordinary chondrites (OC) are divided into subclasses of LL (low-low), L (low), H (high), and HH (high-high); and the enstatite chondrites (EC) into EL (low) and EH (high). Each subclass of chondrites occupies a distinct field in the molar ratio plot of $(Fe + FeS)/Si$ vs. $(FeO + 2Fe_2O_3)/Si$ from the whole rock analysis (Figure IV-1). Apparently, CI and CM were formed under the most oxidizing conditions, and EL and EH under the most reducing ones.

Each chondritic subclass can be further subdivided into various petrologic types according to the general criteria given in Table IV-3. Most meteorites in the CC subclasses belong to petrologic types 1 to 3 and a few to type 4; those in OC and EC subclasses are of types 4 to 6 and a few are types 3 and 7. The OC and EC of petrologic type 3 are often called **unequilibrated chondrites** (UOC and UEC), because the mineral compositions of olivine and pyroxene are quite variable within the same rock. Dodd (1981) proposed that the petrologic types 3 to 7 correspond roughly to the metamorphic temperature ranges of 400–600°C (type 3), 600–700°C (type 4), 700–750°C (type 5), 750–950°C (type 6), and greater than 950°C (type 7) for ordinary and enstatite chondrites.

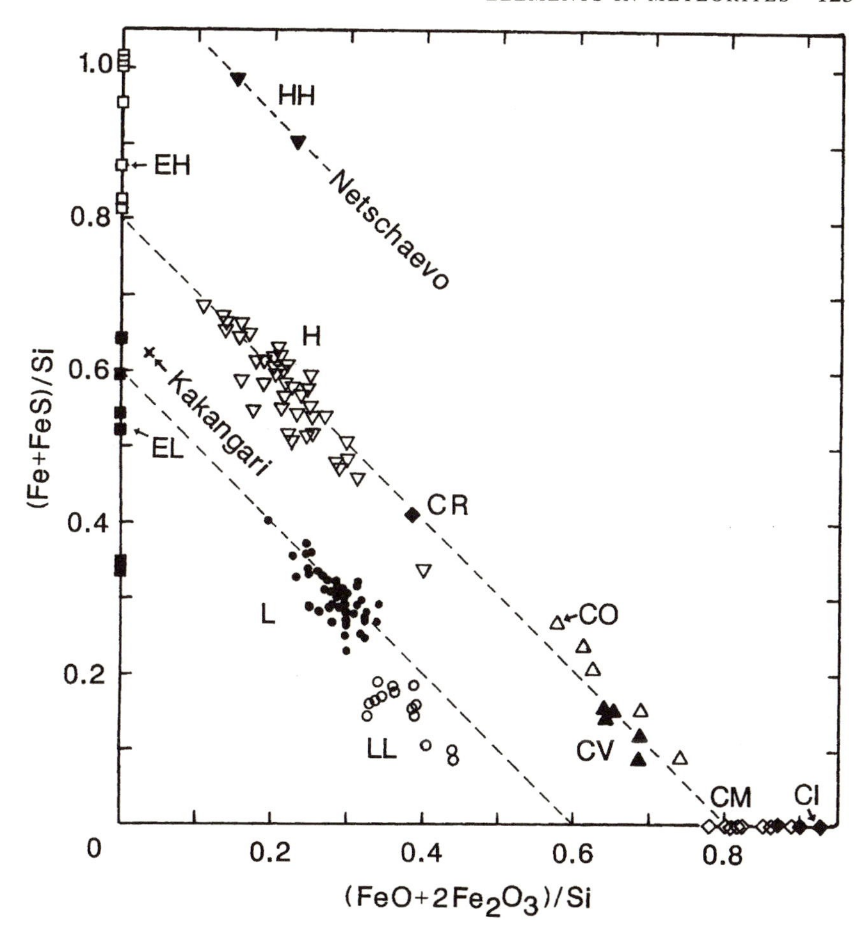

FIGURE IV-1. Plot of molar ratio of (Fe + FeS)/Si vs. (FeO + $2Fe_2O_3$)/Si for various chondrite subclasses. The diagram is mainly based on Bild and Wasson (1977). Additional data are CR (Renazzo) and CK (Kakangari; Mason and Wiik, 1962, 1966), EH and EL (Urey and Craig, 1953; Mason, 1966b) HH (Bild and Wasson, 1977). The dotted lines represent the constant total Fe/Si ratios.

The subclasses of achondrites and silicate fractions of stony irons can be conveniently distinguished from one another in the molar ratio plot of Fe/(Fe + Mg) vs. Ca/(Ca + Mg) as shown in Figure IV-2. These molar ratios were obtained from the analytical data for either whole rocks (eucrites, diogenites, howardites, SNC group, and primitive achondrites) or silicate fractions (ureilites, enstatite achondrites, mesosiderites, and pallasites). One exception is that the eucrite-howardite-diogenite group and SNC group overlap in Figure IV-2. The SNC group members are, however, characterized by

TABLE IV-3

Some criteria for petrologic types of chondrite subclasses

Criterion	Petrologic type						
	1	2	3	4	5	6	7
Dispersion of olivine, pyroxene compositions		>5% mean deviation of Fe		≤5%	Homogeneous		
Polymorph of low-Ca pyroxene		Chiefly monoclinic		Partly monoclinic	Orthorhombic: CaO ≤ 1 wt %		Orthorhombic: CaO > 1%
Secondary feldspar		Absent or minor		Micro-crystalline aggregates	Clear interstitial grains Coarsening from type 5 to type 7		
Chondrule glass		Clear and isotropic		Devitrified	Absent		
Metal phase (Ni wt %)		Taenite (<20) minor or absent	Kamacite and taenite (>20) in exsolution relationship				
Mean Ni in troilite		>0.5 wt%	<0.5 wt %				
Matrix	All fine-grained, opaque	Chiefly fine, opaque	Clastic and minor opaque	Recrystallized Coarsening from type 4 to type 7			
Chondrule-matrix integration	No chondrules	Chondrules very sharply defined		Well defined	Readily delineated	Poorly defined	Relicts only
Carbon wt %	3–5	0.8–2.6	0.2–1	<0.4			
Water wt %	18–22	2–16	0.3–3	<1.5			

Source: Dodd (1981), reprinted by permission of Cambridge University Press.

their relatively young ages (about 1.2 to 1.4 Ga) as compared to all other meteorites (4.5 ± 0.1 Ga). The SNC group meteorites are thought to originate from Mars (e.g., Wasson and Wetherill, 1979; Gratz et al., 1993).

The linear relationship among eucrites, howardites, and diogenites in Figure IV-2 confirms the conclusion that howardites were formed mainly from the mechanical mixing of eucrites and diogenites (Fukuoka et al., 1977). Mesosiderites and some eucrites, which lie on the left-hand side of the eucrite-howardite-diogenite mixing line in Figure IV-2, may indicate an additional end member such as pallasites, and/or the separation of Fe from silicate fractions through impact shock melting. The eucrite-howardite-diogenite achondrites along with mesosiderite and pallasite are often called the AMP group.

The field of the primitive achondrites (PA; including Winona, Mount Morris, Pontlyfni, Acapulco, ALH-77081, Brachina, Lodran, etc.) in Figure IV-2 almost coincides with that of chondrites as one would have expected,

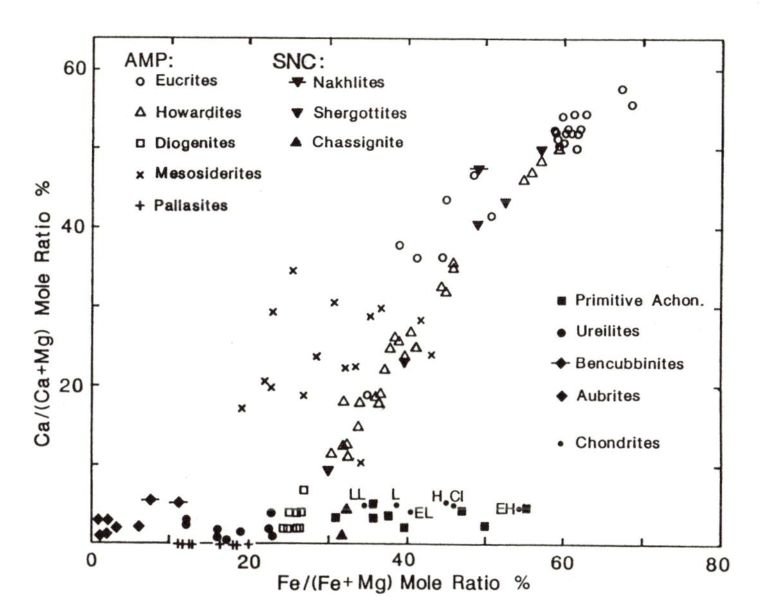

FIGURE IV-2. Plot of molar ratios of Fe/(Fe + Mg) vs. Ca/(Ca + Mg) for various achondrite subclasses, silicate fraction of stony irons, and major chondrite subclasses. Data sources are as follows: eucrites and howardites (Mason et al., 1979), diogenites (Fredricksson et al., 1976), mesosiderites (Simpson and Ahrens, 1977), pallasites (Buseck and Goldstein, 1969), SNC group (Smith et al., 1984; McCarthy et al., 1974), aubrites (Urey and Craig, 1953), ureilites (Takeda, 1986/87), primitive achondrites (Graham et al., 1977; Kallemeyn and Wasson, 1985; Nagahara and Ozawa, 1985), chondrites (Table IV-6 below), and bencubbinites (Kallemeyn et al., 1978).

because by definition, the primitive achondrites (PA) have bulk compositions of major cations very similar to those of chondrites (Table IV-4). The enstatite achondrites (EA; aubrite, bencubbinite) are extremely low in Fe/(Fe + Mg) ratio as compared to other groups of achondrites, again indicating effective separation of Fe from the silicate fraction (Figure IV-2).

The redox conditions during the formation of chondrites and achondrites can be roughly deduced from the mole % content of Fe_2SiO_4 in olivine and of $FeSiO_3$ in low-Ca pyroxene, such as orthopyroxene and pigeonite (Table IV-1). The higher the iron content the higher the oxygen fugacity. If olivine and low-Ca pyroxene coexist and are more or less in equilibrium, the iron contents in both minerals show a positive correlation (Figure IV-3). The high iron contents in low-Ca pyroxene and/or in olivine for eucrites-diogenite and the SNC group plus angrite (Table IV-1) may indicate that they were formed under a similar redox condition to LL chondrites or even more oxidizing. The ureilite field in Figure IV-3 overlaps with those for the ordinary chondrites (LL to HH). According to Berkley et al. (1980), the ureilites with $FeSiO_3$ in low-Ca pyroxene greater than 18 mole % are called group I, 18 to 15% group II, and less than 15% group III. The primitive achondrites (Acapulco, ALHA-77081, Enon, Lodran, Winona, Mount Morris, Pontlyfni) and aubrites cover a wide range of reducing conditions (Figure IV-3).

The classification of iron meteorites and the metals in stony irons according to their Ni, Ga, and Ge contents is shown in Figure IV-4 and Table IV-2. Many iron meteorites may not belong to any of those subclasses. Finally, one more important source of information is the statistics on numbers of meteorites found long after their fall (finds) and those recovered right after their fall to earth (falls) as summarized in Table IV-5 (excluding meteorites found on the Antarctic continent). The statistics on falls are more meaningful than those on finds, because weathering over time causes preferential decomposition of carbonaceous and stony meteorites over iron meteorites. The most abundant meteorite falls are the ordinary chondrites (OC), especially the L and H subclasses. The most common finds of iron meteorites are subclasses IIIAB > IAB > IIAB > IVA and many unclassified. The statistics from 1100 Antarctic meteorites give OC = 88, CC = 2.8, EC = 0.7, achondrites = 5.9, stony irons = 0.7, and irons = 2.3% which are very similar to those for the falls as given in Table IV-5. This similarity is confirmed by Cassidy and Harvey (1991). The implication is that weathering in the Antarctic environment has not been severe enough to bias the statistics greatly. However, there is always some probability that the relative abundance of iron meteorites in the Antarctic collection is underestimated due to the preferential breakup of ordinary chondrites into many pieces.

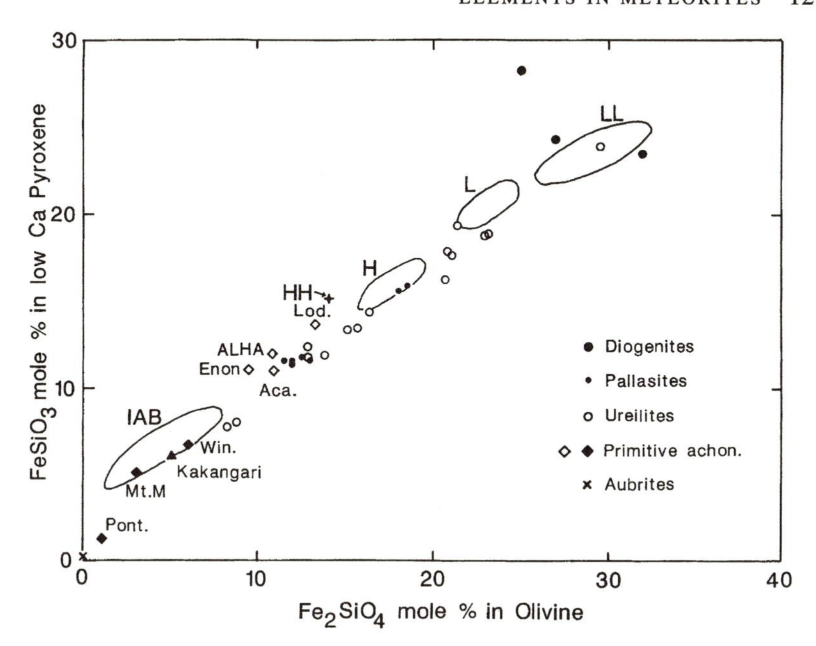

FIGURE IV-3. Plot of mole % Fe_2SiO_4 in olivine and mole % $FeSiO_3$ in low-Ca pyroxene from chondrites, achondrites, and silicate fraction of IAB irons. Data sources are as follows: H, L, LL, and IAB fields, Dodd (1981) and Bild and Wasson (1976); primitive achondrites (including Lodran, ALHA-77081, Enon, Acapulco, Winona, Mount Morris, Pontlyfni), Prinz et al. (1985), Graham et al. (1977), and Kallemeyn and Wasson (1985); pallasites, Buseck (1977); ureilites, Takeda (1986/87); and diogenites, Fredriksson et al. (1976).

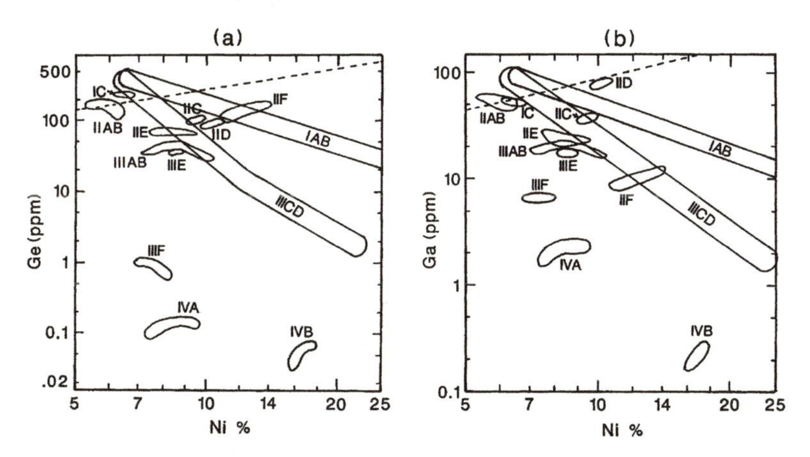

FIGURE IV-4. Classification of iron meteorites and iron fractions of stony irons according to the plots of Ni vs. Ge and Ni vs. Ga (based on Scott and Wasson, 1975; Wasson, 1985).

TABLE IV-4
Enrichment factors (E^i_{AI}) relative to CI for Kakangari chondrite and the primitive achondrites

	Kakangari (1)	*Winona (1)*	*Mount Morris (1)*	*Pontlyfni (1)*	*Acapulco (2)*	*ALH-77081 (3)*	*Brachina (4)*
Si	* 1.20	* 1.26	* 1.16	* 0.90	* 1.20	* 1.30	* 1.29
Ti	* 1.04	* 1.34	0.56	* 0.91	* 1.30		* 1.27
Cr	* 0.79	0.53	0.44	0.57	1.4	* 1.35	* 1.16
Fe	* 0.91	0.6	* 0.71	* 1.18	* 1.11	* 0.90	* 0.84
Mn	* 1.01	0.68	0.4	0.62	* 1.08	* 1.10	* 1.03
Mg	* 1.09	* 1.14	* 1.08	* 0.80	* 1.10	* 1.14	* 1.29
Ca	* 1.00	* 0.70	0.65	* 0.71	* 0.88	0.46	* 1.25
Na	* 1.02	* 0.99	* 0.81	* 0.89	* 0.91	* 1.09	* 0.72
K	* 1.24	* 0.73	0.51	* 0.89	0.54	* 0.86	* 0.92
P	* 1.37	* 0.77	0.67	0.36	3.4		* 1.13

Sources: (1) Graham et al. (1977); (2) Palme et al. (1981); (3) Mason (1978); (4) Floran et al. (1978).
Note: Asterisks indicate the E^i_{AI} values within 1 ± 0.3.

TABLE IV-5
Meteorite classes found (excluding Antarctic) and those recovered shortly after their fall

Class	*Find (no.)*	*Fall (no.)*	*Fall (%)*	*Class*	*Find (no.)*	*Fall (no.)*	*Fall (%)*
Chondrites	(672)	(784)	(86.6)	Stony irons	(57)	(10)	(1.1)
CI	0	5	0.6	Mesosiderite	22	6	0.7
CM	5	18	2	Pallasites	34	3	0.3
CO	2	5	0.6	Lodranites	1	1	0.1
CV	4	7	0.7	Irons	(670)	(42)	(4.6)
H	347	276	30.5	IAB	97	6	0.7
L	286	319	35.2	IC	11	0	
LL	21	66	7.3	IIAB	60	5	0.6
EH	3	7	0.7	IIC	7	0	
EL	4	6	0.7	IID	12	3	0.3
Others		75	8.3	IIE	13	1	0.1
Achondrites	(20)	(69)	(7.7)	IIF	4	1	0.1
Ureilites	6	4	0.4	IIIAB	189	8	0.9
Eucrites	8	25	2.8	IIICD	19	2	0.2
Diogenites	0	9	1.1	IIIE	13	0	
Howardites	3	18	2	IIIF	6	0	
Aubrites	1	9	1.1	IVA	52	3	0.3
SNC	2	4	0.4	IVB	12	0	
				Others	175	13	1.5

Source: Graham et al., (1985); Sears and Dodd (1988).
Note: Numbers in parentheses are the totals for each class.

IV-2. Oxygen Isotopes and Possible Genetic Relationships among Subclasses of Meteorites

The $\delta^{18}O$ of terrestrial water reservoirs ranges from near zero in ocean waters to $-50‰$ in polar snows; and from 5.5 to 7.5‰ in terrestrial ultramafic and basaltic rocks; from 7.0 to 11‰ in granitic rocks; from 15 to 25‰ in deep-sea clays; and from 20 to 35‰ in chert (all relative to SMOW; Garlick, 1974; Faure, 1986). The plot of $\delta^{18}O$ vs. $\delta^{17}O$ for these terrestrial samples forms a straight line $\delta^{17}O = 0.52 \times \delta^{18}O$, which is called the **terrestrial fractionation line** (Clayton et al., 1973). The slope of near 1/2 is as expected if the isotopic fractionation of ^{17}O and ^{18}O relative to ^{16}O is mainly related to isotopic mass differences, i.e., $(^{17}O - {}^{16}O)/(^{18}O - {}^{16}O) = 1/2$. Therefore a fractionation line with a slope of near 1/2 in the $\delta^{18}O$ vs. $\delta^{17}O$ plot is also called the **mass-dependent fractionation line**. The most surprising result for meteorites is that the oxygen isotopes of many bulk chondrite and ureilite samples do not fall on the mass-dependent fractionation line, but instead fall on two to three distinct lines all with slopes near one (Clayton et al., 1973), as is summarized in Figure IV-5. For example, the ordinary chondrite subclasses and two special inclusions from Antarctic meteorite ALHA-76004 (LL3, Mayeda et al., 1980) fall on a straight line of $\delta^{17}O = -1.35 + 1.0 \times \delta^{18}O$. One may call this line the **ordinary chondrite mixing line** or, as Clayton et al. (1991) called it, the equilibrated chondrite line (ECL). The CO, CV, ureilites, and one unusual carbonaceous matrix clast from the Nilpena ureilite (Clayton and Mayeda, 1988) form another well-defined line of $\delta^{17}O = -4.2 + 0.944 \times \delta^{18}O$. The above two straight lines intersect at a point around $\delta^{17}O = -52.2‰$ and $\delta^{18}O = -50.9‰$. A probable third line with a slope of near one is formed by Cumberland Falls (EA fraction), Winona (EA), Pontlyfni (EA), Renazzo (CR) and its olivine and enstatite inclusions, Y790112 (CR, Antarctic meteorite), Kakangari (CK), Weatherford (EA fraction), and Bencubbin (EA fraction). These meteorites are characterized by their high contents of forsterite and enstatite minerals. If one assumes this third line also intersects with the other two lines at the same point, it will have an equation of $\delta^{17}O = -2.67 + 0.974 \times \delta^{18}O$ and may be called the **enstatite meteorite mixing line.** The slope of near one in Figure IV-5 indicates a near constant $^{18}O/^{17}O$ ratio for those meteorites, and the changes in $\delta^{18}O$ and $\delta^{17}O$ are mainly caused by the change in ^{16}O content. In short, the above-mentioned three groups of meteorites were formed by mixing one common ^{16}O-enriched end member ($\delta^{17}O = -52.2‰$, $\delta^{18}O = -50.9‰$) with three ^{16}O-depleted end members. These three ^{16}O-depleted end members were, in turn, probably related to one another through mass-dependent fractionation processes.

The existence of the ^{16}O-enriched end member is clearly shown by additional oxygen isotopic data from Ca-Al-rich inclusions (CAI) in Allende (CV)

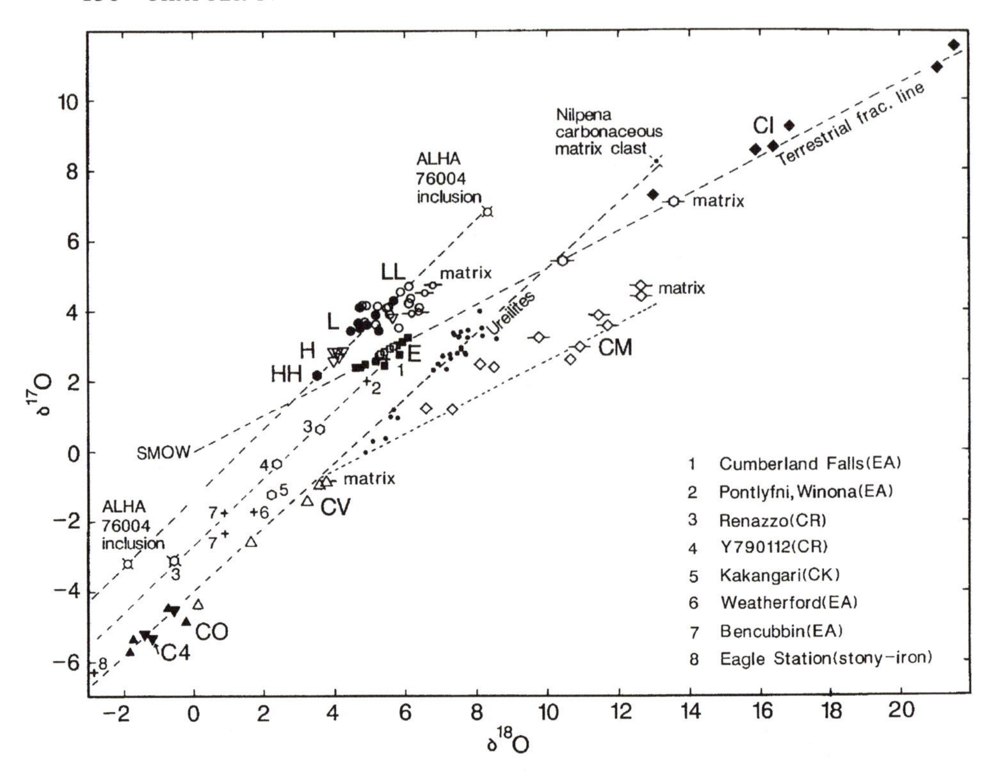

FIGURE IV-5. Oxygen isotopic variations of bulk chondrites, their matrices (symbols with horizontal bar), and enstatite achondrite fractions of some stony irons and Cumberland Falls. Data sources follow: chondrites, Clayton et al. (1976, 1977), Clayton and Mayeda 1977a, 1984a,b, and Halbout et al. (1986); ALHA-76004 inclusions, Mayeda et al. (1980); ureilites, Clayton and Mayeda (1988); stony irons and Cumberland Falls, Clayton et al. (1976), and Clayton and Mayeda (1978b); and matrix of LL3, Grossman et al. (1987).

and Murchison (CM) meteorites (Clayton et al., 1977; Clayton and Mayeda, 1984a; Fahey et al., 1987b; Mayeda et al., 1986) and from interplanetary dust particles (IDP) collected from the Earth's stratosphere (McKeegan, 1987), as is summarized in Figure IV-6. Most of the data points fall on the mixing line for the CV-CO-ureilite bulk rock samples shown in Figure IV-5, but the ranges for $\delta^{17}O$ and $\delta^{18}O$ extend down to around −60‰. This mixing line is often called the **Allende mixing line** (Clayton et al., 1977). A few CAI, samples (EK1-4-1, C1S, and HAL), which fall off to the right of the Allende mixing line, are called **FUN inclusions** (Fractionation and Unknown Nuclear effects; Lee et al., 1980) and will be discussed later.

The ^{16}O-depleted end members are probably related to the regular solar nebular gas reservoir. However, the origin of the ^{16}O-enriched end mem-

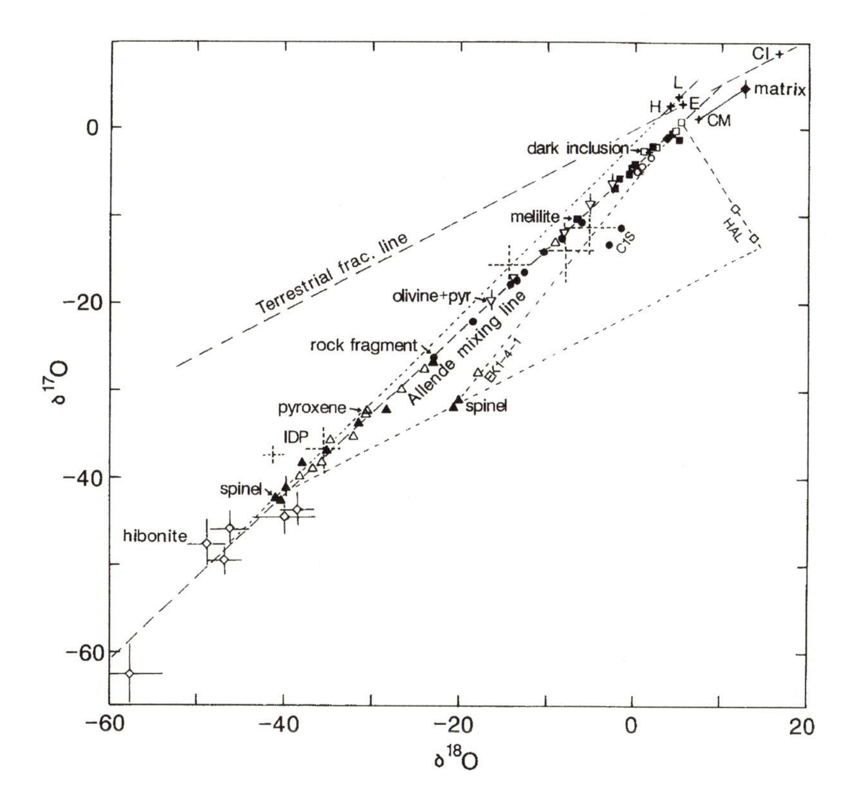

FIGURE IV-6. Oxygen isotopic variations of high-temperature minerals from CAI in Allende (CV) and Murchison chondrites (CM; with vertical bars or crosses); and interplanetary dust particles (IDP; large dotted crosses) collected from the stratosphere. EK1-4-1, C1S, and HAL are FUN inclusions from Allende. Data sources: Clayton et al. (1977), Clayton and Mayeda (1977b, 1984a,b), Fahey et al. (1987a,b), Lee et al. (1980), and McKeegan (1987) for IDP.

ber is still an open question. Clayton et al. (1973) favor an injection of ^{16}O-rich supernova explosion debris into the solar nebula during the accretion of meteorites. Thiemens and Heidenreich (1983) prefer some unusual mass-independent chemical fractionation processes within the solar nebula to explain the origin. Whatever the origin, the systematic variation of oxygen isotopes helps to elucidate the possible genetic relationship among meteorites.

As can be seen in Figure IV-5, the data points for the bulk rock and matrix of Semarkona (LL3) (open circles with horizontal bar), as well as other unequilibrated ordinary chondrites (Clayton et al., 1991; not shown here), tend to be slightly off to the right of the OC mixing line, probably indicating some mass-dependent fractionation effects through aqueous alteration or carbon reduction (Clayton et al., 1991). The oxygen isotopes of enstatite chondrites, aubrites, and lunar basaltic rocks are very similar (Mayeda and Clayton, 1980; Clayton and Mayeda, 1984b) and all fall on the terrestrial fractionation line along with matrices of Renazzo (CR) and Al Rais (CR) (Clayton and Mayeda, 1977a). The implication is that Earth, moon, enstatite chondrites, and enstatite achondrites might all have originated from the same oxygen isotope reservoir that was probably quite low in oxygen fugacity. The CI data points fall slightly above the terrestrial fractionation line, and show a large mass-dependent fractionation effect through aqueous alteration (Clayton and Mayeda, 1984a). Similarly, the oxygen isotopes of CM and their matrix materials could also be produced through aqueous alteration of a precursor similar to CV or ureilites with low $\delta^{18}O$. Because ureilite achondrites

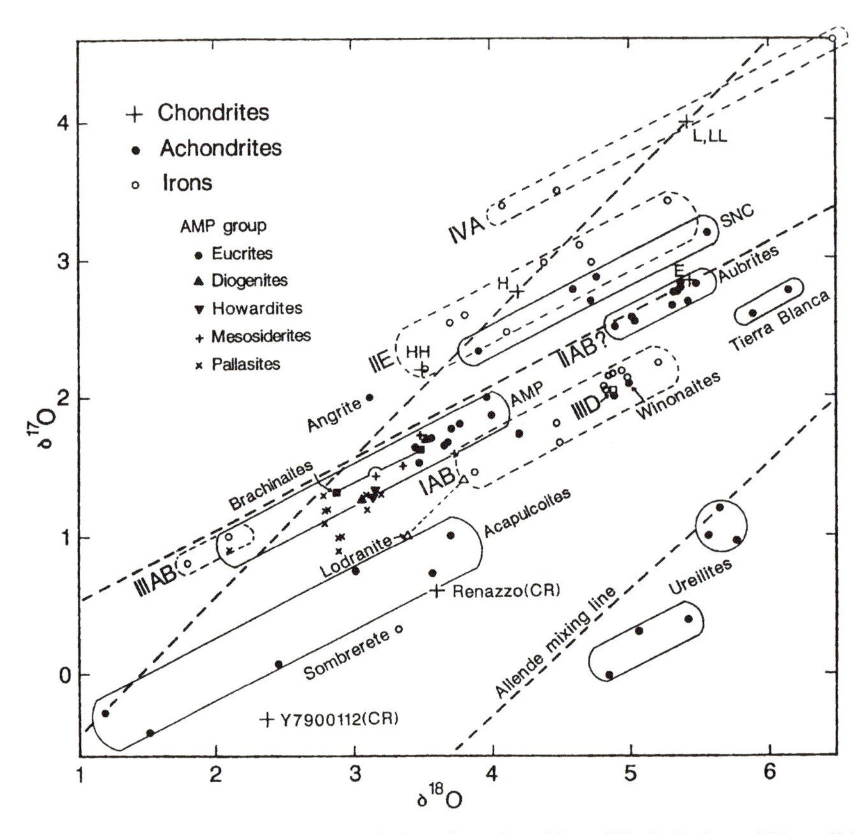

FIGURE IV-7. Oxygen isotopic variations in achondrites (filled circles within solid line enclosures) and silicate inclusions in iron meteorites (open circles within dashed line enclosures). Data sources: Mayeda et al. (1987), Clayton et al. (1976, 1983b, 1986), Clayton and Mayeda (1978a, 1983, 1984b, 1988), and Mayeda and Clayton (1980).

fall on the Allende mixing line, they should have formed from isotopically heterogeneous precursors, and did not undergo significant mass-dependent fractionation after formation (Clayton and Mayeda, 1988).

Except for ureilites, other achondrite subclasses (bulk rocks) and silicate inclusions from iron meteorites all fall on distinct mass-dependent fractionation lines (or fields) in the $\delta^{17}O$ vs. $\delta^{18}O$ plot (Figure IV-7). The relations in Figure IV-7 suggest common oxygen reservoirs (thus genetically related) for the following meteorites: IIIAB–AMP group–brachinaites; SNC group–HH-angrite; IIE-H-HH; IVA-L-LL; IAB-IIID-winonaite; aubrites-moon-Earth-EH-EL. Kelly and Larimer (1977) suggest that IIAB irons may also relate to aubrites-EH-EL, because the Fe/Ni ratio in IIAB irons is almost cosmic, indicating a very reducing environment such that no significant fractionation between Fe and Ni occurred through oxidation of Fe. The precursors for the AMP group, the SNC group, and Acapulcoites might all lie on the ordinary chondrite mixing line (but with higher ^{16}O content than the OC) before subsequent mass-dependent fractionation through melting and crystallization.

IV-3. Bulk Compositions of Chondrites and Factor Analysis

Table IV-6 gives the average compositions of the major subclasses of chondrites as summarized by Wasson and Kallemeyn (1988). A few additions and exceptions are N for CI and EH (Mason, 1979); W for EH, EL (Rambaldi and Cendales, 1980) and LL (Mason, 1979); Mo = 1.2 ppm for CI (average of Mason, 1979; Anders and Ebihara, 1982); and Mo for EH (Imamura and Honda, 1976).

One convenient way to compare the compositions of a sample and a reference material (in our case, CI) is the so-called **enrichment factor** (E_j^i). That is defined as the concentration ratio of element i and the normalizing element $j (X_i/X_j)$ in the sample divided by the same ratio in the reference material, i.e.,

$$E_j^i = \frac{(X_i/X_j)_{\text{sample}}}{(X_i/X_j)_{\text{reference}}}.$$

For example, if Al is chosen as the normalizing element and CI as the reference, then

$$E_{\text{Al}}^i = \frac{(X_i/X_{\text{Al}})_{\text{sample}}}{(X_i/X_{\text{Al}})_{\text{CI}}}$$

Table IV-6
Average composition of chondrites

	CI	*CM*	*CO*	*CV*	*H*	*L*	*LL*	*EH*	*EL*
Ag-47	0.208	0.157	0.097	0.107	0.045	0.065	0.072	0.236	0.023
Al-13	8600	11800	14300	17500	11300	12200	11900	8100	10500
As-33	1.84	1.8	1.95	1.6	2.05	1.55	1.35	3.45	2.2
Au-79	0.144	0.165	0.184	0.144	0.215	0.162	0.14	0.33	0.225
B-5	1.2	0.6		0.3	0.5	0.4			
Ba-56	2.3	3.3	4.3	4.9	4.2	3.7	4.8	2.6	
Be-4	0.027				0.051	0.043	0.051		
Bi-83	0.11	0.075	0.033	0.048	0.017	0.014	0.016	0.088	0.012
Br-35	3.6	2.6	1.3	1.5	0.5	0.8	0.6	2.4	0.8
C-6	32000	22000	4500	5600	1100	900	1200	4000	3600
Ca-20	9200	12700	15800	19000	12500	13100	13000	8500	10100
Cd-48	0.65	0.368	0.008	0.373	0.017	0.011	0.037	0.484	0.027
Ce-58	0.616	0.838	1.02	1.29	0.83	0.9	0.907	0.66	0.3
Cl-17	680	160	240	210	80	76	130	660	210
Co-27	508	575	688	655	810	590	490	840	670
Cr-24	2650	3050	3550	3600	3660	3880	3740	3150	3050
Cs-55	0.183	0.125	0.08	0.095	0.12	0.28	0.18	0.2	0.1
Cu-29	121	115	125	100	82	90	80	185	110
Dy-66	0.245	0.33	0.404	0.475	0.343	0.366	0.351	0.24	0.139
Er-68	0.16	0.218	0.226	0.315	0.226	0.248	0.234	0.166	0.097

Eu-63	0.056	0.076	0.094	0.113	0.073	0.078	0.076	0.054	0.054
F-9	64	38	30	24	32	41	63	238	180
Fe-26	182000	210000	248000	235000	275000	215000	185000	290000	220000
Ga-31	9.8	7.8	7.1	6	6	5.7	5	16	11
Gd-64	0.197	0.276	0.337	0.415	0.299	0.31	0.303	0.214	0.107
Ge-32	33	23	21	17	13	10	9	42	28
H-1	20000	14000	700	28000					
Hf-72	0.12	0.186	0.178	0.194	0.18	0.17	0.15	0.14	0.15
Hg-80	0.39								
Ho-67	0.055	0.077	0.094	0.11	0.073	0.081	0.077	0.05	
I-53	0.5	0.425	0.2	0.188	0.068	0.053		0.15	0.053
In-49	0.08	0.05	0.025	0.033	0.011	0.007	0.012	0.058	0.002
Ir-77	0.46	0.595	0.735	0.76	0.76	0.49	0.36	0.565	0.525
K-19	560	400	345	310	780	825	790	800	735
La-57	0.236	0.317	0.387	0.486	0.295	0.31	0.315	0.235	0.19
Li-3	1.57	1.36	1.2	1.24	1.7	1.8	2.1	2.1	0.58
Lu-71	0.025	0.033	0.04	0.048	0.031	0.033	0.033	0.024	0.024
Mg-12	97000	117000	145000	145000	140000	149000	153000	106000	141000
Mn-25	1900	1700	1650	1450	2320	2570	2620	2200	1630
Mo-42	1.2a	1.5	1.9	2.1	1.7	1.3	1.1	1.7i	

TABLE IV-6 *Continued*

	CI	*CM*	*CO*	*CV*	*H*	*L*	*LL*	*EH*	*EL*
N-7	3180m	1520	90	80	48	43	70	260m	
Na-11	4900	4100	4100	3300	6400	7000	7000	6800	5800
Nb-41	0.27	(0.37)	(0.45)	(0.54)	(0.36)	(0.39)	(0.37)	(0.25)	
Nd-60	0.457	0.631	0.772	0.99	0.628	0.682	0.659	0.46	0.233
Ni-28	10700	12000	14000	13400	16000	12000	10200	17500	13000
Os-76	0.49	0.64	0.79	0.825	0.82	0.515	0.4	0.654	0.589
P-15	1020	900	1040	990	1080	950	850	2000	1170
Pb-82	2.4	1.7	2.2	1.4	0.24	0.37		1.1	0.68*m*
Pd-46	0.56	0.64	0.703	0.705	0.87	0.56	0.53	0.885	0.69
Pr-59	0.093	0.129	0.157	0.2	0.123	0.132	0.122	0.094	
Pt-78	0.99	1.1	1.2	1.25	1.4	1.05	0.85	1.2	1.32m
Rb-37	2.22	1.7	1.45	1.25	2.9	3.1	3.1	2.6	2.5
Re-75	0.037	0.046	0.055	0.065	0.07	0.04	0.03	0.052	0.047
Rh-45	0.134			0.25	0.22				
Ru-44	0.71	0.883	1.09	1.13	1.1	0.75		0.915	0.831
S-16	59000	33000	20000	22000	20000	22000	23000	58000	33000
Sb-51	0.153	0.115	0.105	0.085	0.07	0.068	0.06	0.196	0.09
Sc-21	5.8	8.2	9.6	11.4	7.9	8.6	8.4	5.7	7.4
Se-34	19.6	12.7	7.6	8.3	7.7	9	9.9	25.5	13.5

Si-14	105000	129000	159000	156000	169000	185000	189000	167000	186000
Sm-62	0.149	0.2	0.24	0.295	0.185	0.195	0.2	0.14	0.135
Sn-50	1.72	1.01	0.89	0.9	0.86	0.71			
Sr-38	7.9	10.1	12.7	15.3	10	11.1	11.1	7.2	8.2
Ta-73	0.016	(0.022)	(0.027)	(0.032)	0.023	0.023	(0.022)	(0.015)	
Tb-65	0.036	0.047	0.057	0.065	0.053	0.057	0.048	0.035	
Te-52	2.4	1.91	0.9	1.02	0.26	0.48	0.49	2.23	0.8
Th-90	0.029	0.04		0.06	0.042	0.043	0.043	0.03	0.035
Ti-22	420	580	780	980	600	630	620	450	580
Tl-81	0.142	0.092	0.042	0.046	0.0037	0.002	0.0072	0.103	0.005
Tm-69	0.025	0.033	0.04	0.045	0.039	0.039	0.034	0.025	
U-92	0.0082	0.011	0.013	0.017	0.012	0.013	0.013	0.009	0.01
V-23	55	75	92	96	74	77	75	54	60
W-74	0.1	0.14	0.16	0.19	0.16	0.11	0.08m	0.14r	0.14r
Y-39	1.44	2	2.4	2.4	2.2	2.1	2	1.3	
Yb-70	0.159	0.222	0.27	0.322	0.205	0.22	0.22	0.16	0.165
Zn-30	312	185	100	116	47	50	46	250	17
Zr-40	3.8	8	7.8	8.3	6.3	5.9	5.9	4.9	5.2

Sources: Wasson and Kallemeyn (1988). a: Average between Anders and Ebihara (1982) and Mason (1979); i: Imamura and Honda (1976); m: Mason (1979); r: Rambaldi and Cendales (1980).

Notes: All compositions are given ppm. Values in parentheses are estimated by assuming $E^{i}_{\mathrm{Al}} = 1$.

One can also easily show that

$$E^i_j = E^i_k / E^j_k,$$

where k is any third element. If $E^j_k \approx 1$, then $E^i_j \approx E^i_k$. For example, if $E^{\mathrm{Sc}}_{\mathrm{Al}} \approx 1$ then $E^i_{\mathrm{Sc}} \approx E^i_{\mathrm{Al}}$. An enrichment factor of 1 ($E^i_j = 1$) represents no fractionation of element i in the sample relative to the reference when normalized to element j. A value of greater than 1 implies enrichment, and a value of less than 1, depletion of element i. Sometimes, to avoid confusion, one may also add [sample name/reference name] after E^i_j.

Table IV-7 gives the E^i_{Al} values for various elements in the chondrite subclass with CI as the reference. The elements in the Table are divided into five groups:

1. **Refractory lithophiles**: The E^i_{Al} values are all close to 1 in all subclasses, except in EL where E^i_{Al} for rare earth elements (REE), Ca, V, and Sr is appreciably less than 1.
2. **Moderately volatile lithophiles**: E^i_{Al} values in the ordinary chondrites (H, L, LL) are all close to 1 but in carbonaceous chondrite are less than 1 ($1 > \mathrm{CM} > \mathrm{CO} > \mathrm{CV}$).
3. **Refractory siderophiles**: E^i_{Al} values are correlated to $E^{\mathrm{Fe}}_{\mathrm{Al}}$ among all subclasses. In other words, $E^i_{\mathrm{Fe}} (= E^i_{\mathrm{Al}}/E^{\mathrm{Fe}}_{\mathrm{Al}})$ is always close to 1 (one exception is $E^{\mathrm{Au}}_{\mathrm{Fe}}$ in EH).
4. **Moderately volatile siderophiles**: E^i_{Al} values are correlated to $E^{\mathrm{Fe}}_{\mathrm{Al}}$ among all subclasses but E^i_{Fe} is less than 1 (exceptions are E^i_{Fe} values for P, As, and Ga in EH and EL).
5. **Volatiles**: E^i_{Al} values are all much less than 1, except E^i_{Al} for Cs, Ag, Cl, Se, Te, S, and F in EH. There is no correlation between E^i_{Al} and $E^{\mathrm{Fe}}_{\mathrm{Al}}$.

For refractory lithophiles, moderately volatile Mg, Cr, Mn, Rb, and volatiles (excluding F, Pb, I, C and N) in EH, E^i_{Al} values lie within the range of 1.0 ± 0.3 indicating a very close resemblance between EH and CI, whereas E^i_{Al} values for Si, Li, K, Na, and refractory and moderately volatile siderophiles (excluding Au, P, As) in EH lie within the range of 1.5 ± 0.2, indicating systematic enrichment of those elements in EH relative to CI. The high E^i_{Al} values for As, P, Au, and F (2 to 4) in EH may indicate enrichment of phosphorus minerals yet to be identified. The depletion of refractory lithophiles such as REE, Ca, V, and Sr in EL ($E^i_{\mathrm{Al}} < 1$) was probably caused by the formation of sulfide gas species for those elements in a very reducing nebular gas, so that these elements escaped full condensation.

By comparing Table IV-7 and Table III-5, one can easily see that the refractory lithophiles are the very elements that condense from the solar nebular gas as oxides or solid solution components in perovskite ($CaTiO_3$) with high 50% condensation temperatures ($T_{C,50}$), while the moderately volatile

TABLE IV-7
Classification of elements according to their enrichment factors (E^i_{Al}) relative to CI in various chondritic subclasses along with the 50% condensation temperature from Table III-5

$Z(T_{C,50})$	*CM*	*CO*	*CV*	*H*	*L*	*LL*	*EH*	*EL*
Refractory lithophiles; all $E^i_{Al} \approx 1$, except EL ≤ 1								
Zr(1717)	1.53	1.23	1.07	1.26	1.09	1.12	1.37	1.12
Hf(1690)	1.13	0.89	0.79	1.14	1	0.9	1.24	1.02
Sc(1652)	1.03	1	0.97	1.04	1.05	1.05	1.04	1.05
Al(1650)	1	1	1	1	1	1	1	1
Y(1622)	1.01	1	0.82	1.16	1.03	1	0.96	
Ti(1598)	1.01	1.12	1.14	1.09	1.06	1.07	1.14	1.13
Lu(1598)	0.98	0.98	0.96	0.96	0.95	0.97	1.04	0.8
Er(1598)	0.99	1	0.97	1.08	1.09	1.06	1.1	0.5
Ho(1598)	1.03	1.03	0.99	1.02	1.04	1.02	0.97	
Tb(1598)	0.97	0.97	0.9	1.14	1.13	0.98	1.04	
Tm(1598)	0.97	0.97	0.9	1.2	1.11	0.99	1.07	
Dy(1598)	0.98	0.99	0.95	1.07	1.05	1.04	1.04	0.46
Gd(1597)	1.02	1.03	1.04	1.16	1.11	1.11	1.15	0.44
Th(1598)	1.01	0.93	1.02	1.1	1.05	1.07	1.1	0.99
U(1580)	0.98	0.95	1.02	1.11	1.12	1.15	1.17	1
Nd(1563)	1.01	1.02	1.06	1.05	1.05	1.04	1.07	0.42
Sm(1560)	0.98	0.97	0.97	0.94	0.92	0.97	1	0.74
Pr(1557)	1.01	1.02	1.06	1.01	1	0.95	1.07	
La(1544)	0.98	0.99	1.01	0.95	0.93	0.96	1.06	0.66
Ta(1543)	1	1.01	0.98	1.05	1.01	0.99	1	
Ca(1518)	1.01	1.03	1.01	1.03	1	1.02	0.98	0.9
Yb(1493)	1.02	1.02	1	0.98	0.98	1	1.07	0.85
V(1455)	0.99	1.01	0.86	1.02	0.99	0.99	1.04	0.89
Ce(1440)	0.99	1	1.03	1.03	1.03	1.06	1.14	0.4
Eu(1338)	0.99	1.01	0.99	0.99	0.98	0.98	1.02	0.79
Sr(1217)	0.93	0.97	0.95	0.96	0.99	1.02	0.97	0.85
Ba(1163)	1.05	1.12	1.05	1.39	1.13	1.51	1.2	
Moderately volatile lithophiles; $E^i_{Al} = 1 > \text{CM} \geq \text{CO} > \text{CV}$; $\text{H} \approx \text{L} \approx \text{LL} \approx 1$								
Mg(1340)	0.88	0.9	0.73	1.1	1.08	1.14	1.16	1.19
Si(1311)	0.9	0.91	0.73	1.22	1.24	1.3	1.69	1.45
Cr(1301)	0.84	0.81	0.67	1.05	1.03	1.02	1.26	0.94
Li(1225)	0.63	0.46	0.39	0.82	0.81	0.97	1.42	0.3
Mn(1190)	0.65	0.52	0.38	0.93	0.95	1	1.23	0.7
K(1000)	0.52	0.37	0.27	1.06	1.04	1.02	1.52	1.08
Rb(1000)	0.56	0.39	0.28	0.99	0.98	1.01	1.24	0.92
Na(970)	0.61	0.5	0.33	0.99	1.01	1.03	1.47	0.97

TABLE IV-7 *Continued*

$Z(T_{C,50})$	*CM*	*CO*	*CV*	*H*	*L*	*LL*	*EH*	*EL*
Refractory siderophiles; $E^i_{Al} \propto E^{Fe}_{Al}$ and $E^i_{Fe} \approx 1$								
Re(1818)	0.91	0.9	0.86	1.44	0.76	0.64	1.49	1.04
Os(1812)	0.95	0.97	0.83	1.27	0.74	0.59	1.42	0.98
W(1794)	1.02	0.96	0.93	1.22	0.78	0.58	1.49	1.14
Ir(1603)	0.94	0.96	0.81	1.26	0.75	0.57	1.3	0.93
Mo(1595)	0.91	0.96	0.86	1.21	0.85	0.71	1.51	
Ru(1565)	0.91	0.92	0.78	1.18	0.74		1.37	0.96
Pt(1411)	0.81	0.73	0.62	1.08	0.75	0.62	1.29	1.08
Co(1356)	0.83	0.81	0.64	1.21	0.82	0.7	1.76	1.08
Ni(1354)	0.82	0.79	0.62	1.14	0.79	0.69	1.74	1
Fe(1337)	0.84	0.82	0.63	1.15	0.83	0.74	1.69	0.99
Pd(1321)	0.83	0.76	0.62	1.18	0.71	0.68	1.68	1.01
Au(1225)	0.84	0.77	0.49	1.14	0.79	0.7	2.43	1.28
Moderately volatile siderophiles; $E^i_{Al} \propto E^{Fe}_{Al}$ and $E^i_{Fe} < 1$								
P(1151)	0.64	0.61	0.48	0.81	0.66	0.6	2.08	0.94
As(1135)	0.71	0.64	0.43	0.85	0.59	0.53	1.99	0.98
Cu(1037)	0.69	0.62	0.41	0.52	0.52	0.48	1.62	0.74
Ga(918)	0.58	0.44	0.3	0.47	0.41	0.37	1.73	0.92
Sb(912)	0.55	0.41	0.27	0.35	0.31	0.28	1.36	0.48
Ge(825)	0.5	0.38	0.25	0.3	0.21	0.2	1.35	0.69
Sn(720)	0.43	0.31	0.25	0.38	0.29			
Volatiles; all $E^i_{Al} < 1$, except a few EH								
Cs(1000)	0.5	0.26	0.26	0.5	1.08	0.71	1.16	0.45
Ag(952)	0.55	0.28	0.25	0.16	0.22	0.25	1.21	0.09
Cl(863)	0.17	0.22	0.15	0.09	0.08	0.14	1.03	0.25
F(736)	0.43	0.28	0.18	0.38	0.45	0.71	3.95	2.3
Se(684)	0.47	0.23	0.21	0.3	0.32	0.37	1.38	0.56
Te(680)	0.58	0.23	0.21	0.08	0.14	0.15	0.99	0.27
Zn(660)	0.43	0.19	0.18	0.11	0.11	0.11	0.85	0.05
S(648)	0.41	0.2	0.18	0.26	0.26	0.28	1.04	0.46
Pb(521)	0.52	0.55	0.29	0.08	0.11		0.49	0.23
Bi(472)	0.5	0.18	0.21	0.12	0.09	0.11	0.85	0.09
In(470)	0.46	0.19	0.2	0.11	0.06	0.11	0.77	0.02
Tl(448)	0.47	0.18	0.16	0.02	0.01	0.04	0.77	0.03
Cd(429)	0.41	0.01	0.28	0.02	0.01	0.04	0.79	0.03
Br(350)	0.52	0.22	0.2	0.11	0.15	0.12	0.71	0.18
I	0.62	0.24	0.18	0.1	0.08		0.32	0.09
C	0.5	0.09	0.09	0.03	0.02	0.03	0.13	0.09
N	0.34	0.02	0.01	0.01	0.01	0.02	0.09	
B	0.36		0.12	0.31	0.23			
H	0.51	0.02	0.69					

lithophiles condense along with magnesium silicates and feldspar with moderate $T_{C,50}$. Similarly, the refractory and moderately volatile siderophiles are the very elements that condense as metal alloys with high and moderate $T_{C,50}$, respectively; volatiles condensed as metal alloys, sulfides, or as components of apatite and sodalite at low temperatures. In Table III-5, the $T_{C,50}$ of Cr was calculated assuming that Cr condensed as metal alloy, but according to Table IV-7, Cr behaves more like a moderately volatile lithophile in chondrites. In Table IV-7, the E^{i}_{Al} values for Be, Hg, Nb, and Rh are not given due to missing data; however, Be and Nb are probably refractory lithophiles, Rh a refractory siderophile, and Hg a volatile.

Factor Analysis of Chondritic Data

So far, we have been discussing the characteristics of chondritic subclasses based on their average compositions. However, there is a very useful statistical technique called **factor analysis** that will help us sort out the underlying interrelationships among the variables (in our case, the concentrations of a set of elements for a group of chondrite samples), and the chemical similarities or particularities among the given samples. For details, one should refer to standard textbooks, e.g., Davis (1973), Reyment and Joereskog (1993), and computer programs in Statistical Packages for the Social Sciences (SPSS) or in the Statistical Analysis System (SAS).

The most useful outputs from computer programs for factor analysis are means, standard deviations of variables, **correlation coefficient matrix, eigenvalues** (i.e., variances of new factors), **factor loadings** of variables on new factors, and **factor scores** of new factors for each sample. One may visualize factor loadings as the correlation coefficients between any variable and new factors, and factor scores as the "concentrations" of new factors in each sample. New factors (or new variables) are mutually orthogonal or linearly independent, i.e., the correlation coefficients between any two different factors are always zero. The new factor 1 is the best linear combination of the original variables to account for the largest fraction of the total variance in the whole data set. The factor 2 is the next best linear combination to account for the residual variance, and so on. Any interrelationships among original variables can be best revealed as cluster(s) in one or two two-dimensional plots of factor loadings on factors 1 to 4 (F1 to F4). The chemical similarity among samples can be best revealed as cluster(s) in one or two two-dimensional plots of factor scores of factors 1 to 4 for all samples. Computer programs provide the option to compute both the so-called **principal component** and **varimax** factor analysis. Both are useful and one may choose the one that best illustrates the data for the final presentation. Sometimes it is just a matter of taste.

Figure IV-8 gives examples of factor loading results from principal component factor analysis for the bulk composition data of carbonaceous chondrites (CC), ordinary chondrites (OC), and enstatite chondrites (EC). The axes F1 and F2 always represent the range of factor loading values between −1 and +1, and their intersection is at the zero origin. A group of ele-

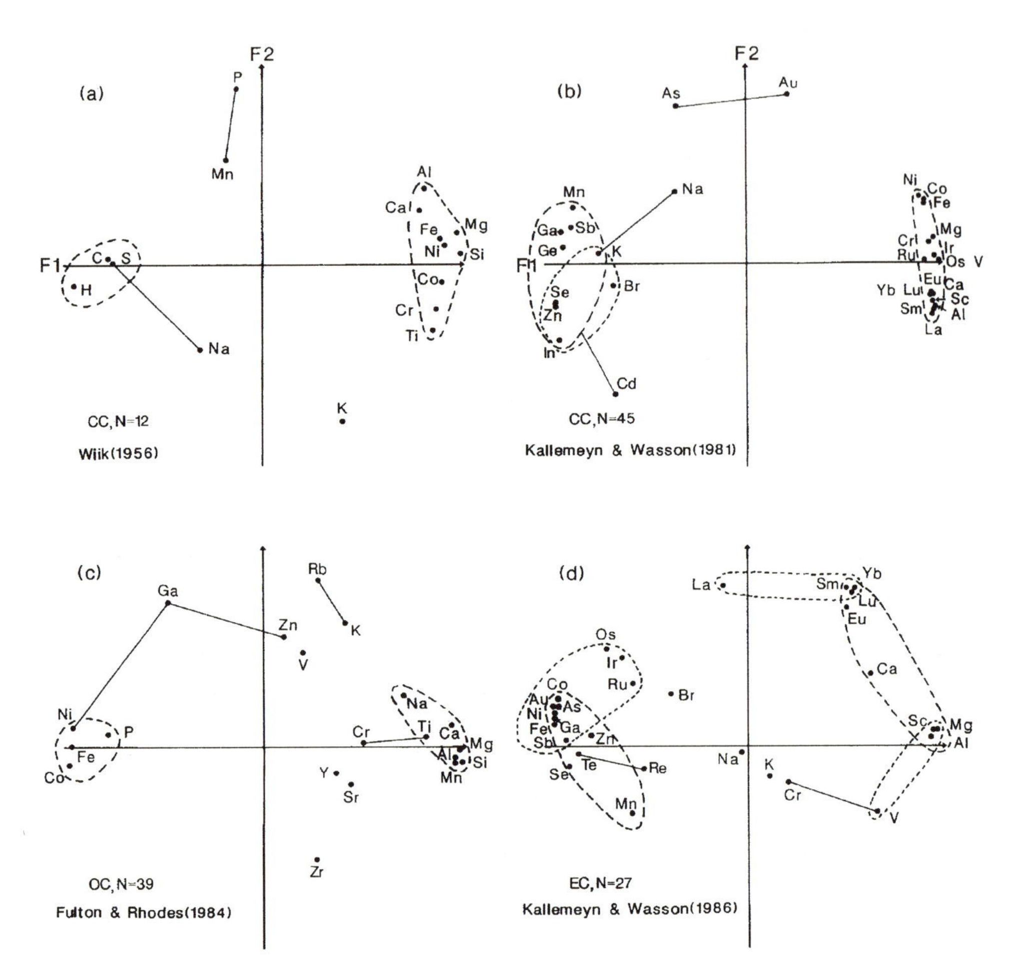

FIGURE IV-8. Plots of factor loadings of elements on factor 1 versus factor 2 for (a) carbonaceous chondrites (CC; Wiik, 1956), (b) carbonaceous chondrites (CC; Kallemeyn and Wasson, 1981), (c) ordinary chondrites (OC; Fulton and Rhodes, 1984), and (d) enstatite chondrites (EC; Kallemeyn and Wasson, 1981). *N* represents the sample numbers used in the factor analysis. The elements enclosed by any dotted line have correlation coefficients of greater than 0.5 between any pair. Any two elements connected by a solid line also have correlation coefficients of greater than 0.5.

ments in Figure IV-8 are enclosed by a dotted loop whenever the correlation coefficients are greater than 0.5 among all pairs (based on the correlation coefficient matrix); thus these elements are closely associated. Any two elements connected by a solid line also have a correlation coefficient of greater than 0.5. This kind of graphic presentation applies to all factor loading plots in this book.

The major feature for carbonaceous chondrites (Figures IV-8a and IV-8b) is that the refractory lithophiles plus Si, Mg, and Cr, and refractory siderophiles form a highly coherent cluster (positive F1), while the moderately volatile lithophiles + siderophiles and volatiles form another cluster (negative F1). These two inversely correlated clusters nicely reflect the fractionation of elements according to their volatility or condensation temperature during the formation of carbonaceous chondrites. This is consistent with the suggestion by Anders (1964) that carbonaceous chondrites are essentially mixtures of high- and low-temperature nebular condensates. The factor 2 in Figures IV-8a and b (P, As, and Au) may again represent the unidentified phosphorus minerals. The carbonaceous chondrite data (Figure IV-8b) given by Kallemeyn and Wasson (1981) were again factor-analyzed by excluding elements with missing data in order to obtain the factor scores. The results of factor loadings and factor scores are shown in Figure IV-9a and IV-9b, respectively. Figures IV-8b and IV-9a are essentially the same. As can be seen in Figure IV-9b, the carbonaceous chondrite subgroups CI, CM, CO, and CV are chemically distinct from one another, revealed mainly through the change in factor 1 scores. For example, CI is high in the elements of low condensation temperatures. The CO and CV subgroups are exactly opposite. Furthermore, CO and CV are separated by factor 2 scores, i.e., CO is high and CV is low in As and Au.

In ordinary chondrites, the refractory siderophiles plus P are fractionated from the refractory lithophiles plus Si, Mg, and Cr (Figure IV-8c). Also, the moderately volatile Na and Mn now join the refractory lithophile group. The refractory-lithophile and -siderophile factors are inversely correlated (Figure IV-8c). Thus, the elements were fractionated in the OC according to their affinity to either silicate or metal phases, instead of their volatility. The implication is that the primary high- and low-temperature nebular condensates were further differentiated into distinct silicate and metal fractions through fusion (this will be discussed later with the chondrule formation processes) and gravitational segregation inside the nebular disk. Aggregation of silicate and metal fractions in variable proportions resulted in the L to HH types of ordinary chondrites. The scatter of V, Y, Sr, and Zr is probably caused by low accuracy of those data.

In enstatite chondrites (Figure IV-8d), the refractory and moderately volatile siderophiles form a highly coherent cluster, while Os, Ir, and Ru form a subcluster within. Zn, Se, Te, Mn, and probably Re form a sulfide group and are closely correlated with siderophiles other than Os, Ir, and Ru. The rare

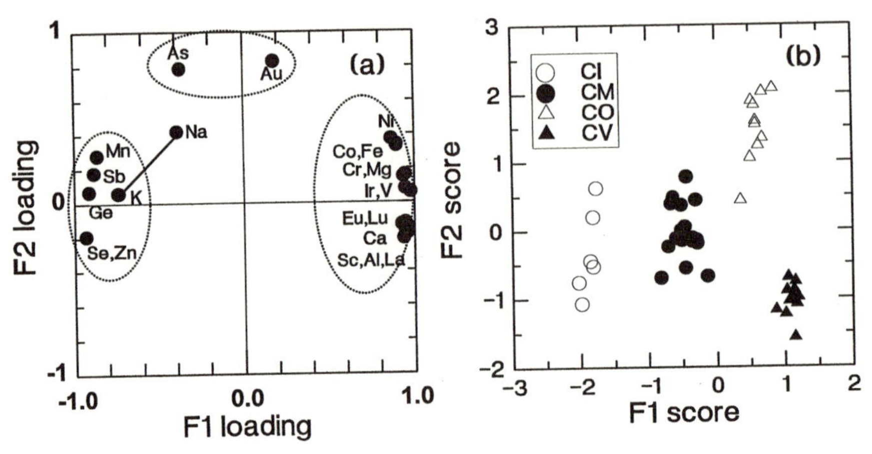

FIGURE IV-9. (a) Similar plot to Figure IV-8b, except excluding elements that are not analyzed in some cases. (b) Plot of factor scores of factors 1 and 2 for carbonaceous chondrite data by Kallemeyn and Wasson (1981).

earth elements are fractionated from the cluster of Al, Sc, Mg, and V, and probably are concentrated in sulfides such as oldhamite (CaS), as discussed earlier. Ca and Eu are related to both rare earth and Al clusters. The Cr and V pair may indicate another sulfide such as daubreelite ($FeCr_2S_4$).

IV-4. Cosmochemical Classification of Elements

The terms **lithophile**, **siderophile**, and **chalcophile**, in addition to **atmophile** (noble gases, H, C, N, and O), were introduced by Goldschmidt (1954) to describe elements that have the tendency to concentrate, respectively, in the silicate + oxide, metal, and troilite phases of the ordinary chondrites. Tables IV-8a and IV-8b provide examples of how the elements distribute among different mineral phases in the Modoc chondrite (L6; Mason and Graham, 1970), in metal from L type chondrites (Rambaldi, 1976, 1977), and between metal and troilite phases of the Cape York iron meteorite (IIIAB, Koeberl et al., 1986). The cosmochemical classification of elements according to Goldschmidt's simple criteria is also shown in the same tables. In general, the classification of elements in Tables IV-8a and IV-8b agrees quite well with that in Table IV-7 for lithophiles and siderophiles. There are, however, some subtle differences: Cu, Ga, and Mo are primarily siderophiles in Table IV-7, but they also show a chalcophile tendency (concentrations higher in troilite than in bulk rock; Table IV-8a). Ga also shows strong lithophile character (high concentration in chromite and plagioclase). P is a moderately volatile siderophile in Table IV-7 and in the Cape York iron meteorite, but

TABLE IV-8A

Concentrations (ppm) of elements in various mineral phases of the ordinary chondrite Modoc (L6)

Z	Class	Bulk 100%	Metal 8.1%	FeS 6.5%	Olivine 46.1%	Hypersthene 23.6%	Plagioclase 10.8%	Chromite 0.8%	Phosphate 0.7%
B	1	0.4L			3		2		
Ba	1	3.8			< 0.5	2	42 (100)	2	11
Br	1	0.7	1 (12)	1 (9)					8 (8)
Cl	1	76	74?	13					12000 (100)
Co	s	800	10000 (100)	38	25	7	1	28	620
Cr	1	3800	<10	140	350	800	370	370000 (78)	340
Cu	s/c	170	620 (30)	220 (8)	14	7	5	460 (2)?	52
Ga	1/s	6.3	22 (28)	6 (6)	2	2	17 (29)	65 (8)	2
In	c	.0002		0.1 (100)					
Mn	1	2600	<10	150	3600 (64)	3600 (33)	230	6000 (2)	580
Mo	s/c	1.3L	7 (44)	3 (15)					
Ni	s	13000	150000 (93)	820	230	46	<20	20	
Sc	1	8.1	1	8 (7)	3	10 (29)	6	2	44 (4)
Se	c	12.9		130 (66)					
Sn	s	0.28	2 (58)						
Sr	1	11			0.4	4	75 (74)	8	87 (6)
Ti	1	700	<50	50	100	1000 (34)	360	17000 (19)	2100 (2)
V	1	69	7	4	9	40	12	4000 (46)	5
Y	1	2.1		1		0.8	<1		160 (53)
Zn	1	46		10	20	17	3	2000 (34)	19
Zr	1	8		3		6	7	5	650 (57)

TABLE IV-8A
Continued

Z	Class	Bulk 100%	Phosphate 0.7%	Z	Class	Bulk 100%	Phosphate 0.7%	Z	Class	Bulk 100%	Diopside 3.1%
La	1	0.34	51 (100)	Ho	1	0.084	8.6 (72)	Cr	1	3800	4100 (1)
Ce	1	0.9L	120 (93)	Er	1	0.23	19 (58)	Mn	1	2600	2000
Pr	1	0.13	17 (92)	Tm	1	0.033	2.4 (51)				
Nd	1	0.63	65 (72)	Yb	1	0.15	18 (84)				
Sm	1	0.2	27 (95)	P	1	1300	185000 (99)				
Eu	1	0.08	2.5 (22)	F	1	41L	880 (15)				
Gd	1	0.33	30 (64)	Hf	1	0.17L	18 (74)				
Tb	1	0.048	4.4 (64)	Th	1	0.059	7 (83)				
Dy	1	0.3	34 (79)	U	1	0.019	2.3 (83)				

Sources: Mason and Graham (1970) and Mason (1979) except for values followed by L (Table IV-6, L column).

Notes: Column class: 1 = lithophile; s = siderophile; c = chalcophile. The value in parentheses represents the percent contribution by the indicated phase to the bulk composition of the indicated element.

TABLE IV-8B

Concentrations of elements in the metal phase of average L chondrites, and metal and troilite phases of the iron meteorite Cape York (IIIAB)

L chondrite				Cape York (IIIAB)							
Z	Class	Bulk 100%	Metal 8.1 wt%	Z	Class	Metal	FeS	Z	Class	Metal	FeS
As	s	1.55	18.6 (97)	Ag*	c	0.001–0.014	0.06	Os	s	3.8	< 0.005
Au	s	0.16	1.77 (89)	As	s	6.1	0.002	P	l/s	1400	0.9–4
Co	s	590	7040 (97)	Au	s	1.4	0.0014	Pb*	c	0.01–0.1	2–10
Cu	s/c	90	708 (64)	Bi*	c	< 0.008	0.06–0.13	Pd	s	2.5	0.036
Fe	s/c/l	215000	844000 (32)	Cd*	c	0.02	0.05–0.5	Pt*	s	9.4	0.7
Ga	l/s	5.7	12 (17)	Cl*	l/c	0.01–0.4	39	Re	s	0.4	< 0.0004
Ge	s	10	125 (100)	Co	s	4800	5.2	Rh	s	1.7	< 0.2
Ir	s	0.49	5.3 (88)	Cr	l/c	51	10^3–5.10^4	Ru	s	4.5	0.016
Mo	s/c	1.3	7.7 (48)	Cu	s/c	189	100	Sb	s	0.28	
Ni	s	12000	148000 (99)	Ga	l/s/c	15	0.2–23	Se	c	0.006	38–92
Os	s	0.52	1.9 (93)	Ge	s	37	0.15	Sn*	s	1.9	
Pt	s	1.05	13 (100)	Hg*	c	0.04	0.17–0.49	Te*	c	0.02–0.09	1.2–7.8
Re	s	0.04	0.59 (100)	I*	c	0.01–0.3	0.02–3.6	Ti	l/c	< 17	24
Ru	s	0.75	7.8 (84)	In	c	< 0.01		Tl*	c	0.0001–0.01	0.002–0.2
Sb	s	0.068	0.88 (100)	Ir	s	4.6	0.003	V	l/c	0.06	3.7
W	s	0.11	1.05 (77)	Mn	l/c	20	200	W	s	1.02	< 0.7
				Mo	s/c	7.2	5.6	Zn	l/c	1.5	8.7

Sources: Bulk of L chondrite (Table IV-5, L column), metals of L Chonidrites (Rambaldi, 1976, 1977) except Mo (Imamura and Honda, 1976). Cape York (Koeberl et al., 1986) except elements with asterisk which are averages or ranges for iron meteorites (Mason, 1971, 1979; Wedepohl, 1969–1978).

Notes: The value in parentheses represents the percent contribution by the indicated phase toward the bulk composition of the indicated element. All concentrations in ppm.

is lithophile in Modoc. The lithophile Cr, Mn, Ti, and V show chalcophile character in IIIAB iron meteorites (Table IV-8b). The apatite phase in Modoc is the most prominent reservoir for rare earth elements, Y, Hf, Zr, Th, U, and P; and is also high in Ba, Sc, Sr, and Ti.

The elements F, Cl, Br, and B are volatiles, but the first three are also concentrated in apatite, and B in olivine and plagioclase (Table IV-8a). Therefore F, Cl, Br, and B along with Cs may be called the volatile lithophiles. However, Cl, I, and probably Br also show a chalcophile tendency in iron meteorites

TABLE IV-8C
Concentrations of elements in bulk samples of five enstatite chondrites (EH), Indarch (EH4), and Abee (EH4); and in their metal, sulfide, and oldhamite (CaS) phases.

Five EH			*Indarch (EH4)*			*Abee (EH4)*		
Z	*Bulk 100 wt%*	*Sulfides 10.7 wt%*	Z	*Bulk 100 wt%*	*Oldhamite 1 wt%*	Z	*Bulk 100 wt%*	*Metal 21 wt%*
Ca	6160	52200 (91)	Ca	8900	523000 (59)	As	3.5	14.5 (87)
Cr	2420	9270 (41)	Cs	0.24	7 (29)	Au	0.34	1.35 (83)
Fe	342000	489000 (15)	Mg	105000	13200 (0.1)	Co	892	3660 (86)
K	650	2050 (34)	Mn	1900	2200 (1)	Cu	186	244 (28)
Mg	121000	27900 (3)	S	51800	431000 (8)	Fe	305000	906000 (62)
Mn	2190	16800 (82)	Sb	0.24	9 (38)	Ga	16.5	71 (90)
Na	6860	6840 (11)	Zn	430	2900 (7)	Ge	55	198 (76)
S	41500	387000 (100)	La	0.21	75 ± 36 (100)	Ir	0.55	2.43 (93)
Ti	721	5470 (81)	Ce	0.59	68 ± 11 (100)	Mo	1.7	4.3 (39)
		Metal 31.3 wt%	Sm	0.14	16 ± 4 (100)	Ni	17800	71000 (97)
			Eu	0.042	7 ± 2 (100)	Os	0.61	2.95 (100)
P	952	1653 (54)	Tb	0.031	0.8 ± 0.5	Pt	1.23	5.76 (98)
Si	176000	20000 (4)	Yb	0.122	75 ± 43 (100)	Re	0.045	0.187 (87)
Co	852	2720 (100)				Ru	0.96	4.94 (100)
Fe	342000	917000 (84)				Sb	0.197	0.803 (86)
Ni	1700	5420 (100)				W	0.138	0.664 (100)
								Oldhamite 0.9 wt%
						Th	0.033	0.91(25)
						U	0.007	0.26(34)

Sources: Five EH average (Easton, 1985); Indarch bulk (Mason, 1979), oldhamite (Larimer and Ganapathy, 1987); Abee (Rambaldi and Cendales, 1980) except Mo (Imamura and Honda, 1976); Th and U (Murrell and Burnett, 1982).

Notes: The value in parentheses represents the percent contribution by the indicated phase toward the bulk composition of the indicated element. All concentrations in ppm. Djerfisherite, $K_3(Na, Cu)(Fe, Ni)_{12}S_{14}$, contains 1.54% Cl, 170 ppm Br, 660 ppm Rb, 570 ppm Se, and 19 ppm Sr; and sphalerite (Zn, Fe)S, 1700 ppm Ga in Qingzhen (EH3) (Woolum et al., 1983).

(Table IV-8b). The volatiles such as Ag, Se, Te, S, Zn, Pb, Bi, In, Tl, and Cd are all concentrated in the troilite phase of iron meteorites, and thus are chalcophile. Except for Ag, the rest are the same elements that condensed from the solar nebular gas as solid solutions with troilite and sometimes partly with metal alloys (Table III-5). Moreover, Zn also acts as a lithophile in the Modoc chondrite.

In the very reducing environments of enstatite chondrites, many lithophile elements such as alkalis, alkaline earths, rare earths, actinides (U, Th, and Pa), Ti, V, Cr, and Mn exist in part as sulfides, thus showing various degrees of chalcophile tendency (Table IV-8c). The siderophiles Sb, Cu, Ga, and probably also Mo show chalcophile tendencies in EC (Sb in oldhamite, CaS; Cu in djerfisherite, $K_3CuFe_{12}S_{14}$; Ga in sphalerite, [(Zn,Fe)S]). Chlorine and Br are also very much concentrated in djerfisherite. Some silicon occurs as metal in solid solution with iron (1 to 3 wt% Si; Keil, 1968). Carbon, Si, and P show siderophile character by the formation of minerals such as cohenite (Fe_3C), schreibersite [$(Fe, Ni)_3P$], and perryite [$(Ni, Fe)_x(Si, P)_y$]. Nitrogen forms minerals such as osbornite (TiN), carlsbergite (CrN), and sinoite (Si_2N_2O).

Figure IV-10 summarizes the cosmochemical classification of elements in a periodic table format based on Tables IV-8a to IV-8c. The siderophile character of radioactive Tc is predicted from the similarity of phase diagrams among Fe-Ru, Fe-Os, Fe-Ru, and Fe-Tc pairs (Massalski, 1986). In the following, a simple thermodynamic model is presented to predict the cosmochemical classification of elements and compare it with the observations (Tables IV-8a to IV-8c).

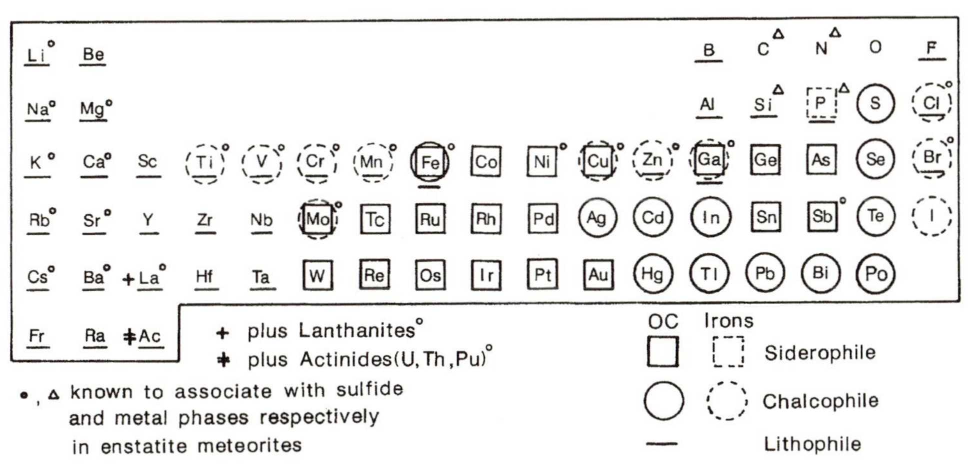

FIGURE IV-10. Cosmochemical classification of the elements in the form of the periodic table, based on data from Tables IV-8a to IV-8c.

Partitioning of Elements among Iron Metal, Troilite, and Iron Silicates at 25°C and 1 Atmosphere

In a hypothetical system where pure metallic iron (Fe), troilite (FeS), and iron silicates ($FeSiO_3$, Fe_2SiO_4) coexist in equilibrium, the fugacities of O_2 and S_2 in the system are fixed or buffered by the following reactions:

$$\mathrm{Fe(s)} + \frac{1}{2}\mathrm{S_2(g)} \leftrightarrow \mathrm{FeS(s)}, \tag{IV-1a}$$

$$\mathrm{Fe(s)} + \frac{1}{2}\mathrm{O_2(g)} + \mathrm{FeSiO_3(s)} \leftrightarrow \mathrm{Fe_2SiO_4(s)}. \tag{IV-1b}$$

Equation IV-1b can be simplified to

$$\mathrm{Fe(s)} + \frac{1}{2}\mathrm{O_2(g)} \leftrightarrow \mathrm{FeO^*(s)},$$

where FeO^* represents the iron oxide component in the silicate phases. For example,

$$\mathrm{FeSiO_3} = \mathrm{FeO^*} \cdot \mathrm{SiO_2}^*, \text{ or } \mathrm{Fe_2SiO_4} = 2\mathrm{FeO^*} \cdot \mathrm{SiO_2}^*.$$

Hereafter, any oxide with an asterisk represents the oxide component in silicate phases (meta- and ortho-). If a minor element M is introduced into the same system, what will be the stable form of the element M, assuming no solid solution formation? For the reaction

$$\mathrm{FeO^*} + \frac{2}{z}M \leftrightarrow \frac{1}{z}M_2O_z{}^* + \mathrm{Fe}, \tag{IV-2a}$$

the Gibbs free energy of reaction at 25°C and 1 atmosphere is

$$\Delta\tilde{G} = \frac{1}{z}\Delta G_f M_2 O_z^* - \Delta G_f \mathrm{FeO^*}, \tag{IV-2b}$$

where ΔG_f are the Gibbs free energies of formation of various phases or components at 25°C and 1 atmosphere pressure. For pure metal, $\Delta G_f = 0$. If $\Delta\tilde{G} < 0$, the above reaction proceeds to the right until all of the metal M is converted into $M_2O_z{}^*$. Similarly, in the reaction

$$\mathrm{FeS} + \frac{2}{z}M \leftrightarrow \frac{1}{z}M_2\mathrm{S}_z + \mathrm{Fe},$$

$$\Delta\tilde{G} = \frac{1}{z}\Delta G_f M_2\mathrm{S}_z - \Delta G_f \mathrm{FeS}. \tag{IV-3}$$

If $\Delta\tilde{G} < 0$, the metal M will be converted to M_2S_z. In the reaction involving oxide and sulfide, i.e.,

$$\text{FeO}^* + \frac{1}{z}M_2\text{S}_z \leftrightarrow \text{FeS} + \frac{1}{z}M_2O_z^*,$$

$$\Delta\tilde{G} = (\Delta G_f\text{FeS} - \Delta G_f\text{FeO}^*) + \frac{1}{z}(\Delta G_f M_2\text{O}_z{}^* - \Delta G_f M_2\text{S}_z). \quad \text{(IV-4)}$$

If $\Delta\tilde{G} < 0$, $M_2O_z{}^*$ in silicate is the stable form instead of M_2S_z. Therefore, in order to predict the stable form of element M among metal M, M_2S_z, and $M_2O_z{}^*$, we need to know $\Delta G_f M_2O_z{}^*$ in silicate first. Other ΔG_f values can be obtained from regular thermodynamic tables.

If one assumes that the Gibbs free energies of formation of oxide components in silicates are additive, then

$$\Delta G_f\text{H}_2\text{SiO}_3(\text{s}) = \Delta G_f\text{SiO}_2{}^* + \Delta G_f\text{H}_2\text{O}^* = -261.1,$$

and

$$\Delta G_f\text{H}_4\text{SiO}_4(\text{s}) = \Delta G_f\text{SiO}_2{}^* + 2\Delta G_f\text{H}_2\text{O}^* = -318.6.$$

Solving the above two equations, one obtains

$$\Delta G_f\text{SiO}_2{}^* = -203.6 \text{ and } \Delta G_f H_2O^* = -57.5,$$

which are essentially the same as the pure phase values, i.e.,

$$\Delta G_f\text{SiO}_2 = -204.6 \text{ and } \Delta G_f\text{H}_2\text{O} = -56.7.$$

Therefore one can safely assume that

$$\Delta G_f\text{SiO}_2{}^* \cong \Delta G_f\text{SiO}_2 = -204.6,$$
$$\Delta G_f\text{H}_2\text{O}^* \cong \Delta G_f\text{H}_2\text{O} = -56.7.$$

ΔG_f values are all in units of kcal/mole. Once $\Delta G_f\text{SiO}_2{}^*$ and $\Delta G_f\text{H}_2\text{O}^*$ are defined, $\Delta G_f M_2O_z{}^*$ in silicate and in hydroxides can be calculated. For example,

$$\Delta G_f M_2\text{O}^* = \Delta G_f M_2\text{SiO}_3 - \Delta G_f\text{SiO}_2{}^* \text{ for } z = 1,$$
$$\Delta G_f M_2\text{O}_2{}^* = \Delta G_f M_2\text{SiO}_4 - \Delta G_f\text{SiO}_2{}^* \text{ for } z = 2.$$

Similarly,

$$\Delta G_f M_2\text{O}^* = 2\Delta G_f M(\text{OH}) - \Delta G_f\text{H}_2\text{O}^* \text{ for } z = 1,$$
$$\Delta G_f M_2\text{O}_2{}^* = 2\Delta G_f M(\text{OH})_2 - 2\Delta G_f\text{H}_2\text{O}^* \text{ for } z = 2.$$

Interestingly, $(\Delta G_f M_2O_z{}^* - \Delta G_f M_2O_z)/z$ values obtained from the known silicate data and from the corresponding hydroxide data are linearly correlated, as shown in Figure IV-11. Because we can easily obtain $(\Delta G_f M_2O_z{}^* -$

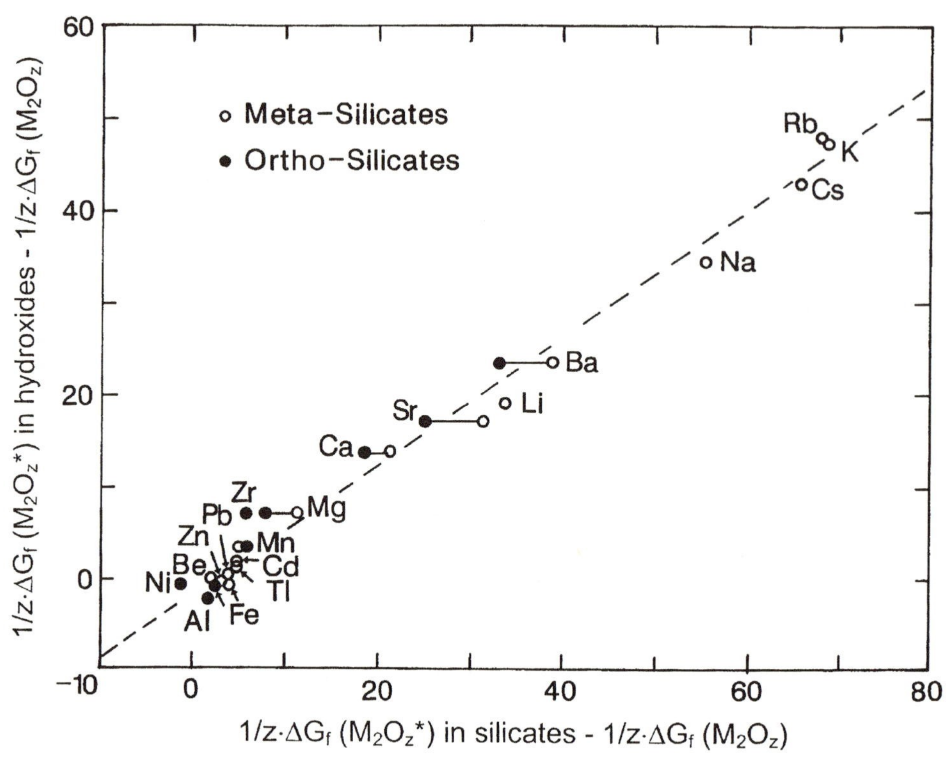

FIGURE IV-11. Plots of $(\Delta G_f M_2 O_z^* - \Delta G_f M_2 O_z)/z$ for silicates versus those for hydroxides (where z is the charge of cation M^{+z}). All values in units of kcal/mol. Data sources: Robie et al. (1978) and Dean (1985).

$\Delta G_f M_2\text{O}_z)/z$ values for most elements in hydroxides, we can predict the same parameters in silicates from the correlation line in Figure IV-11, and thus obtain $\Delta G_f M_2\text{O}_z{}^*$ in silicates. Actually, the predicted $\Delta G_f M_2\text{O}_z{}^*$ values in silicates are not much different from $\Delta G_f M_2\text{O}_z$ for most elements, except alkalis, alkaline earths, and some lanthanides and actinides.

Figure IV-12 is a plot of $-(\Delta G_f M_2\text{S}_z/z - \Delta G_f\text{FeS})$ versus $-(\Delta G_f M_2\text{O}_z/z - \Delta G_f\text{FeO}^*)$ for many elements, and the stability fields for siderophile, chalcophile, and lithophile group elements in an iron silicate–FeS-Fe coexisting system at 25°C and 1 atmospheric pressure are also indicated according to the criteria given in equations IV-2, IV-3, and IV-4. The data points (and thus also the boundary lines among phases) in Figure IV-12 probably have uncertainties of about ±2 kcal/mole in both the x and y directions. Considering these uncertainties, the simple model correctly predicts the cosmochemical classification of most elements. The outstanding exceptions are the volatiles Ag, Hg, Bi, and probably Tl and Pb (borderline cases), which are siderophile accord-

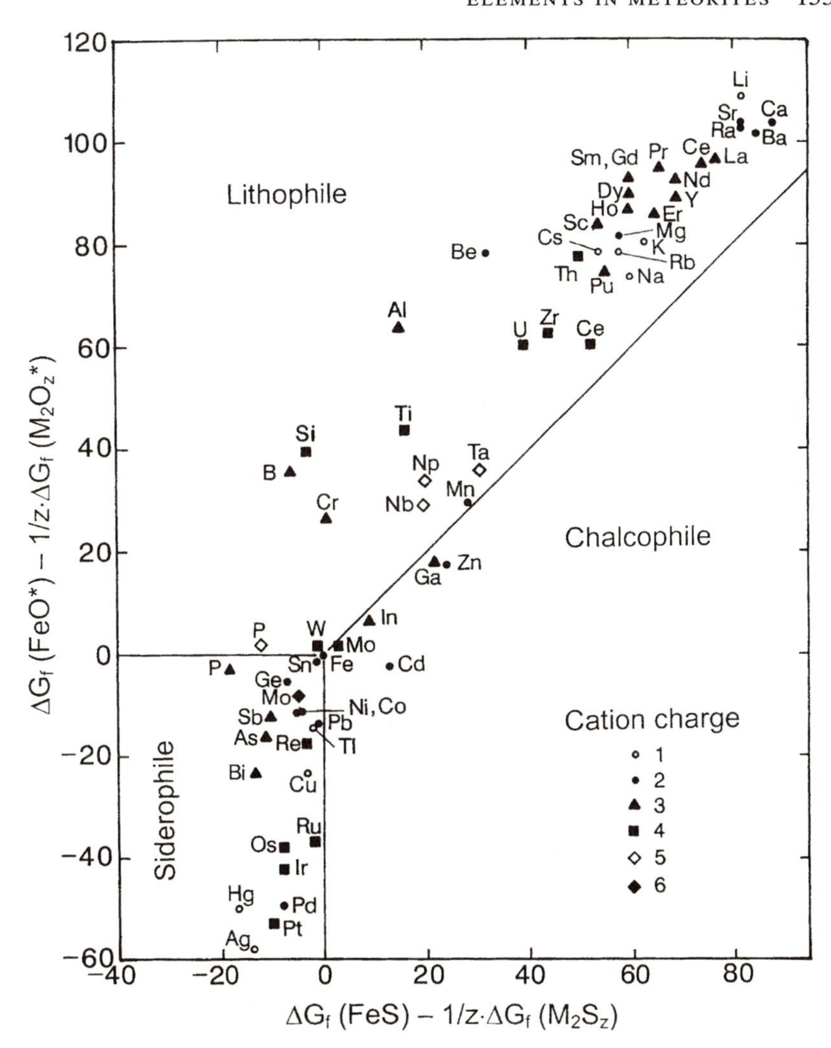

FIGURE IV-12. Plot of $\Delta G_f \text{FeS} - \Delta G_f M_2S_z/z$ versus $\Delta G_f \text{FeO}^* - \Delta G_f M_2 O_z^*/z$ for various cations. All are in units of kcal/mol. The stability fields for silicates + oxides (lithophile), metals (siderophile), and sulfides (chalcophile) are also indicated. Thermodynamic data sources: Robie et al. (1978), Dean (1985), Karapet'yants and Karapet'yants (1970), and Wagman et al. (1982).

ing to this model, but are observed as chalcophile in ordinary chondrites and iron meteorites (Figure IV-10). The contradiction between the model prediction and observation may imply some nonequilibrium processes during the sulfide formation of these elements. For example, repeated cycles of condensation and evaporation in the temperature range of 400 to 1000 K might produce an environment locally high in the fugacities of these volatiles and S_2, thus forming sulfides. Another odd element is Ga, which is enriched in metal and troilite as well as in chromite and plagioclase (Table IV-8a and IV-8b). The condensation model assumes that Ga condenses as the metal in solid solution with Fe-Ni alloy (Wai and Wasson, 1977). However, the simple equilibrium model here predicts that the stable form of Ga is sulfide and probably also oxide (borderline case in Figure IV-12). One possible explanation is that the moderately volatile Ga in the nebular gas condensed with Fe first. As temperatures continued to decrease, part of the Fe and Ga was oxidized and formed chromite (Table IV-8a). At even lower temperatures, Fe and Ga reacted with sulfur to form sulfides.

In the case of an iron silicate–FeS system without metallic iron, where the fugacity ratio of O_2 and S_2 is fixed by the FeO*-FeS pair, the boundary between the lithophile and chalcophile fields (the siderophile field disappears) is the diagonal line extending through Fe in Figure IV-12. The originally siderophile elements such as P, Ge, Sn, W, and probably Mo and Sb (borderline cases) may become lithophile, but all other siderophile elements become chalcophile.

A simple equilibrium model at room temperature was tried earlier by Goldschmidt (1954). However, he used $\Delta G_f M_2 O_z$ of oxides in his model instead of $\Delta G_f M_2 O_z{}^*$ in silicates. Thus, his model calculations predicted Na, K, Rb, Cs, Ca, Sr, Ba, and even some lanthanides as chalcophile in ordinary chondrites, in contradiction to observations.

IV-5. Matrices and Chondrules of Chondrites

Table IV-9a summarizes the relative abundances of major constituents such as matrix, chondrules, amoeboid olivine-rich inclusions (AOI), Ca-Al-rich inclusions (CAI), and others (including individual minerals, lithic fragments, and secondary minerals) in chondrite subclasses. In the following, the mineralogy and chemistry of matrices and chondrules are discussed first.

Matrices

The matrices of chondrites can be defined as mixtures of fine-grained (0.1 to 10 μm in size) phases filling interstitial space among larger (0.2 to 1 mm) chondrules, inclusions (CAI, AOI), isolated mineral grains, and lithic fragments. The major mineral components of matrices are high-Fe phyllosilicates (i.e., hydrous silicates such as serpentine, chlorite, montmorillonite, etc.) and extremely complex organic material in CI chondrites. The opaque matrix of CM chondrites observed under an optical microscope can be distinguished

TABLE IV-9
(a) Relative abundance of major components and (b) relative abundance of various chondrule types in chondrites

(a)

	CI	*CM*	*CO*	*CV*	*UOC*	*EH3*
Matrix	97 ± 2	70 ± 15	35 ± 5	43 ± 7	10 ± 5	$< 3 \pm 2$
Chondrules	0	4 ± 2	38 ± 3	40 ± 5	70 ± 5	17 ± 3
AOI	0	12 ± 6	12 ± 4	6 ± 5	trace	0
CAI	0	trace	2 ± 1	6 ± 4	trace	0
Others	3 ± 2	14 ± 14	13 ± 7	5 ± 5	20 ± 7	80 ± 4

(b)

	CV	*UOC*	*EH3*
high-Fe PO	6	$23 - x$	0
low-Fe PO	79	x	0.2
POP	7	49	4
PP	2	10	77
BO	6	4	trace
RP + G	trace	15	19

Sources: Based on Dodd (1981, his Tables 3.2 and 3.4), Grossman et al. (1988, their Table 9.1.1).

Notes: AOI = ameboid olivine inclusions; CAI = calc-aluminous inclusions; others = isolated larger mineral grains and lithic fragments plus minor opaque phases of Fe-Ni metal, troilite, etc.; PO = porphyritic olivine; POP = porphyritic olivine and pyroxene; PP = porphyritic pyroxene; BO = barred olivine; RP = radial pyroxene; G = glassy. All abundances in volume %.

under a scanning electron microscope as the "clastic matrix," and the "dust mantle" of phyllosilicates coating the coarse-grained components (Metzler et al., 1992). The dust mantles were probably formed by the direct accretion of fine-grained phyllosilicate dust from the solar nebula onto the preformed coarse-grained components, and they were agglomerated into primary CM chondrites. The clastic matrix is essentially the pulverized fine debris of primary CM chondrites formed during impact events of chondrites (Metzler et al., 1992). The matrices of CO chondrites mainly consist of high-Fe olivine (Fa 30-60) and phyllosilicates, and the matrices of CV chondrites of high-Fe olivine (Fa 40-60), phyllosilicates, Fe-Ni metal, sulfides, and small amounts of high-Ca pyroxene, nepheline, and sodalite. The matrix materials for unequilibrated ordinary chondrites (UOC) are, in order of abundance, olivine of variable Fe content (Fa 9-91), enstatite, high-Ca pyroxene, albite, Fe-Ni metal, troilite, magnetite, spinel, chromite, and calcite. In addition, Mg- and Fe-rich fluffy particles (with normative compositions of intermediate olivine + albite ± nepheline or phyllosilicates), and albitic particles or albitic interstitial glasses are also important components of UOC matrices (Nagahara, 1984). Zolensky and McSween (1988), Zolensky et al. (1993, and references therein), and Buseck and Hua (1993) provide more detailed information on the matrices of carbonaceous chondrites (CC).

Fine- to coarse-grained rims of about 10 to 400 μm thickness are often found around chondrules of carbonaceous chondrites and unequilibrated ordinary chondrites. The bulk chemical and mineral compositions of these rims are similar to the matrix materials (Matsunami, 1984; Zolensky et al., 1993). Table IV-10 summarizes the average compositions of matrix + rim and the percent standard deviations ($= 100\times$ standard deviation/mean; in parentheses) for four CI, twenty-three CM, two CO, and four CV chondrite samples (Zolensky et al., 1993). The percent standard deviations for Na, K, Ca, and P are consistently high ($>$40% in all subgroups), probably reflecting random aqueous alteration processes (Zolensky and McSween, 1988). As compared to the bulk chondrite compositions (Table IV-6), Na and Ca are consistently low and K high in the average matrix + rim (see the E^i_{Si} values in Table IV-10), whereas E^i_{Si} for P varies greatly from 0.25 to 2.6.

The factor analysis of matrix + rim data from carbonaceous chondrites (Zolensky et al., 1993) shows the groupings of Si and Mg (and probably Ti, but the correlation coefficient between Ti and Si or Mg is only about 0.38); of Na and K; of S and Ni; and Fe is inversely related to Si + Mg in the factor loading plots of Figure IV-13a. The factor score plots of Figure IV-13b confirm the general chemical similarity of matrix and rim among all subclasses of carbonaceous chondrites (Zolensky et al., 1993). Factor 2 (Na + K) is related to aqueous alteration processes (Zolensky and McSween, 1988). The association of S and Ni may represent some metal + sulfide components. The inverse relationship between Fe and Mg + Si in Figure IV-13a probably reflects the similar relationship observed in serpentines of CM chondrites (Zolenky et al., 1993). Tomeoka and Buseck (1985) proposed that the enrichment of Mg and depletion of Fe in serpentine are directly related to the degree of aqueous alteration.

The elemental interrelationships for the UOC matrix are quite different from those for the CC matrix. For example, Figure IV-13c shows the factor 1 and 2 loadings obtained from matrix data of Antarctic sample ALH-764 (LL3) (Ikeda, 1980). The close association among Na, K, Ca, Al, and Si as factor 1 may represent a feldspathic component; Fe and Mn (negative factor 1) a metal + sulfide component; and Mg, Cr, and Ni (factor 2) an olivine + chromite component. A mixture of these three components can explain the observed compositional variation of the matrix in ALH-764 (LL3) (Figure IV-13d). In contrast, there is no feldspathic component in the matrix material of carbonaceous chondrites.

TABLE IV-10

Average compositions, percent standard deviations (parentheses), and E^i_{Si} values for matrix + rim materials from carbonaceous chondrite subclasses

	CI [4]	*CM [23]*	*CO [2]*	*CV [4]*	E^i_{Si} [matrix/bulk]			
					CI	*CM*	*CO*	*CV*
P	0.035 (41)	0.103 (66)	0.068	0.23 (110)	0.25	1.1	0.8	2.6
Si	14.4 (13)	13.3 (13)	13	13.7 (8)	1	1	1	1
Ti	0.05 (6)	0.049 (44)	0.054	0.048 (40)	0.88	0.81	0.85	0.56
Al	1.23 (14)	1.42 (25)	1.65	1.67 (46)	1.1	1.4	1.2	1
Cr	0.55 (50)	0.31 (31)	0.29	0.31 (19)	1.5	0.99	1	0.98
Ca	0.18 (51)	0.46 (80)	0.72	0.70 (60)	0.14	0.35	0.56	0.42
Mg	11.9 (14)	10.3 (20)	11.5	12.4 (23)	0.89	0.85	0.98	0.98
Fe	17.5 (7)	24.2 (19)	25.3	27.7 (15)	0.7	1.1	1.3	1.3
Mn	0.19 (19)	0.16 (24)	0.15	0.16 (23)	0.74	0.88	1.1	1.3
Ni	1.87 (18)	1.76 (34)	0.25	1.72 (61)	1.3	1.4	0.22	1.5
K	0.13 (90)	0.07 (57)	0.033	0.048 (67)	1.6	1.7	1.2	1.8
Na	0.40 (110)	0.32 (51)	0.18	0.28 (72)	0.59	0.76	0.55	0.95
S	2.02 (24)	2.41 (46)	0.08	0.86 (75)	0.25	0.71	0.05	0.45

Source: Data from Zolensky et al. (1993).

Note: Number in brackets shows number of samples used for averaging.

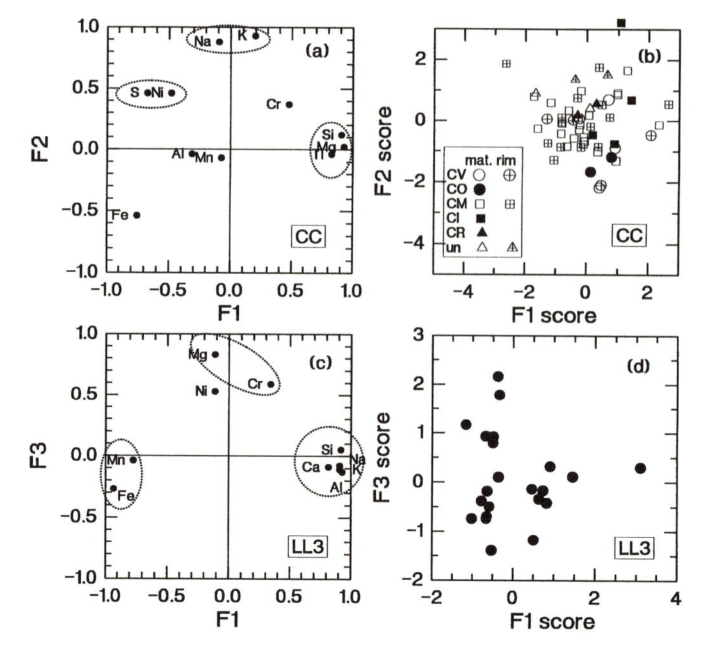

FIGURE IV-13. Plots of factor loadings and factor scores for matrix + rim data from carbonaceous chondrites by Zolensky (1993) [diagrams (a) and (b)], and matrix data from ALH-764 (LL3) by Ikeda (1980) [diagrams (c) and (d)].

Chondrules

As compared to whole rock compositions, average chondrules are generally enriched in lithophile and depleted in siderophile and chalcophile elements (Table IV-11). The depleted siderophile and chalcophile elements should be concentrated in metal and sulfide phases outside the chondrules. The E^i_{Al} values for refractory lithophile elements (Al, Ca, Hf, REE, Sc, Ti, V) in the average chondrules of Allende (CV3), Semarkona (LL3), and Qingzhen (EH3) relative to CI are around one (Table IV-11); exceptions are Eu and V in EH3. Therefore, chondrule precursor materials could not have originated from differentiated igneous materials.

Based on texture and mineralogy, chondrules can be grouped into high-Fe and low-Fe porphyritic olivine (PO), porphyritic olivine–pyroxene (POP), porphyritic pyroxene (PP), barred olivine (BO), barred pyroxene (BP), radial pyroxene (RP), glassy or cryptocrystalline (G), Ca-Al–rich chondrules (CAC), etc. (e.g., Scott and Taylor, 1983; Nagahara, 1983). Figure IV-14 schemat-

TABLE IV-11

Average compositions of the bulk Allende (CV3), Semarkona (LL3), and Qingzhen (EH3) and their chondrules, and the enrichment factors of elements $\left(E_{\mathrm{Al}}^{i}\right)$ for chondrules relative to CI

	Allende		*Semarkona*		*Qingzhen*		E_{Al}^{i}		
	CV3		*LL3*		*EH3*		*CV3*	*LL3*	*EH3*
	Bulk	*Chondrules*	*Bulk*	*Chondrules*	*Bulk*	*Chondrules*			
Lithophiles									
Hf	0.21		(0.15)	0.18	0.13	0.2		0.8	1.04
Sc	11.2	12.8	(8.4)	10.9	4.4	8.3	0.87	1	0.89
Al	17900	21800	13200	16200	8200	13900	1	1	1
Ti	980	1120	660	840	479	574	1.05	1.06	0.85
Lu	0.052		(0.033)	0.046	0.021	0.039		0.98	0.97
Sm	0.29	0.39	(0.2)	0.24	0.11	0.24	1.02	0.87	0.99
La	0.46	0.64	(0.32)	0.38	0.19	0.37	1.07	0.86	0.97
Yb	0.3	0.44	(0.22)	0.29	0.2	0.25	1.08	0.96	0.98
Eu	0.12	0.12	(0.076)	0.1	0.037	0.059	0.87	0.97	0.65
Ca	18100	19000	16400	17000	9800	12300	0.82	0.98	0.83
V	100	135	(75)	101	55	58	0.97	0.98	0.65
Mg	147000	206000	168000	218000	125000	217000	0.84	1.19	1.38
Si	160000	194000	211000	199000	171000	243000	0.73	1.01	1.43
Cr	3650	4310	3800	4550	2390	2400	0.64	0.91	0.56
Mn	1440	1470	2740	3920	2050	1960	0.31	1.1	0.64
K	295	394	910	799	353	235	0.28	0.76	0.26
Na	3310	4670	7100	6730	5900	10300	0.38	0.73	1.3

TABLE IV-11 *Continued*

	Allende		*Semarkona*		*Qingzhen*		E^i_{Al}		
	CV3		*LL3*		*EH3*		*CV3*	*LL3*	*EH3*
	Bulk	*Chondrules*	*Bulk*	*Chondrules*	*Bulk*	*Chondrules*			
Siderophiles and chalcophiles									
Os	0.86	0.91	0.45	0.26	0.75	0.1	0.73	0.28	0.13
Ir	0.84	0.78	0.26	0.14	0.67	0.06	0.67	0.16	0.08
Ru			0.76	0.53	1.04	0.26		0.4	0.23
Co	662	446	505	157	880	55	0.35	0.16	0.07
Ni	13800	9330	12000	3380	16900	1110	0.34	0.17	0.06
Fe	233000	131000	153000	111000	279000	39900	0.28	0.32	0.14
Au	0.15	0.1	0.12	0.05	0.28	0.18	0.27	0.2	0.77
As	1.6	0.83	1.4	0.65	3.7	0.37	0.18	0.19	0.12
Ga	6.2	3	4	2.4	18.3	3.7	0.12	0.13	0.23
Se	8.4	10.4	12.9	4.8	21	12.6	0.21	0.13	0.4
Zn	102	65	40	13	94	94	0.08	0.02	0.19
Br	1.7	2.6	(0.6)	0.6	3	9.2	0.28	0.09	1.58

Sources: Allende (Rubin and Wasson, 1987), Semarkona (Grossman and Wasson, 1983, 1985), Qinzhen (Grossman et al., 1985). Additional data in parentheses and for bulk rocks are based on Huss et al. (1981), Wang and Xie (1977), and Table IV-5.

Notes: The elements are listed in decreasing order of $T_{C,50}$ in each group. All compositions in ppm.

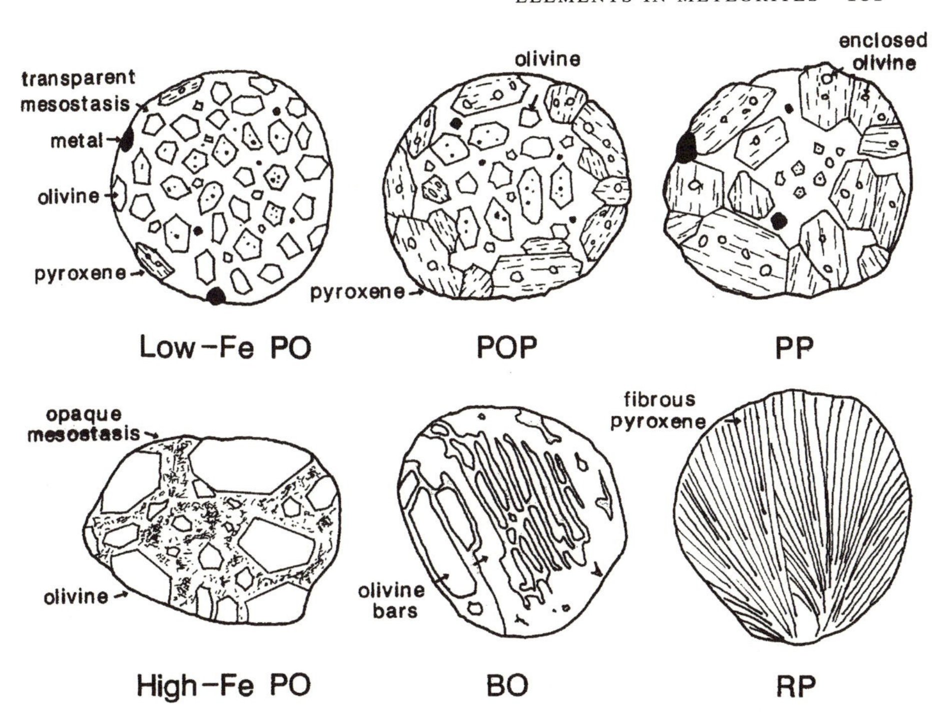

FIGURE IV-14. Schematic sketches of low-Fe porphyritic olivine (PO), porphyritic olivine pyroxene (POP), porphyritic pyroxene (PP), high-Fe porphyritic olivine (PO), barred olivine (BO), and radial pyroxene (RP) chondrules. The sketches are mainly based on Scott and Taylor (1983) and Nagahara (1983).

ically shows some of those chondrule types. High-Fe porphyritic olivine (PO) chondrules contain large euhedral olivines with high-Fe contents (Fa 15-50) embedded in translucent to opaque mesostasis (i.e., the glassy to crypocrystalline interstitial material). Pyroxene phenocrysts are absent, and opaque Fe-Ni metal is sparsely distributed. Zoning within individual olivines is common, with rims enriched in Fe, Ca, and Na. Low-Fe porphyritic olivine (PO) chondrules contain euhedral olivines (Fa 0-10) embedded in transparent mesostasis. Opaque Fe-Ni metal and troilite droplets are common in both olivine and mesostasis. A few low-Ca pyroxenes may be located at the chondrule's edges. Occasionally, mesostasis becomes sparse and olivine anhedral. Porphyritic pyroxene (PP) chondrules contain pyroxenes which enclose rounded olivine grains (i.e., poikilitic texture). Small olivines also exist as euhedral phenocrysts in the core of chondrules. Porphyritic olivine-pyroxene (POP) chondrules are an intermediate type between PO and PP. Radial pyroxene (RP) chondrules contain fanlike arrays of fibrous pyroxene crystals and glassy mesostasis radiating inward from one or more points near the edge of the chondrule. Barred olivine (BO) consists of one to a few sets of olivine lamellae or bars. Translucent glass fills in the volume between the olivine lamellae. Each set of olivine lamellae has the same optical orientation. The barred olivine (BO) chondrules show the whole range of Fe content exhibited by low-Fe and high-Fe porphyritic olivine chondrules in both CC and OC. As reviewed by Rubin (1986), the mesostasis of chondrules is of feldspathic composition and very much enriched in Al, Ca, Ti, Na, and K, as compared to the olivine and pyroxene of chondrules and bulk chondrules (Table IV-12). Alexander (1994) further showed that lithophile Rb, Cs, Sr, Ba, Y, REE, Hf, Zr, Nb, and P are also enriched in feldspathic mesostasis. The mineral assemblages of the Ca-Al-rich chondrules (CAC) are similar to those of CAI (usually anorthite, spinel, mellilite, diopside, olivine, etc.) and will be discussed later along with CAI. The relative abundances of various chondrule types in CV, OC, and EH are summarized in Table IV-9b.

Many experiments have been conducted to simulate the formation of chondrule textures. These experiments (well summarized by Hewins, 1988) show that the barred olivine (BO) and radial pyroxene (RP) chondrule textures can be reproduced by complete melting of the chondrule precursor material at about zero to 20°C above the liquidus. Meanwhile the porphyritic olivine (PO), porphyritic olivine-pyroxene (POP) and porphyritic pyroxene (PP) chondrule textures can be reproduced by incomplete melting at about zero to 30°C below the liquidus, and by subsequent cooling of the melts at rates about 100 to 2000°C/hr (Grossman et al., 1988; Alexander, 1994). The liquidus temperatures of chondrule precursor materials range from about 1500 to 1600°C (the higher the Fe and Na content, the lower the liquidus temperature). Because glassy chondrules (which can easily be produced by superheating of melts) are rare in nature, transient heating events for chondrule formation should rarely exceed a temperature above 1600°C. Also, the finite cooling rate of 100 to 2000°C/hr indicates that there were sufficient numbers of chondrules formed per unit volume of space to retard radiative cooling (the radiative cooling rate of a molten droplet in vacuum would be about 10^6°C/hr). The precursor materials of CAI composition melt at a temperature range similar to regular chondrules but should be cooled at a very slow rate of about 10°C/hr to reproduce the mineral texture of CAC (Ca-Al-rich chondrules). Therefore, CAC had to be formed in quite a different environment from regular chondrules. Because the porphyritic olivine, porphyritic olivine-pyroxene, and porphyritic pyroxene chondrules (PO, POP, and PP) were not completely molten, those chondrules may contain some chondrule precursor materials that survived the melting processes (i.e., so-called relict grains). For example, some olivine crystals in low-Fe porphyritic olivine (PO) chondrules have numerous small metal blebs around their centers and appear "dusty" in transmitted light. These dusty olivine

TABLE IV-12

Average compositions of bulk chondrules; and olivine, low-Ca pyroxene, and mesostasis (glass) from chondrules of Allende (CV), Semarkona (LL3), and Qingzhen (EH3)

	Si	*Ti*	*Al*	*Cr*	*Fe*	*Mn*	*Mg*	*Ca*	*Na*	*K*
Allende										
chondrule	194	1.12	21.8	*4.3	*131	*1.47	206	19	4.7	0.39
olivine	193	0.32	0.53	1	37.4	0.8	*323	2.2		
pyroxene	*266	1.3	6.4	*5.5	23.6	1	221	8.2	0.2	
mesostasis	*214	*3.4	*101	2.8	39.3	0.9	79.6	*72.2	*34.7	*3.2
Semarkona										
chondrule	199	0.84	16.2	*4.5	*111	*3.9	218	17	6.7	0.8
olivine	191	< 0.24	< 0.21	3.7	60.3	2.9	*302	1.1		
pyroxene	*268	0.5	3	*5.2	42.5	2.5	201	4.4	(0.6)	
mesostasis	*278	*2.4	*105	0.9	25.3	2.4	16.7	*67.1	*24.8	*2.7
Qingzhen										
chondrule	243	*0.57	13.9	*2.4	*40	*2.0	217	*12.3	10.3	0.24
olivine	202	< 0.24	< 0.21	1.2	6.4	0.9	*333	1.6		
pyroxene	*277	0.3	1.4	*2.1	6.2	1	233	2.9		
mesostasis	*307	0.3	*78.9	0.8	13	0.4	28.5	*12.2	*60.6	*1.6

Sources: Chondrules from Table IV-11; others are summarized by Rubin (1986).

Notes: Asterisks represent the highest concentration value(s) among the bulk chondrule, olivine, pyroxene, and mesostasis. All compositions in mg/g.

grains are often considered as the relict grains. Smaller dusty olivine grains are also commonly poikilitically enclosed in pyroxene of porphyritic olivine-pyroxene (POP) and in porphyritic pyroxene (PP) chondrules (Grossman et al., 1988). The heating event(s) of chondrule precursors might be related to some kind of nebular shock wave in the early solar system (Boss, 1996; Hewins, 1997).

On the basis of normative plagioclase compositions of chondrules in the CC and OC, Ikeda (1980, 1982, 1983) classified the chondrules into IP (intermediate plagioclase, An 30-80), SP (sodic plagioclase, An 0-30), N (nepheline and sodic plagioclase), and G (glassy) types. Figures IV-15a to IV-15f summarize the factor analysis results for these chondrule types and matrix data from Antarctic chondrites ALH-77003 (CO3) and ALH-764 (LL3). There are three distinct factors in ALH-77003 (CO3): Al + Na + Ca + Si (factor 1), Fe (negative factor 1), and Mg (factor 2), representing, respectively, feldspathic mesostasis, metal + sulfides, and olivine components in the chondrules. Except for a few chondrule samples, the chemistry among IP and SP chondrules and matrix are quite distinct (Figure IV-15b).

For ALH-764 (LL3), there are four distinct groups of elements, i.e., Al+ Na + K, Mg, Fe, and Si + Cr (Figure IV-15c and IV-15e), representing, respectively, feldspathic mesostasis, olivine, metal + sulfides, and pyroxene components in the chondrules, as also shown in Table IV-12. A mixture of these four components can explain the compositional changes among chondrules and matrices in ALH-764 (LL3). These four components might have preexisted before or during the aggregation of chondrules (Rubin and Wasson, 1988). The IP and SP type chondrules and matrices again occupy distinctive fields in Figures IV-15d and IV-15f. The G (glassy) type chondrules are indistinguishable from the SP type, while N (nepheline) type chondrules are very much enriched in the feldspathic component and depleted in the Fe component (Figures IV-15d and IV-15f). The IP and SP types roughly correspond to the Fe-poor type I and the Fe-rich type II chondrules of McSween (1977) and Jones (1994, 1996). Additional factor analysis of chrondrule data from Allende (CV3) and Semarkona (LL3) (Rubin and Wasson, 1988; Grossman and Wasson, 1983, 1985) indicates that REE, Sc, and V also belong to the feldspathic mesostasis component; Os, Ir, Ni, Co, Se, Zn, Ga, and probably As and Au belong to metal + sulfide components; and Mn belongs to pyroxene.

Oxygen Isotopes and Origin of Chondrules

The oxygen isotopic data of chondrules from unequilibrated ordinary chondrites (UOC; Clayton et al., 1983a, 1985, 1991), enstatite chondrites (EC; Clayton and Mayeda, 1985), and Allende (CV3) along with its chondrule rims (Clayton et al., 1983a, 1987) are summarized in Figure IV-16. Each group of

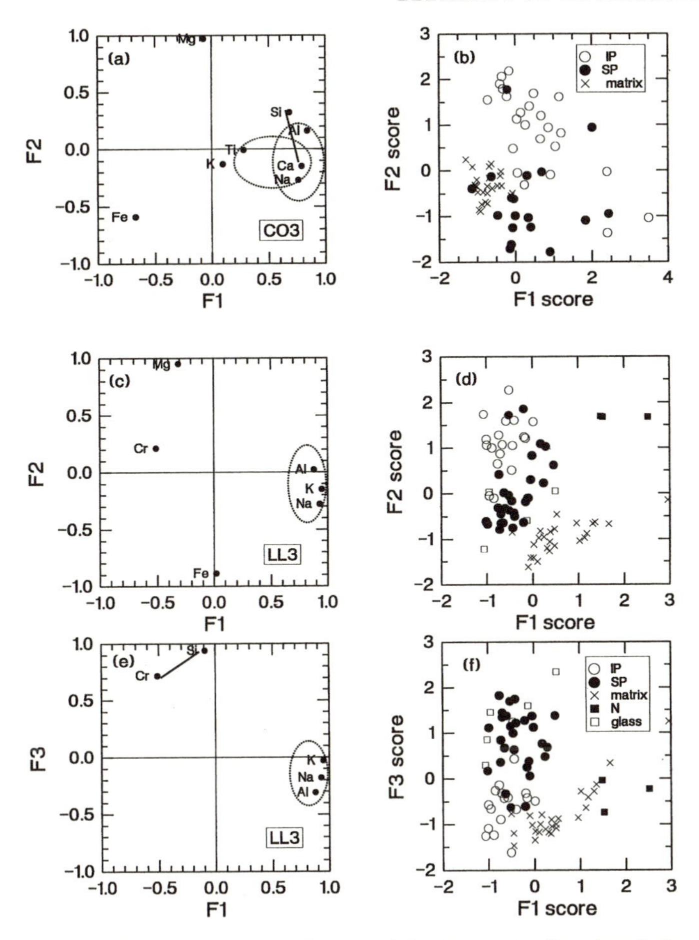

FIGURE IV-15. Plots of factor loadings and factor scores for chondrules (IP, SP, N, and G types) and matrix data from the Antarctic chondrite ALH-77003 (CO3) [diagrams (a) and (b)], and ALH-764 (LL3) [diagrams (c) to (f)] by Ikeda (1980, 1982, and 1983).

chondrules occupies a distinct field in Figure IV-16. The obvious conclusion is that the chondrules in each group must have formed in a reservoir with a distinct oxygen isotopic composition, and that there were no discernible mixing among chondrules produced in different reservoirs. Therefore, any apparent similarity of texture, mineralogy, and chemistry among chondrules from different groups may not necessarily imply their common origin.

The data points for UOC chondrules fall mostly on the ordinary chondrite mixing line, and some fall slightly to the right, as in the case for the bulk ordinary chondrites (compare Figure IV-16 and Figure IV-5). However, there

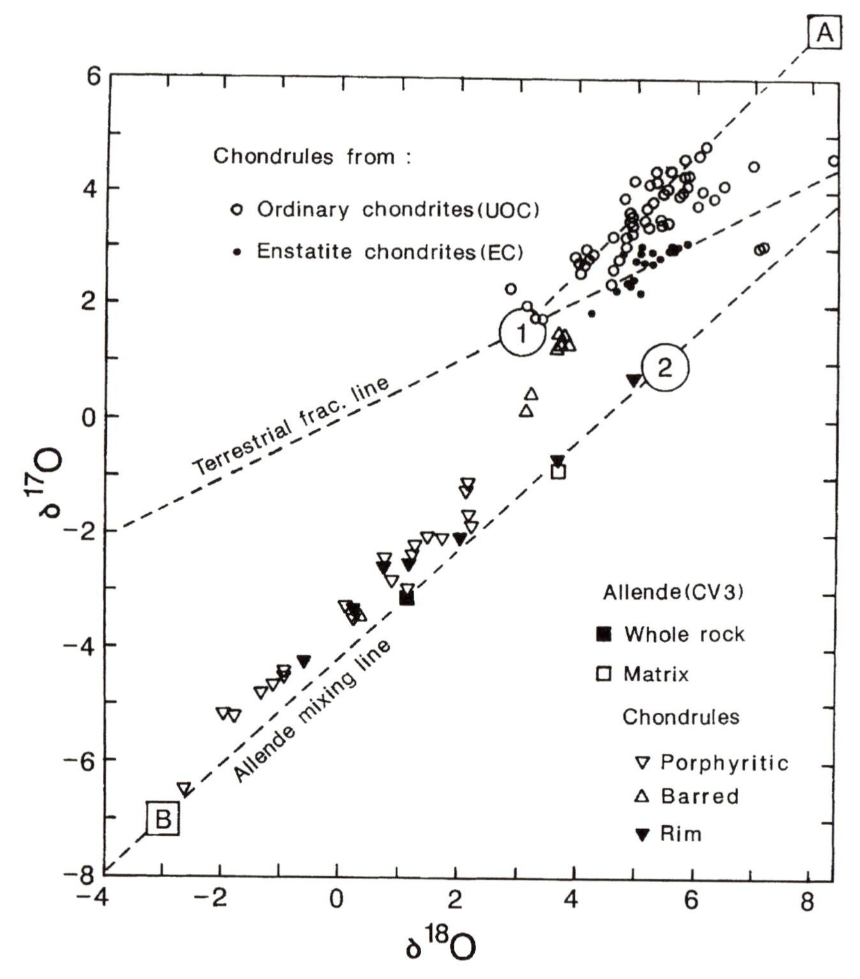

FIGURE IV-16. Oxygen isotope variations of various components from Allende (CV3) and chondrules from ordinary and enstatite chondrites. Big circles and squares represent the possible gas and solid end members, respectively. Data sources: Clayton et al. (1981, 1983a, 1987), Clayton and Mayeda (1985), and Mayeda et al. (1987).

is no distinction among H, L, and LL chondrules with regard to their oxygen isotope composition, whereas the bulk H, L, and LL chondrites show distinct different average oxygen isotope compositions. The implication is that the H, L, and LL chondrites derived their chondrules from a common pool but incorporated different amounts of ^{16}O-enriched components (Clayton et al., 1991). The chondrules of enstatite chondrites occupy the same oxygen isotope field as the bulk enstatite chondrites. In contrast, the chondrules from Allende all lie slightly above the Allende mixing line (Figure IV-16). Note that the barred chondrules (indicative of complete melting) approach the UOC chondrule field. Because melting is supposed to enhance the isotopic exchange between gas and solid or liquid phases, the oxygen isotopes of the barred chondrules of Allende may suggest the existence of a gas reservoir with oxygen isotopic composition around number 1 in Figure IV-16.

The variation of oxygen isotopes in the UOC chondrules can be explained by various degrees of isotopic exchange between chondrule precursors, which might have isotopic compositions similar to an anomalous inclusion in ALH-76004 (LL3; point A in Figure IV-16), and the gas reservoir number 1 during the chondrule formation event. For the formation of Allende chondrules, one needs the additional gas reservoir number 2, which is around the convergent point among the Allende and FUN inclusion mixing lines and the CM matrix fractionation line (Figures IV-5, and IV-6). First, the chondrule precursor materials, which might have a hypothetical isotopic composition at point B in Figure IV-16, exchanged oxygen isotopes with gas reservoir number 1 during the chondrule formation event. Then the newly produced chondrules came into contact with gas reservoir number 2. Another minor heating event might cause isotopic exchange between gas reservoir number 2 and the chondrules as well as chondrule rims. As shown by Clayton et al. (1987), the coarse-grained rims of Allende chondrules are always ^{16}O depleted relative to the enclosed chondrules, and form a line intersecting with gas reservoir number 2 (Figure IV-16).

The oxygen isotopic composition of gas reservoir number 1, in principle, could be formed by a proper mixture between the ^{16}O-enriched end member $\delta^{17}O = -52.2‰$ and $\delta^{18}O = -50.9‰$ (Figure IV-6) and the end member A in Figure IV-16. Meanwhile, gas reservoir number 2 and the hypothetical chondrule precursor reservoir B could be formed by mixing the ^{16}O-enriched end member and another end member similar to Nilpena carbonaceous matrix clasts (Figure IV-5). Of course, the possibility also exists that the hypothetical chondrule precursor reservoir B was the ^{16}O-enriched end member itself. One should be aware that the above mixing model is only a possible scenario, and there is no unique solution to the formation of various chondrules.

Additional reviews on chondrules are given by Hewins et al. (1996).

IV-6. Ca-Al-Rich Inclusions in Chondrites

Mineralogical Classification

Ca-Al-rich inclusions (CAI) are mineral aggregates with diameter on the order of mm and cm, and are highly irregular to nearly spheroidal in shape. They all contain high-condensation-temperature minerals such as hibonite $CaAl_{12}O_{19}$, spinel $MgAl_2O_4$, melilite $Ca_2[MgSi, Al_2]SiO_7$, perovskite $CaTiO_3$, fassaite (diopside enriched in Al and Ti), anorthite $CaAl_2Si_2O_8$, and their alteration products such as nepheline $NaAlSiO_4$, sodalite $Na_4Al_3Si_3O_{12}Cl$, grossular $Ca_3Al_2Si_3O_{12}$, anorthite, andradite $Ca_3Fe_2Si_3O_{12}$, and hedenbergite $CaFeSi_2O_6$. The oxygen isotopes of these mineral components all fall more or less on the Allende mixing line, and a few fall on the FUN inclusion mixing lines (Figure IV-6).

According to the grain size of mineral aggregates, CAI can be divided into the fine-grained (μm range) and coarse-grained (mm to cm range) classes (Grossman and Ganapathy, 1976; Grossman et al., 1977). The common fine-grained CAI are the amoeboid olivine-rich inclusions (AOI) and the spinel-rich inclusions (FSI). AOI are irregularly shaped aggregates of micron-size olivine (Fo 60-90), high-Ca pyroxene, and feldspathoids, and contain spheroidal to irregularly shaped nodules of 5 to 300 μm in size. The nodules consist of spinel, melilite, fassaite, perovskite, diopside rim, and their alteration products. For example, the gehlenitic melilites ($Ca_2Al_2SiO_7$) are often readily converted into grossular ($Ca_3Al_2Si_3O_{12}$), anorthite ($CaAl_2Si_2O_8$), and nepheline ($NaAlSiO_4$) (Hashimoto and Grossman, 1987) by

$$2Ca_2Al_2SiO_7 + 3SiO_2(g) \rightarrow Ca_3Al_2Si_3O_{12} + CaAl_2Si_2O_8,$$
$$3Ca_2Al_2SiO_7 + Na_2O(g) + 5SiO_2(g) \rightarrow 2NaAlSiO_4 + 2Ca_3Al_2Si_3O_{12}; \quad \text{(IV-5a)}$$

fassaite into phyllosilicates and ilmenite ($FeTiO_3$); spinel ($MgAl_2O_4$) into olivine (Mg_2SiO_4) and nepheline (or phyllosilicates) by

$$MgAl_2O_4 + Na_2O(g) + \frac{5}{2}SiO_2(g) \rightarrow \frac{1}{2}Mg_2SiO_4 + 2NaAlSiO_4; \quad \text{(IV-5b)}$$

perovskite ($CaTiO_3$) into ilmenite ($FeTiO_3$), wollastonite ($CaSiO_3$), and hedenbergite ($CaFeSi_2O_6$) by

$$CaTiO_3 + FeO(g) + SiO_2(g) \rightarrow FeTiO_3 + CaSiO_3,$$
$$CaSiO_3 + FeO(g) + SiO_2(g) \rightarrow CaFeSi_2O_6; \quad \text{(IV-5c)}$$

and diopside ($CaMgSi_2O_6$) into hedenbergite and olivine or clinopyroxene ([Ca, Mg, Fe]SiO_3) by

$$CaMgSi_2O_6 + FeO(g) + \frac{1}{2}SiO_2(g) \rightarrow CaFeSi_2O_6 + \frac{1}{2}Mg_2SiO_4,$$
$$CaMgSi_2O_6 + FeO(g) + SiO_2(g) \rightarrow (Ca, Mg, Fe)(SiO_3)_3. \quad \text{(IV-5d)}$$

All the above reactions indicate the addition of Si, Na (as well as K), Fe, and H_2O from the gas phase to solid phases after the formation of Al-Ca-rich nodules.

The fine-grained spinel-rich inclusions (FSI) consist of irregularly shaped spinel grains which are often rimmed with Ca pyroxene and dispersed in a porous matrix of grossular, anorthite, and feldspathoids (i.e., alteration products of melilite). Spinel grains may intergrow with melilite, hibonite, and perovskite. Other rarer fine-grained inclusions are porous aggregates of hibonite crystals ($CaAl_{12}O_{19}$).

The common coarse-grained CAI are the melilite-rich type A, fassaite-rich type B, anorthite-rich type C, and Ca-Al-rich chondrules (CAC). The type A CAI is characterized by abundant melilite with some spinel ± perovskite ± hibonite, and only minor fassaite (MacPherson and Grossman, 1984). Melilite can be a porous network ("fluffy" type) or compact patches ("compact" type). The "fluffy" type A CAI may never have been totally molten after their formation. The inferred crystallization sequence from petrography for the "compact" type A is in the order of spinel and then melilite.

Type B CAI is characterized by abundant fassaite, melilite, and primary anorthite, with some spinel ± perovskite ± hibonite. Type B can be subdivided into types B1, B2, and B3 (Wark and Lovering, 1982; Wark et al., 1987). In general, the abundance of melilite decreases while the **akermanite** component ($Ca_2MgSi_2O_7$) in the melilite increases from type B1 to type B3. The TiO_2 content in fassaite decreases from type B1 to type B3. The inferred crystallization sequence from petrography is in the order of spinel, melilite, anorthite, and fassaite (or the order of anorthite and fassaite is sometimes reversed) for most of type B1 and B2. For type B3, the sequence is in the order of spinel, fosterite, fassaite, ±melilite, ±anorthite (Wark et al., 1987).

Type C CAI contain 30 to 60 vol% anorthite and the rest consists of fassaite, melilite, and spinel (Wark, 1987). The crystallization sequence is spinel, anorthite, melilite, and fassaite (or the order of melilite and fassaite is reversed). The Ca-Al-rich chondrules (CAC) in carbonaceous chondrites are characterized by sodic plagioclase laths (An 47-95), spinel ± pyroxene, and interstitial olivine. Fassaite is absent or minor. The crystallization sequence is spinel, plagioclase, and olivine (Wark, 1987).

In summary, the inferred crystallization sequence for coarse-grained CAI is always spinel first, then melilite, which is in the reverse order predicted by the equilibrium condensation model (Figure III-10). As discussed by Stolper (1982), the observed crystallization sequences strongly suggest that the precursors of those CAI underwent partial to total melting.

Chemical Compositions

Composition data of CAI type A (Wark, 1981), types B1 to B3 (Wark and Lovering, 1982; Wark et al., 1987), type C (Wark, 1987) and Ca-Al-rich chondrules (CAC; Bischoff and Keil, 1983), and FSI (Mason and Taylor, 1982) from the Allende (CV3) chondrite; and AOI (Ikeda, 1982) from

ALH77003 (CO3) chondrite are all plotted in the ternary diagrams of Figures IV-17a and IV-17b. All elements and oxide concentrations were converted to mole %. The mole % of SiO_2 is approximately corrected for SiO_2 added during the formation of secondary minerals by the formula $SiO_2(\text{corrected}) = SiO_2 - FeO - 4Na_2O$, where SiO_2, FeO, and Na_2O are the measured mole % of those species in the bulk CAI. As already shown in equations IV-5a to IV-5d, addition of one mole of FeO is always associated with the addition of one mole SiO_2, and one mole of Na_2O with 2.5 to 5 moles SiO_2 (here 4 is taken as an average). This correction is only signifi-

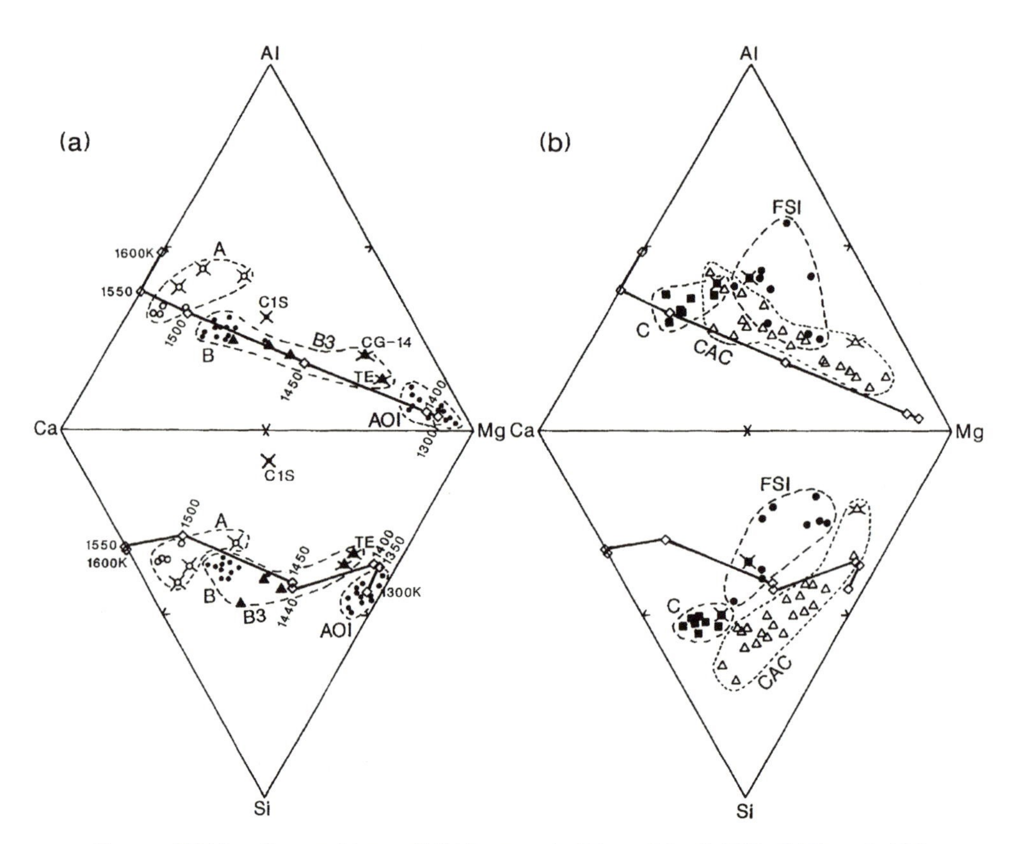

FIGURE IV-17. Compositions of CAI types A, B1 to B3, C, FSI, CAC, and AOI are plotted in the Al-Ca-Mg and Si-Ca-Mg ternary diagrams (all in mole %). The thick solid line in each diagram represents the compositional change of equilibrium condensates from a gas of the average solar composition at $P = 10^{-3}$ atm as a function of temperature (open diamonds) (Grossman and Clark, 1973; Grossman, 1975). Symbols with crosses represent CAI samples that have high modal content of spinel and some are FUN inclusions (C1S, CG-14, and TE).

cant for CAC and type A CAI that have a high content of FeO and Na_2O. Another reasonable assumption is that any CAI as a whole does not gain or lose appreciable CaO, Al_2O_3, and MgO during the alteration process. In Figures IV-17a and IV-17b, the equilibrium condensate compositions from a gas of solar composition at $P = 10^{-3}$ atm as a function of temperature (Grossman and Clark, 1973; Grossman, 1975) are also shown as a thick solid line connecting open diamond symbols. The most interesting result is that the compositional fields occupied by types A, B1 to B3, and AOI are almost identical to the equilibrium condensation trend in both the Al-Ca-Mg and Si-Ca-Mg ternary diagrams (Figure IV-17a). One may conclude that the precursors of the type A and B CAI and AOI most likely represent a series of equilibrium solid condensates from the solar nebula at different temperatures, and underwent various degrees of partial melting within the turbulent solar nebula. However, the assumption that CaO, Al_2O_3, and MgO do not gain or lose much during the alteration process needs further study.

Symbols with a cross above the equilibrium condensation line in the Al-Ca-Mg ternary diagram (Figure IV-17a) are extra-Al-rich, and have high modal spinel contents (>15%) in the spinel-gehlenite-anorthite ternary diagram of Stolper (1982). These CAI belong either to group II of REE abundance patterns (Martin and Mason, 1974), as will be discussed later, or to the FUN inclusions (C1S, CG-14, and TE).

Type C CAI and Ca-Al rich chondrules (CAC) are parallel to the equilibrium condensation trend in the Al-Ca-Mg diagram but mostly enriched in both Al and Si relative to the equilibrium condensates (Figure IV-17b). The fine-grained spinel-rich inclusions (FSI) are mostly enriched in Al but depleted in Si relative to the equilibrium condensates (Figure IV-17b). The implication is that the precursors of type C CAI, CAC, FSI, and even some type A and B CAI with extra-high Al contents were condensed from gases whose compositions were different from the average solar nebula. The local heterogeneity of solar nebular composition might be produced by some complicated evaporation and recondensation cycles of early condensates within the turbulent convective cells of the solar nebula, as proposed by Morfill et al. (1985). For example, when the condensates in convection cells moved toward the Sun or high-temperature regions, the volatile components would evaporate back into the gas phase. When the gas phase moved toward cooler regions, the less volatile components would condense again. The convection cells also might allow differential movement between the gas phase and solid condensates.

REE and Trace Elements

According to rare earth element abundance patterns, Martin and Mason (1974) classified CAI from the Allende chondrite into three major groups (I, II, III) as shown in Table IV-13 and Figure IV-18a. In Figure IV-18a, the moderately volatile lithophiles, refractory lithophiles, and retractory siderophiles, are arranged separately according to the 50% concentration temperature, except for the REE, which are arranged according to atomic weight. Mason and Taylor (1982) later introduced additional groups V and VI. However, groups V and VI are essentially small variations of group I. As shown in Figure IV-18a, group I (including V and VI here) is characterized by an unfractionated flat REE pattern relative to the bulk Allende or CI, except for occasional small positive anomalies of Eu and/or Yb. Also, all other refractory lithophiles and ultrarefractory siderophiles (Ru, Ir, W, Os, and Re except Mo) are not fractionated ($E^i_{\mathrm{La}} \approx 1$). Group III is similar to group I except for a marked deficiency in the low-$T_{C,50}$ Eu and Yb as well as in other low-$T_{C,50}$ refractory lithophiles such as Ba, Sr, Nb, Ca, and U, and high-$T_{C,50}$ Mo, which will be discussed later. Therefore, group III CAI was probably formed at higher temperatures than group I CAI. Group II is characterized by relatively unfractionated light lanthanides, $E^i_{\mathrm{La}} \approx 1$ for La to Sm) but various degrees of deficiency in high-$T_{C,50}$ heavy lanthanides (Gd to Tm and Lu), and even in other high-$T_{C,50}$ refractory lithophiles and siderophiles, as well as in low-$T_{C,50}$ Ba, Sr, Nb, Ca, U, Eu, and Yb ($E^i_{\mathrm{La}} < 1$). Group II CAI were probably condensed from gas(es) already depleted in high-$T_{C,50}$ elements. Therefore, one can expect to find other kinds of CAI that are highly enriched only in high-$T_{C,50}$ heavy lanthanides and other high$T_{C,50}$ lithophiles and siderophiles in Allende chondrite. The core of CAI-3643 from Allende (Wark, 1986) fits this description and is termed an ultrarefractory CAI.

The variance of groups II and III CAI is exemplified by a hibonite-rich CAI named HAL in the Allende chondrite (Davis et al., 1982): HAL is a member of the FUN inclusions. Its hibonite shows group II and its friable rim group III REE patterns, except that both are also very much depleted in Ce as well as V (Figure IV-18b). The extreme depletion of Ce and V suggests that HAL was condensed from a hot oxidizing gas, because Ce and V are highly volatile under oxidizing condition (Boynton, 1975; Davis et al., 1982). One way to produce a hot oxidizing gas locally is to vaporize chondritic material (of CI composition) into a vacuum at high temperature. In CAI, Mo is often depleted relative to other ultrarefractory siderophiles (i.e., $E^{\mathrm{Mo}}_{\mathrm{Ir}}$; Figure IV-18a), indicative of highly oxidizing environments, with a fugacity of oxygen 10^3 to 10^4 times greater than that of the normal solar nebula (Fegley and Palme, 1985).

Most of the coarse-grained CAI have group I REE patterns. However, they are either group II or FUN inclusions whenever their modal spinel contents are high, as discussed earlier. The fine-grained spinel-rich inclusions (FSI) are mostly group II and a few group III (Martin and Mason, 1974; Mason and Taylor, 1982).

Table IV-13

Average compositions of whole rock, matrix, chondrules and CAI (groups I to III) from Allende (CV3)

	Whole rock	Matrix	Chondrules	CAI I	CAI III	CAI II		Whole rock	Matrix	Chondrules	CAI I	CAI III	CAI II
Al %	1.79	1.32	2.18	15.1	21.5	18.3	S %	2.2					
As	1.6	2	0.83				Sc	11.2	7	13			
Au	0.15	0.16	0.1				Se	8.4	8	10			
Ba	4.4			57	34	18	Si %	16		19.4	14	11.1	13.4
Ca %	1.81	1.73	1.9	19.8	12.5	8.86	Sr	(15.3)			136	35	37
Co	662	660	450				Th	0.06			0.45	0.87	0.46
Cr	3650	3100	4300				Ti	980		1100	7190	8750	3950
Fe %	23.3	24.6	13.1	1.32	2.23	5.44	U	(0.017)			0.09	0.09	0.045
Ga	6.2	6	3				V	100	91	135			
Hf	0.19			1.3	2.7	0.3	W	(0.19)					
Ir	0.84	0.6	0.78	8.9	12	0.56	Y	3.1			23	37	1.8
K	295	250	390	170	415	996	Zn	102	130	65			
Mg %	14.7	15.9	20.6	5.97	7.18	7.66	Zr	(8.3)			56	99	4.2
Mn	1440	1700	1500				REE						
Mo	2			11	7	< 0.7	La	0.44	0.4	0.64	4.1	6.7	8.3
Na %	0.33	0.2	0.47	0.28	0.54	1.71	Ce	1.25			11	14	21
Nb	(0.54)			3.7	4.3	1.4	Pr	0.2			1.6	2.5	3.3
Ni %	1.38	1.5	0.93				Nd	0.91			6.8	12.4	15.5
Os	0.86		0.91	8.2	11	0.41	Sm	0.29	0.19	0.39	2.2	3.7	4.6
P	990						Eu	0.11	0.085	0.12	1	0.72	0.33
Pb	1.4			3	1.1	1.3	Gd	0.43			3.1	5.1	2.4
Pd	(0.71)			1.4	1.2	0.48	Tb	0.07			0.53	0.95	0.38
Pt	(1.25)			14	15	0.78	Dy	0.42			3.7	6.7	1.9
Rb	1.3			0.33	1.2	3.2	Ho	0.12			0.94	1.8	0.15
Re	(0.065)			0.71	0.94		Er	0.31			2.8	5.2	0.22
Rh	(0.25)			1.7	1.2	0.23	Tm	0.05			0.38	0.74	0.72
Ru	(1.13)			13	16	1.2	Yb	0.32			2.6	2.92	0.75

Sources: Allende whole rock, Martin and Mason (1974); matrix and chondrules, Rubin and Wasson (1987); and CAI Mason and Martin (1977), Mason and Taylor (1982).

Notes: All compositions in ppm unless noted otherwise.

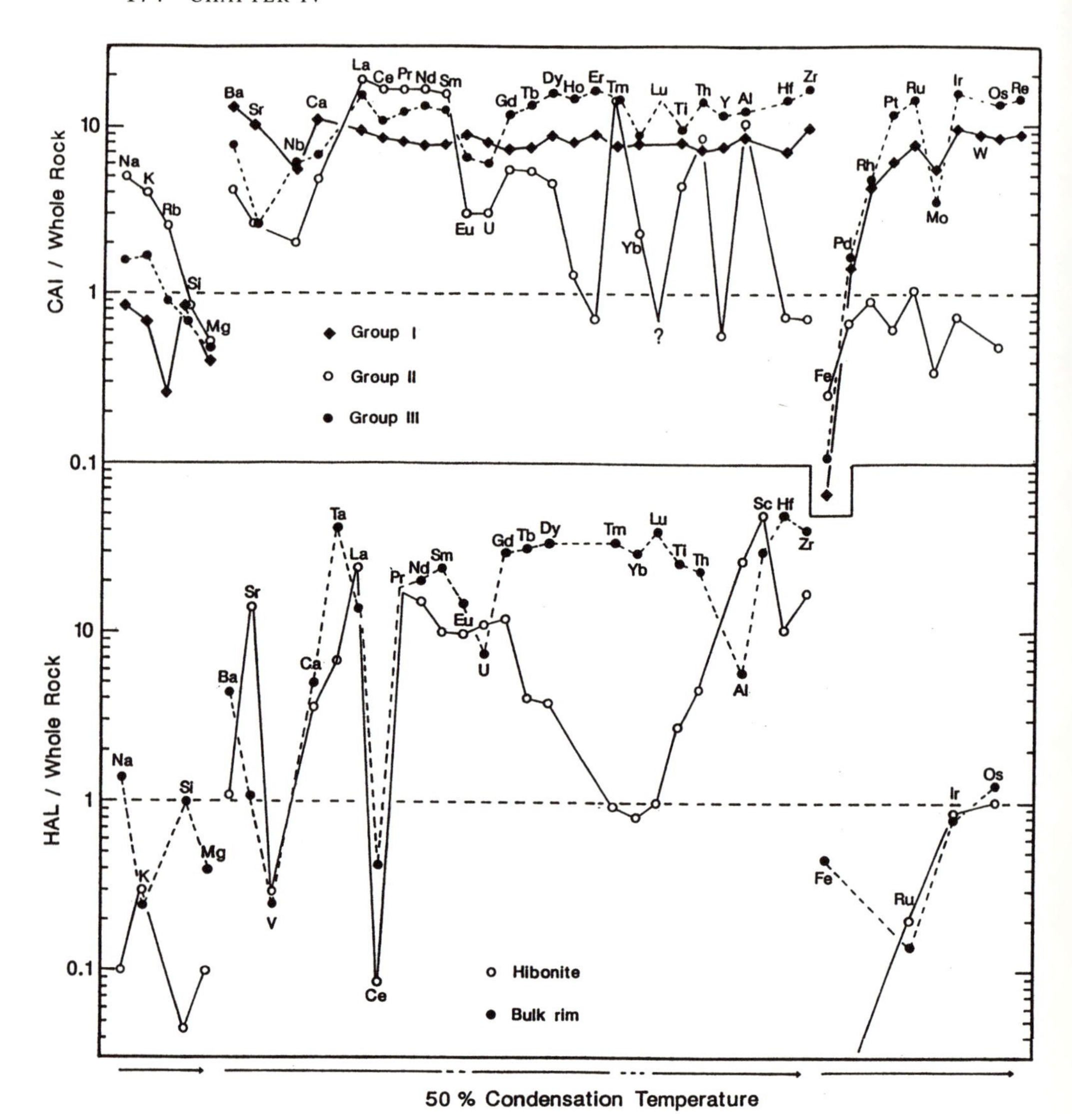

FIGURE IV-18. (a) Concentrations of elements for the groups I to III CAI from the Allende chondrite divided by those for the bulk Allende (Mason and Martin, 1977; Mason and Taylor, 1982). The elements are arranged roughly in increasing order of their 50% condensation temperature (Table III-5) except for the REE, which are in order of atomic weight. (b) A similar plot except the samples are hibonite and its rim separated from the Allende HAL inclusion (Davis et al., 1982).

Evaporation and Recondensation of Early Condensates

Besides oxygen, silicon and magnesium also have three stable isotopes (^{28}Si, ^{29}Si, and ^{30}Si; ^{24}Mg, ^{25}Mg, and ^{26}Mg), and calcium has six stable isotopes. It is instructive to see how these isotopes also vary in CAI and FUN inclusions.

Figure IV-19 summarizes the plots of $\delta^{29}Si$ vs. $\delta^{30}Si$ data (relative to the terrestrial quartz standard NBS-28) for chondrules, CAI and FUN inclusions in Allende given by Clayton et al. (1985). In contrast to oxygen isotope data, most of the silicon isotopic data fall on a mass-dependent fractionation line (i.e., $\delta^{29}Si = 0.5 \times \delta^{30}Si$), except for CG-14 and C1S FUN inclusions, which tend to have slight excess ^{29}Si (Figure IV-19). Also, the extent of fractionation increases in the order of chondrules < CAI < FUN inclusions.

The observed high $\delta^{29}Si$ and $\delta^{30}Si$ values for the FUN inclusions and some CAI could be produced by **Rayleigh evaporation** of preexisting silicate solids. In the Rayleigh evaporation process, the vapor is always preferentially enriched in lighter isotopes relative to the solid, and is continuously

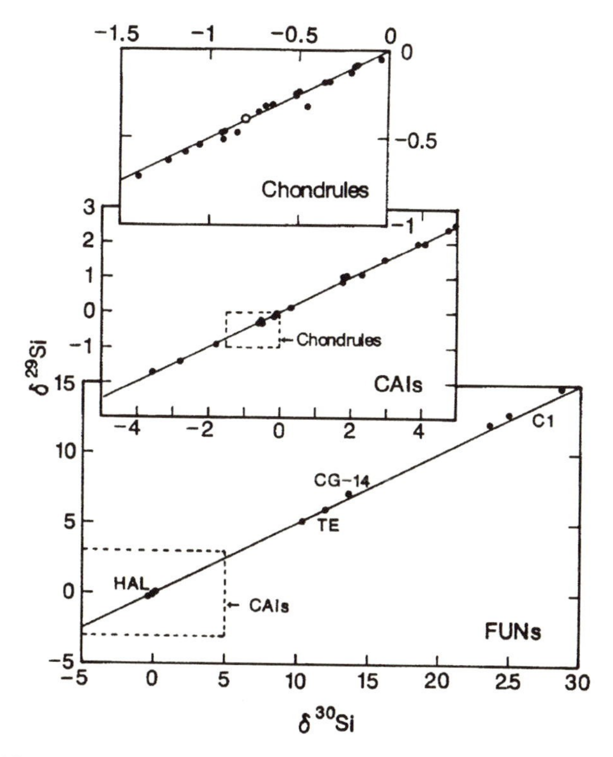

FIGURE IV-19. Variations of silicon isotopes for chondrules, CAI, and FUN inclusions in the Allende chondrite (modified after Clayton et al., 1985).

isolated from the rest of the system as soon as it is formed. Thus, the residual solid becomes steadily enriched in heavier isotopes (Figure IV-20). However, Rayleigh evaporation alone would predict an inverse relationship between silica content and δ^{29}Si (or δ^{30}Si) values in CAI, inconsistent with observation (Clayton et al., 1985). Therefore, the high δ^{29}Si and δ^{30}Si for CG-14 and C1S FUN inclusions must represent the condensates (not residue) from vapors which were already high in δ^{29}Si and δ^{30}Si at the end of Rayleigh evaporation processes as shown in Figure IV-20. The lighter δ^{29}Si and δ^{30}Si for some chondrules and CAI relative to bulk Allende could result from Rayleigh condensation of vapor into solid condensates, i.e., continuous separation of condensates from the vapor or the cessation of interaction between condensates and vapor as soon as the condensates are formed (Figure IV-20).

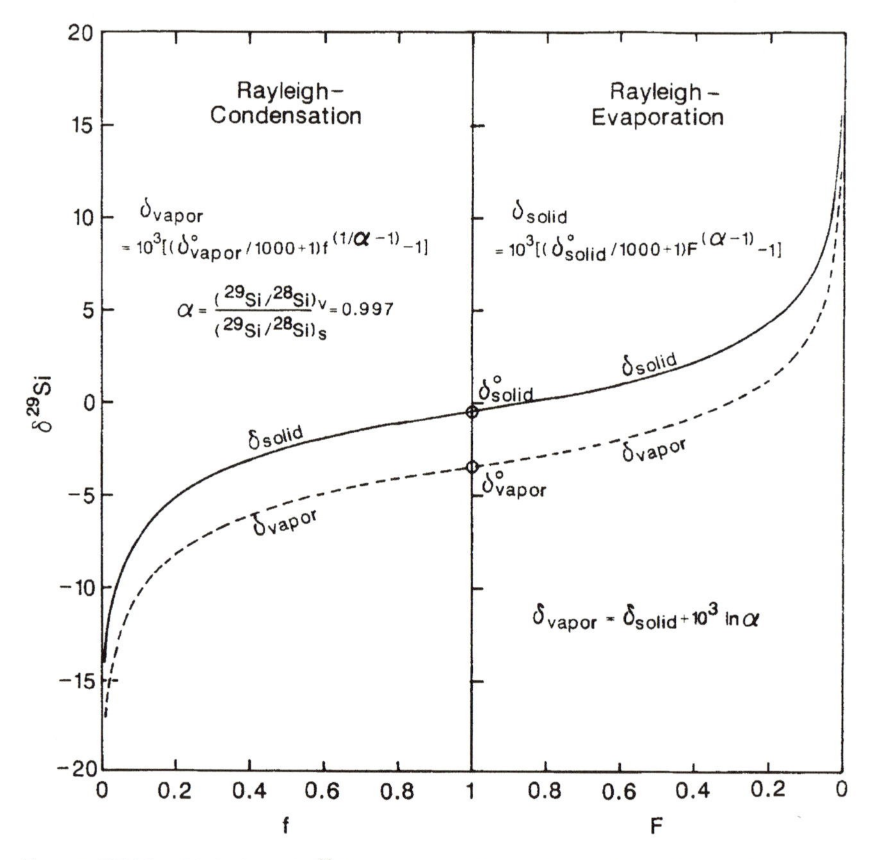

FIGURE IV-20. Variation of δ^{29}Si for solid and gas phases due to Rayleigh evaporation (right) and condensation (left). The fractionation factor α is assumed to be 0.997 and the initial δ^{29}Si for solid and vapor is respectively, −0.04‰ and −3.04‰. f and F are the fractions of vapor and solid remaining.

The derivation of Rayleigh evaporation and condensation equations shown in Figure IV-20 is given in Appendix 1. The Mg and Ca isotopes also all show similar mass-dependent fractionation patterns with some even containing excess ^{26}Mg due to decay of ^{26}Al (Wasserburg et al., 1977; Clayton and Mayeda, 1977b; Niederer and Papanastassiou, 1984; Hinton et al., 1988). The important conclusion is that the solar nebular condensates underwent complicated evaporation and recondensation cycles before finally aggregating into chondrules and CAI (Clayton et al., 1985; Morfill et al., 1985).

Other Features

CAI are often coated by thin multilayered rims (20 to 30 μm in thickness). The rims may consist of (from innermost to outermost layers)

1. spinel + perovskite ± hibonite ± fassaite
2. melilite and its alteration products
3. inner fassaite to outer diopside
4. hedenbergite ± andradite or olivine ± diopside

In reality, rims seldom contain all of these layers (see MacPherson et al., 1988, for a more detailed review). How these CAI rims were formed is still debatable (e.g., Boynton and Wark, 1987). They represent either a second episode of condensation of CAI material or another heating episode of the CAI.

Other unusual objects found in CAI are the so-called **Fremdlinge** (German, meaning "strangers"). Their size is typically 10 to 100 μm. The chemical and mineral compositions of the well-studied Fremdlinge Willy (Armstrong and El Goresy, 1985), Zelda (Armstrong et al., 1987), and A37A (Bischoff and Palme, 1987) are summarized in Tables IV-14a and IV-14b. Armstrong and El Goresy (1985), Armstrong et al. (1987), and Bischoff and Palme (1987) postulated that Fremdlinge formed in the complex environment of the early solar nebula, where various phases of Fremdlinge formed under quite different temperatures and fugacities of O_2 and S_2 (temporally and spatially). They were then mixed together before incorporation into CAI. The main problem with this model is that many low-temperature mineral assemblages in Fremdlinge could not have survived the melting event(s) of the CAI. Also, the observed sharp phase boundaries among Pt-rich, Ru + Os + Re–rich, and Ni+Fe-rich metallic phases could not be maintained during the subsequent cooling of the CAI under any plausible cooling rates (Blum et al., 1989). Blum et al. (1988, 1989) provided the following less complicated explanation for the origin of Fremdlinge. The precursors of Fremdlinge were metal alloy droplets that were formed inside the CAI. The condition of formation was that the CAI cooled down to about 770 K and came in contact with a gas phase that had fugacities of S_2 and O_2 about one and six orders of magnitude above the normal solar gas composition. Thus, the sulfidation and oxidation of metal alloys in

TABLE IV-14
(a) Mineral Compositions (volume %) of Fremdlinge Willy, Zelda, and A37A from Allende (CV3); (b) Average Chemical Compositions of Willy, Zelda and A37A

(a)

Minerals	*Willy*	*Zelda*	*A37A*
Ni-Fe metal (Ni 53–62%)	40	5	44.8
Os nuggests (Os 46–77%)	1	5.1	15.3
Pt-Ir nuggets (Pt 31%; Ir 12%)		2.2	
V-magnetite $FeO{\cdot}(Fe,V)_2O_3$	50	20.5	24.3
Pyrrhotite $Fe_{1-x}(Ni,Co)_xS$	5	28.2	9.9
Pentlandite $(Fe,Ni)_9S_8$		31	
Molybdenite MoS_2	<1	5.2	
Scheelite $Ca(Mo,W)O_4$	3		4.7
Apatite $Ca_5(PO_4)_3Cl$	2		
Whitlockite $Ca_9MgNa(PO_4)_7$		2.4	
Mo-lath (MoO_3 48%; MoS_2 28%)			0.8

(b)

	Willy	*Zelda*	*A37A*
Al	0.31	0.25	0.09
Ca	0.86	0.82	0.43
Co	1.33	0.38	1.06
Cr	0.59	1.46	0.27
Fe	45.7	42.9	29
Ir	0.13	1.49	4.61
Mg	0.29	0.08	0.028
Mo	0.19	2.91	0.64
Na	0.04	0.05	
Ni	30.9	8.56	27
Os	0.92	2.55	2.53
P	0.35	0.48	
Pt	0.24	1.71	1.14
Re	0.07	0.25	2
Ru	0.94	1.59	2.51
S	1.38	23.4	1.8
Si	0.51	0.28	0.043
Ti	0.01	0.08	0.03
V	1.44	3.85	0.74
W	1.79	<0.1	0.74

Source: Willy, Armstrong and El Goresy (1985); A37A, Bischoff and Palme (1987); and Zelda, Armstrong et al. (1987).

Notes: Mineral compositions in volume % and chemical compositions in weight %.

the CAI caused the exsolution of noble metal alloys, Ni-rich taenite, etc., and the formation of various oxides and sulfides in Fremdlinge. In this model the question remaining to be answered is how permeable CAI were with respect to the gases. Certainly further studies are needed.

Finally, primitive carbonaceous chondrites contain a few ppm of fine-grained graphite (1-6 μm), diamond (0.002 μm), silicon carbide (SiC; 0.2–10 μm), and sometimes small titanium carbide (TiC) crystals inside graphite grains as well as a few corundum (Al_2O_3) grains (Ott, 1993, and references therein). All these grains have anomalous isotopic compositions of O, C, N, noble gases, Mg, Si, REE, etc., as compared to the average solar nebula (Ott, 1993; Amari et al., 1993, and references therein). These carbon compounds could have formed only under very reducing conditions ($C/O > 1$) outside the solar system. The most probable places are in circumstellar atmospheres of mass-losing red giant stars, novae, massive supergiant stars (so-called Wolf-Rayet stars), and carbon-rich shells from supernovae (Ott, 1993). These grains became interstellar materials and are thought to have been incorporated into primitive meteorites intact during the formation of the solar system, providing important information on stellar nucleosynthesis. Research on these new materials is advancing rapidly (Nittler et al., 1994; Arnett and Bazan, 1997) and is beyond the scope of this book.

IV-7. Igneous Differentiation in Achondrites and Iron Meteorites

As shown in Table IV-4, the primitive achondrites such as Acapulco and ALH-77081 are more or less unfractionated in their major element compositions relative to CI ($E^i_{Al} \approx 1$). This also holds true for other refractory and moderately volatile lithophiles + siderophiles (Kallemeyn and Wasson, 1985). In contrast, the high-Ca achondrites such as the AMP group and the SNC group are believed to be the end products of igneous magmatic differentiation in asteroidal-size bodies and Mars, respectively (reviewed by McSween, 1989). The low-Ca achondrites such as ureilites and aubrites also indicate significant separation of Fe from achondrite precursors (see Figure IV-2, and McSween, 1989). The following examples give some highlights of these igneous differentiation processes.

Eucrites-Diogenites-Pallasites

The compositions of Juvinas (eucrite) and Johnstown (diogenite) are summarized in Table IV-15 and plotted in Figure IV-21 The lithophile, chalcophile, and siderophile group elements are again arranged separately according to

TABLE IV-15
Compositions of achondrites Juvinas (eucrite), Johnstown (diogenite),and Kenna (ureilite)

	Eucrite (Juvinas) [1]	*Diogenite (Johnstown)* [2]	*Ureilite (Kenna)* [3]		*Eucrite (Juvinas)* [1]	*Diogenite (Johnstown)* [2]	*Ureilite (Kenna)* [3]
Ag-47	0.1	0.011	0.0052[j]	Mo-42	0.015		
Al-13	68800[m]	6500[m]	1220	Na-11	3490[m]	200[m]	225
As-33			0.5[w]	Nd-60	4	0.11	0.011
Au-79	0.0071	0.00087	0.032[j]	Ni-28	1.1	28	1180
Ba-56	31	2.5[ma]	0.054	Os-76	18×10^{-6}	0.00074	0.70[j]
Bi-83	0.0035		0.00043[j]	P-15	402[m]	600[m]	
Br-35	0.16	0.11[ma]	0.15[j]	Pd-46	0.0004	0.001	0.073[j]
Ca-20	74200[m]	10400[m]	9290	Pt-78		0.019	
Cd-48	0.029	0.021	0.013[j]	Rb-37	0.25	0.027	0.004[j]
Ce-58	7.25	0.13	0.047	Re-75	9.7×10^{-6}	60×10^{-6}	0.069[j]
Cl-17	18[ma]	13[ma]		Ru-44			0.84[w]
Co-27	3.3	24	164	S-16	1900[ma]	4300[ma]	1790[g]
Cr-24	2330[m]	5900[m]	5060	Sb-51	0.042	0.0054	0.0098[j]
Cs-55	0.005	0.0008	37×10^{-6}[j]	Sc-21	28[ma]	14[ma]	8.5
Dy-66	3.03	0.14	0.013	Se-34	0.077	0.22	0.49[j]
Er-68	1.85	0.14	0.025	Si-14	231000[m]	251000[m]	200000
Eu-63	0.62	0.0089	0.00033	Sm-62	1.74	0.08	0.0017
F-9	19[ma]	3.5[ma]		Sr-38	78	2.1[ma]	0.16
Fe-26	132000[m]	126000[m]	159000	Ta-73	0.2	0.008	
Gd-64	2.55		0.0054	Tb-65	0.4	0.021	0.0023
Ge-32	0.004	0.047	18	Te-52	0.001	0.005	0.053[j]
Hf-72	1.3[ma]	0.05[ma]		Th-90	0.37[ma]		
Ho-67	0.59	0.036	0.0032	Ti-22	3820[m]	720	
I-53	0.039[ma]	0.025[ma]		Tl-81	0.0011	0.001	62×10^{-6}[j]
In-49		0.0032	0.0021[j]	Tm-69	0.28	0.021	0.0052
Ir-77	28×10^{-6}	0.00067	0.56	U-92	0.057	0.002	0.0001[j]
K-19	340[m]	20[m]	1.3	V-23	96[ma]	147[ma]	100
La-57	2.53	0.044	0.015	W-74	0.027	0.006[ma]	
Li-3	5.1	2.2		Y-39	17[ma]	1.2[ma]	
Lu-71	0.23	0.033	0.008	Yb-70	1.5	0.15	0.046
Mg-12	43800[m]	156000[m]	200000	Zn-30	2.5	0.86[ma]	233[j]
Mn-25	4330[m]	3870[m]	2840	Zr-40	46	1.3	

Sources: [1] Morgan et al. (1978); [2] Wolf et al. (1983); [3] Boynton et al. (1976); [j]Janssens et al. (1987); [m]McCarthy et al. (1972); [ma]Mason (1979); [w]Wasson et al. (1976).
Note: Compositions in ppm.

their 50% condensation temperature, except for the REE, which are arranged according to atomic weight. The noble metal contents in Johnstown are quite variable but only their lowest values are listed in Table IV-15. It is evident from Figure IV-21 that the $E^{i}_{\rm Al}$ values of siderophiles Re, Os, Ir, Co, Ni, Pd, and Ge and chalcophiles Se and Te in the eucrite are in the range of 10^{-4} to 10^{-5}; of siderophiles W, Mo, Au, Ga, and Sb and chalcophiles S, Ag, Zn, Tl, Cd, Cl, Br, and Fe, 0.05 to 10^{-3}; of moderately volatile lithophiles Mg, Si, Cr, P, Mn, K, Na, and Fe, 0.05–0.3; and of refractory lithophiles except V and Sc, all around one (i.e., unfractionated). One possible scenario to produce the above-mentioned pattern was suggested by Takahashi (1983) from melting

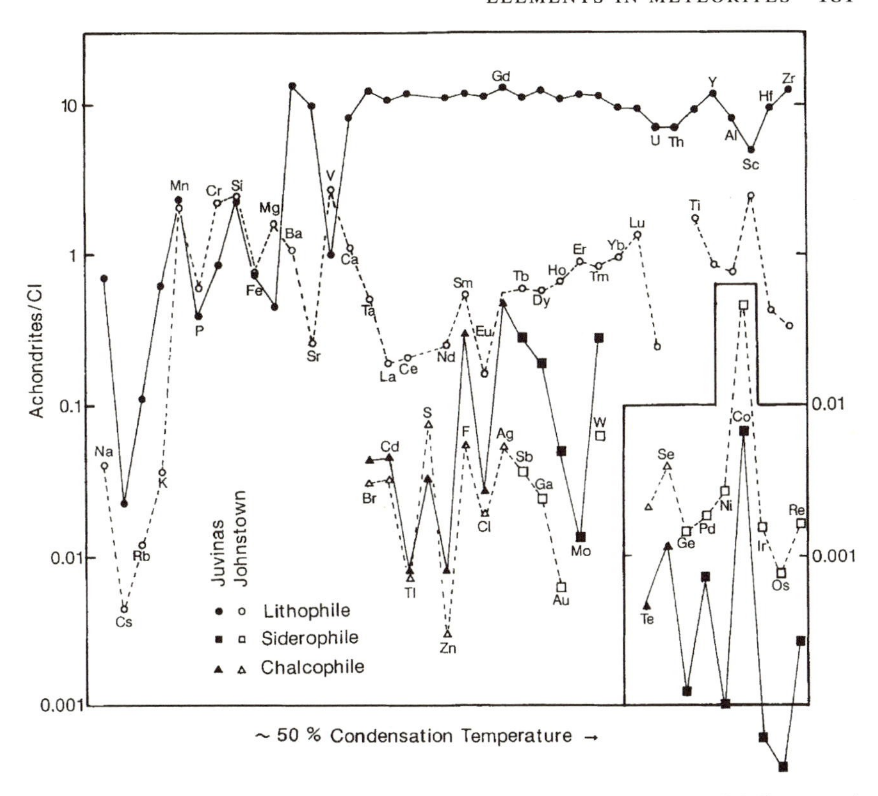

FIGURE IV-21. Compositions of Juvinas (eucrite; connected by solid lines) and Johnstown (diogenite; connected by dotted lines) relative to CI. The lithophile, chalcophile, and siderophile group elements are listed separately in order of increasing volatility, except for the REE, which are arranged according to atomic weight. Data are from Table IV-15.

experiments with chondrites. At a temperature not too far above solidus, the chondritic precursor, which was already low in moderately volatile lithophile concentrations, started to melt partially, resulting in a eucritic silicate melt (which could dissolve up to 0.2 wt% S), metal + sulfide melt, and olivine residues. The olivine residues and metal + sulfide melt became the pallasites, and the silicate melt became eucrites after some segregation and solidification processes. Note that W, Mo, Au, Ga, Co, and Sb are not as much depleted as the other refractory siderophiles, such as Re, Os, Ir, Ni, Pd, and Ge, in the eucrite. As also shown by Kloeck and Palme (1987) and Jones and Drake (1986) in a basaltic melt system, the degree of enrichment of the former elements in the metal + sulfide phase is much less than for the latter elements. According to Figure IV-21, chalcophiles Se and Te are more depleted than

S, and behave more like noble metal siderophiles during the separation of metal + sulfide melts.

The depletion of V and Sc in eucrites is complemented by the enrichment of these elements in diogenites. Also, the refractory lithophiles in diogenite are quite fractionated relative to CI (Figure IV-21). Therefore diogenite was most likely a crystal cumulate from the eucritic silicate melt (McSween, 1989). Figure IV-22 shows the partition ratios of elements between eucrite and diogenite, based on Table IV-15, as a function of ionic radius. The solid lines connect lithophile and chalcophile + siderophile ions with the same charge separately. As a rule, the partition ratios of each group of cations increase with increase of ionic radius in Figure IV-22, as one would expect from Goldschmidt's rule (1954). The rule states that the higher the ionic radius, the higher the tendency for a cation to stay in the melt rather than to be incorporated into crystals. However, the partition ratios become flat or even decrease toward the high end of the ionic radius. One possible explanation is that diogenite was contaminated by the last bit of residual melt that was very much enriched in cations with high ionic radii. One more important

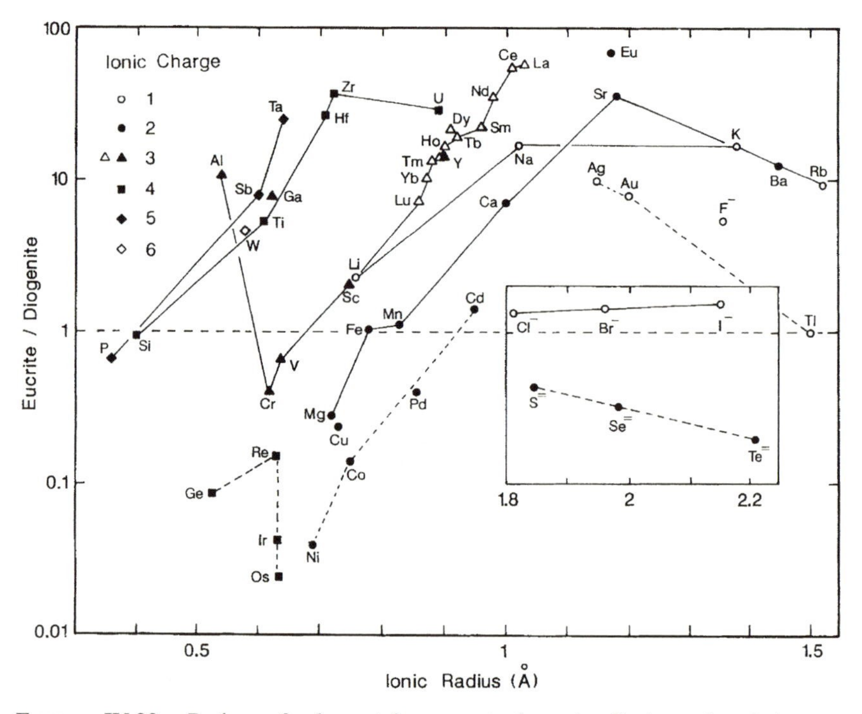

FIGURE IV-22. Ratios of elemental concentrations in Juvinas (eucrite) over Johnstown (diogenite) as a function of ionic radius. The solid and dotted lines connect ions with the same charge and either all lithophile or chalcophile plus siderophile.

exception is Al, which has a smaller ionic radius but a higher partition ratio than Cr. As will be discussed in Chapter V, Goldschmidt's rule needs to be revised, i.e., for ionic radii below a certain point, the cation also prefers to stay in the melt.

Ureilites

Unlike most achondrites, the oxygen isotopes of ureilites do not fall on a mass-dependent fractionation line (compare Figures IV-5 and IV-7), and the fO_2 (fugacity of oxygen) of ureilites has a wide range (Figure IV-3; low fO_2 for group III to high fO_2 for group I ureilites). These facts strongly indicate that group I to III ureilites are not related by any igneous differentiation processes (Clayton and Mayeda, 1988). Figure IV-23 further supports this conclusion: eucrites-diogenites-howardites as well as terrestrial igneous rocks and SNC group achondrites, all of which underwent igneous differentiation, show positive correlations between their MnO and FeO contents in low-Ca pyroxene (Takeda, 1986/87). In contrast, the MnO and FeO contents in low-Ca pyroxene are negatively correlated among group I to III ureilites

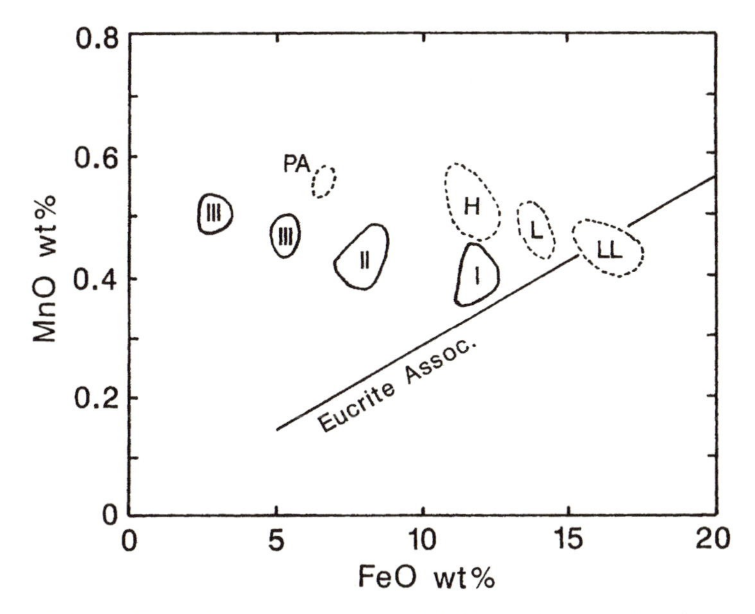

FIGURE IV-23. Plot of FeO (weight %) vs. MnO (weight %) in low-Ca pyroxene from ordinary chondrite (LL, L, H), primitive achondrite (PA), and ureilites group I to III. The solid line represents the correlation trend for the AMP and SNC group achondrites. Data from Takeda (1986/87).

as well as among ordinary chondrites (LL, L, H) and primitive achondrites (PA). However, the ratios of elemental concentrations in the Kenna ureilite (group I; Table IV-15) over that in carbonaceous chondrite (assumed to be the precursor of ureilites, since ureilites are also rich in carbon) again show the nice inverse relationship to ionic radius (Figure IV-24). Therefore, the formation of ureilites should also involve some moderately high-temperature events that would produce a small amount of Fe-FeS eutectic melts and silicate partial melts enriched in Al and cations with large ionic radius. The small amount of silicate melt was subsequently separated from the carbonaceous chondritic precursor (Takeda, 1986/87). The ureilites as residuals would still retain the original heterogeneous signals of oxygen isotopes and MnO/FeO ratios.

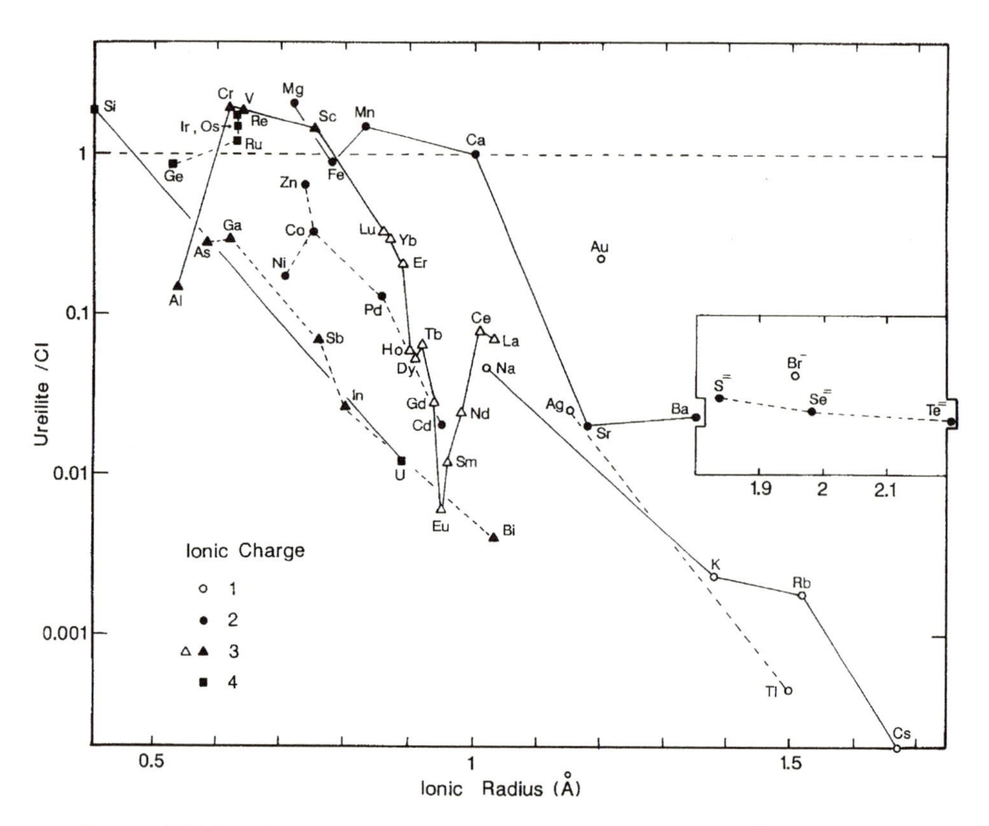

FIGURE IV-24. Elemental concentrations in Kenna (a group I ureilite) relative to CI as a function of ionic radius. The solid and dotted lines connect ions with the same charge and either all lithophile or chalcophile plus siderophile. Data are from Table IV-15.

The increase of partition ratios from Eu to La (Figure IV-24) may also suggest a contamination of ureilite by a last bit of melt that was enriched in the light REE (Boynton et al., 1976). The high temperature melting events might be related to shock heating through asteroidal collisions, which would explain the occurrence of micron-sized diamonds and another high pressure polymorph of carbon (lonsdaleite) in ureilites (Berkley et al., 1976, 1980).

Iron Meteorites

As already shown in Figure III-9, siderophiles with $T_{C,50}$ much higher than that of Fe (i.e., Rh, Pt, Ru, Mo, Ir, W, Os, and Re) were concentrated in the early metal-alloy condensates. At a temperature around 1480 K (in the case of $P = 10^{-3}$ atm), they were virtually completely condensed, and iron along with Pd, Ni, and Co (their $T_{C,50}$ values are similar) started to condense rapidly. Thus, the concentrations of high $T_{C,50}$ siderophiles in metal alloys decreased rapidly around 1480 K, and then slowly as the nebular temperature continued to decrease. In contrast, the concentrations of siderophiles with much lower $T_{C,50}$ than Fe (i.e., Ge, Sb, Ga, Cu, As, P, Au) were extremely low at 1480 K in the metal alloy, but increased rapidly as the temperature decreased to around 800 K, at which point all siderophiles were condensed. Therefore, at 800 K, siderophiles in the metal alloy should have cosmic or solar abundance.

The thick solid curves in Figures IV-25a and IV-25b depict how the concentrations of Ni and Ir (high $T_{C,50}$) decrease and how the concentration of Au (low $T_{C,50}$) increase as the condensation of nebular gas proceeds from the time when Fe and Ni condense rapidly down to the complete condensation of siderophiles as calculated by Kelly and Larimer (1977) for the case of $P = 10^{-5}$ atm. However, iron meteorite data do not fit into this simple condensation path except for group IVB, as is evident in Figure IV-25. The wide changes in Ir concentration (about four orders of magnitude), and the linear relationship between the logarithmic concentrations of Ni versus Ir and Au (Figure IV-25), suggest that iron meteorite groups were products of large-scale magmatic differentiation in asteroid-sized bodies (Kelly and Larimer, 1977).

During or soon after accretion, the asteroidal bodies might have become totally or partially molten by yet undetermined heat sources (e.g., extinct ^{26}Al, intense solar wind heating). By gravitational segregation, the molten Fe and FeS along with associated siderophile and chalcophile elements would tend to sink toward the center of an asteroidal body. If a molten iron core was finally formed, the relative abundance of siderophiles in the molten core should be again close to the cosmic abundance, except for Fe, which might exist in part as a cation in the silicate mantle if fO_2 was relatively high.

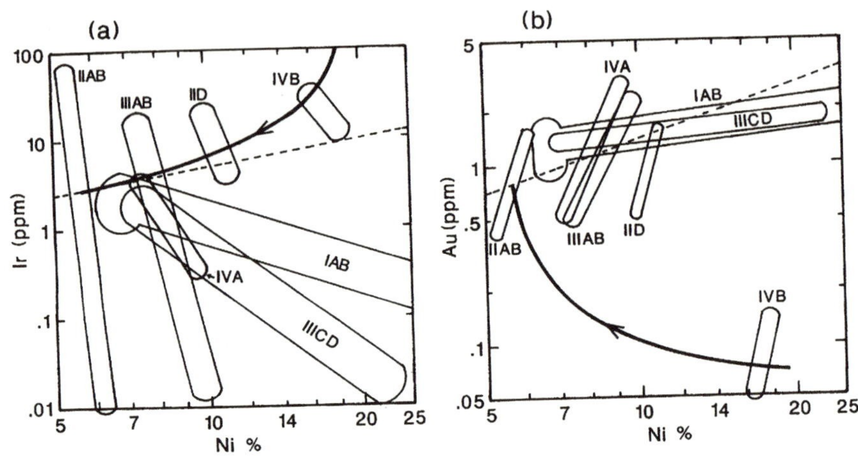

FIGURE IV-25. Compositional fields occupied by different iron meteorite groups in the plots of Ni vs. Ir and Ni vs. Au. The thick solid lines represent the compositional change of metal-alloy condensates as the nebular gas temperature decreases at $P = 10^{-6}$ atm (Kelly and Larimer, 1977). The dashed straight lines with slope of one represent the cosmic Ir/Ni and Au/Ni ratios.

When the temperature of the molten core had fallen to the liquidus temperature, the Fe-Ni alloy called taenite would start to crystallize and precipitate. If taenite solids were continuously segregated from the melt as soon as they were formed, the concentrations of siderophiles in sequential taenites would differ greatly. This is the so-called **Rayleigh fractional crystallization**.

Mathematically,

$$X_i = D_i \cdot C_i^0 \cdot f^{(D_i - 1)} \tag{IV-6a}$$

where X_i = concentration of element i in the solid; C_i^0 = initial concentration of element i in the melt; f = fraction of melt remaining; and D_i = distribution coefficient = X_i/C_i, which is assumed to be constant in this case. The derivation is given in Appendix 1, and the models will be discussed in more detail in Chapter V. Similarly,

$$X_{\mathrm{Ni}} = D_{\mathrm{Ni}} \cdot C_{\mathrm{Ni}}^0 \cdot f^{D_{\mathrm{Ni}} - 1} \text{ for Ni.} \tag{IV-6b}$$

Eliminating f from equations IV-6a and IV-6b, one obtains

$$\ln X_i = A + B \cdot \ln X_{\mathrm{Ni}}, \tag{IV-6c}$$

where $A = \ln(C_i^0 \cdot D_i) - B \cdot \ln(C_{Ni}^0 \cdot D_{Ni}) = \text{constant}$, and

$$B = (D_i - 1)/(D_{Ni} - 1).$$

Therefore a plot of $\ln X_{Ni}$ versus $\ln X_i$ should form a straight line with a slope of B and an intercept of A.

Another effective way to fractionate the siderophiles is by **Rayleigh partial melting**. During partial melting of an asteroid-size body, an Fe-FeS eutectic melt was continuously segregated from the solid phase as soon as it was formed. If each parcel of separated melt resolidified before having a chance to accumulate into a molten metal core, every resolidified melt parcel would have different siderophile concentrations. Mathematically,

$$C_i = X_i^0/D_i \cdot F^{(1/D_i - 1)}, \qquad \text{(IV-7a)}$$

where C_i = concentration of i in the resolidified melt, X_i^0 = initial concentration of i in the solid phase, F = fraction of the solid phase remaining, and D_i = distribution coefficient = X_i/C_i, and

$$\ln C_i = A' + B' \cdot \ln C_{Ni} \qquad \text{(IV-7b)}$$

where A' and B' are again constants and $B' = (D_i - 1) \cdot D_{Ni}/[(D_{Ni} - 1) \cdot D_i]$.

Therefore a plot of $\ln C_{Ni}$ vs. $\ln C_i$ also forms a straight line but with a slope of B'. D_{Ni} in the Fe-Ni binary system is about 0.9 but increases to about one and 1.3 when the dissolved sulfur in the melt increases to 15 and 25 wt%, respectively. D_i values for other siderophiles such as Ga, Ge, Au, W, Ir, and P are also strong functions of S and P content in the melt (Jones and Drake, 1983). For illustration purpose, the effective average D_{Ni} in the iron meteorite system is taken to be 0.9 . Therefore, if $D_i > 1$, the slopes B and B' are both negative, as in the case of Ir in Figure IV-25, and Co, Rh, Pt, Ru, W, Os, Re, and Cr in figures given by Scott (1972), Wasson (1985), and Pernicka and Wasson (1987). If $D_i < 1$, the slopes B and B' are positive, as in the case for Au in Figure IV-25 and Sb, As, P, Pd, and Mo in figures by Scott (1972). As shown in Figure IV-4, D_{Ga} and D_{Ge} must range from slightly less than one to slightly larger than one for most of the iron groups, except for groups IAB and IIICD, whose D_{Ga} and D_{Ge} are definitely larger than one. Kelly and Larimer (1977) demonstrated that the metal data for groups IAB and IIICD can be best explained by the Rayleigh partial melting model, and others by Rayleigh fractional crystallization. Jones and Drake (1983) also showed that iron meteorite data for groups other than IAB and IIICD can best fit into the Rayleigh fractional crystallization model, allowing D_i to be a function of S and P content.

IV-8. Concluding Remarks

Mineralogical, chemical, and isotopic compositions of major constituents of chondrites (e.g., CAI, chondrules, and matrices) reveal their complex physicochemical conditions of formation. The CAI (including FUN inclusions) are a mixture of various precursors that condensed from the nebular gases at high temperatures, and at times under highly oxidizing conditions. The crystallization sequences of high-temperature minerals in CAI often indicate partial to total melting of CAI. The occurrence of multilayered rims around CAI and Fremdlinge in CAI suggest, respectively, some episodic heating events and locally unusual fO_2 and fS_2 conditions in the nebular disk. The observed large mass-dependent fractionation of Si, Ca, and Mg isotopes in CAI further indicates complex evaporation-recondensation cycles of early condensates within turbulent convective cells of the solar nebula.

The chondrules and matrices of ordinary chondrites are mixtures of feldspathic mesostasis, olivine, pyroxene, and metal + sulfide components. These components must have formed from earlier condensates through partial to total melting and gravitational sorting within the turbulent nebular disk. The formation of chondrules and rims around chondrules should be related to complex episodic heating events. The distinct oxygen isotope compositions for various meteorites indicates that the solar nebula was heterogeneous in the distribution of oxygen isotopes. Anomalous isotopic compositions of fine-grained graphite, diamond, and silicon carbide in primitive meteorites may suggest their origin as interstellar dust that survived intact during the formation processes of meteorites.

Factor analysis of chondrite data indicates that carbonaceous chondrites are a mixture of high- and low-temperature condensates. For the formation of ordinary chondrites, these condensates were first differentiated into distinct silicate, sulfide, and metal phases through melting and gravitational sorting, which were eventually aggregated in different proportions to form the L to H type ordinary chondrites. The very existence of enstatite chondrites indicates greatly reducing conditions in some region of the nebular disk.

In short, the condensation model of the solar nebula provides an important thermodynamic reference point, but the actual formation processes of meteorites, and thus of the whole solar system, had to be more complex than a simple unidirectional cooling of the solar nebula.

Chapter V

IGNEOUS ROCKS AND THE COMPOSITION OF THE EARTH

INTRODUCTION

TO FAMILIARIZE readers with various igneous rock types mentioned in this chapter, the classification scheme for igneous rocks is introduced first. This chapter discusses how the average composition of the Earth's crust, mantle, and core are estimated, and presents probable scenarios for the formation of a stratified Earth. We must explain both (1) the observed partition patterns of elements among the continental crust, oceanic crust, mantle, and core, and (2) the systematic changes in chemical composition of various igneous rock types. Partial melting and fractional crystallization models are presented, with special focus given to the concept of elemental partitioning between solid and liquid phases as described by the distribution coefficient. The heterogeneity of the mantle with regard to isotopic and elemental composition is illustrated by examples.

V-1. CLASSIFICATION SCHEME FOR IGNEOUS ROCKS

Appendix Table A-4 summarizes the common minerals found in igneous rocks (Deer et al., 1992). Additional mineral information can be found in Appendix Table A-3.

The plutonic igneous rocks are classified according to the relative abundance of actual mineral assemblages, namely, Q = quartz + tridymite + cristobalite; A = alkali feldspar, including albite (An_0–An_5; i.e., anorthite content of zero to 5 mole % in plagioclase); P = plagioclase (An_5–An_{100}); F = feldspathoids or foids (nepheline, leucite, sodalite, etc.); and M = mafic minerals (e.g., mica, amphibole, pyroxene, olivine, etc.)+ accessory minerals (zircon, apatite, sphene, etc.). Q and F mineral assemblages, in principle, do not coexist in the same rock.

For rocks with M less than 90%, the parameters A, P, and Q, or A, P, and F are recalculated at 100% and plotted into the double **QAPF triangle diagram** (Figure V-1a; Le Maitre, 1989) to determine their nomenclature.

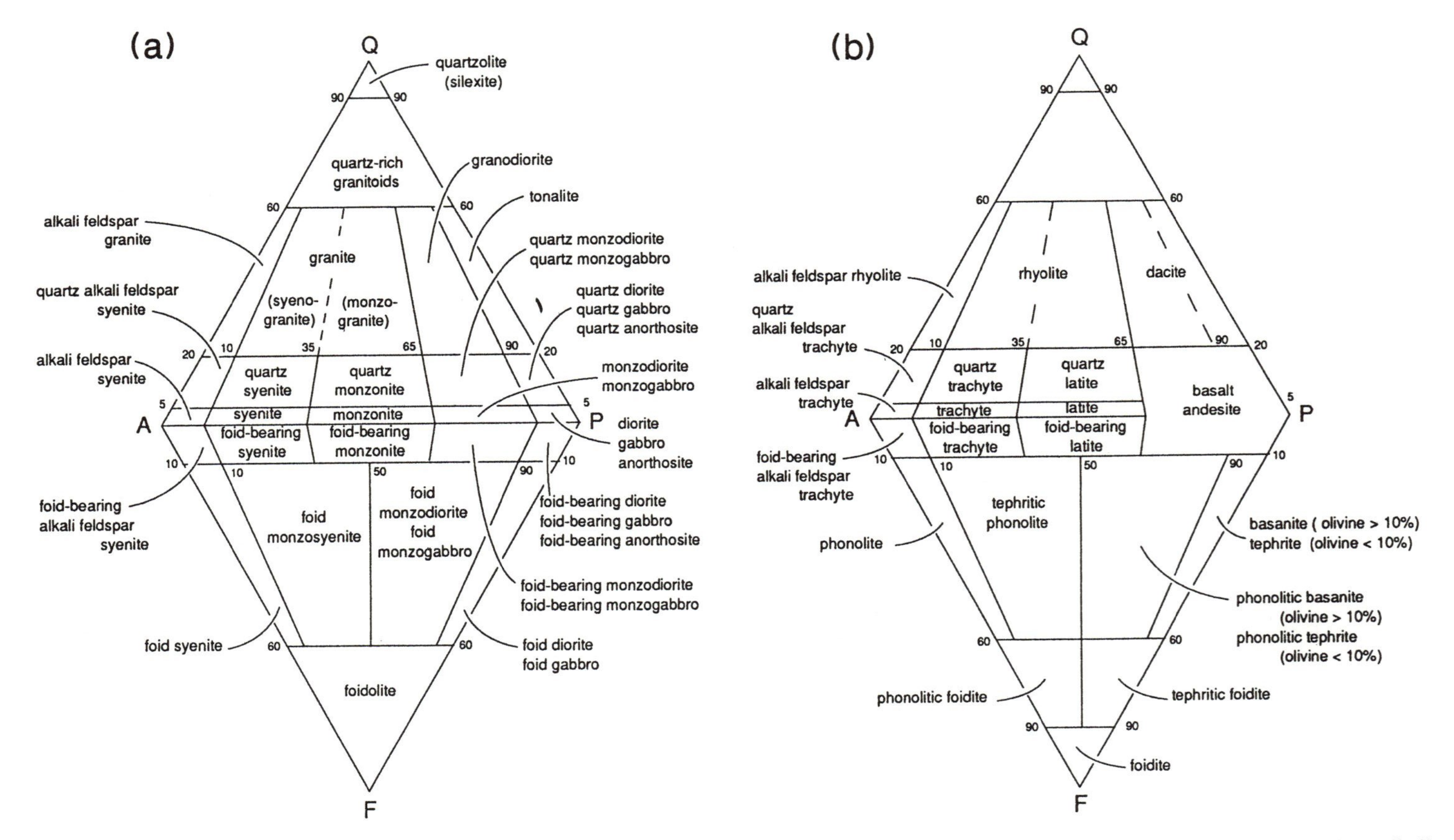

FIGURE V-1. Classification of (a) plutonic igneous rocks and (b) volcanic rocks according to the relative abundance of Q = quartz, A = alkali feldspar, P = plagioclase, and F = feldspathoid, when the mafic minerals (M) are less than 90% (Le Maitre, 1989). Reprinted by permission of Blackwell Science Ltd.

Though diorite, gabbro, and anorthosite fall in the same field in Figure V-1a, they are characterized, respectively, by plagioclase of more than An_{50}, less than An_{50}, and M fraction of less than 10%. The plutonic rocks with $M = 90$ to 100% are called **ultramafic rocks** and are classified according to the relative abundance of olivine (ol), clinopyroxene (cpx), orthopyroxene (opx), and hornblende (hbl) as shown in Figure V-2 (Le Maitre, 1989). If plagioclase or spinel or garnet is present in peridotites, one may add this mineral as a prefix to the rock type, e.g., spinel lherzolite, garnet harzburgite, etc.

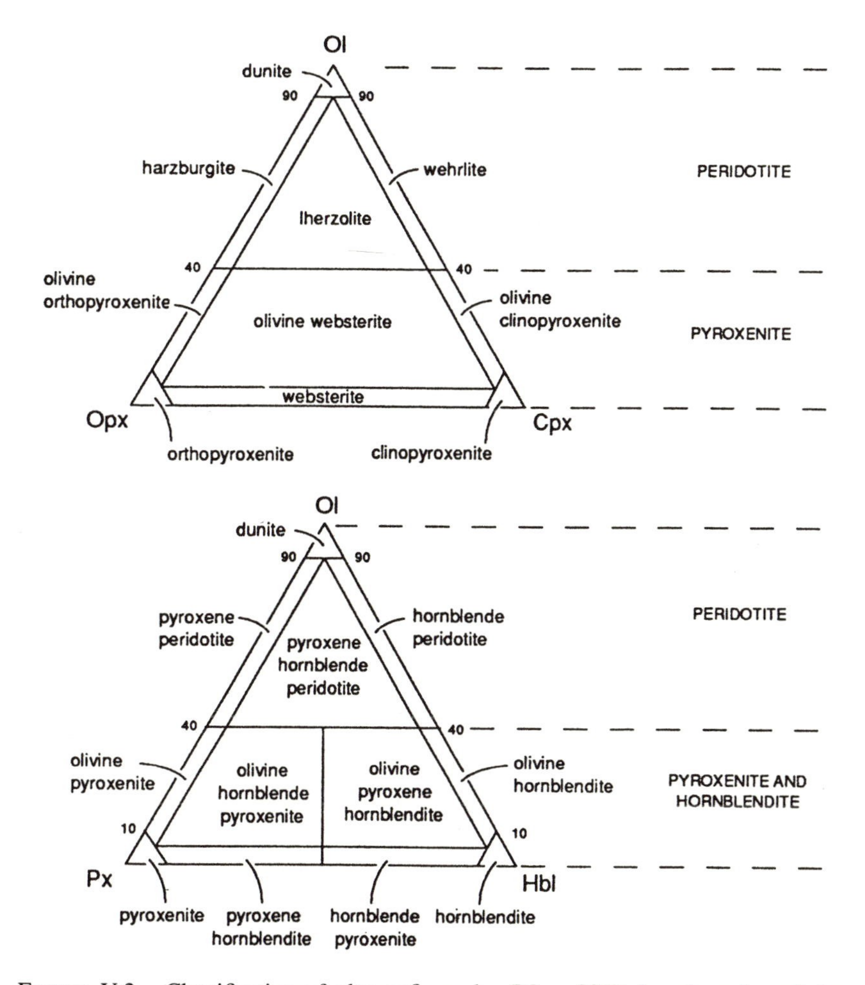

FIGURE V-2. Classification of ultramafic rocks ($M > 90\%$) based on the relative abundance of olivine (ol), orthopyroxene (opx), clinopyroxene (cpx), and hornblende (hbl) (Le Maitre, 1989). Reprinted by permission of Blackwell Science Ltd.

The volcanic rocks can also be classified according to the QAPF diagram as shown in Figure V-1b (Le Maitre, 1989), if the relative abundance of actual mineral assemblages can be determined. However, more often than not, the volcanic rocks consist mostly of glassy matrices; thus, they are classified according to their bulk chemical compositions. If the MgO content is greater than 18 weight %, SiO_2 < 53%, and $Na_2O + K_2O$ < 2%, the volcanic rock is called **picritic rock**. Picritic rocks are further subdivided into **picrites** ($Na_2O + K_2O$ > 1%) and **komatiites** ($Na_2O + K_2O$ < 1% and TiO_2 < 1%). If the MgO content is less than 18%, the volcanic rocks are classified according to the **total alkali silica (TAS) diagram** as shown in Figure V-3 (Le Bas et al., 1986; Le Maitre, 1989). Using the chemical composition of volcanic rock, one may also calculate the hypothetical assemblage of standard minerals known as the norm (e.g., apatite, ilmenite, orthoclase, albite, anorthite, olivine, hypersthene, diopside, and quartz or nepheline + leucite, etc.) following a standard procedure (e.g., the CIPW sequence; for details, see Cox et al., 1979). The volcanic rock is called **silica oversaturated** or **silica undersaturated** if it contains normative quartz or nepheline, respectively.

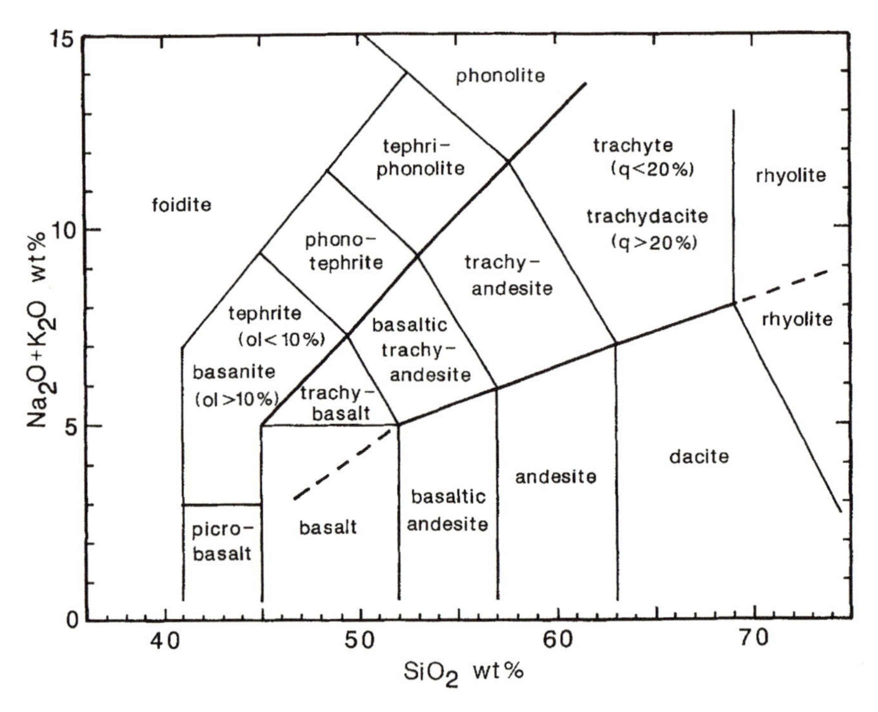

FIGURE V-3. Classification of volcanic rocks based on the weight content of SiO_2 and $Na_2O + K_2O$, when MgO is less than 18% (Le Maitre, 1989).

The volcanic rock types above the thick boundary line with the dotted extension in Figure V-3 are called the **alkali magma series**, and below, the **subalkali magma series**. In general, the subalkali magma series is silica oversaturated, and the alkali magma series above the other thick boundary line is silica undersaturated. The subalkali volcanic rock series can be further subdivided into an iron-depleted **calc alkali series** and an iron-enriched **tholeiitic series**. Miyashiro (1974) showed that if $SiO_2 > 6.4 \sum FeO/MgO + 42.8$ (where elements and their oxides are in weight %, and $\sum FeO$ is the total iron expressed as FeO), the rock belongs to the calc alkali series; otherwise to the tholeiitic series. The basalts and andesites of the calc alkali series contain 16–20% Al_2O_3, while their tholeiitic counterparts contain only 12 to 16%. The calc alkali series is mainly associated with subduction-related plate tectonic settings, i.e., island arcs and active continental margins. The tholeiitic series occurs mainly in young immature island arcs and the oceanic islands. Tholeiitic basalts with low potassium content are mostly associated with mid-ocean ridge basalt (MORB) and continental flood basalt provinces. However, the tholeiitic series can also occur in other tectonic settings.

The alkali magma series occurs mainly in the oceanic islands and continental rift zones. The trachybasalt, basaltic trachyandesite, and trachyandesite in the alkali magma series (Fig V-3) can be called hawaiite, mugearite, and benmoreite, respectively, provided that $(Na_2O - 2) \geq K_2O$. When $(Na_2O - 2) < K_2O$, they are called potassium trachybasalt, shoshonite, and latite. The spinel lherzolites often occur as xenoliths in basaltic rocks and basanites. The MgO-rich komatiite lavas (up to 33% MgO) have been found only within the basalt sequences of the Archean greenstone belts. The younger komatiite lavas are less MgO rich (22% > MgO > 18%). For more detailed classification of igneous rocks and geological occurrences one should refer to Le Maitre (1989) and Wilson (1989).

V-2. Earth's Structure and Mineral Composition

According to seismic compressional and shear velocity changes (Figure V-4), the Earth is stratified into crust (depth of about 50 km for continental crust and 10 km for oceanic crust), upper mantle (depth from 10 or 50 to 670 km), lower mantle (670 to 2890 km), liquid outer core (2890 to 5150 km), and solid inner core (5150 to 6370 km). Their relative masses are 0.47 (including 0.1% oceanic crust), 18.5, 48.5, 30.8, and 1.7%, respectively, with a total Earth mass (m_E) of 6×10^{27} g. The basal 200 km of the mantle has low seismic velocity gradient, is called the **D″ region**, and may have a chemical composition quite different from the normal mantle.

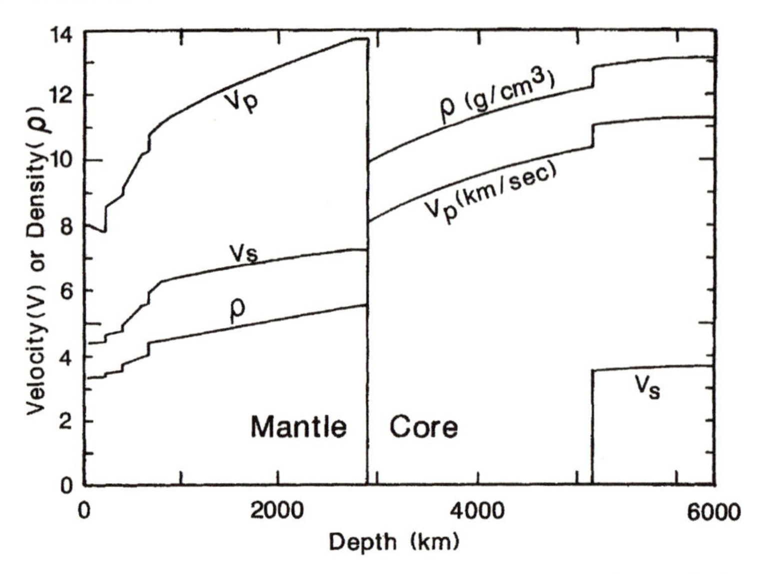

FIGURE V-4. Seismic compressional velocity (V_p; km/sec), shear velocity (V_s; km/sec), and density (ρ; g/cm^3) as a function of depth according to the Preliminary Reference Earth Model (PREM) of Dziewonski and Anderson (1981).

Figure V-5a summarizes the phase relationships of dry peridotite in the upper mantle (Takahashi, 1986; Takahashi and Ito, 1987) along with the oceanic geotherm (Sato et al., 1988). The subsolidus phase changes along the oceanic geotherm include the conversion of plagioclase to spinel by

$$\underset{\text{plagioclase}}{CaAl_2Si_2O_8} + \underset{\text{olivine}}{2Mg_2SiO_4} \rightarrow \underset{\text{spinel}}{MgAl_2O_4} + \underset{\text{opx}}{2MgSiO_3} + \underset{\text{cpx}}{CaMgSi_2O_6},$$

and spinel to garnet by

$$\underset{\text{spinel}}{MgAl_2O_4} + \underset{\text{opx}}{4MgSiO_3} \rightarrow \underset{\text{garnet}}{Mg_3Al_2Si_3O_{12}} + \underset{\text{olivine}}{Mg_2SiO_4},$$

where opx and cpx are orthopyroxene and clinopyroxene, respectively. As the pressure increases, olivine changes into the denser β-spinel (at 150 kbar) and γ-spinel (at 190 kbar) phases without change in chemical composition, while

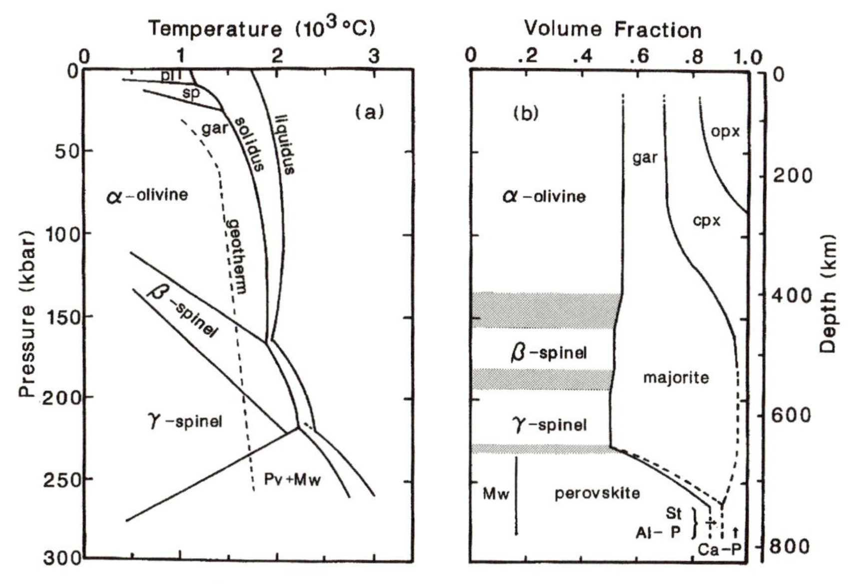

FIGURE V-5. (a) Mineral phase relationship of dry peridotite (KLB-1) as a function of pressure (or depth) and temperature, where pl = plagioclase, sp = spinel, gar = garnet, Pv = perovskite, and Mw = magnesiowuestite. (b) Relative abundance of mineral phases (in volume fraction) along the oceanic mantle geotherm (dashed line in Fig. V-5a), where opx = orthopyroxene, cpx = clinopyroxene, St = stishovite, Al-P = Al_2O_3-rich phase(s), and Ca-P = CaO-rich phase(s). After Takahashi and Ito (1987).

pyroxene (opx and cpx) reacts extensively with garnet to form majorite with the general formula of $(Mg, Fe, Ca)_3[(Mg, Fe)Si, Al_2]Si_3O_{12}$. At pressures greater than 240 kbar the majorite changes into an assemblage of perovskite, unidentified "CaO-rich" and "Al_2O_3-rich" phases, and stishovite, whereas γ-spinel at 240 kbar dissociates into magnesiowuestite ([Mg, Fe]O), and perovskite. The general formula for perovskite is $(Mg, Ca, Si)(Si, Fe, Al)O_3$. The relative abundance of various phases along the geotherm is summarized in Figure V-5b (Ito and Takahashi, 1987). From the relative mineral abundance, one can predict the compressional (V_p) and shear (V_s) velocities as functions of depth in the mantle. The close agreement between the observed and predicted velocities (Weidner and Ito, 1987) strongly supports the contention that the upper mantle is composed of dry peridotite, and that the seismic discontinuities at 400 and 670 km depths mainly reflect phase changes (Ringwood, 1975; Yoder, 1976).

The agreement between the seismologically determined density profiles in the lower mantle and the measured densities for perovskite under lower mantle pressure and a temperature of about 2500 ± 1000 K further suggests that perovskite is also the predominant phase in the lower mantle (Williams et al., 1989). However, whether the upper and lower mantles have exactly the same chemical composition is still hotly debated (Anderson, 1989). If there is no chemical stratification, thermal convection cells, as postulated by plate tectonic theory, would extend through the entire mantle (Figure V-6b).

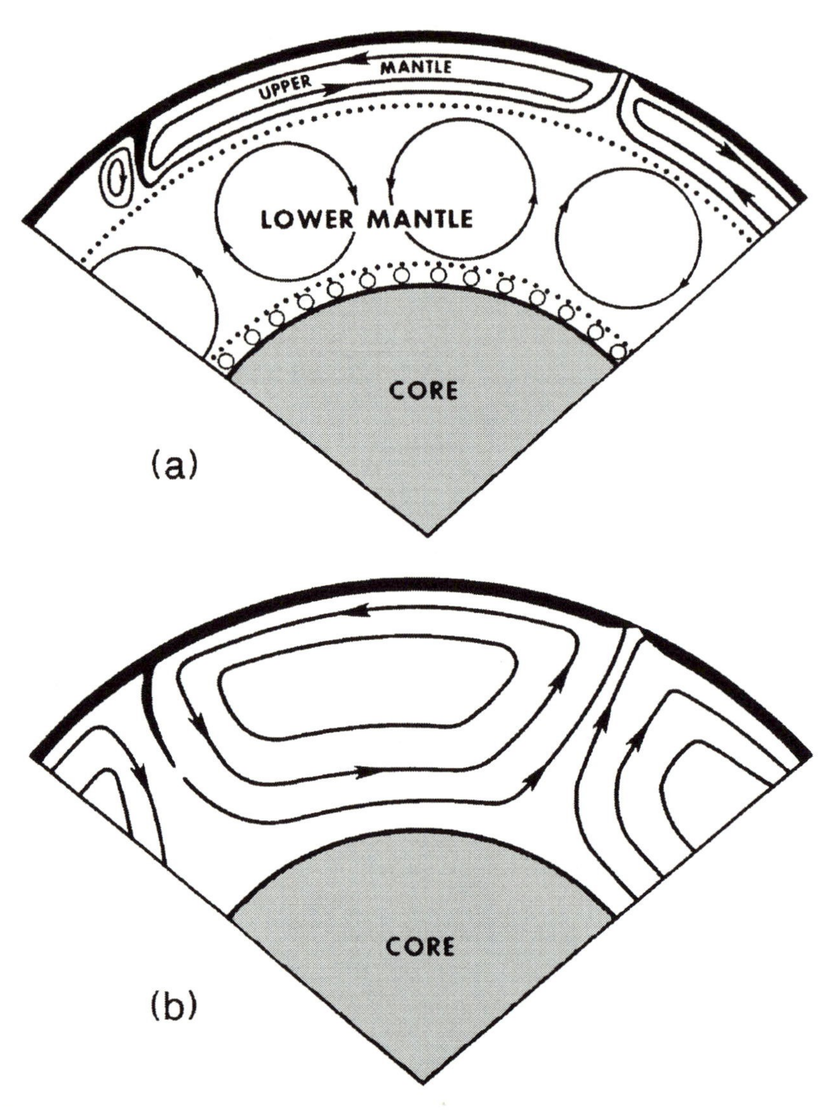

FIGURE V-6. Models of the thermal convection in the mantle. (a) Convection cells confined within the upper and lower mantles. (b) Convection cells extending through the entire mantle (after Richter, 1979; Basaltic Volcanism Study Project, 1981).

Otherwise, they would circulate in the upper and lower mantles separately (Figure V-6a; Richter, 1979). Using improved analytical methods for seismological data, van der Hilst et al. (1991) demonstrated that the subducted slabs of Pacific plate beneath the Japan and Izu Bonin island arcs are deflected at the upper/lower mantle boundary, whereas those below the Kuril-Kamchatka and Mariana arcs do sink into the lower mantle. These results strongly suggest that the upper/lower mantle boundary does not totally shut off material exchange between the upper and lower mantles. Kesson et al. (1994) also demonstrated in the laboratory that at the upper/lower mantle boundary condition, mid-ocean ridge basalt can be transformed into a new mineral assemblage which is denser than an estimated lower mantle density. Therefore, the subducted slab should have no problem moving into the lower mantle. Van der Hilst et al. (1997) provided additional evidence for mantle-wide convective flow.

According to Knittle and Jeanloz (1991), the D″ region may represent reaction products between the mantle perovskite and the outer core iron at the core-mantle boundary (a heterogeneous mixture of the metallic alloy Fe-Ni-O-Si and nonmetallic SiO_2 and $MgSiO_3$). From the observed low shear-wave and compressional-wave velocities and strong scattering of seismic waves, Vidale and Hedlin (1998) also suggested the presence of partial melt and small-scale convection at the core-mantle boundary. Tromp and Dziewonski (1998) emphasized that the large-scale heterogeneity in the lower mantle (in temperature and composition) may cause apparent three-dimensional variations in wave speeds in the lowermost mantle. Other possible explanations for the D″ layer will be discussed later.

The high density of Earth's core (Figure V-4) indicates a Fe-Ni alloy composition. However, the core may also contain some light elements such as S (Anderson, 1989; Anderson et al., 1989), O (Ohtani and Ringwood, 1984; Ohtani et al., 1984; Knittle and Jeanloz, 1991), H (Fukai and Suzuki, 1986), and C (Wood, 1993). However, recent high-T and P experimental results favor S over O and C (Li and Agee, 1996; Jana and Walker, 1997a–c). The presence of these elements would explain the $\sim$10% density deficiency as compared to that of pure Fe metal. How and when the Earth differentiated into the core, mantle, and crust will be discussed later.

V-3. Partial Melting and Fractional Crystallization Models of Igneous Rocks

Simple batch (or equilibrium) and Rayleigh partial melting/fractional crystallization models for the formation of various igneous rocks were first introduced by Gast (1968) and later clarified by Shaw (1970). These models relate the concentrations of trace elements in liquid and solid phases during the melting of igneous rocks or the crystallization of molten magma. For simplicity, it is assumed that the partitioning of any element between the liquid and solid phases represents a kind of chemical equilibrium, i.e.,

$$\frac{X}{C} = D, \tag{V-1}$$

where X and C are the concentrations of a given element in the solid and liquid phases, respectively, and D is the **bulk distribution coefficient** of the element.

Partial Melting Model

In a batch partial melting model of rocks, an isolated reservoir of rocks starts to melt. However, the liquid produced separates from the solid only when the fraction of solid in the isolated system is reduced from one to, say, F; then, by mass balance, one obtains

$$X^0 = FX + (1 - F)C, \tag{V-2a}$$

where X^0 = the original concentration of a given element in a solid before any melting occurs; F = the fraction of the remaining solid; and $(1 - F)$ = the fraction of liquid produced or melted solid. To avoid confusion, the character f will be used later to designate the fraction of the remaining melt in the fractional crystallization model.

Let M^0 represent the original mass of solid before any melting, and M and L represent the masses of solid and liquid after the melting. Then by mass balance,

$$M^0 = M + L,$$

and

$$F = M/M^0, \text{ as well as } 1 - F = L/M^0.$$

Rearranging the above equations, one also gets

$$M = FM^0 = LF/(1 - F). \tag{V-2b}$$

Combining equations V-1 and V-2a, one obtains

$$\frac{X}{X^0} = \frac{D}{DF + (1 - F)}, \tag{V-3a}$$

or

$$\frac{C}{X^0} = \frac{1}{DF + (1 - F)}. \quad \text{(V-3b)}$$

If the solid consists of a mineral assemblage, then

$$X = (m_1 x_1 + m_2 x_2 + \cdots + m_i x_i)/M$$
$$= \sum a_i x_i, \quad \text{(V-3c)}$$

where m_i and $a_i (= m_i/M)$ are, respectively, the mass and fraction of mineral i in the solid phase; $M = \sum m_i$ by definition; and x_i is the concentration of a given element in mineral i. Thus,

$$D = \frac{X}{C} = \frac{\sum a_i x_i}{C} = \sum a_i d_i, \quad \text{(V-4a)}$$

where $d_i = x_i/C$ = **mineral distribution coefficient** of a given element between mineral i and the common liquid pool (L); D is constant only if a_i and d_i are constant during the melting process.

Letting $m_i^0 = m_i + l_i$, and by substituting m_i with $(m_i^0 - l_i)$ and M with equation V-2b in equation V-3c, one obtains

$$X = [(m_1^0 - l_1)x_1 + (m_2^0 - l_2)x_2 + \cdots]/M$$
$$= \frac{\sum a_i^0 x_i - (1 - F)\sum b_i x_i}{F},$$

where m_i^0 is the original mass of mineral i before melting; $a_i^0 = m_i^0/M^0$; and l_i and $b_i (= l_i/L)$ are, respectively, the liquid mass and fraction of liquid mass contributed by melting of mineral i. Therefore,

$$D = \frac{X}{C} = \frac{\sum a_i^0 d_i - (1 - F)\sum b_i d_i}{F}$$
$$= \frac{D^0 - (1 - F)B}{F}, \quad \text{(V-4b)}$$

where $D^0 = \sum a_i^0 d_i$ and $B = \sum b_i d_i$. Equation V-4b was first introduced by Shaw (1970) and can be substituted into equation V-3a or V-3b. Here one needs to keep track of b_i instead of a_i. D^0 and B are constant only if b_i and d_i are constant. Although the d_i are complicated functions of temperature, pressure, and composition of solid and melt, one may treat d_i as constant within a small range of changes for these parameters. If the melt represents the eutectic composition within a certain range of partial melting, B can be treated as a near constant. Once any one mineral phase is exhausted, B will change accordingly.

In a Rayleigh partial melting model of rocks, the liquid is continuously separated from the solid as soon as it is formed. Mathematically, the instantaneous change rates of the mass of solid (M) and the mass of an element in the solid (MX) can be given by

$$\frac{dM}{dt} = -kM, \tag{V-5a}$$

$$\frac{d(MX)}{dt} = -CkM, \tag{V-5b}$$

where k is the hypothetical first-order removal rate constant of liquid from the solid. By substituting $C = X/D$ and equation V-5a into equation V-5b, and expanding the left-hand side of equation V-5b, one obtains

$$M\frac{dX}{dt} + X\frac{dM}{dt} = \frac{X}{D}\frac{dM}{dt},$$

i.e.,

$$\frac{dX}{X} = \left(\frac{1}{D} - 1\right)\frac{dM}{M}.$$

Integrating the above with initial conditions of $X = X^0$, $M = M^0$, and assuming D to be constant, one obtains

$$\frac{X}{X^0} = \left(\frac{M}{M^0}\right)^{(1/D-1)} = F^{(1/D-1)}, \tag{V-6a}$$

or

$$\frac{C}{X^0} = \frac{F^{(1/D-1)}}{D}. \tag{V-6b}$$

If all the melts produced during Rayleigh partial melting are collected in a magma chamber and well mixed, then the average concentration of an element in the magma chamber is

$$\overline{C} = \frac{1}{1-F}\int_F^1 C dF = \frac{1}{1-F}\int_F^1 \frac{X^0}{D}F^{(1/D-1)}dF$$
$$= \frac{X^0(1 - F^{1/D})}{1-F}. \tag{V-6c}$$

It is evident that equations V-6c (Figure V-7h) and V-3b (Figure V-7i) are very similar as functions of F and D. Also, the variations of X/X^0 are similar in the batch and Rayleigh partial melting models for D greater than 1 (Figures V-7j and V-7g).

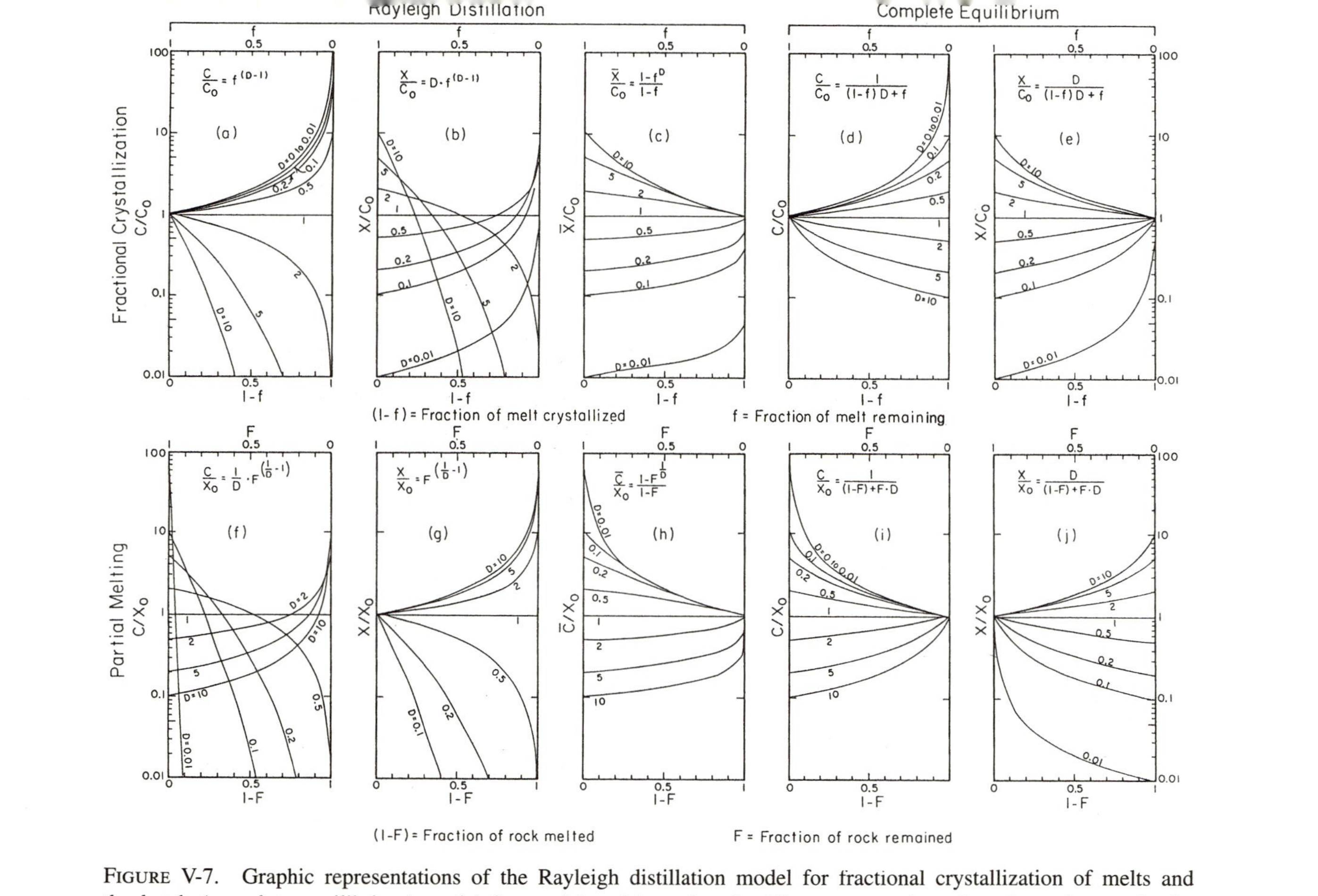

FIGURE V-7. Graphic representations of the Rayleigh distillation model for fractional crystallization of melts and the batch (complete equilibrium) model for partial melting of rocks. The concentration of a given element in the remaining melt (C) or remaining rock (X) is expressed as a function of bulk distribution coefficient (D) and degree of crystallization $(1 - f)$, and degree of partial melt $(1 - F)$.

Fractional Crystallization Model

In a batch fractional crystallization model of magma, an isolated reservoir of liquid starts to crystallize. However, the crystals precipitated separate from the liquid only when the fraction of liquid in the system is reduced from 1 to, say, f. In a Rayleigh fractional crystallization model, the crystals are separated from the liquid or cease to equilibrate with the liquid as soon as they are formed. The mathematical derivations and expressions for the fractional crystallization models are similar to those for partial melting. The results are summarized in Figure V-7. If minerals precipitate at a eutectic point during the fractional crystallization, a_i in equation V-4a is near constant and so is the bulk distribution coefficient D. If the system shifts to another eutectic point, D will change accordingly.

The Rayleigh fractional crystallization model equation $C/C^0 = f^{(D-1)}$ (Figure V-7a) can be transformed into $\log f = (\log C/C^0)/(D-1)$. For all elements in the same rock, f is the same, and therefore

$$\log f = \frac{\log(C_1/C_1^0)}{D_1 - 1} = \frac{\log(C_2/C_2^0)}{D_2 - 1},$$

or

$$\log C_2 = \left(\log C_2^0 - \frac{D_2 - 1}{D_1 - 1}\log C_1^0\right) + \frac{D_2 - 1}{D_1 - 1}\log C_1, \qquad \text{(V-7)}$$

where the subscripts 1 and 2 represent any two different elements and the superscript zero represents the initial or time-zero value.

A plot of $\log C_2$ against $\log C_1$ should form a straight line with a slope of $(D_2 - 1)/(D_1 - 1)$, if a suite of volcanic rocks was formed from the Rayleigh fractional crystallization process and D_1 and D_2 are nearly constant. This is the case for the alkali-basalt suite from the Massif Central shown in Figure V-8. The break in the slope for many elements indicates at least three stages of crystallization. The first stage may represent the fractional crystallization of olivine and clinopyroxene; the second stage, of magnetite, ilmenite, and clinopyroxene; and the last stage, of hornblende, biotite, and alkali feldspars to form trachyte magmas (Villemant et al., 1981).

Distribution Coefficients

The d_i values for elements can be calculated from the elemental composition data of minerals and the glassy or microcrystalline matrix (representing the liquid phase) from any volcanic rock. Table V-1a (see page 204) summarizes d_i values for various minerals in an alkali-basalt suite of Massif Central, France (Villemant et al. 1981). Figure V-9a (see page 206) shows

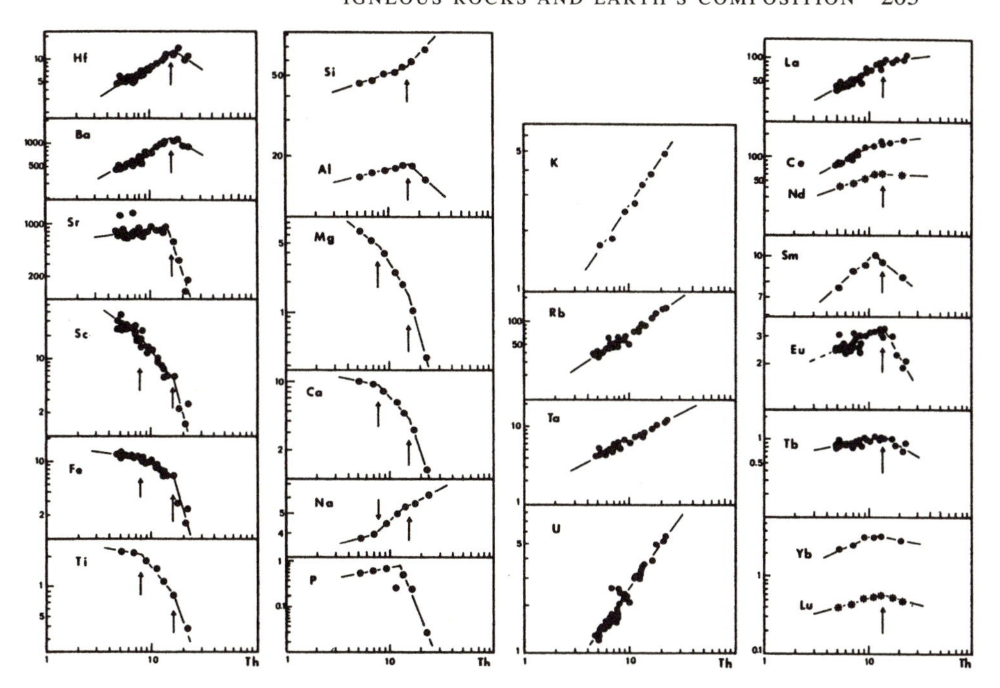

FIGURE V-8. Log-log plots of concentrations for various elements against thorium in the alkali basalt suite of Massif Central, France (Villemant et al., 1981; with permission of Elsevier Science). The arrows indicate changes in slope.

the d_i values from Table V-1a as a function of ionic radius of cations in different minerals. Table V-1b provides additional d_i values of REE for various minerals in average basaltic and rhyolitic rocks (Arth, 1976), and high-silica rhyolites (Mahood and Hildreth, 1983). These data are also plotted in Figure V-9b (see page 207). In general, the d_i values of REE for a given mineral are the highest in the high-silica rhyolites, followed by average rhyolites. They are the lowest in average basaltic rocks (compare orthopyroxene, clinopyroxene, plagioclase, etc., in Figure V-9b). The silicate melts of high-silica rhyolites are highly polymerized due to depletion of silicate-network modifiers (mainly major cations). Thus, they are poor in octahedral sites for accommodating various cations. The end result is the observed generally high d_i values in high-silica rhyolites (Mahood and Hildreth, 1983). The d_i values decrease systematically from Lu to Ce or La for olivine, orthopyroxene, garnet, and zircon in average basaltic and rhyolitic rocks (Figure V-9b). This is consistent with the fact that the ionic radius of Lu^{+3} (0.86 Å), is closer than La^{+3} (1.03 Å) to that of the major cations of these minerals (e.g., $Mg^{+2} = 0.72$; $Fe^{+2} = 0.78$; $Zr^{+4} = 0.72$ Å). Similarly, d_i increases systematically from Lu to La (except for unusually

TABLE V-1A
Mineral distribution coefficients (d_i) of various cations in the alkali-basalt suite of Massif Central, France

Ion	Radius Å	Olivine	Clinopyroxene	Plagioclase	Magnetite	Ilmenite	Hornblende	Feldspar	Biotite
Na^{+}	1.02	0.005	0.17	*1.2	0.04	0.04	0.54	0.99	0.2
K^{+}	1.38			0.17	0.003		0.29	*1.9	*1.9
Rb^{+}	1.52	0.04	0.04	0.13	< 0.06	(0.14)	*1.9	0.3	*1.9
Cs^{+}	1.67		0.04	0.13	0.08		0.5	0.14	*1.6
Mg^{+2}	0.72	*9.5	*5.4	0.04	*1.5	*7	*11		*35
Ca^{+2}	1.00	0.04	*3.4	*2.0			*3.4	0.28	0.05
Sr^{+2}	1.18	0.02	0.16	*2.7	0.1	*(4.6)	0.3	*10	0.7
Ba^{+2}	1.35	0.03	0.04	0.56	0.14	(0.4)	*6.4	*3.6	*10
Ni^{+2}	0.69	*34	*2.5	0.04	*1.9	*(30)	0.8	0.5	*1.3
Co^{+2}	0.75	*5.1	1.0	0.07	*4.8	*(38)	*16	0.5	*23
Fe^{+2}	0.78	*1.9	0.99	0.04	*1.4	*20	*3.6	0.08	*7.3
Mn^{+2}	0.83	*3.0	*1.2		*7.4	*63	*1.6		*5
Al^{+3}	0.54	0.002	0.48	*1.6		0.01	0.75	*1.3	0.92
Cr^{+3}	0.73	*2.8	*5.3	0.08	*3.3	*(82)	*2.9	0.6	*5.4
Sc^{+3}	0.75	0.22	*3.0	0.04	*1.4	*(15)	*6	0.1	*8.3
Tb^{+3}	0.92	0.03	0.73	0.11	0.1	*(8.2)	0.9	0.1	*1.1
Eu^{+3}	0.95	0.03	0.63	0.5	0.06	*(7.5)	0.76	*1.2	*1.1
La^{+3}	1.03	0.03	0.12	0.2	0.13	*(4)	0.6	0.24	0.7
Si^{+4}	0.40	0.78	0.91	1.0	0.002		0.66	0.96	0.52
Ti^{+4}	0.61	0.05	*1.1	0.05	*12	*100	*5.0	0.05	*13
Hf^{+4}	0.71	0.04	0.48	0.05	0.2	*(12)	0.92	0.13	*1.8
Zr^{+4}	0.72	0.06	0.27	0.13	0.4	*(18)	*1.2	0.27	*2.5
U^{+4}	0.89	0.04	0.05	0.06	0.08	(0.6)	0.15	0.1	0.13
Th^{+4}	0.94	0.03	0.04	0.05	0.14	*(1.2)	0.11	0.09	0.12
Sb^{+5}	0.60		0.1	0.18	0.1	(1)	0.15	0.12	0.04–0.8
Ta^{+5}	0.64	0.03	0.06	0.04	0.3	*(2.7)	0.62	0.08	0.56

Source: Villemant et al. (1981)
Note: The asterisks are d_i values of greater than one (compatible); values in parentheses represent only one determination.

TABLE V-1B

Selected mineral distribution coefficients (d_i) of REE in basaltic rocks (B), rhyolitic rocks (R); and high-silica rhyolite (H)

	Olivine		*Orthopyroxene*			*Clinopyroxene*			*Plagioclase*		*Garnet*	*Hornblende*	
	(B)	(H)	(B)	(R)	(H)	(B)	(R)	(H)	(B)	(R)	(B)	(B)	(B)
La		23			18			24					
Ce	0.0069	24	0.024	0.15	16	0.15	0.50	22	0.120	0.27	0.028	0.20	1.5
Nd	0.0066	14	0.033	0.22	14	0.31	1.11	17	0.081	0.21	0.068	0.33	4.3
Sm	0.0066	8.4	0.054	0.27	9.3	0.50	1.67	13	0.067	0.13	0.29	0.52	7.8
Eu	0.0068	5.8	0.054	0.17	3.3	0.51	1.56	11	0.340	2.15	0.49	0.59	5.1
Gd	0.0077		0.091	0.034		0.61	1.85		0.063	0.097	0.97	0.63	10
Tb		2.9			6.3			6.1					
Dy	0.0096	3.3	0.15	0.46	4.4	0.68	1.93	4.8	0.055	0.064	3.17	0.64	13
Er	0.0110		0.23	0.65		0.65	1.80		0.063	0.055	6.56	0.55	12
Yb	0.0114	2.0	0.34	0.86	2.6	0.62	1.58	5.5	0.067	0.049	11.5	0.49	8.4
Lu	0.0160	2.2	0.42	0.90	3.3	0.56	1.54	7.2	0.060	0.046	11.9	0.43	5.5

	K-feldspar		*Phlogopite*		*Biotite*		*Zircon*		*Apatite*	*Sph.*	*Mag.*	*Ilm.*	*All.*
	(R)	(H)	(B)	(R)	(R)	(H)	(R)	(H)	(R)	(R)	(H)	(H)	(H)
La		0.110				3.6		27			32	1.30	2830
Ce	0.044	0.095	0.034	0.23	0.32	3.5	2.6	24	35	53	26	1.20	2490
Nd	0.025	0.093	0.032	0.34	0.29	2.7	2.2	22	57	88	18	0.96	1840
Sm	0.018	0.046	0.031	0.39	0.26	1.8	3.1	18	63	102	13	0.68	980
Eu	1.130	2.2	0.030	0.50	0.24	0.87	3.1	12	30	101	7.2	0.40	100
Gd	0.011		0.030	0.35	0.28		12		56	102			
Td		0.022				1.20		37			3.7	0.36	310
Dy	0.006	0.023	0.030	0.20	0.29	0.92	46	95	61	81	2.5	0.37	150
Er	0.006		0.034	0.17	0.35		135		37	59			
Yb	0.012	0.018	0.042	0.17	0.44	0.69	270	490	24	37	1.3	0.55	37
Lu	0.006	0.014	0.046	0.21	0.33	0.80	320	635	20	27	1.4	0.74	44

Source: B and R, Arth (1976); H, Mahood and Hildreth (1983).

Notes: sph. = sphene, mag. = magnetite, ilm. = ilmenite, and all. = allanite.

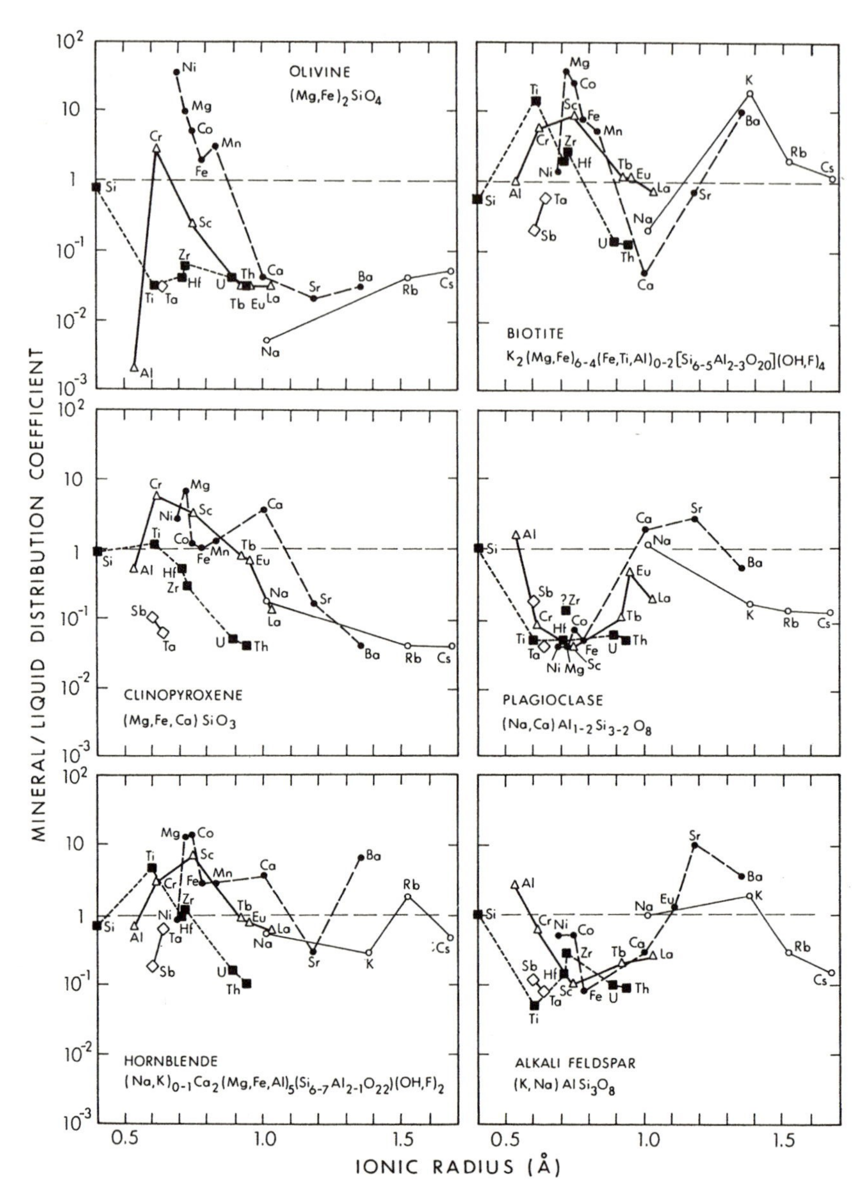

FIGURE V-9A. Mineral distribution coefficients (d_i) of various elements in various minerals obtained from the alkali basalt suite of Massif Central, France (Table V-1a).

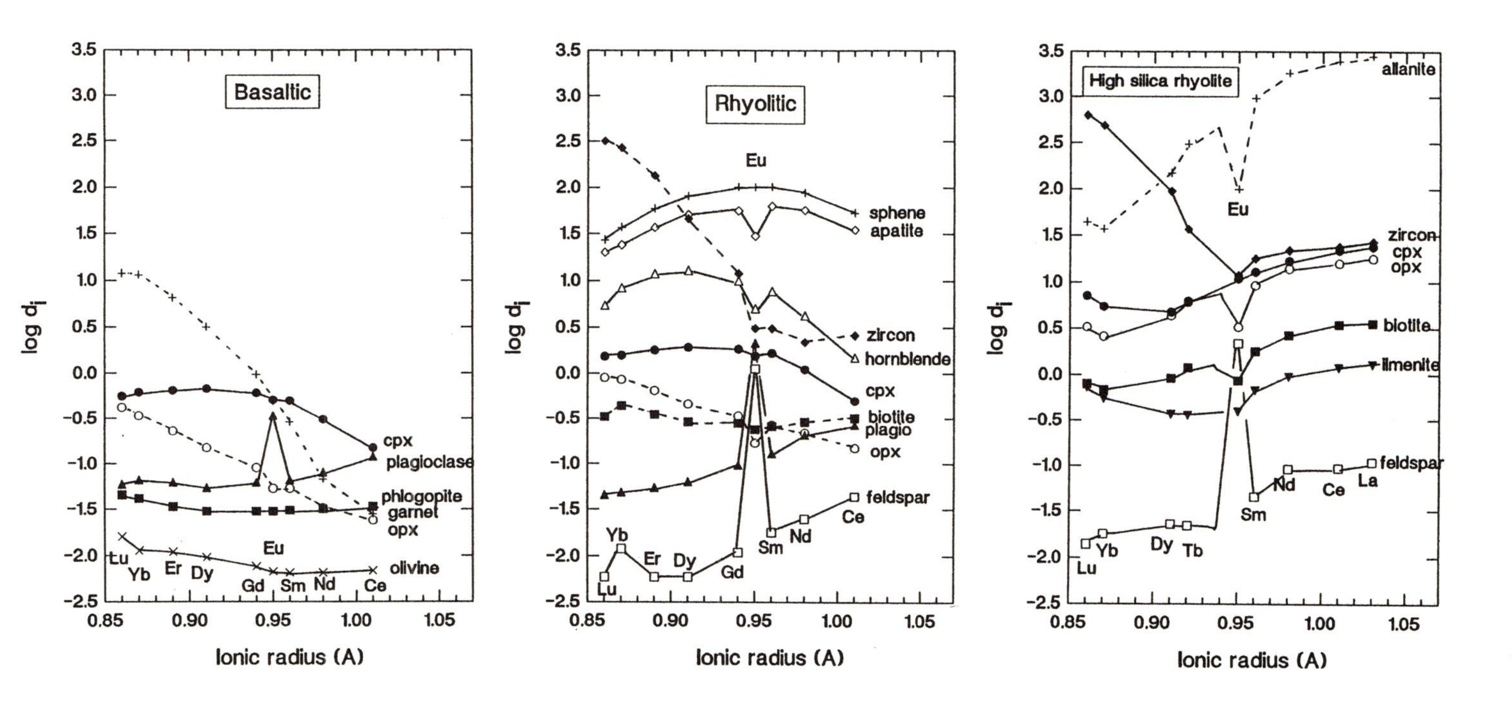

FIGURE V-9B. Mineral distribution coefficients of REE in average basaltic and rhyolitic rocks and high silica rhyolites (Table V-1b).

high values for Eu) for plagioclase and feldspar, because the ionic radius of La^{+3} is closer than Lu^{+3} to that of the major cations of plagioclase and feldspar (e.g., $Ca^{+2} = 1.00$; $Na^{+} = 1.02$; $K^{+} = 1.35$ Å). The exceptionally high d_i for Eu is due to the similarity in both charge and ionic radius among Eu^{+2}(1.17 Å), Ca^{+2}, Na^{+}, and K^{+}. In short, the relative magnitude of d_i (in average basaltic and rhyolitic rocks) is mainly controlled by the crystal chemistry.

In contrast, the d_i values for Sm to La increase with increasing ionic radius for all minerals in high-silica rhyolites (Figure V-9b). The probable explanation is that the limited octahedral sites of silicate melts may have sizes equal to or smaller than 0.95 Å. Thus, Sm^{+3} to La^{+2} (0.96 to 1.03 Å) are harder to fit into these sites and force d_i to become larger. Therefore, the degree of polymerization of silicate melts along with the availability of metal complexing agents (e.g. H_2O, F, Cl, and CO_2) in melts are additional important factors that affect the relative magnitude of d_i (Mahood and Hildreth, 1983). Additional data and discussion on d_i are given by Irving and Frey (1984), Nash and Crecraft (1985), Lemarchand et al. (1987), and McKay (1989).

From the a_i and d_i of Th for various minerals in alkali basalts and trachytes (from Massif Central), D of Th is estimated to be 0.03 and 0.1, respectively. Therefore, from the slopes in Figure V-8, one can estimate D for other elements (see equation V-7). The results are summarized in Figure V-10 as a

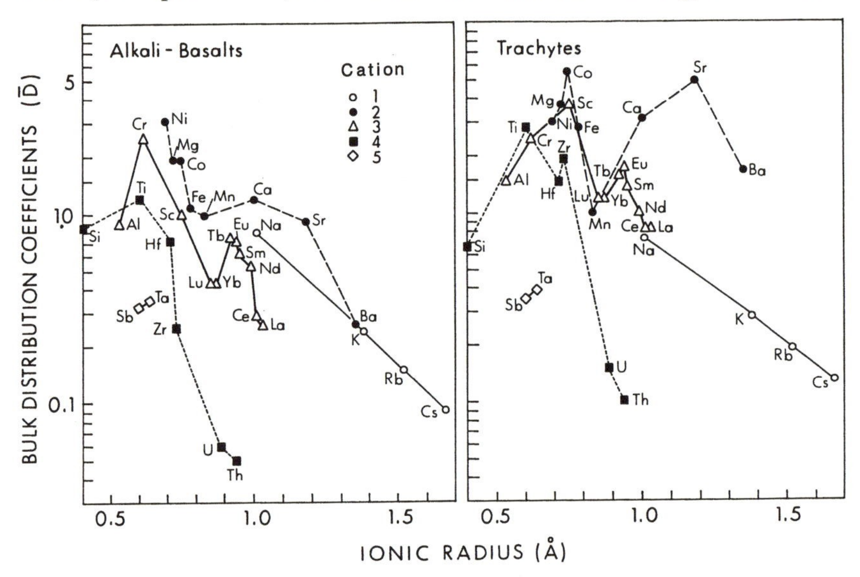

FIGURE V-10. Bulk distribution coefficients for various elements in alkali basalts and trachytes from the volcanic suite of Massif Central, France (data from Villemant et al., 1981).

function of ionic radius. It is apparent from Figure V-10 that the bulk distribution coefficients of elements, D, are closely related to the ionic radii of the elements (Goldschmidt, 1954) and are in general higher in trachytes than in alkali basalts. The elements with D or d_i less than 1 are often called **incompatible**, and those with D or d_i greater than one, **compatible** elements. For example, the elements Ni, Mg, Co, Fe, Mn, and Cr are compatible with olivine in Figure V-9a; whereas additional Sc, Ti, and Ca become compatible with clinopyroxene; and Zr, Hf, heavy rare earths, K, Rb, Cs, and Ba are compatible with biotite. In Figure V-10, Ni, Mg, Co, Fe, Mn, Cr, and to a lesser degree Sc, Ti, and Ca behave compatibly in the alkali basalt, representing removal of these elements from the magma by olivine and minor clinopyroxene.

V-4. Deduction of the Primitive Upper Mantle Composition

Spinel lherzolite xenoliths in basanite hosts are believed to be upper mantle materials that may have already undergone various degrees of partial melting (Jagoutz et al., 1979). Other useful volcanic rocks that can be used to deduce the upper mantle composition are the **Archaean komatiites** (Nesbitt and Sun, 1976), which may represent either partial melts of low degree ($F < 10\%$) in the upper mantle at high pressure (equivalent to depths of 150 to 200 km; Takahashi and Scarfe, 1985; Takahashi, 1986) or partial melts of high degree ($F > 30\%$) at low pressure (around 45 km; Jaques and Green, 1980). If one can correct for the effect of partial melting on these rock types, one can deduce the original composition of the undifferentiated upper mantle. The following provides an example using ultramafic xenoliths.

The concentration of elements in the spinel lherzolites are shown in Figure V-11. Data from Jagoutz et al. (1979) are shown as solid and open triangles and data from the Basaltic Volcanism Study Project (hereafter BVSP, 1981) as solid and open circles in Figure V-11a. Data on the left side of Figure V-11b are from the BVSP (1981) and on the right side from Jagoutz et al. (1979). The magnesium concentration was chosen as an indicator of the degree of partial melting, i.e., the lower the magnesium concentration in the ultramafic nodules, the lower the degree of partial melting (or the closer to the undifferentiated upper mantle concentration). The elements shown in Figure V-11a, and Se and Ag in Figure V-11b follow the prediction of the partial melting model nicely (Figures V-7g and V-7j); namely, the concentrations of elements with D less than 1 decrease smoothly and of elements with D equal to or greater than 1 stay the same or increase slightly when the Mg concentration in the ultramafic nodules (or the degree of partial melting) increases. Figure V-11a also shows good agreement between the data sets

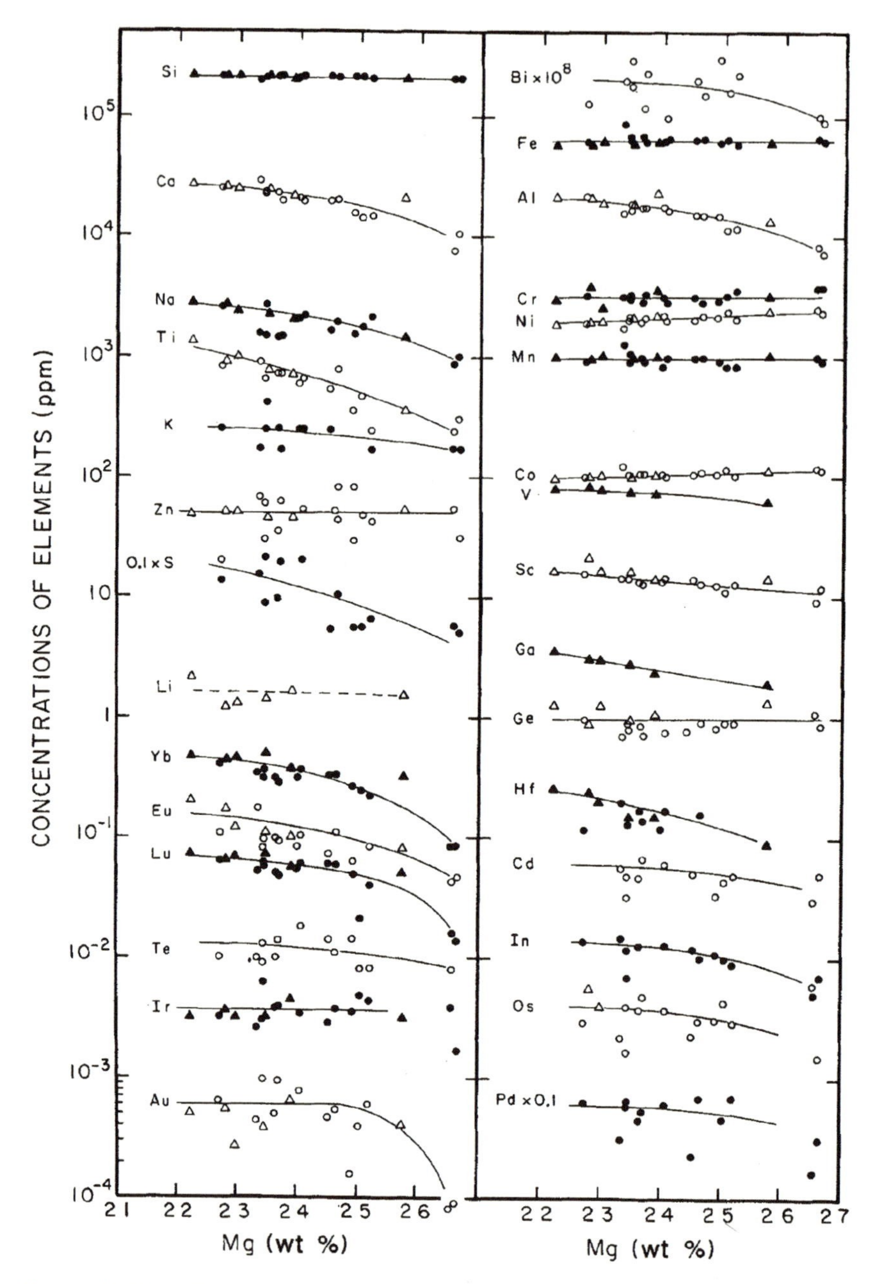

FIGURE V-11A. Concentration plots of various elements versus magnesium in spinel lherzolites. The solid and open circles are data from the Basaltic Volcanism Study Project (1981), and solid and open triangles are from Jagoutz et al. (1979).

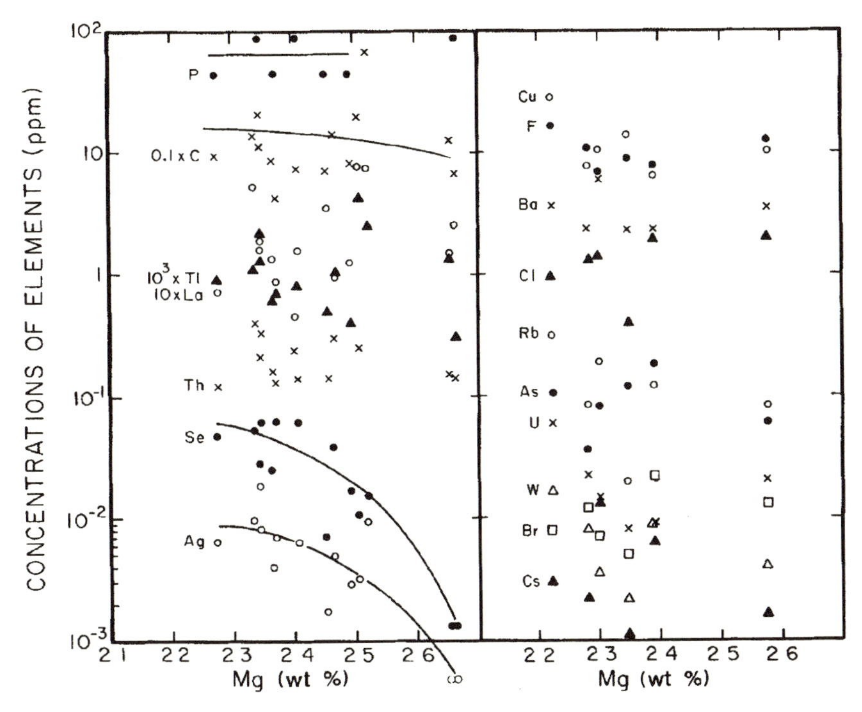

FIGURE V-11B. Data in the left half are from the Basaltic Volcanism Study Project (1981) and in the right half from Jagoutz et al. (1979).

given by Jagoutz et al. (1979) and the BVSP (1981). Considering that these ultramafic nodule samples were obtained from different continents and one sample even from the island of Hawaii, the smooth variation of the concentration data in Figure V-11a is remarkable.

The solid lines in Figures V-11a and V-11b represent approximate fits to the data points, ignoring occasional outliers. The concentrations of elements in the least differentiated (or primitive) upper mantle can be easily read off from the solid lines at Mg concentrations of about $22.2 \pm 0.5\%$. The results are tabulated in Table V-2 (indicated by reference numbers 1 and 2). One can see from Table V-2 that the abundances of rare earth elements Lu, Yb, Tb, and Eu in the upper mantle are in chondritic proportions. The rare earth data of the most primitive ultramafic nodule (SC-1) given by Jagoutz et al. (1979) also show chondritic proportions, except for La and Ce, which may be slightly depleted. Therefore the concentrations of other rare earth elements in the primitive upper mantle are probably also chondritic and were estimated by rare earth concentrations of type 1 carbonaceous chondrite (CI) multiplied by 2.69, where 2.69 is the weight ratio of Lu in the primitive upper mantle and CI. The CI composition data are from Table II-1. It is worthwhile

TABLE V-2

Concentrations of elements in the Earth's primitive mantle, and the enrichment factors (E^i_{Al}) relative to carbonaceous chondrite CI

Z	*Upper Mantle*	E^i_{Al}	*Ref.*	*Z*	*Upper Mantle*	E^i_{Al}	*Ref.*
Ag-47	0.008	0.016	2	Mo-42	0.05	0.019	10
Al-13%	2.2	1	1 to 4	N-7	*10	0.001	5, 6
As-33	*0.1	0.021	1	Na-11%	0.26	0.2	1 to 4
Au-79 ppb	1	0.003	1, 2, 8	Nb-41	0.6	1	4
B-5	*0.3	0.12	14	Nd-60	1.23	1.1	7
Ba-56	6	1	7	Ni-28	2000	0.071	1 to 4
Be-4	0.06	1	7	Os-76 ppb	3.7	0.003	1, 2, 8
Bi-83	0.002	0.007	2	P-15	80	0.026	2, 3
Br-35	*0.005	0.0006	11	Pb-82	0.156	0.025	15
C-6	*150	0.0017	2	Pd-46 ppb	4.3	0.003	2, 5, 8
Ca-20%	2.6	1.2	1 to 4	Pr-59	0.25	1.1	7
Cd-48	0.06	0.04	2, 6	Pt-78 ppb	7.6	0.003	8
Ce-58	1.7	1.1	7	Rb-37	0.54	0.089	9
Cl-17	*0.5	0.0003	11	Re-75 ppb	0.28	0.003	2, 8
Co-27	100	0.078	1, 2, 4	Rh-45 ppb	1	0.003	8
Cr-24	3200	0.48	1 to 4	Ru-44 ppb	5.5	0.003	8
Cs-55	0.013	0.03	12	S-16	180	0.0012	2
Cu-29	*28	0.09	1	Sb-51	*0.003	0.0083	2
Dy-66	0.66	1.1	7	Sc-21	17	1.1	1 to 6
Er-68	0.43	1.1	7	Se-34	0.05	0.001	1, 2, 5
Eu-63	0.15	1.1	1, 2	Si-14%	21	0.78	1 to 6
F-9	*16	0.11	1	Sm-62	0.4	1.1	7
Fe-26%	6.4	0.19	1, 2, 3	Sn-50	0.15	0.034	13

Ga-31	3.4	0.13	1	Sr-38	20	1	3, 4, 7
Gd-64	0.53	1.1	7	Ta-73	0.04	1	7
Ge-32	1	0.012	1, 2	Tb-65	0.1	1.1	1, 3
Hf-72	0.26	1	2, 3, 7	Te-52	0.011	0.002	2
Hg-80	*0.01	0.016	5, 6	Th-90	0.074	1	3, 6, 7
Ho-67	0.15	1.1	7	Ti-22	1100	1	1 to 4
I-53	*0.004	0.004	11	Tl-81	0.006	0.017	12
In-49	0.014	0.069	2	Tm-69	0.066	1.1	7
Ir-77 ppb	3.7	0.003	1, 2	U-92	0.02	1	6, 7
K-19%	0.025	0.18	2	V-23	82	0.57	1, 4
La-57	0.64	1.1	7	W-74	0.016	0.068	1, 12
Li-3	1.6	0.42	1, 6	Y-39	4.1	1.1	3, 7
Lu-71	0.066	1.1	1, 3	Yb-70	0.45	1.1	1, 2, 3
Mg-12%	22.2	0.9	1, 2	Zn-30	50	0.063	1 to 4
Mn-25	1000	0.2	1, 2, 3	Zr-40	10	1	4, 7

Sources: (1) Jagoutz et al. (1979) and Waenke et al. (1984): spinel lherzolites; (2) Basaltic Volcanism Study Project (1981): spinel lherzolites; (3) Basaltic Volcanism Study Project (1981): Archean komatities; (4) Nesbitt and Sun (1976): Archean komatities; (5) Turekian and Wedepohl (1961); (6) Wedepohl and Muramatsu (1979); (7) the concentration in CI multiplied by 2.69 for REE (2.69 = weight ratio of Lu in spinel lherzolites over CI) or by 2.53 for others (2.53 = Weight ratio of Al in lherzolites over CI); (8) 0.003 ± CI (0.003 = weight ratio of Ir in lherzolite over CI); (9) Rb/Sr = 0.027 ± 0.003 obtained from the $^{87}Rb/^{86}Sr$ ratio of 0.078 ± 0.009 for the primitive mantle at present (Faure, 1986); (10) Newson and Palme (1984); (11) Waenke et al. (1984); (12) Sun (1982); (13) Hofmann and White (1983); (14) assuming E^i_{Al} for B is 0.12 as in the CV chondrite; (15) U/Pb = 0.128 obtained from the $^{238}U/^{204}Pb$ ratio of 8.0 for the primitive mantle at present (Faure, 1986).

Notes: Asterisks indicate mostly volatile elements and data are less reliable. Concentrated of elements in ppm unless indicated otherwise.

to mention that S, Se, and Te also behave nicely according to the partial melting model (Figures V-11a and V-11b), and their relative abundances are also nearly chondritic. The relative abundances for Au, Ir, Pd, Os, and Re in Table V-2 are also chondritic, as is the case for the other platinum group elements (e.g., Chou et al., 1983). Therefore, the concentrations of the platinum group elements are estimated from the chondritic concentrations multiplied by 0.0077, which is the weight ratio of Ir in the primitive upper mantle and CI.

In Figure V-11b, the data points for many elements are too scattered or too few to discern any systematic trend. For As, Cu, and F, the values in the most primitive nodule sample SC-1 (Jagoutz et al., 1979) were adopted. The Cl and Br data in Table V-2 are from sample SC-1 leached with hot water (Waenke et al., 1984). The concentrations of Mo and W in the upper mantle are estimated from the Mo/Nd and W/U weight ratios of 0.043 and 0.41, respectively, for terrestrial rock samples (Newsom and Palme, 1984). Rb and Pb concentrations are estimated from $Rb/Sr = 0.027 \pm 0.003$ and $U/Pb = 0.128$, which are obtained from the $^{87}Rb/^{86}Sr$ ratio of 0.078 ± 0.009 and $^{238}U/^{204}Pb$ ratio of 8.0 for the bulk Earth at present (Faure, 1986). Some additional data in Table V-2 are adopted from data for the ultramafic rocks (Turekian and Wedepohl, 1961; Wedepohl and Muramatsu, 1979; Abbey, 1983; Chou et al., 1983). The recent estimate of the primitive mantle composition by McDonough and Sun (1995) agrees with Table V-2 within $\pm 10\%$ or better for most of the elements. However, they give much higher values for Br (0.05 ppm), Cl (17 ppm), and I (0.01 ppm); and a lower value for N (2 ppm). Certainly, better constrains are needed for these volatile elements.

If the spinel lherzolite xenoliths are really representative of present upper mantle samples, the upper mantle should have undergone various degrees of partial melting in space and in time (Figure V-11a). The wide scatter of the incompatible elements such as Tl, Th, U, La, W, Cs, and Rb in Figure V-11b may also suggest the importance of **metasomatic fluids** in modifying the composition of the upper mantle. The metasomatism may include the percolation of H_2O and/or CO_2-rich vapors (or supercritical solutions) and/or infiltration of volatile-rich silicate melts. These vapors and melts were formed within the mantle and are enriched in incompatible elements (e.g., Harte, 1987).

The results of factor analysis for the spinel lherzolite data by BVSP (1981) are summarized in Figure V-12 (excluding the abnormal sample from Nunivak Island). As before, any pair of elements within a dotted enclosure in Figure V-12 always has a correlation coefficient of greater than 0.5. Negative factor 1 (Mg, Ni, Ge, and Cr) represents the elements with D_i greater than 1; the factor 2 (Fe, Mn, Co, and Zn) with D_i near 1; and factor 1 (Ca, Al, Ti, Sc, REE, S, Se, Re, and In) with D_i less than 1 (compare Figure V-12 with Figures V-11a and V-11b). Ag, Te, Pd, and Na may be partly related to factor 1 also. The close association of C, La, Tl, and Th as factor 3 may represent their common metasomatic origin. Factor 4 (Ir, Au, Os, Bi) represents

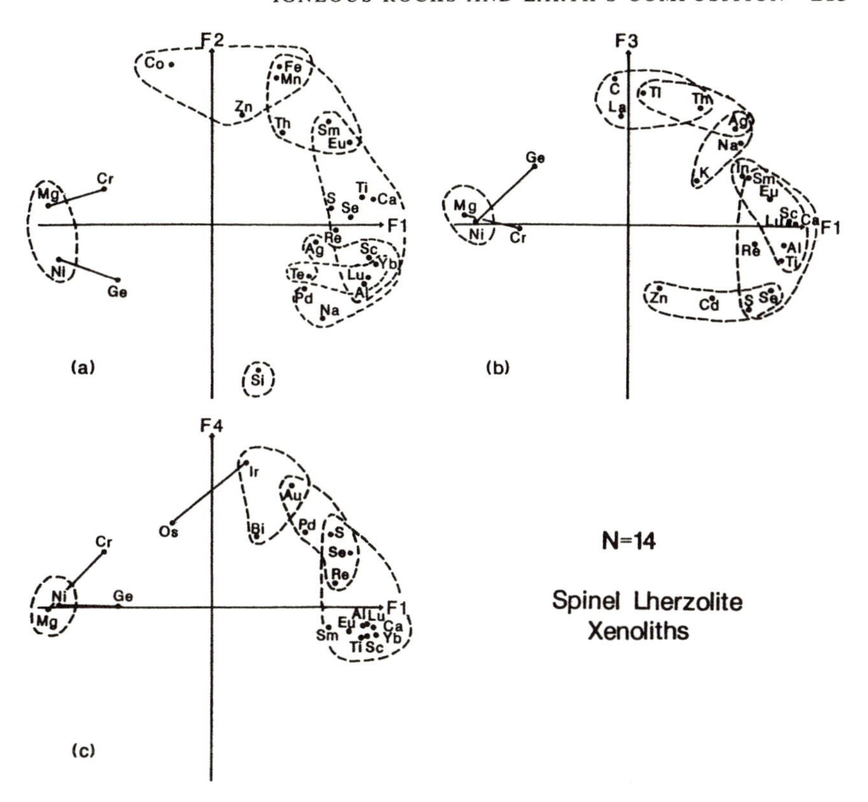

FIGURE V-12. Factor loadings (F1 to F4) for the compositional data of 14 spinel lherzolite xenoliths given by the Basaltic Volcanism Study Project (1981).

unknown noble metal phases. Si does not correlate to any other elements but inversely correlates to the Fe-Mn-Co-Zn group (Figure V-12). Sample numbers are smaller than the number of elements; therefore the factor scores cannot be extracted.

V-5. Partition of the Elements between Mantle and Core

The enrichment factors (E^i_{Al}) for elements in the primitive upper mantle relative to CI carbonaceous chondrite are summarized in Table V-2 and plotted in Figure V-13 as a function of the 50% condensation temperature. The E^i_{Al} values for the refractory lithophiles other than V (open circles in Figure V-13) are all close to 1. These chondritic relative abundances of refractory lithophiles in the primitive upper mantle strongly suggest that the

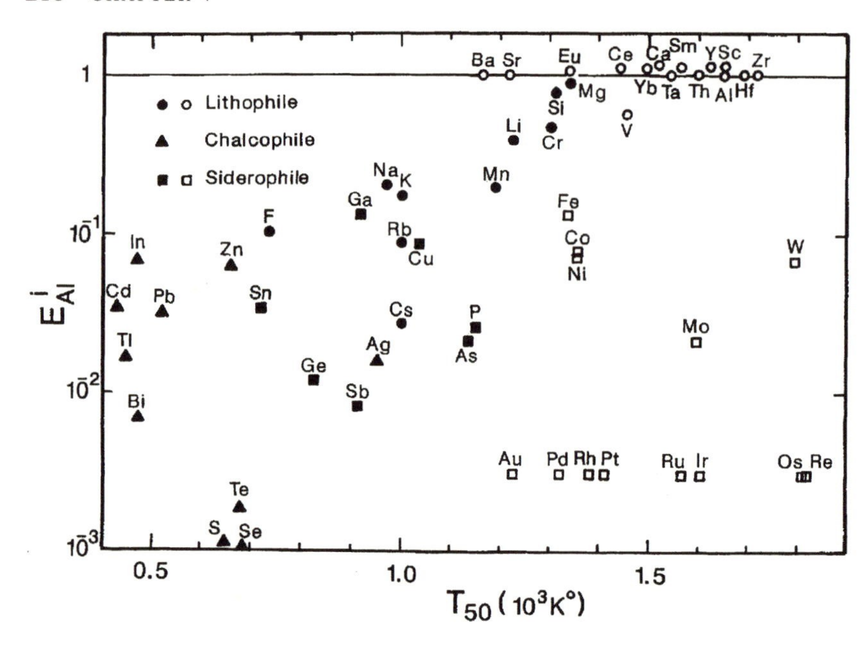

FIGURE V-13. Enrichment factors ($E^i_{\rm Al}$) for elements in the primitive upper mantle (Table V-2) relative to CI carbonaceous chondrite (Table IV-6) as a function of the 50% condensation temperature (Table III-5). The open circles and squares are the refractory lithophile and refractory siderophile elements, respectively. Others are moderately volatile and volatile elements.

primitive lower mantle should also have similar chondritic abundances in order to keep a mass balance of refractory lithophiles (Kato et al., 1988). The $E^i_{\rm Al}$ values for the moderately volatile and volatile lithophiles (solid circles) are all less than 1. This pattern is very similar to those found for the CO and CV carbonaceous chondrites. However, the concentrations of volatile elements such as C, N, Br, Cl, and I are much lower in the primitive mantle than in CO and CV chondrites, even using the higher Br, Cl, and I values of McDonough and Sun (1995; compare Table V-2 and Table IV-6). These volatiles could be already low in the Earth's accreting materials and/or degassed and stripped from the Earth during the accretion process.

The $E^i_{\rm Al}$ values for the siderophiles and chalcophiles are all less than 1 (squares and triangles in Figure V-13) and even less than those for all subclasses of chondrites (compare Tables IV-7 and V-2). The obvious implication is that the siderophile and chalcophile elements must be highly concentrated in the Earth's core, especially the platinum group elements plus Au and S group elements during the iron core formation processes. If 90 ± 3 weight

% of the iron core is iron and the rest S and Ni, then the Fe/Al weight ratio for the Earth's mantle + core system will be about 22.5 ± 0.7 as compared to 21.9 ± 1.0 for CI chondrites (i.e., $E_{\mathrm{Al}}^{\mathrm{Fe}} \approx 1$). McDonough and Sun (1995) gave an Fe/Al ratio of 20 ± 2. Therefore it is quite probable that the relative abundances of both refractory lithophiles and siderophiles for the bulk Earth are very similar to those for the CI.

If the bulk composition of the Earth's mantle plus core system for the refractory siderophiles is really similar to CI, then

$$X_0/\mathrm{Al}_0 = X_{\mathrm{CI}}/\mathrm{Al}_{\mathrm{CI}}, \tag{V-8a}$$

where X_0 and X_{CI} are, respectively, the average concentrations of refractory siderophile element X in the mantle + core system and in CI chondrite. Also, by mass balance, one obtains

$$X_0 = X_m f + X_c(1 - f), \tag{V-8b}$$

$$\mathrm{Al}_0 = \mathrm{Al}_m f, \tag{V-8c}$$

where X_m and X_c are, respectively, the concentrations of element X in the mantle and the core; Al_c is near zero, and f is the fraction of mantle mass in the mantle + core system = 0.675.

Solving equations V-8a–c, one obtains

$$\frac{X_c}{X_m} = \frac{\left(\dfrac{\mathrm{Al}_m X_{\mathrm{CI}}}{\mathrm{Al}_{\mathrm{CI}} X_m} - 1\right) f}{1 - f} = 5.27 \frac{X_{\mathrm{CI}}}{X_m} - 2.08, \tag{V-8d}$$

and the fraction of element X in the core, $X_c(1 - f)/X_0$, is equal to $(1 - 0.395 X_m/X_{\mathrm{CI}})$. The X_c/X_m and $X_c(1 - f)/X_0$ values for refractory siderophiles as calculated from the above equations are summarized in Table V-3 along with the distribution coefficient D ([metal + sulfides]/liquid silicates) determined from a naturally reduced basalt (Kloeck and Palme, 1987) and from synthetic basalts (Jones and Drake, 1986). It is evident that values for X_c/X_m and D are quite different. X_c/X_m and D are not directly comparable, because X_c/X_m may represent the distribution coefficient between the core and the peridotitic mantle under the P-T conditions at the core-mantle boundary, whereas D represents the distribution coefficients between metal + sulfides and basaltic silicate melts at 1 atmosphere pressure and 1200 ± 100°C.

As shown by recent studies (Schmitt et al., 1989; Jana and Walker, 1997a–c; Li and Agee, 1996; Walker et al., 1993), D(metal melt/silicate melt) is a sensitive function of oxygen fugacity, S and C content, and temperature and pressure. Walker et al. (1993) demonstrated that the distribution coefficients of liquid metal/liquid silicate for siderophile elements Fe, Ge, and Ni converge to unity with increasing temperature, up to about 3000 K. Therefore, the X_c/X_m values in Table V-3 may reflect a chemical equilibrium between the mantle and the core during their differentiation. In order to explain the observed relatively high and near chondritic abundances of PGE (platinum group elements) plus Au in the mantle, some researchers postulated an incomplete segregation of metal melt droplets from the mantle into the core (Jones and Drake, 1986). Other possible explanations are the mixing of the outer core material back to the mantle by the mantle plumes, which originated at the core-mantle boundary after the core formation (Snow and Schmidt, 1998), or the addition of about 1% chondritic accreting material to the mantle after the separation of the proto-core and -mantle (Schmitt et al., 1989, and references therein). The latter scenario is also called the **heterogeneous accretion model**. Certainly, further experiments on determining distribution coefficients as functions of composition, T, P, and oxygen and sulfur fugacity are badly needed.

Mg and Si are moderately volatile lithophiles, and the Mg/Si ratio of the solar nebular disk might change with heliocentric distance (Ringwood, 1989). Therefore the slight depletion of Mg and Si in the upper mantle relative to CI (Figure V-13) may not necessarily imply the enrichment of Mg and Si in the lower mantle. Another possibility is that the relative abundance of Mg, Si, and Al for the bulk earth is chondritic, but the depleted Mg and Si in the

TABLE V-3

Fractions of refractory siderophile elements in the Earth's core and distribution coefficients in various systems

	$\frac{X_c(1-f)}{X_0}$	$\frac{X_c}{X_m}$	*D(LM/LS)* (1)	*D(LM/LS)* (2)	*D(SM/LS)* (2)
Re	0.998	683		200	10^5
Ir	0.997	683		20000	9×10^5
Au	0.997	683		10000	10^4
W	0.932	29	25	1	20
Mo	0.979	96	1000	1300	3100
Fe	0.867	14	10		
Co	0.921	27	230	130	310
Ni	0.928	27	>5200	5000	62000

Sources: Kolek and Palme (1987) at $T = 1175$°C, log $fO_2 = -13.9$, and S = 30%; (2) Jones and Drake (1986) at $T = 1260$°C, log $fO_2 = -12$ to 13, S = 25%.

Notes: X_0, X_c, X_m are, respectively, the concentration of elements in the bulk Earth, core and, mantle; f = fraction of the mantle; LM = liquid mental + sulfide, SM = solid metal, LS = liquid silicate.

upper mantle is balanced by the dissolution of Mg and Si in the molten outer core at high pressure (Knittle and Jeanloz, 1991). Then by mass balance, the outer core should contain about 0.7% Mg and 3% Si. The E^i_{Al} of less than 1 for refractory lithophile V along with Cr and Mn (Figure V-13) may indicate extraction of some V, Cr, and Mn into the core (Dreibus and Waenke, 1979), probably as sulfides, because these elements are very much enriched in the troilite phase of iron meteorites (Table IV-8b). Rammensee et al. (1983) also demonstrated that *D*(liquid metal/liquid silicates) values for V and Cr become greater than 1 under reducing conditions ($\log fO_2 < -11$). If the metal melt contains some sulfur, the *D* for V, Cr, and Mn might become much greater than 1.

If the Earth's core contains 7% sulfur (with a possible range of 6 to 12%; Ahrens, 1979), the E^i_{Al} values for the sulfur group elements in the bulk earth relative to CI would be about 0.2, which is again very similar to those in the CO and CV chondrites.

A possible scenario for the formation of a stratified Earth was discussed by Stevenson (1981), Fukai and Suzuki (1986), and Sasaki and Nakazawa (1986). In brief, as shown in Figure V-14, (a) the proto-Earth started as a cold planetesimal of chondritic composition (with regard to refractory lithophile and siderophile elements). (b) When it grew in size to greater than about 10^{26} g (for comparison, the mass of our moon is about 0.74×10^{26} g), it captured gravitationally any gas released by the impact of accreting material and formed an atmosphere. (c) Because of the blanketing effect of the atmosphere, the impact energy of accreting material was trapped and caused the surface temperature to exceed the eutectic melting point of Fe-FeS and some silicates. The molten Fe-FeS was gravitationally segregated from the partially molten silicate. (d) As the molten core of the proto-Earth grew larger, a kind of physical instability eventually set in and broke up the cold chondritic core ($\approx 2\%$ of the Earth's mass), and (e) the proto-core, -mantle, and -atmosphere were in place. The cold broken chondritic core might represent the so-called D″ layer at the core-mantle interface. If the scenario for the origin of the Moon by the impact of a Mars-sized or larger body onto the proto-Earth were real (Section III-2), the impact could easily heat up the interior of the proto-Earth. The surface temperature of the proto-Earth could easily reach 16000 K, vaporize the surface rocks, and eject the entire proto-atmosphere into outer space without invoking the proto-Sun's T-Tauri event (Benz and Cameron, 1990). At the high pressure of the lower mantle, the density difference between melt and coexisting crystals might be very small and/or the viscosity of the melt might be very high (Williams et al., 1989). Thus, the partially molten lower mantle might not undergo extensive gravitational segregation before its complete solidification, and might retain its proto-mantle composition (chondrite minus metal core). In contrast, a portion of the partially molten upper mantle did eventually differentiate into the atmosphere,

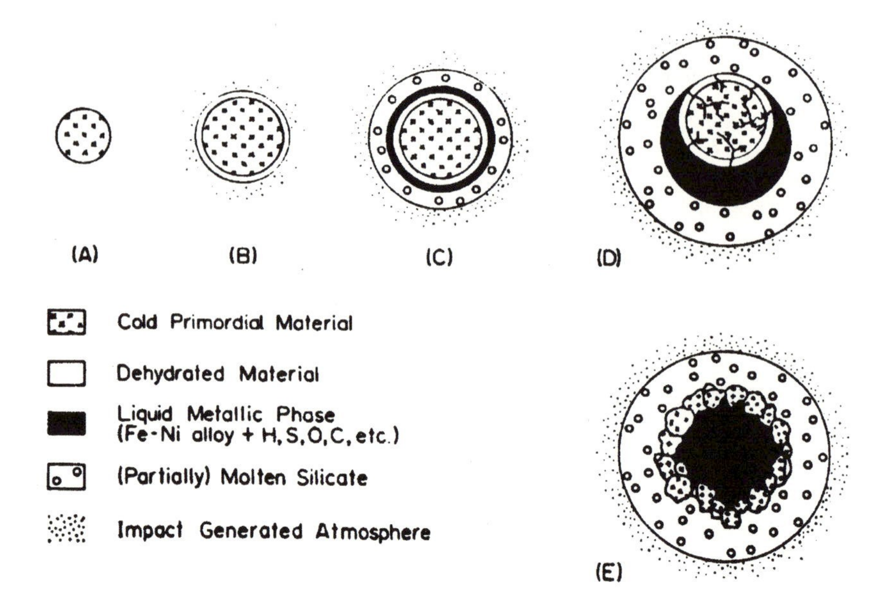

FIGURE V-14. Schematic of the formation processes of Earth's core (Fukai and Suzuki, 1986). See the text for details.

ocean, crust, and the so-called depleted upper mantle. The energy released from the solidification of the molten Fe in the proto-core into the inner core probably has sustained the Earth's magnetic field (Stevenson, 1981).

The above-mentioned scenario is certainly not the only possible one. For example, Agee (1990, and references therein) proposed that the molten proto-mantle might differentiate into innermost magnesiowuestite, perovskite, and komatiite, and outmost peridotite layers. A large portion of the magnesiowuestite layer might dissolve into the Fe-Ni-S proto-core to form a molten Fe-Ni-S-O outer core and the rest become the D″ layer. Considering large variations in distribution coefficients for refractory lithophile elements between majorite, garnet, perovskites, and coexisting liquid melt, the differentiation of the proto-mantle would produce a nonchondritic abundance pattern for these elements among different layers (Kato et al., 1988). This is contrary to observation. One needs to invoke a subsequent solid-state convective mixing to homogenize different layers again. However, the effectiveness of this mechanism in a short period of time ($\approx 0.3 \times 10^9$ y) is in serious doubt (Kato et al., 1988). Certainly the model for the evolution of the core-mantle-crust system will change when new constraints are introduced.

V-6. Continental and Oceanic Crusts

Basaltic and granitic rocks are the most abundant rock types in the present upper continental crust (Mason and Moore, 1982). Therefore, the average composition of the present upper continental crust may be reproduced by the proper mixture of basaltic and granitic rocks. Mead (1914) first showed that the major elemental composition of the upper continental crust (as represented by the average shale) can be reproduced by a mixture of two parts granite and one part basalt.

The average compositions of basalt and granite given by Taylor (1964; his Table 3) are summarized in Table V-4a, except that the major elements (Al, C, Ca, Fe, K, Mg, Mn, Na, P, Si, and Ti) and some other elements are replaced with new estimates by Le Maitre (1976), Marowsky and Wedepohl (1971), and Wedepohl (1969–1978). For comparison, the composition data of basaltic geostandard W1 and granitic geostandard GA (Govindaraju, 1989, 1994) are also given in Table V-4a. The geostandards W1 and GA are chosen here among many other geostandards, because the concentrations of major elements in W1 and GA are very similar to those in average basalt and granite, respectively. Interestingly, the concentrations of trace elements in WI and GA are also in close agreement with those in average basalt and granite, respectively (see Table V-4a). Because the geostandards W1 and GA can reproduce the average basalt and granite so well, their rare earth data are chosen here to represent those for average basalt and granite.

The average composition of the modern upper continental crust as calculated by mixing one part basalt (B) and two parts granite (G) agrees quite well with the estimate given by Taylor and McLennan (1985) and Condie (1993) (Table V-4a). However, Taylor and McLennan (1985) give lower values for Cr, Ni, and Se and a higher value for Nb (Table V-4a). Condie's (1993) estimate was based on relative proportions and average compositions of major rock types in the upper continental crusts as a function of age. His estimates on the average composition of juvenile upper continental crusts at different periods are also summarized in Table V-4b. It is apparent from Table V-4b that the post-Archean upper crusts are enriched in incompatible elements, (such as Ti, K, P, Rb, Ba, Th, U, Zr, Hf, Nb, Ta, Y, heavy REE, Sc, and V) and are depleted in the compatible elements Mg, Co, Cr, and Ni, especially the latter two, as compared to the Archean upper crust.

According to Taylor and McLennan (1985), about 75% of the continental crust was in place by the end of the Archean. The rest was added afterward through igneous activity in orogenic regions (e.g., island arcs). Because the average composition of igneous rocks in orogenic regions is andesitic, the bulk composition of the post-Archean continental crust may be represented by the present-day island arc andesites as summarized in Table V-5 (Taylor and McLennan, 1985). Interestingly a mixture of two parts basalt and one

Table V-4A
Average compositions of basalts, granites, and the Earth's upper crust

	Basalts (B)		*Granites (G)*			*Upper continental crust*	
	Ave. (1)	*W1* (2)	*Ave.* (1)	*GA* (2)	$B/3 + 2G/3$	(3)	(4)
Ag-47	0.1	(0.06)	0.04	0.016L	0.06	0.05	
Al-13%	8.33	7.94	7.58	7.67	7.8	8	7.9
As-33	1.5w	2.2	1.6w	1.7	1.6	1.5	
Au-79 ppb	3.6w	4.3	1.7w		2.3	2s	
B-5	5	13	15	26	12	15	
Ba-56	250	162	730w	840	570	550	630
Be-4	0.5	0.76	4.5w	3.6	3.2	3	
Bi-83	0.031m	(0.048)	0.065m		0.054	0.13	
Br-35	3.6	(0.36)	1.3		2.1		
C-6	300	(160)	200w	300	230	300s	
Ca-20%	6.8	7.86	1.32	1.75	3.2	3	2.57
Cd-48	0.13m	0.15	0.09m	(0.13)	0.1	0.1	
Ce-58		23.5		76	58	64	58
Cl-17	60	205	200	(250)	150	100s	
Co-27	48	47	1	5	17	10	18
Cr-24	200	119	4	12	69	35	100
Cs-55	1	0.96	5	6	3.7	3.7	
Cu-29	90w	113	13w	16	39	25	
Dy-66		4		3.3	3.5	3.5	

Er-68		2.3		1.9	2	2.3	
Eu-63		1.12		1.1	1.1	0.88	1.1
F-9	400	220	850	500	700	500s	
Fe-26%	8.3	7.81	2.11	1.98	4.2	3.5	3.7
Ga-31	17w	17	19w	16	18	17	
Gd-64		4		3.8	3.9	3.8	4.2
Ge-32	1.5	(1.65)	1.6	(1.7)g	1.5	1.6	
Hf-72	3.4	2.5	4	4	(4)	5.8	4.3
Hg-80	0.08	0.2	0.08	0.03f	0.08		
Ho-67		0.81		0.7	0.74	0.8	
I-53	0.5	(0.054)	0.5		0.5		
In-49	0.06w	(0.064)	0.04w	(0.03)	0.05	0.05	
Ir-77 ppb		(0.3)			*0.05	0.02s	
K-19%	0.91	0.53	3.4	3.34	2.6	2.8	2.3
La-57	10	11	40	40	30	30	28
Li-3	10	13	30	90	23	20	
Lu-71		(0.32)		0.3	0.32	0.32	0.32
Mg-12%	4.05	3.99	0.43	0.57	1.6	1.3	1.5
Mn-25	1550	1290	390	700	770	600	
Mo-42	1.2w	(0.75)	1.8w	0.5	1.6	1.5	
N-7	20	(23)	20		20		
Na-11%	2.16	1.6	2.73	2.63	2.54	2.89	2.6

TABLE V-4A
Continued

	Basalts (B)		*Granites (G)*			*Upper continental crust*	
	Ave. (1)	*W1* (2)	*Ave.* (1)	*GA* (2)	$B/3 + 2G/3$	(3)	(4)
Nb-41	10w	9.9	18w	12	15	25?	9.8
Nd-60		14.6		27	23	26	26
Ni-28	150	75	0.5?	7	55	20	56
Os-76 ppb		(0.2)			*0.05		
P-15	1530	570	520	520	860	650s	520
Pb-82	5	(7.5)	23w	30	17	20	17
Pd-46 ppb		12L		0.2L	*1	0.5	
Pr-59		3.2		8.3g	6.6	7.1	
Pt-78 ppb		(13)?			*1		
Rb-37	30	21.4	150	175	110	110	83
Re-75 ppb		(0.5)			*0.4	0.5	
Rh-45 ppb							
Ru-44 ppb		(0.2)			0.07		
S-16	1000w	(130)	300w	(80)	530		
Sb-51	0.2	1.04	0.2	(0.2)	0.2	0.2	
Sc-21	33w	35	5	7	14	11	13
Se-34	0.13w	(0.12)	0.14w		0.14	0.05	
Si-14%	23	24.5	33.3	32.7	30	30.8	30.9
Sm-62		3.68		5	4.5	4.5	4.6
Sn-50	2w	2.7	4w	2.7g	3.3	5.5	
Sr-38	470	186	290	310	350	350	290

Ta-73	0.5	0.48	2.0w	1.3	1.5	2.2	0.79
Tb-65		0.63		0.6	0.6	0.64	0.66
Te-52		(0.0074)	0.0013	0.0013	0.003		
Th-90	2w	2.4	15w	17	11	11	8.6
Ti-22	6000w	6400	1900	2300	3300	3000	3300
Tl-81	0.1	0.11	0.75	(0.008)?	0.53	0.75	
Tm-69		(0.34)		0.31	0.32	0.33	
U-92	0.5w	0.57	4w	5	2.8	2.8	2.2
V-23	270w	257	72w	38	140	60	86
W-74	0.8w	(0.46)	1.5w	(1.5)g	1.3	2	
Y-39	25	26	33?	21	22	22	24
Yb-70		2.03		2	2	2.2	1.9
Zn-30	100	84	50w	80	67	71	
Zr-40	150	99	180	150	170	190	160

Sources: (1) Taylor (1964) except major ions by Le Maitre (1976); (2) Govindaraju (1989); (3) Taylor and McLennan (1985); (4) Condie (1993) (f) Flanagan et al. (1982); (g) Govindaraju (1994); (L) Loss et al. (1983); (m) Marowsky and Wedepohl (1971); (s) Shaw et al. (1967, 1976); (w) Wedephol (1969–1978).

Notes: Question marks: doubtful data. Values with asterisks are adopted from the average shale (Table VI-5a). Values in parentheses for WI and GA are uncertified; thus they may be less reliable. Compositions in ppm unless indicated otherwise in the first column.

TABLE V-4B

Average compositions of juvenile upper continental crusts as a function of age and the present-day average composition of the upper continental crust

	Juvenile upper crust			*Present upper crust*		*Juvenile upper crust*			*Present upper crust*
	>3.5–2.5	2.5–0.8 (Ga)	0.8–0.2			>3.5–2.5	2.5–0.8 (Ga)	0.8–0.2	
SiO_2 %	66.25	65.92	65.69	66.21	Nb ppm	8.4	11.3	11.6	9.8
TiO_2	0.44	0.63	0.67	0.55	Ta	0.7	0.88	0.9	0.79
Al_2O_3	14.8	14.98	15.18	14.96	Y	18	30	30	24
FeO	4.5	4.77	4.99	4.7	La	29.5	28.5	25.5	28.4
MgO	3	2.15	2.23	2.42	Ce	56.7	59.3	56.3	57.5
CaO	3.55	3.5	3.82	3.6	Nd	25	26.7	25.3	25.6
Na_2O	3.66	3.29	3.36	3.51	Sm	4	5.25	5	4.59
K_2O	2.53	3.01	2.83	2.73	Eu	0.98	1.17	1.02	1.05
P_2O_5	0.11	0.14	0.14	0.12	Gd	3.53	4.9	4.76	4.21
					Tb	0.54	0.77	0.75	0.66
Rb ppm	73	93	92	83	Yb	1.53	2.25	2.31	1.91
Sr	294	290	267	289	Lu	0.26	0.36	0.41	0.32
Ba	569	692	695	633	Sc	10.9	15.9	15.6	13.3
Pb	18	17	16	17	V	72	97	108	86
Th	8	9.3	8.9	8.6	Cr	186	61	51	104
U	2.1	2.4	2.4	2.2	Co	21	17	16	18
Zr	148	173	168	160	Ni	105	33	28	56
Hf	3.9	4.9	4.5	4.3					

Source: Modified after Table 3 by Condie (1993).

TABLE V-5

Average compositions of the mixture of $2B/3 + G/3$, island arc andesites, Archean upper crust

	2B/3 + G/3	*Island arc andesites* (1)	*Archean upper crust* (1)	*Archean bulk crust* (1)	*Bulk continental crust* (1)	*Oceanic crust (MORB)* (2)
Ag-47	0.08				0.08	0.032h
Al-13%	8.1	9.5	8.1	8	8.4	8.1
As-33	1.5				1	
Au-79 ppb	3				3	*1.0h
B-5	8.3				10	
Ba-56	410	350	265	220	250	14
Be-4	1.8	1.5			1.5	
Bi-83	0.042				0.06	0.007h
Br-35	2.8					
C-6	270					140
Ca-20%	5	5.4	4.4	5.2	5.3	8.1
Cd-48	0.12				0.098	0.14h
Ce-58	41	38	42	31	33	12
Cl-17	110					
Co-27	32	25	25	30	29	47
Cr-24	134	55	180	230	185	300e
Cs-55	2.3	1.7			1	0.024
Cu-29	64	60		80	75	74

Table V-5
Continued

	2B/3 + G/3	Island arc andesites (1)	Archean upper crust (1)	Archean bulk crust (1)	Bulk continental crust (1)	Oceanic crust (MORB) (2)
Dy-66	3.8	3.7	3.4	3.6	3.7	6.3
Er-68	2.2	2.3	2.1	2.2	2.2	4.1
Eu-63	1.1	1.1	1.2	1.1	1.1	1.3
F-9	550					
Fe-26%	6.2	5.8	6.2	7.5	7.1	8.1
Ga-31	18	18			18	17e
Gd-64	3.9	3.6	3.4	3.2	3.3	5.1
Ge-32	1.5				1.6	1.5h
Hf-72	3	3	3	3	3	3
Hg-80	0.08					
Ho-67	0.77	0.82	0.74	0.77	0.78	1.3
I-53	0.5					
In-49	0.05				0.05	0.071h
Ir-77 ppb	~0.2				0.1	*0.046h
K-19%	1.7	1.3	1.5	0.75	0.91	0.088
La-57	20	19	20	15	16	3.9
Li-3	17	10			13	9e
Lu-71	0.31	0.3	0.31	0.33	0.3	0.59
Mg-12%	2.8	2.1	2.8	3.6	3.2	4.6
Mn-25	1200	1100	1400	1500	1400	1300e
Mo-42	1.4				1	0.46n
N-7	20					
Na-11%	2.35	2.6	2.45	2.2	2.3	2

Nb-41	13	11			11	3.5e
Nd-60	19	16	20	16	16	11
Ni-28	105	30	105	130	105	150
Os-76 ppb						* < 0.05h
P-15	1200					700e
Pb-82	10	10			8	0.49
Pd-46 ppb	~1				1	* < 0.89h
Pr-59	4.9	4.3	4.9	3.7	3.9	2.1
Pt-78 ppb	~1					
Rb-37	64	42	50	28	32	1.3
Re-75 ppb	0.4				0.5	1.1h
Rh-45 ppb						
Ru-44 ppb	~0.1					
S-16	770					800m
Sb-51	0.2				0.2	0.006h
Sc-21	24	30	14	30	30	41
Se-34	0.13				0.05	0.2h

TABLE V-5
Continued

	$2B/3 + G/3$	*Island arc andesites* (1)	*Archean upper crust* (1)	*Archean bulk crust* (1)	*Bulk continental crust* (1)	*Oceanic crust (MORB)* (2)
Si-14%	26.4	27.1	28.1	26.6	26.8	23.6
Sm-62	4.1	3.7	4	3.4	3.5	3.8
Sn-50	2.7				2.5	1.4
Sr-38	410	400	240	215	260	110
Ta-73	1				1	0.19
Tb-65	0.62	0.64	0.57	0.59	0.6	0.89
Te-52	0.005					*0.009h
Th-90	6.3	4.8	5.7	2.9	3.5	0.19
Ti-22	4800	4800	5000	6000	5400	9700
Tl-81	0.32				0.36	0.011h
Tm-69	0.3	0.32	0.3	0.32	0.32	0.62
U-92	1.7	1.3	1.5	0.75	0.91	0.071
V-23	204	175	195	245	230	290e
W-74	1				1	0.028n
Y-39	24	22	18	19	20	36
Yb-70	2.1	2.2	2	2.2	2.2	3.9
Zn-30	93				80	79h
Zr-40	160	100	125	100	100	100

Sources: (1) Taylor and McLennan (1985); (2) Hofmann (1988); $2B/3 + G/3$ from Table V-3. e: Engel et al. (1965); h: Hertogen et al. (1980), Laul et al. (1972); m: Moore and Fabbi (1971); n: Newsom et al. (1986).

Note: Asterisks represent the concentration ranges for Au: 0.06–6; Ir: 0.001–0.5; Os: <0.0001–0.09; Pd: <0.1–7; Te: 0.002–0.017.

part granite (2B/3 + G/3 in Table V-5) can also reproduce the andesitic composition quite well (within a factor of 1.5 or better), except for Cr and Ni. The Archean volcanic assemblages in the Canadian shield are similar to those of developed island arcs (Goodwin, 1977; mostly tholeiitic and calc-alkali magma series). The average compositions of the Archean upper and bulk continental crust given by Taylor and McLennan (1985) are also summarized in Table V-5. The agreement between the compositions of the Archean upper crust by Taylor and McLennan (1985) and the mixture of 2B/3 + G/3 is again remarkable (within a factor of 1.5 or better). The only exceptions are Ba, Sc, and Sr which are high in 2B/3 + G/3. Condie's (1993, Table V-4b) Archean upper crust gives much higher Ba, Ce, La, Pb, Th, and U and much lower Ca, Fe, Mg, Ti, and V concentrations as compared to those by Taylor and McLennan (1985). The causes of the differences are not obvious.

The estimate of the average composition of the whole continental crust summarized in Table V-5 is given by Taylor and McLennan (1985), who assume a mixture of three parts Archean crust and one part post-Archean crust. The mixture of 2B/3 + G/3 again represents the bulk continental crust quite well, except for K, Rb, Cs, Sr, Ba, Th, and U, which tend to be too high in the 2B/3 + G/3 mixture. Wedepohl (1995) provided another estimate of the average composition of the whole continental crust. However, there is no really independent way to check whose estimate is better.

The oceanic crust can be represented by the mid-oceanic ridge basalts (MORB) and its average composition is summarized in Table V-5. The concentrations of platinum group elements such as Os, Ir, Ru, Pd, and Au in MORB are highly variable, and their concentration ranges are also given in the footnote of Table V-5.

V-7. Relationship between the Compositions of Mantle and Crust

The concentrations of elements in the continental crust (0.75 × Archean crust + 0.25 × island arc; Table V-5) divided by those in the primitive mantle (Table V-2) are shown in Figure V-15 in decreasing order of the ratios. For comparison, similar data for the oceanic crust (MORB) are also plotted in Figure V-15. As noted by Hofmann (1988), the depletion of elements from Cs to Hf in the oceanic crust relative to the continental crust suggests the following. The oceanic crust was differentiated from a mantle that was already depleted in those elements due to earlier differentiation of the continental crust from the primitive mantle. As mentioned earlier, the concentrations of Au, Pd, Os, Ir, and Ru in MORB are highly variable and highly depleted as compared to the continental crust or to the primitive mantle (Figure V-15). The possible explanation is that MORB liquids were sulfur saturated during the partial melting of the depleted mantle, and that immiscible sulfide phases, which were enriched in these elements, were retained in various degrees in the mantle residue (Hamlyn et al., 1985; Bruegmann et al., 1987). According to Hofmann's (1988) two-stage model of crust-mantle differentiation, the **continental crust** was extracted from a portion of the **primitive mantle** (PM) by partial melting processes during the first stage. The residual mantle became the **depleted mantle** (DM1). The **oceanic crust** (MORB) was the partial melt from a portion of the depleted mantle, and the residual became the **extra-depleted mantle** (DM2). By mass balance, one obtains

$$X_0 = X_1(1 - F_1) + C_1 F_1, \tag{V-9a}$$

$$X_1 = X_2(1 - F_2) + C_2 F_2, \tag{V-9b}$$

where X_0, X_1, and X_2 are the concentrations of a given element in PM, DM1, and DM2, respectively; C_1 and C_2 are the concentrations of the given element in the continental and oceanic crusts; F_1 and F_2 are continental and oceanic crusts as fractions of the PM and DM1 portions involved in partial melting, respectively.

Letting

$$\frac{X_1}{C_1} = D_1, \text{ and } \frac{X_2}{C_2} = D_2, \tag{V-9c}$$

then, from equations V-9a–c, one obtains

$$\frac{C_1}{X_0} = \frac{1}{D_1(1 - F_1) + F_1}, \tag{V-10a}$$

$$\frac{C_2}{X_1} = \frac{1}{D_2(1 - F_2) + F_2}, \tag{V-10b}$$

or

$$\frac{C_2}{X_0} = \frac{D_1}{[D_1(1 - F_1) + F_1][D_2(1 - F_2) + F_2]}. \tag{V-10c}$$

For highly incompatible elements, D_1 is negligibly small as compared to F_1; then equation V-10a reduces to

$$F_1 \approx 1/(C_1/X_0).$$

As shown in Figure V-15, the largest C_1/X_0 value among the highly incompatible Rb, Cs, Tl, Sb, and W is probably in the range of 60 ± 10. Therefore,

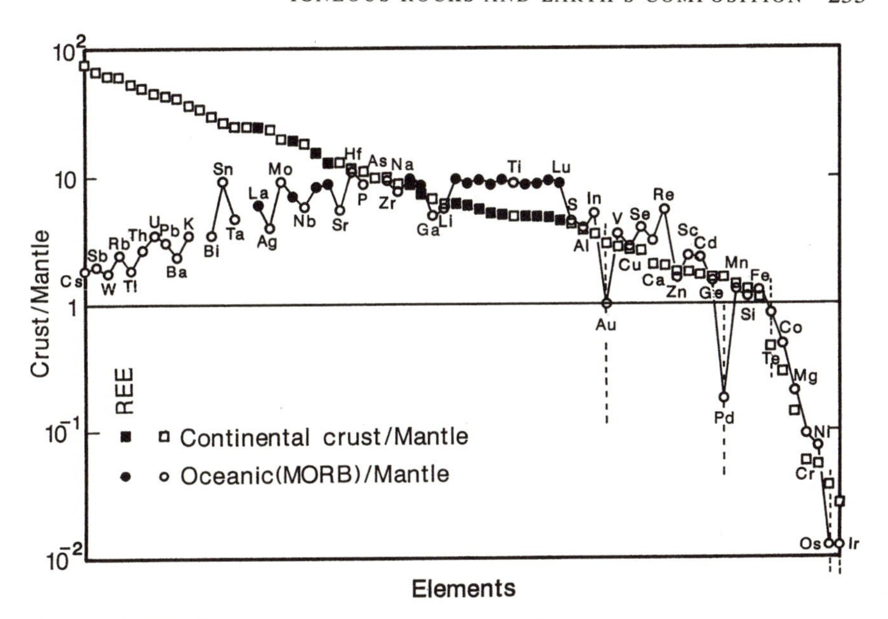

FIGURE V-15. Ratios of the concentrations of elements in the continental crust (0.75 × Archean crust + 0.25 × Island arc; Table V-4) over those in the primitive mantle (Table V-2) plotted in decreasing order (squares). For comparison, similar data for the oceanic crust (MORB, circles; Table V-5) are also plotted according to the same decreasing order for the continental crust. The solid squares and circles are rare earth elements.

F_1 is about 0.017 ± 0.003. Once F_1 is fixed and C_1/X_0 values are known, D_1 for all elements can be calculated from the rearranged equation V-10a, i.e.,

$$D_1 = \frac{1 - F_1 C_1/X_0}{(1 - F_1)C_1/X_0} \quad \text{(V-11)}$$

and X_1/X_0 follows from $D_1 C_1/X_0$.

Furthermore, if one assumes $D_1 = D_2$ for all elements, then F_2 can be roughly estimated from equation V-10c using C_2/X_0 for Sn or Mo, which is representative of the average maximum C_2/X_0 value in Figure V-15. The best estimate for F_2 is 0.045. Once F_2 is known, and D_2 is assumed to be equal to D_1 for all elements, then C_2/X_0 and X_2/X_0 can be calculated from equations V-10c and V-9c, respectively.

The results for C_1/X_0, X_1/X_0, C_2/X_0, and X_2/X_0 as a function of D_1 ($= D_2$) are summarized in Figure V-16a as solid curves. Interestingly, the

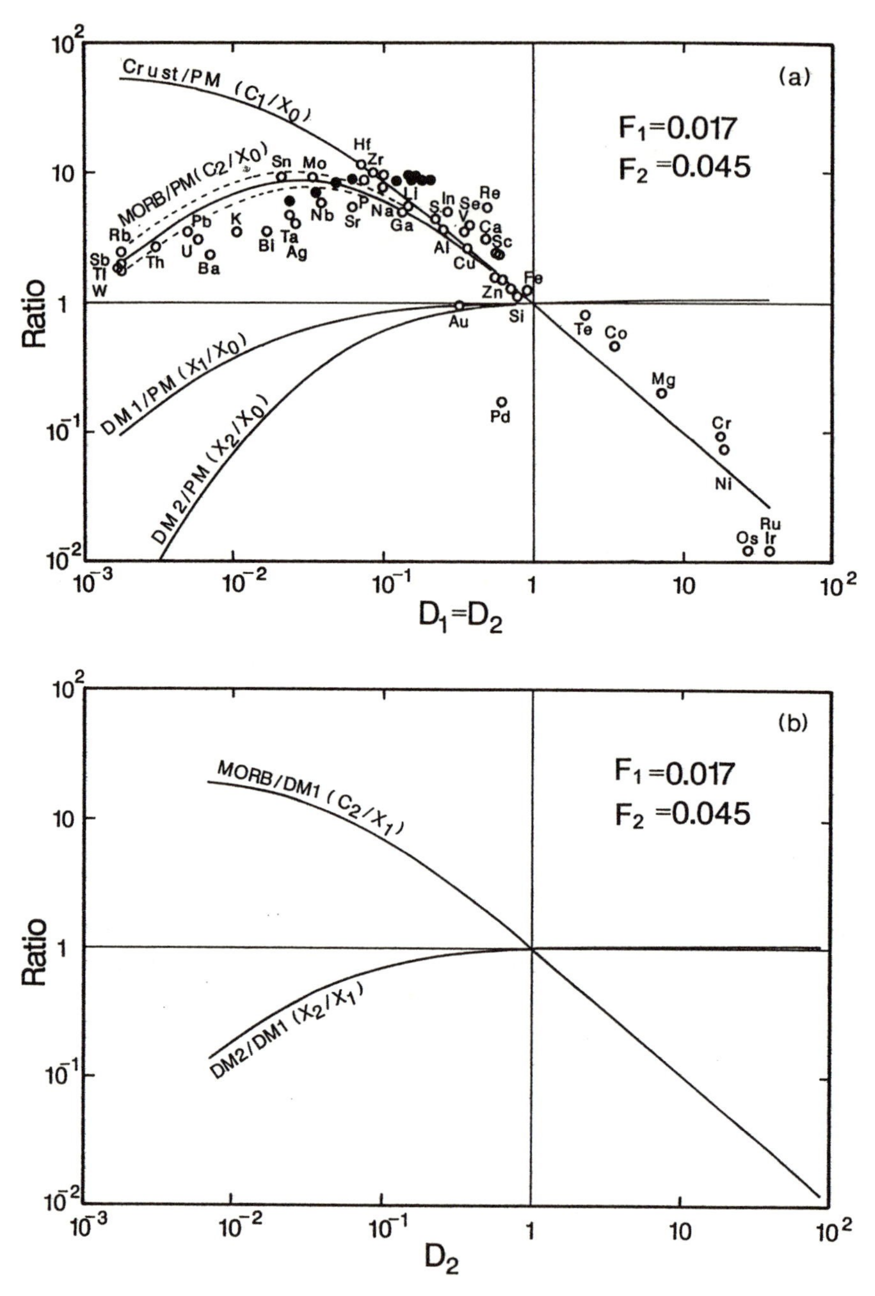

FIGURE V-16. (a) Predicted concentrations of elements in the continental crust (C_1), oceanic crust (C_2), depleted mantle (X_1), and extra-depleted mantle (X_2) relative to the primitive mantle (X_0) as a function of bulk distribution coefficients D_1 ($= X_1/C_1$) according to equations V-9c, V-10a, and V-10c, assuming $F_1 = 0.017$, $F_2 = 0.045$, and $D_2 = D_1$. The dashed curves represent C_2/X_0 for $F_2 = 0.035$ (top) and $F_2 = 0.055$ (bottom). (b) Predicted concentrations of elements in the oceanic crust (C_2) and the extra-depleted mantle (X_2) relative to the depleted mantle (X_1) as a function of D_2 ($= X_2/C_2$), assuming $F_1 = 0.017$ and $F_2 = 0.045$.

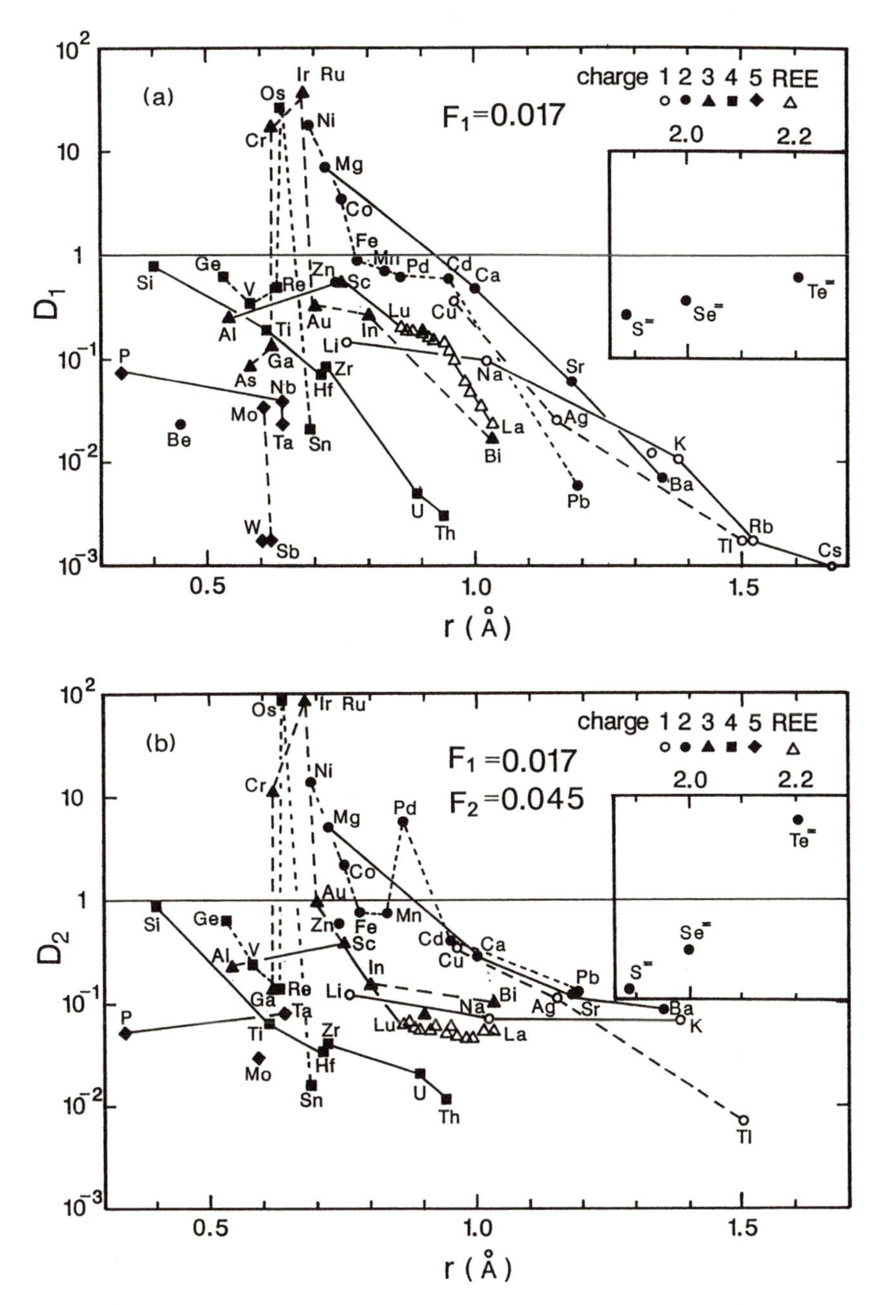

FIGURE V-17. Bulk distribution coefficients of elements, D_1 (depleted mantle/continental crust) and D_2 (extra-depleted mantle/oceanic crust) as a function of ionic radii. The elements with the same charge and geochemical type (lithophile or chalcophile + siderophile) are connected by a solid or dashed line.

general trends for the predicted (solid line) and observed C_2/X_0 (circles) values as a function of D_1 agree quite well except for Au and platinum group elements. Therefore the two-stage model of crust-mantle differentiation by Hofmann (1988) is quite an adequate approximation.

One also can relax the constraint of $D_1 = D_2$. Any observed C_2/X_0 point falling below (or above) the predicted line in Figure V-16a may indicate $D_2 > D_1$ (or $D_2 < D_1$). Actually, D_2 can be calculated from equation V-10c by adopting F_1, F_2, and D_1 values obtained earlier and knowing C_2/X_0. Once D_2 is known, X_2/X_1 can be calculated from equation V-9c. These are plotted against D_2 in Figure V-16b along with C_2/X_1 (equation V-10b). The D_1 (from equation V-11) and D_2 (equation V-10c) values are also plotted against the ionic radii in Figures V-17a and V-17b (see page 235). Even though the exact oxidation states (and thus ionic radii) for some elements are uncertain (e.g., tri- or tetravalent for Ir, Rh, Os, Re, and V; penta- or hexavalent for Mo and W), the general trend is that D_1 and D_2 decrease with increasing ionic radii for elements with the same charge and geochemical type (lithophile versus chalcophile + siderophile). A few exceptions are smaller cations such as Be, Al, As, and Ga, whose trend is reversed. The platinum group elements, Au, Te, and the incompatible elements tend to have much higher D_2 than D_1 values.

The total mass of the continental crust is about $0.0037m_E$. Thus with $F_1 = 0.017 \pm 0.003$, the mass of the depleted mantle (DM1) should be about $(0.22 \pm 0.06)m_E$ [$= 0.0037(1 - F_1)/F_1$], which is comparable to the present upper mantle mass of $0.185m_E$. Similarly, with an oceanic crust of $0.001m_E$ and $F_2 = 0.045 \pm 0.010$, the mass of the DM2 type mantle differentiated from DM1 should be about $(0.022 \pm 0.005)m_E$ [$= 0.001(1 - F_2)/F_2$]. The implication is that the upper mantle may consist mostly of the DM1 and DM2 type depleted mantle and the lower mantle of the undepleted primitive mantle type. It is also probable that the undepleted and depleted mantle types are randomly distributed within the whole mantle, if there have been finite material exchanges between the upper and lower mantles as discussed in Section V-2. The recycled crust materials in the mantle will be discussed in the next section.

V-8. Isotopic Heterogeneity of the Mantle

As discussed in Section V-4, the present upper mantle as represented by spinel lherzolite xenoliths is quite heterogeneous with regard to the concentrations of incompatible elements. Because many radioactive parent-daughter pairs are themselves incompatible elements, it is not unexpected to find extreme heterogeneity of isotopic compositions of those parent-daughter pairs in the mantle-derived magma and xenoliths. Zindler and Hart (1986), Hart (1988), and Hofmann (1997) provided useful reviews on this subject.

The decay of a radioactive parent A into a radiogenic stable daughter B^* in a homogeneous reservoir can be related by these equations:

$$-\frac{dA}{dt} = \frac{dB^*}{dt} = \lambda A,$$

or in integrated form

$$A_2 = A_1 e^{-\lambda(t_2 - t_1)}, \tag{V-12a}$$

$$B_2^* = B_1^* + A_1(1 - e^{-\lambda(t_2 - t_1)}), \tag{V-12b}$$

where A_1 and A_2 are the numbers of parent atoms per unit mass of the reservoir at times t_1 and later t_2, respectively (similarly for the daughter atoms B_1^* and B_2^*); λ is the decay constant of parent A ($= \ln 2/t_{1/2}$), where, $t_{1/2}$ is the half-life of parent A.

Dividing equations V-12a and V-12b by B (the numbers of a nonradiogenic isotope of B^* per unit mass of the reservoir), one obtains

$$\left(\frac{A}{B}\right)_2 = \left(\frac{A}{B}\right)_1 e^{-\lambda(t_2 - t_1)}, \tag{V-13a}$$

$$\left(\frac{B^*}{B}\right)_2 = \left(\frac{B^*}{B}\right)_1 + \left(\frac{A}{B}\right)_1 (1 - e^{-\lambda(t_2 - t_1)}). \tag{V-13b}$$

If t_2 represents the time at present, then $(t_2 - t_1)$ is a time before the present, i.e., the age τ_1; thus $\tau_1 = 0$ is the present (P) and equations V-13a and V-13b become

$$\left(\frac{A}{B}\right)_P = \left(\frac{A}{B}\right)_{\tau_1} e^{-\lambda\tau_1}, \tag{V-14a}$$

$$\left(\frac{B^*}{B}\right)_P = \left(\frac{B^*}{B}\right)_{\tau_1} + \left(\frac{A}{B}\right)_{\tau_1} (1 - e^{-\lambda\tau_1}) \tag{V-14b}$$

$$= \left(\frac{B^*}{B}\right)_{\tau_1} + \left(\frac{A}{B}\right)_P (e^{\lambda\tau_1} - 1). \tag{V-14c}$$

Because $(B^*/B)_P$ and $(A/B)_P$ at present are measurable, $(B^*/B)\tau_1$ can be estimated as a function of the age τ_1. If $(B^*/B)_P$ and $(A/B)_P$ for at least two different mineral phases in a given rock are measured, then both $(B^*/B)\tau_1$ and τ_1 can be estimated from more than two simultaneous equations similar to equation V-14c, assuming $(B^*/B)\tau_1$ is the same for all minerals. For chondrites, τ_1 is estimated to be 4.55 to 4.57 Ga, which is also the age of the

TABLE V-6
Isotopic ratios of chondritic Earth at present time (p) and 4.57 billion years ago (E) along with the half-life ($t_{1/2}$) of the radioactive parent (A). $^*B$ is the radiogenic daughter isotope and B is its nonradiogenic isotope

A	$^*B$	B	$t_{1/2}(10^9 y)$	$(A/B)_p$	$(^*B/B)_p$	$(^*B/B)_E$
^{238}U	^{206}Pb	^{204}Pb	4.468 ± 2	8.0	17.56 ± 1	9.3066 ± 4
^{235}U	^{207}Pb	^{204}Pb	0.7038 ± 5	8/137.88	15.46 ± 1	10.293 ± 1
^{232}Th	^{208}Pb	^{204}Pb	14.01 ± 10	34	38.1 ± 1	29.475 ± 1
^{87}Rb	^{87}Sr	^{86}Sr	48.8 ± 4	0.086	0.7047 ± 7	0.6989 ± 1
^{147}Sm	^{143}Nd	^{144}Nd	106 ± 2	0.187	0.5126 ± 1	0.5068 ± 1
^{176}Lu	^{176}Hf	^{177}Hf	35.7 ± 8	0.0332	0.2828 ± 1	0.2797 ± 1
^{187}Re	^{187}Os	^{188}Os	42.3 ± 13	0.4224	0.1288 ± 3	0.0960 ± 10

Note: The uncertainty ($\pm$) is expressed in the magnitude of the last significant number.

Earth. We may designate $\tau_E = 4.57 \times 10^9$ years. The estimated $(B^*/B)_E$ values for many parent-daughter pairs of incompatible elements in chondrites, which may also represent the Earth's primitive mantle values 4.57 billion years ago, are summarized in Table V-6 along with $(A/B)_P$ and $(B^*/B)_P$ for the primitive mantle. The $(^{147}Sm/^{143}Nd)_P$ and $(^{176}Lu/^{176}Hf)_P$ ratios are again based on the chondrite data (Table IV-6). Once $(A/B)_P$, $(B^*/B)_E$, and τ_E are given, $(B^*/B)_P$ can be calculated from equation V-14c. $(A/B)_P$ and $(B^*/B)_P$ for other parent-daughter pairs in the primitive mantle are estimated by fitting the observed $(B^*/B)\tau_1$ for many terrestrial mafic rocks with known ages to equation V-14c (e.g., Zindler, 1982; Faure, 1986).

If at t_2 a portion of the reservoir (e.g., the primitive mantle) differentiated into reservoir I (e.g., the depleted upper mantle) and reservoir II (e.g., the continental crust), then at t_3 ($> t_2$) one obtains relationships similar to equations V-13a and V-13b for reservoir I:

$$\left(\frac{A}{B}\right)_3^{\mathrm{I}} = \left(\frac{A}{B}\right)_2^{\mathrm{I}} e^{-\lambda(t_3-t_2)}, \tag{V-15a}$$

$$\left(\frac{B^*}{B}\right)_3^{\mathrm{I}} = \left(\frac{B^*}{B}\right)_2^{\mathrm{I}} + \left(\frac{A}{B}\right)_2^{\mathrm{I}} (1 - e^{-\lambda(t_3-t_2)}), \tag{V-15b}$$

where the superscript I represents isotopic ratios for the reservoir I. Also, for the original undifferentiated reservoir

$$\left(\frac{A}{B}\right)_3 = \left(\frac{A}{B}\right)_1 e^{-\lambda(t_3-t_1)}.$$

At t_2 one also assumes

$$\left(\frac{B^*}{B}\right)_2 = \left(\frac{B^*}{B}\right)_2^{\mathrm{I}} = \left(\frac{B^*}{B}\right)_2^{\mathrm{II}}. \tag{V-15c}$$

Therefore if t_3 represents the time at present and $t_3 - t_1 = \tau_1$ and $t_3 - t_2 = \tau_2$, one obtains

$$\left(\frac{B^*}{B}\right)_P^{\mathrm{I}} = \left(\frac{B^*}{B}\right)_{\tau_1} + \left(\frac{A}{B}\right)_P (e^{\lambda\tau_1} - e^{\lambda\tau_2}) + \left(\frac{A}{B}\right)_P^{\mathrm{I}} (e^{\lambda\tau_2} - 1) \qquad \text{(V-16)}$$

by substituting equations V-13b, V-15a, and V-15c into V-15b. For reservoir II, one simply substitutes the superscript I with II in equations V-15a, V-15b, and V-16. If the bulk distribution coefficient of parent over that of daughter elements is greater than one ($D_A/D_B > 1$), which is the case for Sm/Nd and Lu/Hf (Figures V-17a and V-17b), then

$$\left(\frac{A}{B}\right)_P^{\mathrm{I}} > \left(\frac{A}{B}\right)_P > \left(\frac{A}{B}\right)_P^{\mathrm{II}}.$$

If $D_A/D_B < 1$, which is the case for Rb/Sr, U/Pb, Th/Pb, and Re/Os (Figures V-17a and V-17b), then the above order is reversed. Figures V-18a and V-18b provide examples of a two-stage Earth evolution model for ^{143}Nd/^{144}Nd and ^{87}Sr/^{86}Sr, assuming $\tau_2 = 3$ Ga.

Figures V-19a to V-19d summarize the present ranges of ^{87}Sr/^{86}*Sr*, ^{143}Nd/^{144}Nd, ^{206}Pb/^{204}Pb, ^{207}Pb/^{204}Pb, and ^{208}Pb/^{204}Pb ratios for the mid-oceanic ridge basalts (MORB) and the oceanic island basalts (OIB), along with marine sediments, Lashaine granulite xenoliths from East Africa, and other rock types as references. Because the ages of the MORB and OIB samples are mostly young (< 50 Ma), their isotopic ratios do not change much and can represent those for their mantle sources at present time. The isotopic ranges for MORB and OIB fall mostly within the four extrema in Figures V-19a to V-19d designated as DM1 (depleted mantle), EM1 (enriched mantle), EM2, and HIMU (high μ). DM1 is the depleted upper mantle; EM1 is represented by Walvis Ridge basalts (Richardson et al., 1982) and is probably related to the primitive mantle source (PM); EM2 is representative of the oceanic crust mixed with small amount of continent-derived sediments; and HIMU is represented by OIB of Mangaia (Palacz and Saunders, 1986). HIMU members are characterized by high ^{206}Pb/^{204}Pb and ^{208}Pb/^{204}Pb ratios (i.e., also high ^{238}U/^{204}Pb and ^{232}Th/^{204}Pb ratios in the source region), but low ^{87}Sr/^{86}Sr and intermediate ^{143}Nd/^{144}Nd (i.e., also low ^{87}Rb/^{86}Sr and moderate ^{147}Sm/^{144}Nd ratios in the source region). HIMU may represent the subducted ancient oceanic crust (Zindler et al., 1982; Hofmann, 1997) from which Rb, Sr, Pb, and REE but not U and Th were effectively removed by hydrothermal solutions (Chen et al., 1986) during the subduction processes.

Some MORB samples from the Indian ocean, especially sample D5 (Hamelin and Allègre, 1985), require an additional component such as Lashaine granulites from East Africa (Cohen et al., 1984), which may represent subcontinental reservoirs at the upper mantle and lower crust boundary. As is evident

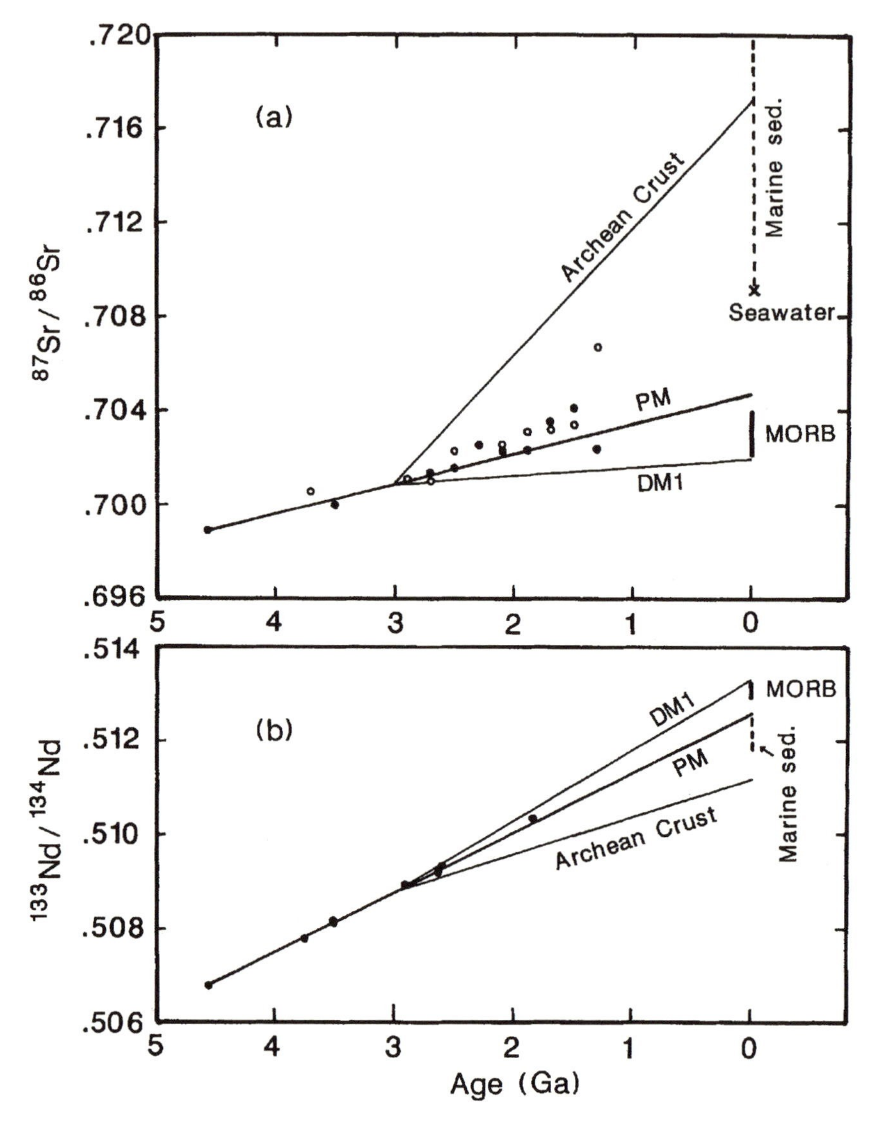

FIGURE V-18. Two-stage evolution model of the Earth for (a) $^{87}Sr/^{86}Sr$ and (b) $^{133}Nd/^{134}Nd$, assuming fractionation of the primitive mantle (PM) into Archean crust and the depleted mantle (DM1) at 3 Ga. The open and solid circles are data given by the Basaltic Volcanism Study Project (1981, Chapter 7.4) for Sr isotopes and by Zindler (1982) for Nd isotopes.

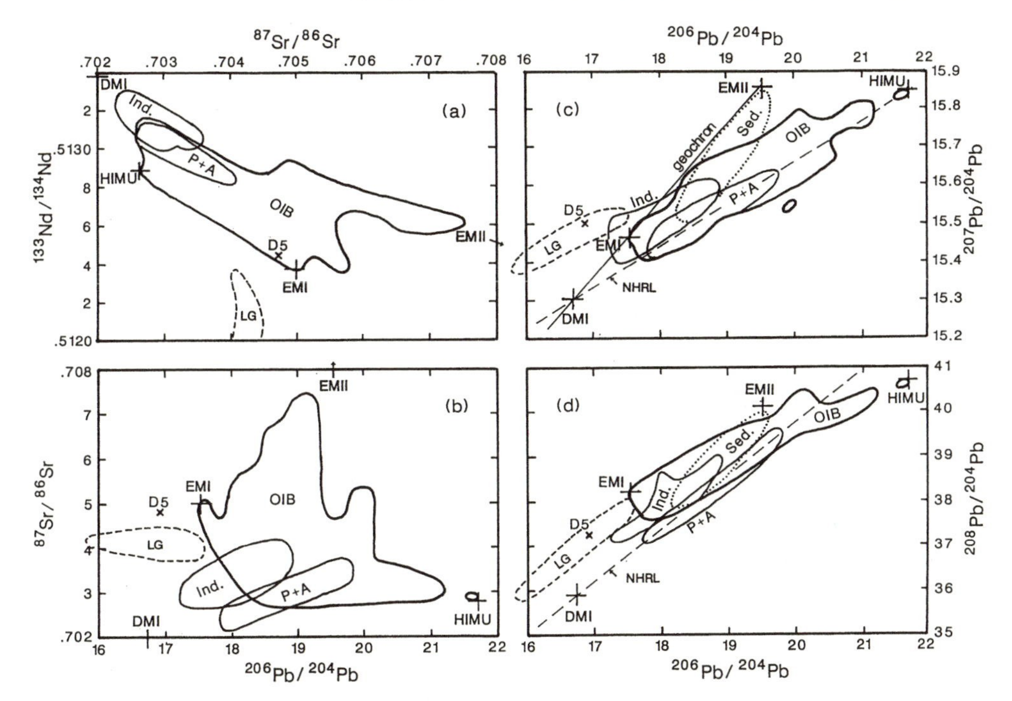

FIGURE V-19. Correlation plots of various isotopes show the fields for Indian (Ind; Mahoney et al., 1989) and Pacific + Atlantic (P + A; Wilson, 1989) mid-oceanic ridge basalts, the oceanic island basalts (OIB; Hart, 1988), Lashaine granulites (LG; Cohen et al., 1984), and marine sediments (sed; Othman et al., 1989). The point D5 is a MORB from the Indian Ocean (Hamelin and Allègre, 1985). See the text for explanation of the end members DM1, EMI, EMII, and HIMU. EMII in (a) has values of $^{87}Sr/^{86}Sr \approx 0.722$ and $^{143}Nd/^{144}Nd \approx 0.512$.

from Figures V-19a to V-19d, once the Lashaine granulite type component is introduced, the EM1 component may not be required to explain the observed large-scale isotope anomaly for MORB and OIB in the southern hemisphere (Hart, 1984). If Lashaine granulites are representative of lower continental crust material, the radiogenic Pb isotope reservoir in the lower crust is still not sufficiently radiogenic poor to balance the radiogenic-rich upper mantle and upper crust reservoirs (Rudnick and Goldstein, 1990). The missing radiogenic-poor reservoir for Pb isotopes must be found in the extra-depleted mantle DM2. Recent studies on osmium isotopes (Walker et al., 1995; Widom and Shirey, 1996) further suggest that the observed high $^{187}Os/^{188}Os$ ratio in some mantle plume magmas may result from the entrainment of the outer core material by mantle plumes at the core-mantle boundary. Certainly, osmium isotope studies will shed more light on the origin of those end members.

V-9. Case Studies of Elemental Association in Igneous Rocks

Mid-Ocean Ridge Basalts

The mid-Atlantic ridge basalts from 29 to 73°N were extensively studied by Schilling et al. (1983). The basalts from 29 to 53°N and 70 to 73°N are called group B basalts, and from 53 to 70°N group A basalts. In general, the former is enriched in Na, K, REE, and P as compared to the latter. Both groups consist of normal mid-ocean ridge basalts (n-MORB), which have E^i_{Lu} values of less than one for light REE relative to chondrite; plume mid-ocean ridge basalts (p-MORB), which have E^i_{Lu} of greater than one for light REE and have relatively high $^{87}Sr/^{86}Sr$ ratios as compared to n-MORB; and transitional zone basalts, whose REE distribution pattern is between those of n- and p-MORB (Schilling et al., 1983). As discussed in Sections V-7 and V-8, the n-MORB are thought to have originated from the depleted upper mantle (DM1) through various degrees of partial melting and subsequent fractional crystalllization, whereas the p-MORB are related to input from enriched mantle sources.

For illustration purposes, only the factor analysis results from group A are presented here (Figure V-20). For convenience, group A basalts are subdivided according to latitude into zone 3 (53–61°N), transitional zone (61–63.6°N), Iceland hot spot (63.6–66°N), and zone 4 (66–70°N). Zone 3 and 4 basalts are tentatively identified as n-MORB, and Iceland basalts as mostly p-MORB by Schilling et al. (1993). The close association of Mg, Cr, Ni, Al, and Ca (as negative factor 1 in Figure V-20a) suggests that olivine (Mg_2SiO_4), clinopyroxene ($MgCaSi_2O_6$), Cr-spinel ($MgCr_2O_4$), and plagioclase ($CaAl_2Si_2O_8$) are the most probable crystallizing minerals during the fractional crystallization of both n- and p-MORB magmas (Schilling et al, 1983). The incompatible Fe, V, Mn, and Na as factor 1 are concentrated in the residual magma during the fractional crystallization processes (Figure V-20a). The highly incompatible K, P, Ti, and light REE (La, Ce, Eu, Sm) as factor 2 may represent inputs from the fertile mantle as metasomatic fluids or partial melts of low degree. As is evident from the factor score plot (Figure V-20b), most of the Iceland basalts (mainly p-MORB) are characterized by high factor 2 scores. In contrast, the basalts from zones 3 and 4 (mainly n-MORB) can be explained by a change in factor score 1 (i.e., fractional crystallization). The transitional zone basalts and a few Iceland basalts can mostly be produced by mixing n- and p-MORB (Figure V-20b). However, a few of the transitional zone basalts are in the field of p-MORB and one in n-MORB, suggesting some interfingering between n- and p-MORB in the transitional zone (61–63.6°N).

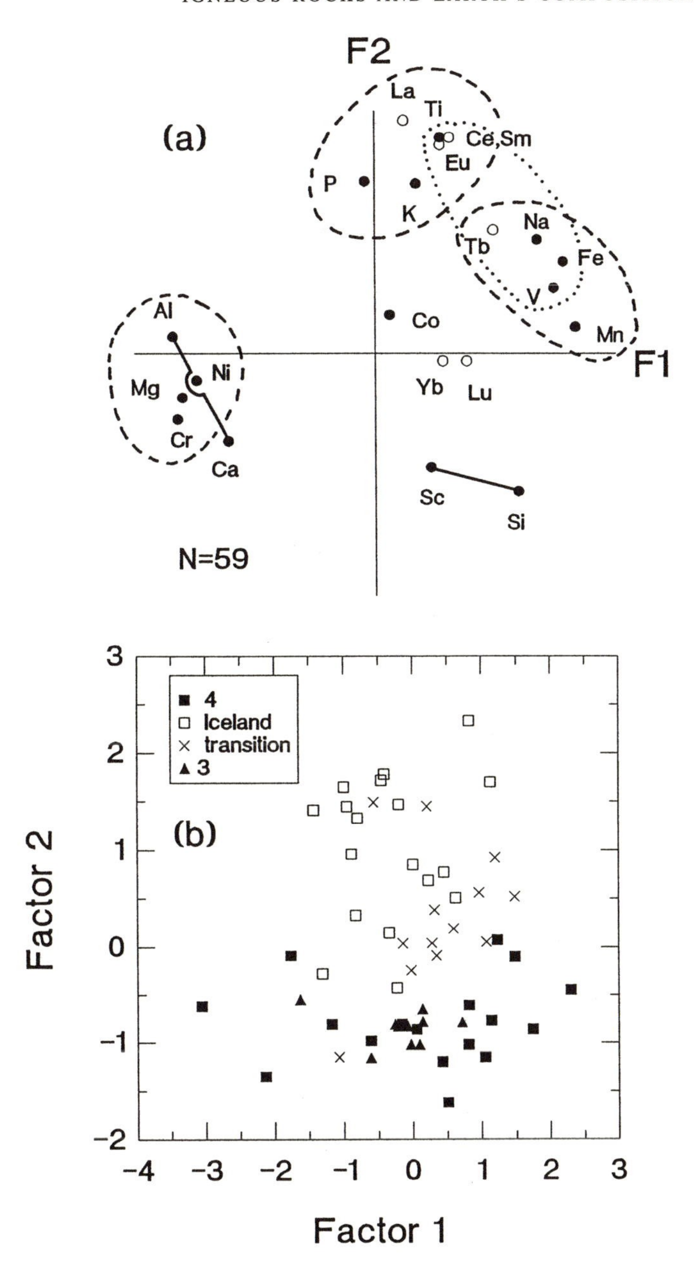

FIGURE V-20. (a) Factor loadings and (b) factor scores of factors 1 and 2 for the group A mid-Atlantic ridge basalts (Schilling et al., 1983). Elements within each dotted enclosure or connected by the solid line in (a) have correlation coefficients of greater than 0.5 between any pair.

Some selected so-called **Harker diagrams** (Figure V-21) confirm the conclusion given above. For example, the rapid decrease of Ni, Ca, and Al with decreasing MgO content for the n-MORB and transitional zone basalts (Figure V-21a–c) reflects the fractional crystallization of olivine, plagioclase, and clinopyroxene. One sample with MgO = 13% is probably a cumulate of olivine crystals, because this sample has very high Ni (Figure V-21a) and Cr, but low Al and Ca (Figures V-21b and V-21c).

The systematic enrichment of La (Figure V-21d) as well as of other factor 2 elements (Ti, P, K, light REE, etc.) in p-MORB as compared to n-MORB at any given MgO content indicates that the p-MORB parental magma was already enriched in these incompatible elements before the fractional crystal-

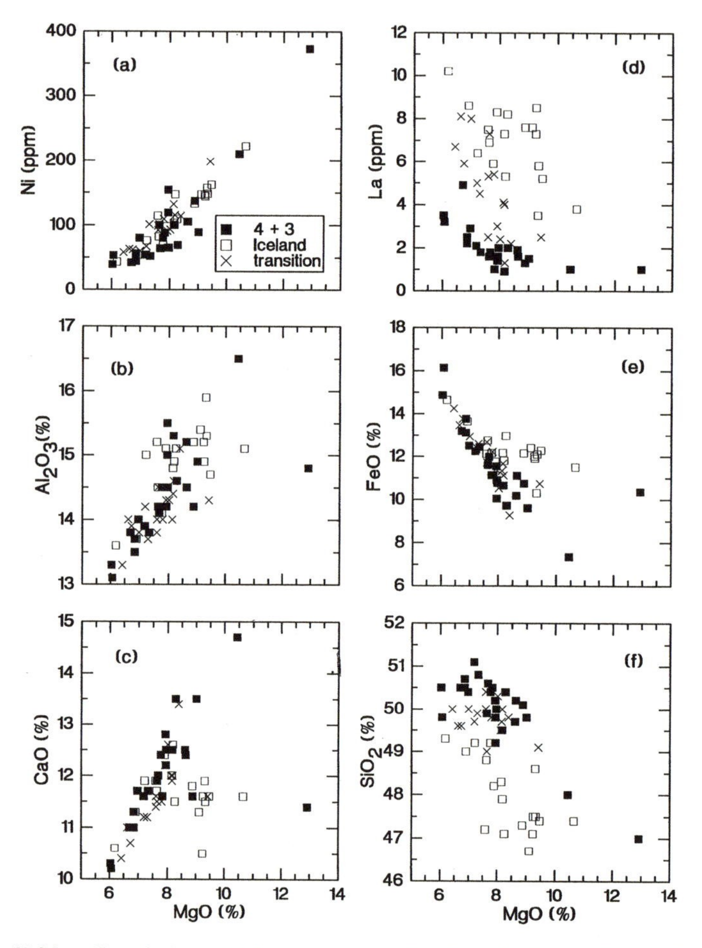

FIGURE V-21. Correlation of MgO versus (a) Ni, (b) Al_2O_3, (c) CaO, (d) La, (e) FeO, and (f) SiO_2 in group A mid-Atlantic ridge basalts.

lization. In contrast, the concentration differences between n- and p-MORB are small for FeO (Figure V-21e) and other factor 1 elements (Mn, V, and Na). Finally, the systematically low silica content (Figure V-21f) as well as Sc in Iceland p-MORB relative to n-MORB may indicate that the parental magma of p-MORB was produced at higher pressure than that of n-MORB (Schilling et al., 1983).

The concentration ratios of pairs of highly incompatible elements in magma, such as Rb/K, Ba/Sr, Ba/K, La/Ce, La/Sm, Ta/Hf, Nb/Zr, and Th/U, whose bulk solid/liquid distribution coefficients are similar and small, will not be changed much by magmatic differention processes. Also, the stable isotopic ratios of Sr, Pb, and Nd in magma are not affected by differentiation. Therefore, two-dimensional plots of isotopic data and ratios of incompatible elements provide a useful means to distinguish sources of magma. For example, the plots between $^{87}Sr/^{86}Sr$, $^{206}Pb/^{204}Pb$, and La/Sm in Figures V-22a and V-22b show distinct sources for p- and n-MORB, whereas transition type MORB can be produced by mixing p- and n-MORB, as mentioned earlier.

Oceanic Island Basalts

The Haleakala Volcano on the island of Maui, Hawaii, is divided into three major units, i.e., the Honomanu unit of tholeiitic and alkalic basalts introduced during the shield building stage (> 0.8 Ma); the Kula unit of alkalic basalts and evolved hawaiites and mugearites formed during the post-shield stage (0.36–0.8 Ma); and the Hana unit of alkalic basalts and basanites extracted during the post-erosional stage (< 0.36 Ma) (Chen et al., 1990, and references therein).

The plot of $^{87}Sr/^{86}Sr$ vs. $^{206}Pb/^{204}Pb$ (Figure V-22c) shows distinct sources for Honomanu, Kula, and Hana unit magmas. West and Leeman (1987) also suggested a three-component mixing model (EM1, DM1, HIMU in Figure V-19b) for Hawaiian volcanic rocks. The plot of $^{87}Sr/^{86}Sr$ versus Nb/Zr (Figure V-22d) shows two-end-member mixing, but is consistent with the above explanation.

The factor analysis results and some Harker diagrams of chemical data from the Haleakala volcano (Chen et al., 1990, 1991) are summarized in Figures V-23 and V-24. The association of Ni, Cr, Mg, and Co as negative factor 1 (Figure V-23a), as well as the rapid decrease of Ni (as well as Cr) and the moderate decrease of Co with decreasing MgO content from 19 to 5% (Figures V-24a and V-24b), indicate the removal of Cr-spinel and olivine in the early stage of fractional crystallization. The association of Fe, V, Ca, Sc, and Co plus Ti as factor 2 (Figure V-23a), and the rapid decrease of these elements below a MgO content of 5% (Figures V-24b and V-24c) indicate the

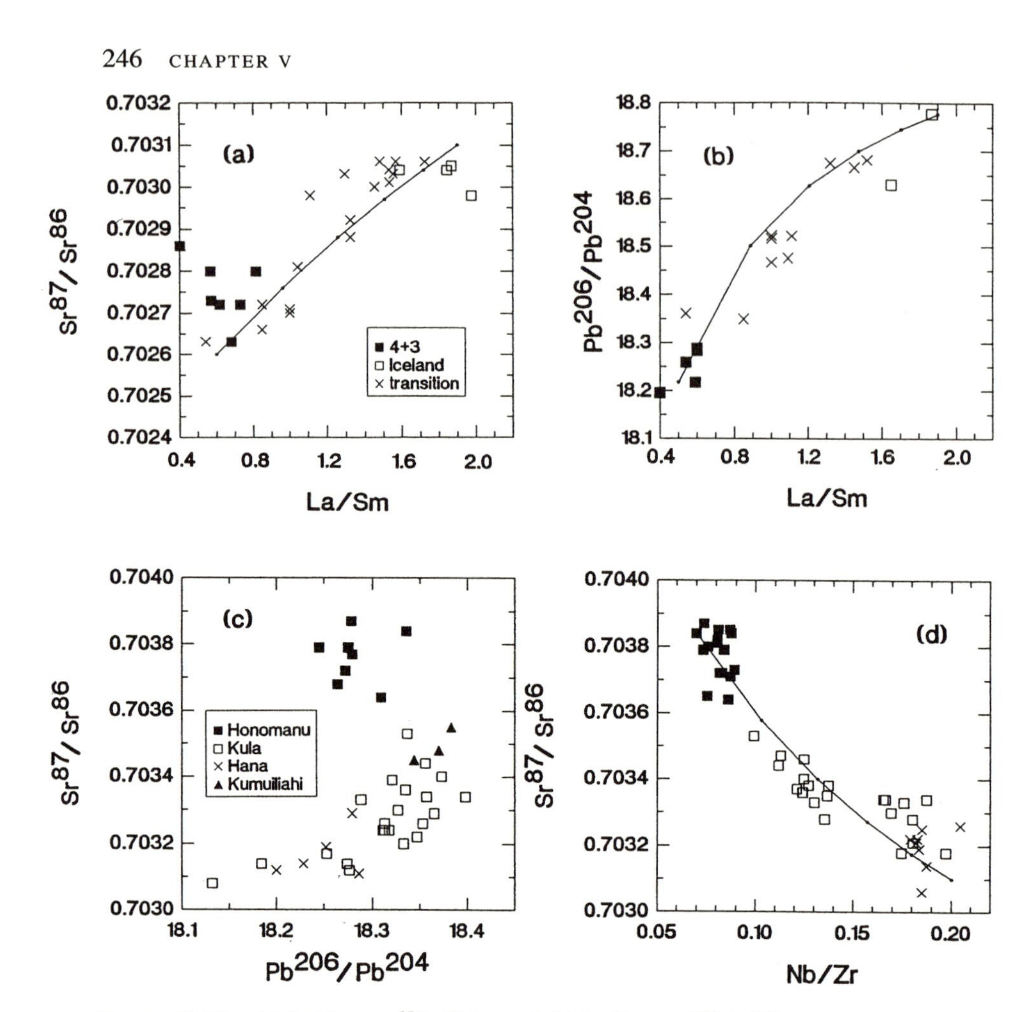

FIGURE V-22. (a) La/Sm vs. $^{87}Sr/^{86}Sr$, and (b) La/Sm vs. $^{206}Pb/^{204}Pb$ for group A mid-Atlantic ridge basalts. (c) $^{206}Pb/^{204}Pb$ vs. $^{87}Sr/^{86}Sr$, and (d) Nb/Zr vs. $^{87}Sr/^{86}Sr$ for the oceanic island basalts from Haleakala Volcano, Maui Island. The solid lines represent two-end-member mixing.

fractional crystallization of clinopyroxene and Fe-Ti oxides. There is no fractional crystallization of apatite here. Therefore, P and related elements such as Sr, Ba, and median REE also become incompatible elements of factor 1 (Figure V-23a). Unlike the MORB, there is not much fractional crystallization of plagioclase in OIB. Therefore Al behaves incompatibly (Figures V-23a and V-24e). Silica behaves incompatibly in the Kula unit (Figure V-24f) but shows no trend in the two other units.

As summarized in Figure V-23b and Figures V-24a–f, the alkali basalts (MgO content between 5 and 11%) could have evolved from tholeiitic parental

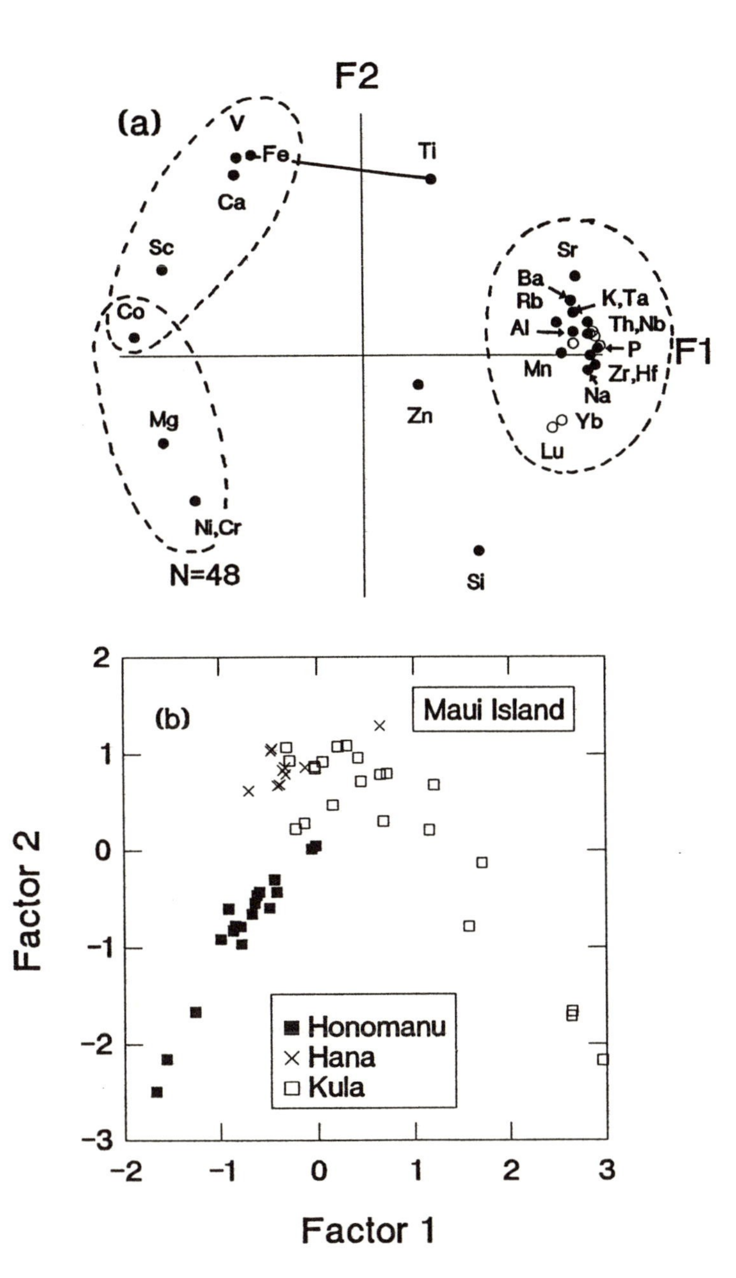

FIGURE V-23. (a) Factor loadings and (b) factor scores of factors 1 and 2 for the oceanic island basalts (OIB) from Haleakala Volcano, Maui Island.

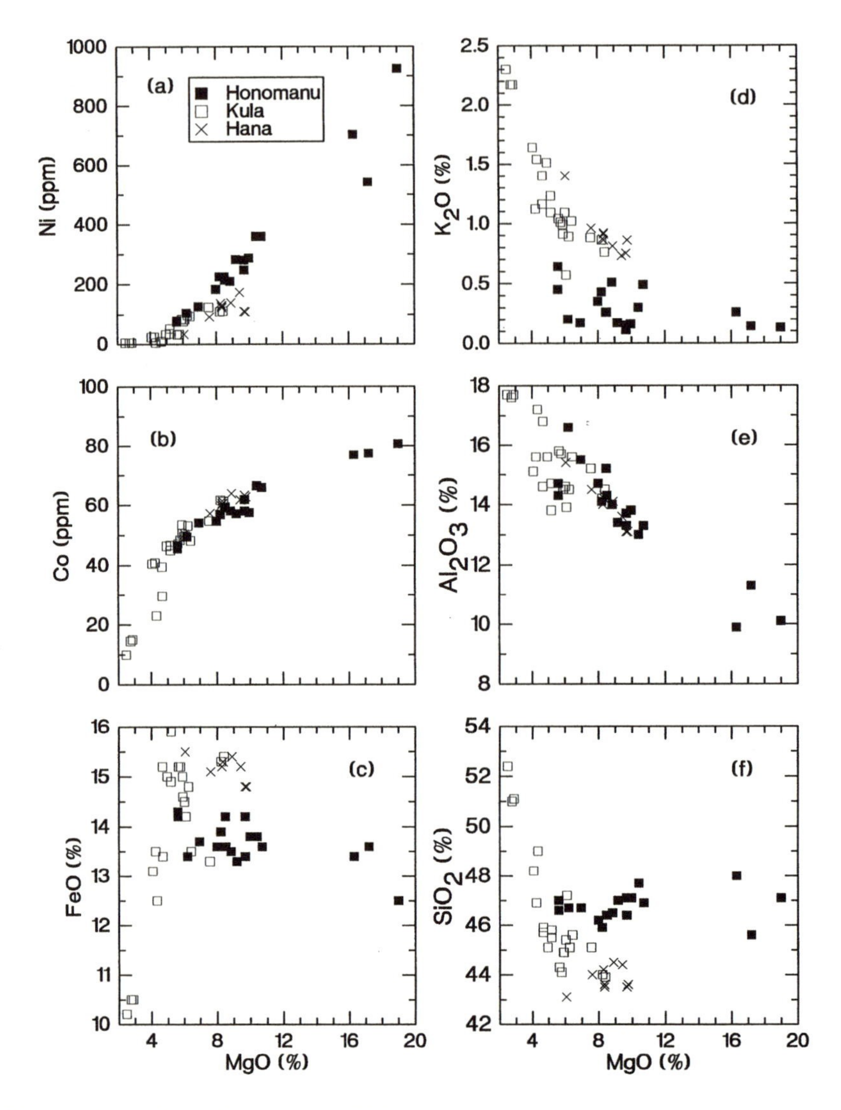

FIGURE V-24. Correlation plots of MgO versus (a) Ni, (b) Co, (c) FeO, (d) K_2O, (e) Al_2O_3, and (f) SiO_2 concentrations in the Honomanu, Kula, and Hana units of Haleakala Volcano.

magma similar to the Honomanu unit (MgO about 19%) through fractional crystallization of olivine. However, the tholeiitic parental magma of the Kula and Hana units should be systematically high in Fe (Figure V-24c) and low in Si (Figure V-24f) content as compared to that of the Honomanu unit. The hawaiites and mugearites evolved from the Kula unit could have been produced by fractional crystallization of clinopyroxene and Fe-Ti oxide minerals.

Two-Mica (Biotite-Muscovite) Granites

Some European granites of Hercynian age (275–350 Ma) and Appalachian granites of Ordovician to Devonian ages (450–350 Ma) are two-mica granites. Their precursors are probably of sedimentary rock origin. Two-mica granites are often closely associated with the mineralization of W, Sn, U, F, Be, Cs, Rb, and Li (Shaw and Guilbert, 1990).

The factor analysis of European granite and randomly selected Appalachian granite data given by Shaw and Guilbert (1990) shows that factor 1 represents the group of elements Mg, Ca, Sr, Mn, Zn, Al, Ti, V, Zr, S, and P, which is inversely correlated to Si (Figure V-25a). These elements are mostly related to hornblende and biotite minerals, and Si to quartz in granites. Factor 2 represents the mineralizing elements W, F, U, Sn, Be, Li, K, Rb, and Cs, which are probably associated with K-feldspar, muscovite, and other minor pegmatite minerals such as beryl, tourmaline, and cassiterite. As shown in Figure V-26b, the high W concentrations (as well as other mineralizing elements) occur mostly in granite with SiO_2 greater than 71 wt%, and especially in European granites. The mineralizing elements are probably carried by metasomatic or pegmatitic solutions. Factor 3 represents the heavy REE and Y, and factor 4 the light REE, Th, and Ba (Figure V-25b). As shown in Figures V-26c and V-26d, La (light REE) and Lu (heavy REE) inversely correlate to Si for the European granite data but not for the Appalachian granite data. Finally, Figure V-25c shows the distinct fields of factor scores occupied by European granites and Appalachian granites. In general, factor 2 (representing mineralization) is higher in European than in Appalachian granites. A few Appalachian granites also tend to have unusually high content of light REE or heavy REE (Figures V-25d, V-26c, and V-26d).

V-10. Concluding Remarks

The relative abundances of refractory lithophiles (except V) in the primitive mantle (PM) are chondritic. The siderophile and chalcophile elements are mostly concentrated in the Earth's metal core. The bulk composition of the Earth is probably similar to that of the CO and CV carbonaceous chondrites.

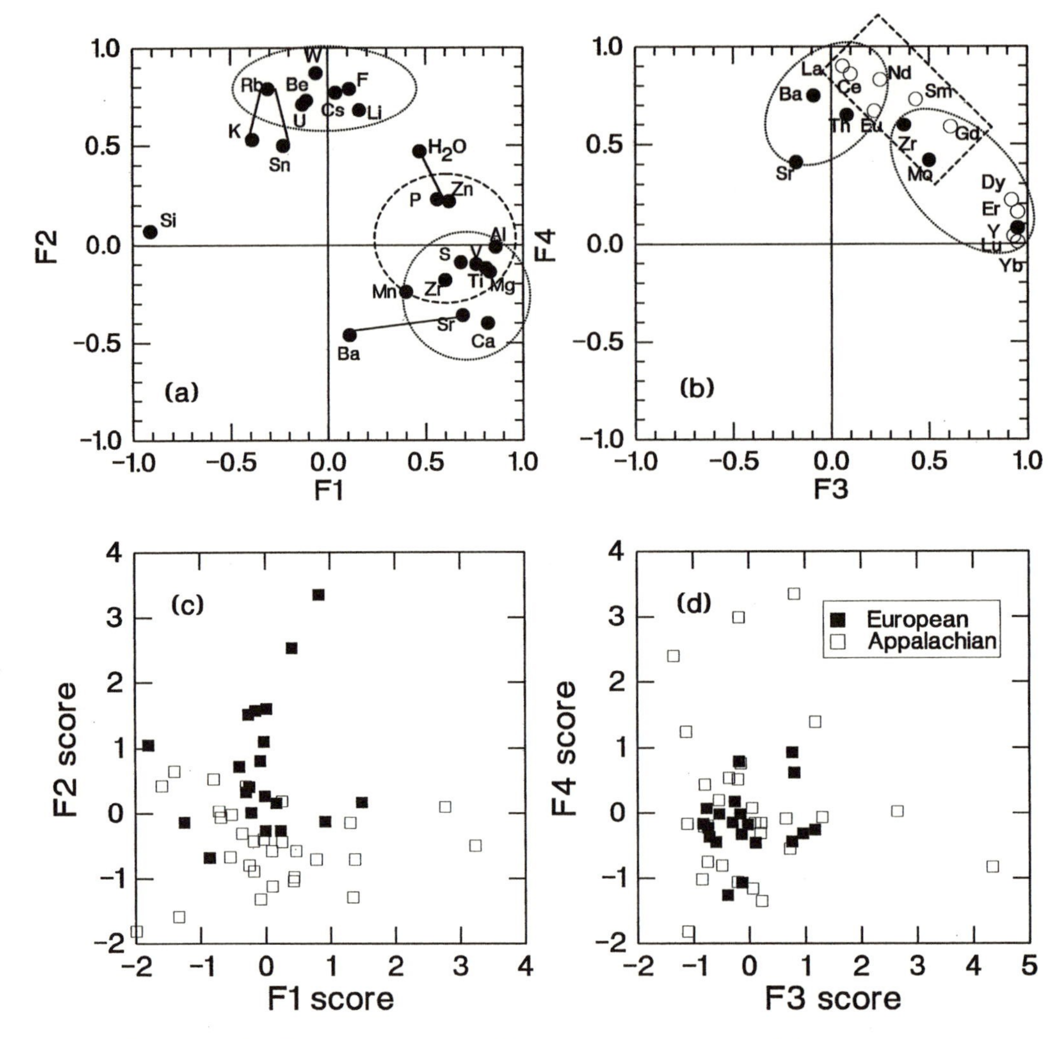

FIGURE V-25. Factor loadings (a and b) and factor scores (c and d) from factors 1 to 4 for European and Appalachian granites.

Through partial melting processes, the chondritic Earth probably first differentiated into the primitive mantle (PM) and the metal core. A part of the primitive mantle, in turn, differentiated into the continental crust and the depleted mantle (DM1); and a part of the depleted mantle differentiated further into the oceanic crust and the extra-depleted mantle (DM2). The subduction of crustal materials back to the mantle through plate tectonic processes has caused the mantle to become even more heterogeneous, especially with

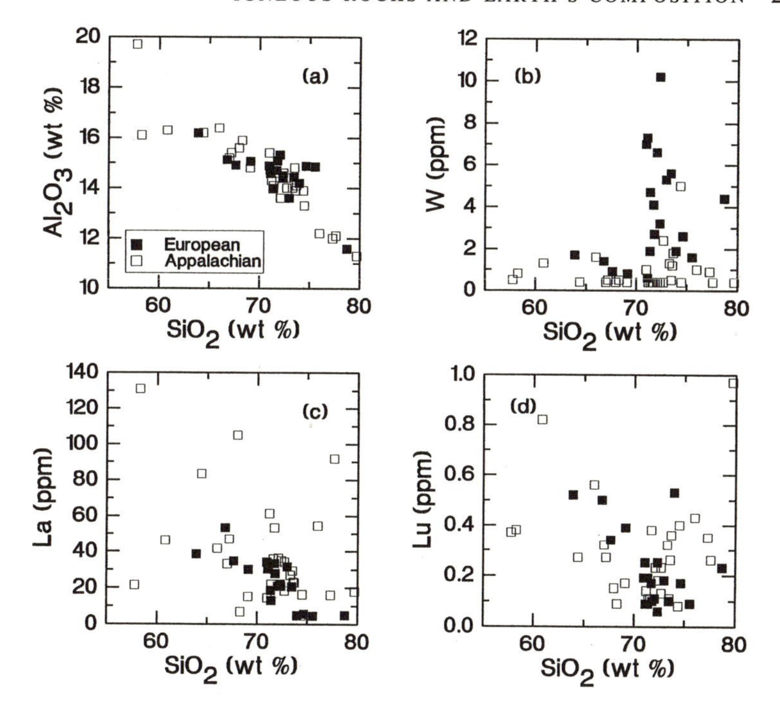

FIGURE V-26. Correlation plots of SiO_2 versus (a) Al_2O_3, (b) W, (c) La, and (d) Lu concentrations in European and Appalachian granites.

regard to the concentrations of incompatible elements and their radioactive parent-daughter pairs.

The partition of elements among various reservoirs is mainly controlled by the distribution coefficients of elements, which are complicated functions of T, P, compositions of melts and associated mineral assemblages, and ionic radii of elements. Additional determinations of distribution coefficients as functions of those variables under controlled conditions are badly needed. The partial melting and fractional crystallization models are the most useful conceptual models to explain the change in the chemical compositions of igneous rock series. Application of radioactive parent-daughter pairs, isotopic ratios of daughter elements, ratios of highly incompatible element pairs, and factor analysis techniques provides additional constraints on the sources and differentiation processes of magmas.

Chapter VI

WEATHERING AND SEDIMENTARY ROCKS

INTRODUCTION

THE DIFFERENTIATION of Earth's crust from the mantle was accompanied by the degassing of magmatic volatiles (H_2O, CO_2, H_2S, HCl, H_2, N_2, and others) from the interior of the Earth to its surface. These **primary magmatic volatiles** reacted with the crustal igneous rocks to produce the presently observed seawater and various sedimentary rock types (Rubey, 1951; Goldschmidt, 1954; Garrels and Mackenzie, 1971; Li, 1972). The total mass of the sedimentary rocks is estimated to be about $2.1 \pm 0.5 \times 10^{24}$ g, which is about 7.5% of the crust and 0.035% of the Earth by mass, and corresponds to a 1.7-km-thick sedimentary layer around the whole Earth (assuming an average density of 2.5 g/cm^3). The total sedimentary mass is directly proportional to the total amount of degassed volatiles (Li, 1972).

According to the magmatic ocean model of Earth's formation (Stevenson, 1981) as discussed in Section V-5, the differentiation of the Earth's crust and the degassing of volatiles were probably accomplished early in the Earth's history (Azbel and Tolstikhin, 1990). The discovery of (4.1–4.3)-Ga-old detrital zircons, which represent the remnants of an ancient felsic continental crust, in early to mid-Archean metasediments (3–3.7 Ga old; Maas and McCulloch, 1991, and references therein) supports the above contention.

Currently, the production rate of new sedimentary rocks through the weathering of crustal igneous rocks is thought to be balanced more or less by the destruction rate of sedimentary rocks through metamorphism and melting at subducting plate boundaries (Li, 1972). The remelting of sedimentary rocks releases the volatiles back to the atmosphere and restarts chemical weathering. Mantle convection was probably much more intensive during the early history of the Earth as compared to post-Archean time (2.5 Ga ago to the present), such that sedimentary rocks of older ages were mostly destroyed by remelting and weathering (Garrels and Mackenzie, 1971).

This chapter first presents the weathering processes of igneous rocks using specific examples and then discusses compositions of soils, river sediments, and various sedimentary rock types as weathering end products. In order to explain the observed partition pattern of elements between suspended particles and water in rivers, a chemical adsorption model is introduced at the end of the chapter.

VI-1. Weathering of Igneous Rocks

Weathering processes include both physical and chemical aspects. **Physical weathering** breaks up an originally massive rock into smaller fragments through physical processes, e.g., thermal expansion-contraction, volcanic explosion, gravitational sliding, and mechanical grinding during riverine and glacial transportation. High relief and rain storms also facilitate physical weathering. **Chemical weathering** includes dissolution, hydration, oxidation, and acid titration. Acid titration is probably the most important weathering process. It involves the replacement of metal cations in igneous rocks by hydrogen ions from natural acids such as organic acids and carbonic acid, and it causes the reconstruction of rock materials into so-called **secondary minerals** like clay minerals and metal hydroxides and oxides. Physical weathering enhances chemical weathering and vice versa.

According to Goldich (1938), the stability of the common minerals in igneous rocks during chemical weathering is roughly in the increasing order of

$$\left.\begin{array}{r}\text{olivine} \rightarrow \text{pyroxene} \rightarrow \text{hornblende} \rightarrow \text{biotite}\\ \text{calcic plagioclase} \rightarrow \text{sodic plagioclase} \rightarrow \text{K-feldspar}\end{array}\right\} \rightarrow \text{muscovite} \rightarrow \text{quartz} + \text{zircon}.$$

Interestingly, this order is similar to Bowen's (1928) crystallization sequences of rock-forming minerals from the magma.

The weathering of calcic plagioclase can be schematically represented by the following reactions of acid titration (Drever, 1988):

$$\underset{\text{Ca-plagioclase}}{4.33\text{CaAl}_2\text{Si}_2\text{O}_8} + 8\text{H}^+(\text{aq}) + 3.33\text{H}_2\text{O} \rightarrow \underset{\text{Ca-smectite}}{\text{Ca}_{0.33}\text{Al}_{4.67}\text{Si}_{7.33}\text{O}_{20}(\text{OH})_4} + \underset{\text{kaolinite}}{0.67\text{Al}_2\text{Si}_2\text{O}_5(\text{OH})_4} + \underset{\text{gibbsite}}{2.67\text{Al(OH)}_3} + 4\text{Ca}^{+2}(\text{aq}).$$

The Ca-smectite can be further weathered into kaolinite which in turn may be altered into gibbsite by the following reactions:

$$3\text{Ca}_{0.33}\text{Al}_{4.67}\text{Si}_{7.33}\text{O}_{20}(\text{OH})_4 + 2\text{H}^+(\text{aq}) + 23\text{H}_2\text{O} \rightarrow 7\text{Al}_2\text{Si}_2\text{O}_5(\text{OH})_4 + \text{Ca}^{+2}(\text{aq}) + 8\text{H}_4\text{SiO}_4(\text{aq}),$$

$$\text{Al}_2\text{Si}_2\text{O}_5(\text{OH})_4 + 5\text{H}_2\text{O} \rightarrow 2\text{Al(OH)}_3 + 2\text{H}_4\text{SiO}_4(\text{aq}).$$

The $H_4SiO_4(aq)$ produced may precipitate as chalcedonic quartz (an aggregate of minute quartz crystals with submicroscopic pores) when the weathering solution becomes saturated with respect to quartz. The end result is consumption of H^+, release of major cations and silica in solution, and introduction of bicarbonate ion in solution from the atomsphere.

Similarly, sodic plagioclase, olivine, and pyroxene can all be transformed into smectite group minerals such as Na-smectite $Na_{0.67}Al_{4.67}Si_{7.33}O_{20}(OH)_4$, nontronite $Fe_4Al_{0.67}Si_{7.33}O_{20}(OH)_4$, and saponite $Mg_6Al_{0.67}Si_{7.33}O_{20}(OH)_4$. Intensive weathering produces kaolinite, gibbsite, goethite (FeOOH), and related minerals such as halloysite $Al_2Si_2O_5(OH)_4{\cdot}2H_2O$, diaspore α-AlOOH, boehmite γ-AlOOH, limonite $FeOOH{\cdot}nH_2O$, hematite Fe_2O_3, and allophane (an amorphous hydrous aluminum silicate gel with highly variable composition).

K-feldspar $KAlSi_3O_8$ and muscovite $K_2Al_8Si_6O_{20}(OH, F)_4$, can easily be converted into illite $K_xAl_{4+x}Si_{8-x}O_{20}(OH)_4$, where $x = 1$ to 1.5, through weathering. Muscovite, biotite, hornblende, and illite also can change into vermiculite, i.e.,

$$(K_2, Na_2, Mg, Ca)_{0.7}(Mg, Fe, Al)_6(Al, Si)_8O_{20}(OH)_4{\cdot}8H_2O.$$

Titanium in ilmenite, $FeTiO_3$, may end up as extremely fine-grained anatase (TiO_2) in the soil.

Figure VI-1 shows how clay mineral compositions of the soils forming on granitic and basaltic bed rocks change with mean annual rainfall

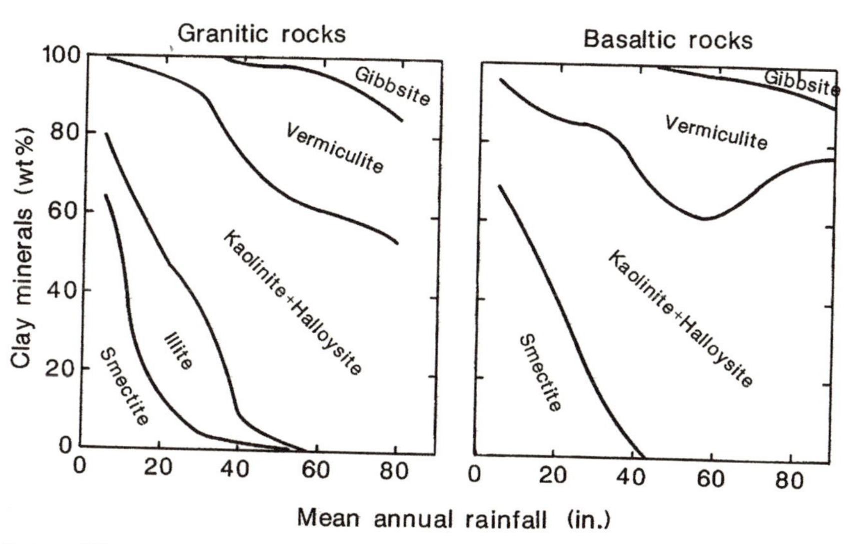

FIGURE VI-1. Cumulative weight% of secondary clay minerals in the soils forming on granitic and basaltic bed rocks, California, as a function of the mean annual rainfall (in inches), modified after Barshad (1966).

(Barshad, 1966). In high-rainfall areas, the main end products are kaolinite + halloysite, vermiculite, and gibbsite, whereas smectites are the main products in drier areas. Illite occurs only in granitic rock areas. Figure VI-2a provides another example of how the mineralogy changes with depth in soils developed over latite bed rock (Craig and Loughnan, 1964). The primary mineral composition of latite is K-feldspar 46%, plagioclase 20%, augite (pyroxene) 20%, magnetite 8%, chlorite 6%, and quartz 1%. The sequence of disappearance of the primary minerals in soil profiles is chlorite → pyroxene → plagioclase → K-feldspar. Secondary minerals such as chalcedonic quartz and kaolinite are abundant near the top and smectites near the bottom of soil profiles. These surface soil samples are depleted in Mg, Ca, Na, K, and Si and are slightly enriched in Fe and Ti as compared to the least weathered bottom sample (Figure VI-2b, Table VI-1a). According to the E^i_{Al} values in Figure VI-2b, the **relative mobility of elements** during the weathering of latite is Mg > Ca > Na > K ≫ Si > Al > Fe > Ti.

The order of relative mobilities, however, is not invariable, as noted by Colman (1982). For example, Figure VI-3 shows the changes in the E^i_{Al} values relative to unweathered basalt (Kauai Island) as a function of depth in a thick saprolite (or laterite) layer. The saprolite layer was formed by intensive leaching of the basalts by downward-percolating water, preserving

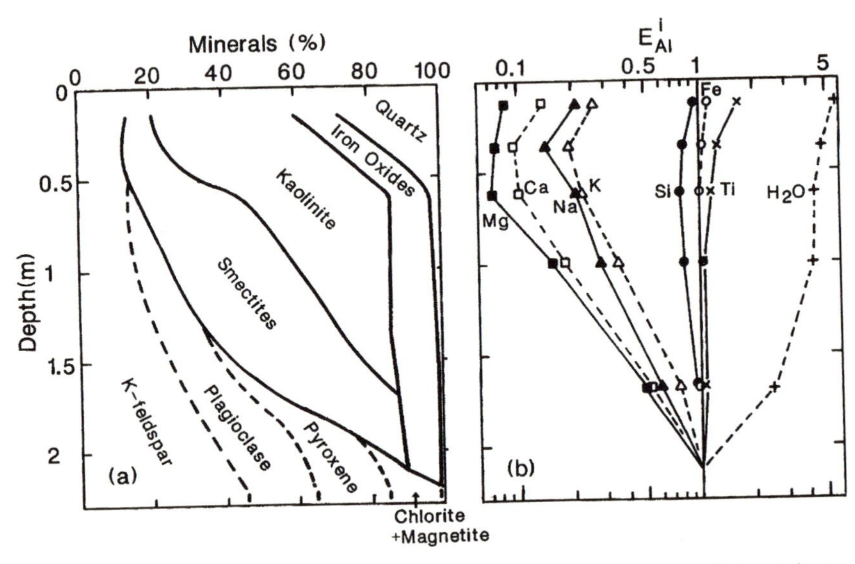

FIGURE VI-2. (a) Depth dependence of the relative abundances of the primary (dashed lines) and secondary (solid lines) minerals in the soil developed over a latite bed rock, New South Wales, Australia, and (b) Enrichment factors (E^i_{Al}) for major elements relative to the least weathered bottom-most samples as a function of depth. Figure VI-2a is modified after Craig and Loughnan (1964).

TABLE VI-1

Chemical compositions (weight %) of (a) weathered latite soil profile from Kiama, Australia and (b) saprolite soil profile from Kauai Island, Hawaii

(a)

Depth (ft)	SiO_2	Al_2O_3	Fe_2O_3	CaO	MgO	Na_2O	K_2O	TiO_2	P_2O_5	MnO	H_2O
0.33	52.1	16.3	10.7	0.9	0.3	0.8	1.1	1.7			15.2
1.17	51.4	18.4	11.2	0.7	0.3	0.6	0.9	1.5			14.2
2	51.8	19.2	11.2	0.8	0.3	0.9	1.1	1.4			13.2
3.25	52	18.5	11.2	1.3	0.6	1.2	1.7	1.2			12.1
5.5	54.8	16.9	9.4	3.3	1.9	2.3	3.3	1.1			6.8
7	54.3	15.8	9.2	6.3	3.4	3.6	4.1	1			2.5

(b)

Depth (ft)	SiO_2	Al_2O_3	Fe_2O_3	CaO	MgO	Na_2O	K_2O	TiO_2	P_2O_5	MnO	H_2O
0–2	16.7	23.7	32.7	0.09	0.09	0.06	0.14	6.2	0.21	0.08	16.9
2–4	16.8	23.6	33.6	0.09	0.14	0.05	0.1	6.2	0.27	0.1	16.9
4–9	15.7	25.7	33.3	0.07	0.12	0.04	0.06	5.3	0.3	0.1	18.3
9–14	15.8	29.4	29.6	0.07	0.18	0.03	0.03	5	0.47	0.19	18.8
14–19	26.1	26.9	25.4	0.06	0.58	0.04	0.02	4	0.4	0.45	16.1
19–24	23	27.7	26.3	0.07	0.47	0.07	0.02	4.4	0.6	0.28	16.8
24–29	23.2	28.1	25.9	0.06	0.52	0.09	0.02	4.3	0.5	0.48	16.8
29–35	25.9	27.2	25.4	0.06	0.43	0.06	0.02	4.2	0.44	0.26	16.4
35–35.5	42.1	14	15.8	6.4	7.9	0.76	0.62	2.4	0.33	0.26	10.2
basalt	42.3	10.9	13.9	11.7	12.8	2.85	0.83	2.7	0.69	0.13	1.3

Sources: (a) Craig and Loughnan (1964); (b) Patterson (1971, sample Ki-64).

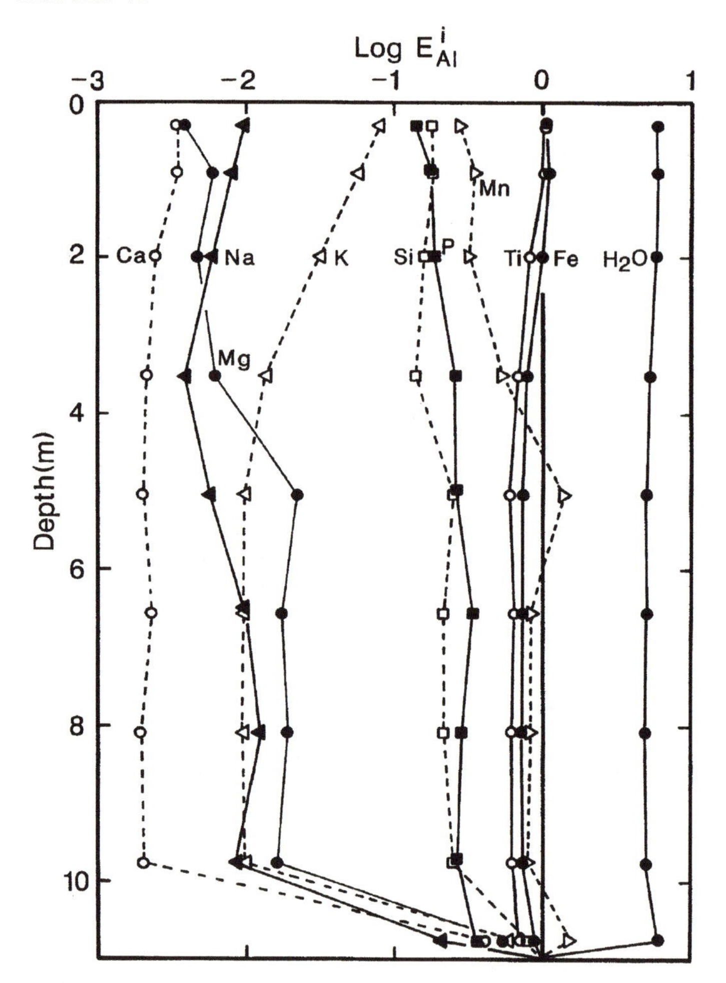

FIGURE VI-3. Depth dependence of the enrichment factors for major elements relative to unweathered basalt in thick saprolite layer, Kauai Island.

the original volume, structure, and texture of the parental rocks. The major minerals of saprolite are gibbsite and goethite. The original chemical data by Patterson (1971) are summarized in Table VI-1b. The high E^i_{Al} for K and low E^i_{Al} for Mn in the top meter relative to mid-depth (4 to 10 m) are probably related to the occurrence of about 2 to 3% organic matter, which contains K and also causes the reduction of manganese oxides to soluble Mn^{+2}.

The relative mobilities of other trace elements in the top layer of saprolite on basalt (Kauai Island) and in residual bauxite on nepheline syenite (Arkansas) are summarized in Table VI-2. Elements with E^i_{Al} values greater than 1 ± 0.2 include B, Cr, Ga, Mo(W), Nb(Ta), Th, Zr(Hf), and probably V.

TABLE VI-2
Compositions of residual deposits (saprolite and bauxite) and their parental rocks (basalt and nepheline syenite) and the enrichment factors of elements (E^i_{Al}) in the residual deposits relative to their parental rocks

	Hawaii		*Arkansas*		E^i_{Al}	
	Basalt(1)	*Saprolite(1)*	*Syenite(2)*	*Bauxite(2)*	*Saprolite*	*Bauxite*
Ag-47	0.14	~0			~0	
Al-13	70000	70000	102000	300000	1	1
B-5	~0	3			>1	
Ba-56	410	8	710	13	0.02	0.006
Be-4	0.9	0.5	1.7		0.6	
Ca-20	66000	40	10100	330	0.0006	0.01
Co-27	90	18			0.2	
Cr-24	400	560	2	43	1.4	7.2
Cu-29	290	33	5	13	0.11	0.87?
Fe-26	100000		27000	58000		0.72
Ga-31	13	15	20	63	1.2	1.1
K-19	3000	~0	52000	500	~0	0.003
La-57	30	3	300	120	0.1	0.13
Mg-12	100000	1200	3800	53	0.012	0.005
Mn-25	1200	450	970	830	0.38	0.29
Mo-42	2.8	3.5	8.5	30	1.3	1.2
Na-11	12000	~0	56000	600	~0	0.004
Nb-41	15	15	130	500	1	1.3
Ni-28	840	250			0.3	
P-15	3000	950			0.32	
Pb-82	12	0.5	20	10	0.04	0.17
Sc-21	18	3	5	5.3	0.17	0.35
Si-14	200000	36500	270000	21000	0.17	0.026
Sr-38	720	2	270	35	0.003	0.043
Th-90			19.3	52		0.9
Ti-22	24000	16500	5100	7300	0.69	0.48
U-92			4.9	6.2		0.42
V-23	210	250	47	60	1.2	0.43
Y-39	60	0.1	130	57	0.002	0.15
Zr-40	190	60	500	1200	0.32	0.80

Sources: (1) Patterson (1971); (2) Gordon et al. (1958), except Th and U data are from Adams and Richardson (1960).

Note: Compositions are in ppm.

They are equally or less mobile than Al. These elements probably exist in weathering-resistant minerals such as titanomagnetite, tourmaline, chromate, zircon, cassiterite, and monazite. Also, La(REE), Y, and U are quite mobile ($E^i_{Al} < 1$) under intensive leaching, as is shown by McLennan and Taylor (1979) and Nesbitt (1979). Other mobile trace elements include Ag, Cu, Co, Ni, Pb, and Sc, in addition to the alkali and alkaline earth elements.

VI-2. Dissolved Products of Chemical Weathering

Dissolved weathering products such as major cations, silica, and bicarbonate eventually end up in rivers and groundwater. The Hawaiian islands are an ideal place to study chemical weathering processes, because the weathering materials are mostly well-defined basaltic rocks and ashes, and the major sources for dissolved chemicals in Hawaiian rivers and groundwater are the chemical weathering of basaltic rocks and seasalt aerosol inputs (Li, 1988). The concentrations of major cations and silica versus bicarbonate in Hawaiian rivers and groundwater after correcting for seasalt inputs are plotted in Figure VI-4. The seasalt contributions in water samples are estimated as $[Cl^-]$ in the water sample multiplied by the $[M]/[Cl^-]$ ratio in seawater, where $[M]$ and $[Cl^-]$ are the concentrations of dissolved species M and chlorine, respectively. The observed linear correlation between bicarbonate and major cations (a slope near one in Figure VI-4; the greater scatter for Na^+ and K^+ is caused in part by large seasalt corrections) implies that the hydrogen ion of carbonic acid must be the major chemical weathering agent. H^+ ions interacting with basaltic rocks to produce secondary minerals, dissolved cations, and bicarbonate. The major sources of carbonic acid must be microbial oxidation of organic matter in the soil column and root respiration, even though the ultimate source of organic carbon is atmospheric CO_2. The light $\delta^{13}C$ values for bicarbonate in Hawaiian rivers and groundwater (mostly within the range of $-18 \pm 2‰$; Hufen, 1974) are consistent with this conclusion. Dissolved silica concentrations do not increase as fast as bicarbonate (slope of less than one), probably indicating some removal processes for silica. Also, the silica concentration in groundwater tends to be higher than in rivers at a given bicarbonate concentration.

The partial pressure of CO_2 (P_{CO_2} in units of atmosphere), pH, and bicarbonate activity ($a_{HCO_3^-}$) in water samples are related by

$$pH = -\log K_{CO_2} \cdot K_1 \cdot P_{CO_2} + \log a_{HCO_3^-}. \qquad \text{(VI-1)}$$

The above equation is derived from combination of the following definitions of equilibrium constants after taking the logarithms of the expressions:

$$K_{CO_2} = a_{H_2CO_3}/(P_{CO_2} \cdot a_{H_2O}) = 10^{-1.47},$$

and

$$K_1 = a_{H^+} \cdot a_{HCO_3^-}/a_{H_2CO_3} = 10^{-6.35}$$

(Drever, 1988). Equation VI-1 is shown as the solid lines in Figure VI-5a at different given P_{CO_2} values. The pH and $a_{HCO_3^-}$ values for the same water

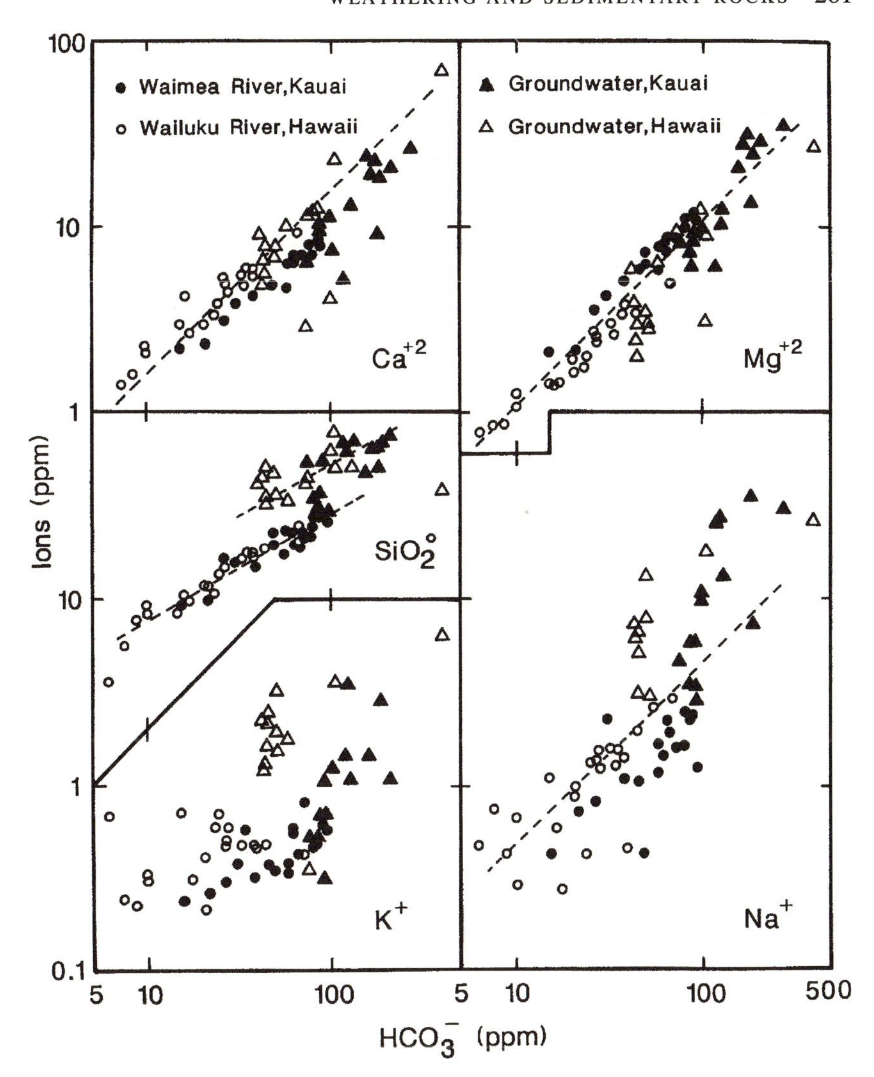

FIGURE VI-4. Seasalt-corrected concentrations of major cations and silica against the bicarbonate concentration in the Hawaiian rivers and groundwaters.

samples shown in Figure VI-4 and additional samples from an upstream station of Waimea River, Kauai, are also plotted in Figure VI-5a. It is apparent that the P_{CO_2} in Hawaiian rivers and groundwaters is always much higher than the atmospheric P_{CO_2} (about $10^{-3.5}$ atm), again indicating the continuous input of carbonic acid by microbial oxidation of organic matter.

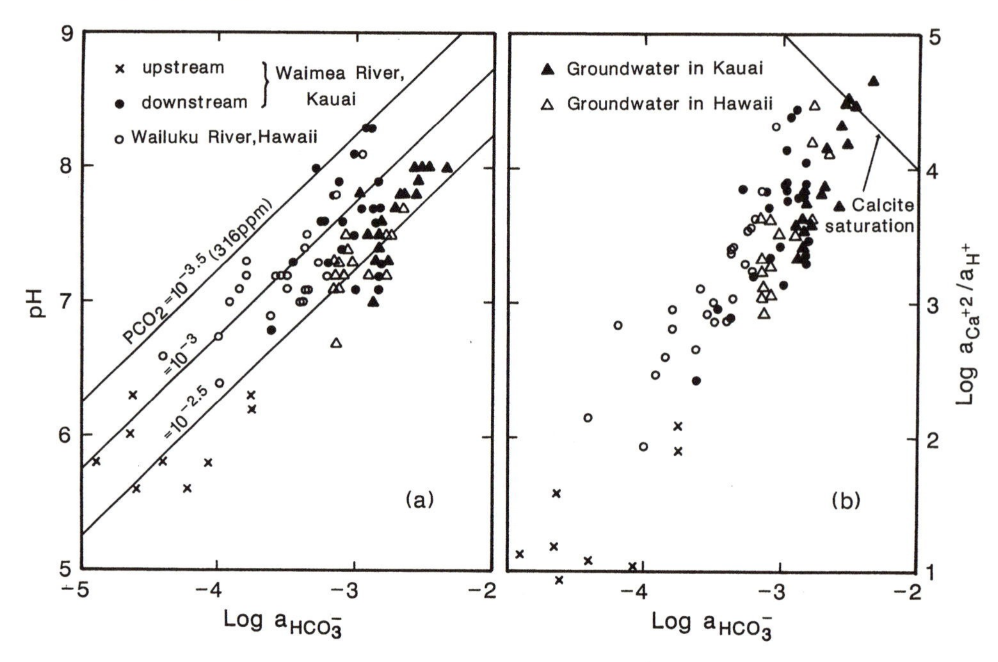

FIGURE VI-5. $\log a_{HCO_3^-}$ against (a) pH, and (b) $\log a_{Ca}^{+2}/a_H^+$ for rivers and groundwaters from Hawaii and Kauai islands.

If a parcel of natural water is in equilibrium with calcite, then

$$\log \frac{a_{Ca^{+2}}}{a_{H^+}} = \log \frac{K_{sp}}{K_2} - \log a_{HCO_3^-}. \qquad \text{(VI-2)}$$

The above equation is derived from the definition of the following equilibrum constants:

$$K_{sp} = a_{Ca^{+2}} \cdot a_{CO_3^{-2}} = 10^{-8.42} \quad \text{and } K_2 = a_{H^+} \cdot a_{CO_3^{-2}}/a_{HCO_3^-} = 10^{-10.33}$$

(Drever, 1988). Equation VI-2 is again shown as a solid line in Figure VI-5b. The observed $a_{Ca^{+2}}/a_{H^+}$ and $a_{HCO_3^-}$ values for the same samples shown in Figure VI-5a are also plotted in Figure VI-5b. The water samples steadily approach saturation with respect to calcite from upstream to downstream and to groundwater reservoirs, as also shown by Garrels (1967) in other areas.

The chemical data from the same water samples are also plotted in various **activity vs. activity phase diagrams** (Drever, 1988) as shown in Figure VI-6. The diagrams in Figure VI-6 indicate that the most stable secondary minerals should be gibbsite in the upstream portion of Waimea River (where rainfall is plentiful and the vegetation coverage is thick), kaolinite in the

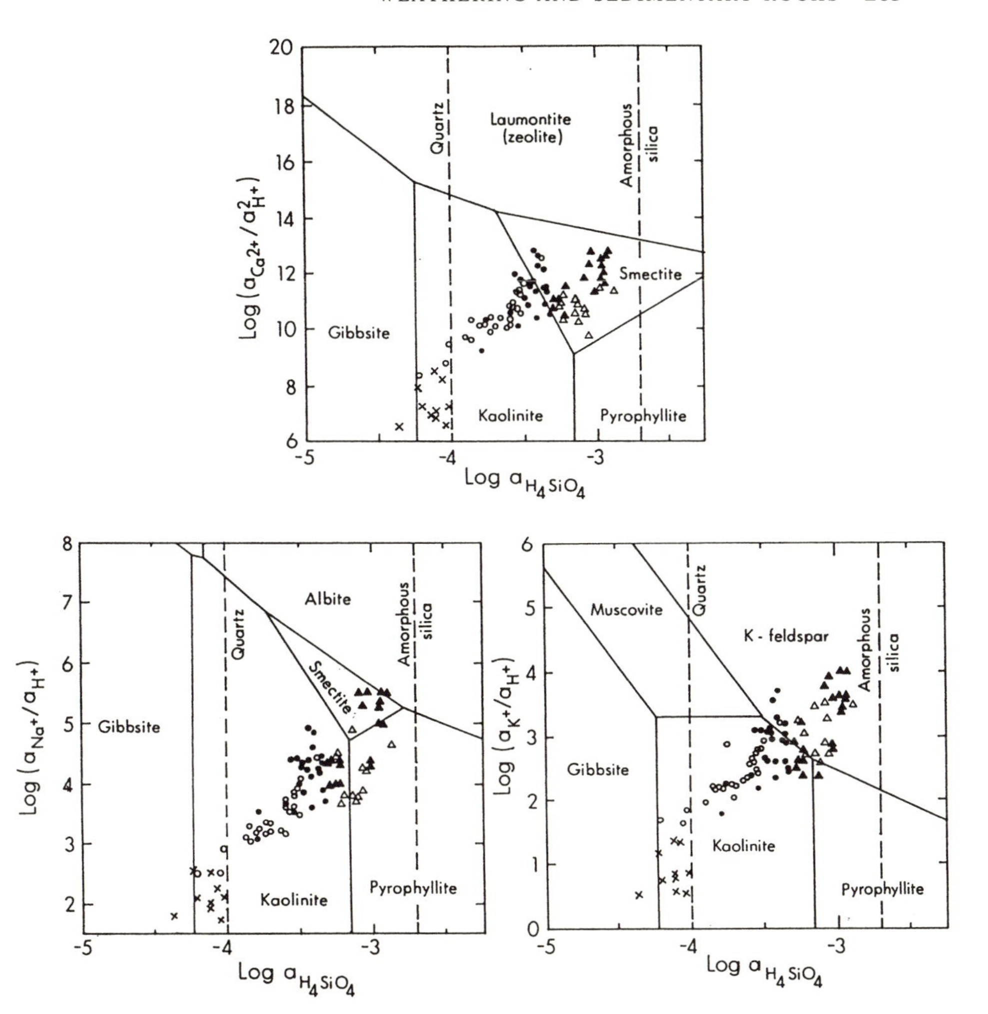

FIGURE VI-6. Plot of log $a_{H_4SiO_4}$ against various log activity ratios for rivers and groundwaters from Hawaii and Kauai islands (the data symbols are the same as in Figure VI-5). Stability fields of various secondary minerals given by Drever (1988) are superimposed.

downstream portion, and smectites and authigenic feldspar in groundwater reservoirs. Sherman (1952) also demonstrated that the clay mineral compositions in Hawaiian soils change systematically from smectites to kaolinite and to gibbsite + goethite as the average annual rainfall of the areas studied increases steadily. In general, an increase of contact time between water and solids or a high solids/water ratio increases the concentrations of dissolved ions and favors the formation of smectite and feldspar.

VI-3. Major Classes of Sedimentary Rocks

Solid weathering products are transported by rivers, glaciers, winds, and currents, and are settled in various depositional environments to form clastic sedimentary rocks after some diagenetic changes. The dissolved weathering products are mainly carried by rivers and groundwaters and are precipitated through chemical and/or biological processes in water basins to form massive nonclastic sedimentary rocks. As summarized in Table VI-3, the clastic sedimentary rocks are classified mainly by the grain size of the sedimentary particles, whereas the nonclastic sedimentary rocks are classified by mineral or chemical compositions.

The size distribution patterns of some clastic sediments and sedimentary rocks are shown in Figure VI-7a as histograms and in Figure VI-7b

Table VI-3
Classification of selected sedimentary rocks

<table>
<tr><th>Particle size or mineral composition</th><th colspan="3">Rock name</th></tr>
<tr><td colspan="4">Clastic</td></tr>
<tr><td>gravel > 2 mm</td><td colspan="3">Conglomerate and breccia</td></tr>
<tr><td>sand 2 to 1/16 mm</td><td colspan="2">Sandstone (Arenite)</td><td rowspan="3">Wack</td></tr>
<tr><td>silt 1/16 to 1/256 mm</td><td>Siltstone</td><td rowspan="2">Shale or Mudstone</td></tr>
<tr><td>clay $< 1/256$ mm</td><td>Claystone</td></tr>
<tr><td colspan="4">Nonclastic</td></tr>
<tr><td>calcite, $CaCO_3$</td><td>Limestone</td><td colspan="2" rowspan="2">Carbonates</td></tr>
<tr><td>dolomite, $CaMg(CO_3)_2$</td><td>Dolomite</td></tr>
<tr><td>gypsum, $CaSO_4 \cdot 2H_2O$; anahydrite, $CaSO_4$
halite, NaCl and others</td><td colspan="3">Evaporite</td></tr>
<tr><td>opal, quartz, SiO_2</td><td colspan="3">Chert</td></tr>
<tr><td>apatite, $Ca_5(PO_4)_3(OH, F, Cl)$</td><td colspan="3">Phosphorite</td></tr>
<tr><td>hematite, Fe_2O_3; magnetite, Fe_3O_4
siderite, $FeCO_3$; ankerite, $Ca(Fe, Mg, Mn)CO_3$
iron silicates (greenalite, stilpnomelane, minesotaite)
pyrite, FeS_2; quartz and chalcedony, SiO_2</td><td colspan="3">Iron-formation</td></tr>
<tr><td>carbon, hydrocarbon</td><td colspan="3">Coal, Oils</td></tr>
</table>

Notes: Ideal formula for greenalite = $(Fe, Mg)_6\,Si_4O_{10}\,(OH)_8$; stilpnomelane = $K_{0.6}\,(Mg, Fe)_6\,(Si, Al)_6\,(O, OH)_{22}$; and minnesotaite = $Fe_3Si_4O_{10}\,(OH)_2$.

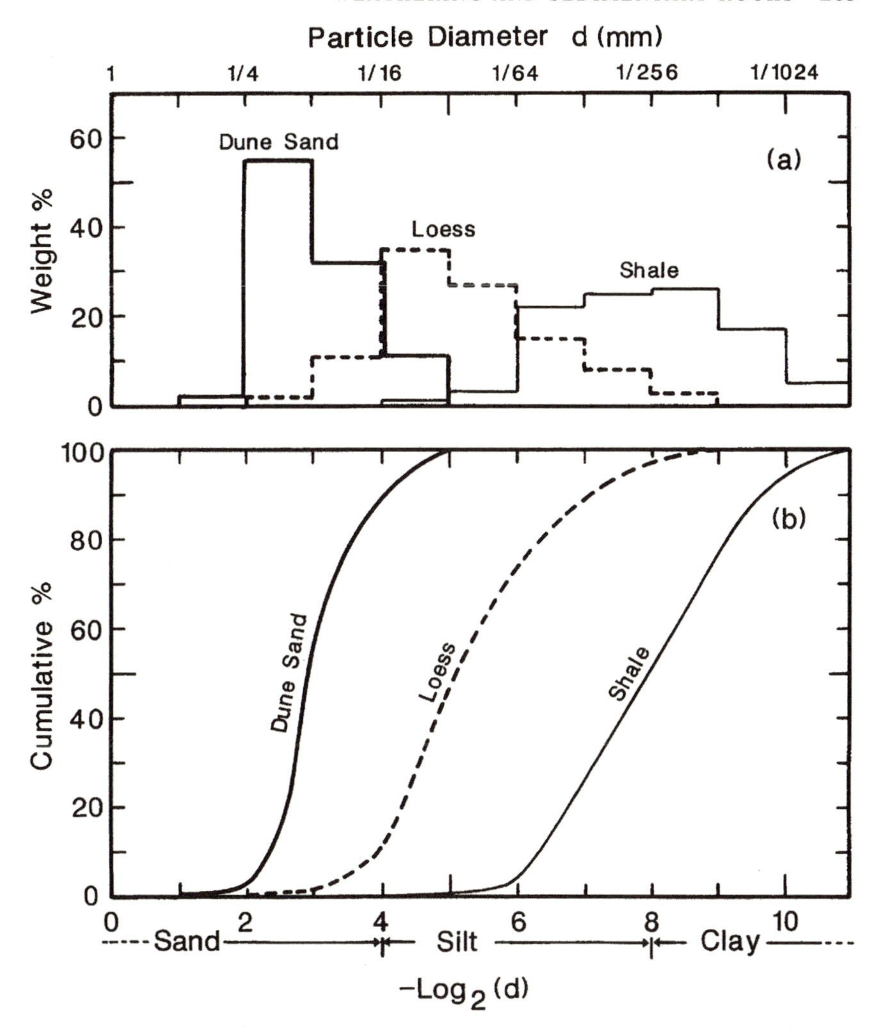

FIGURE VI-7. Grain size distribution patterns of typical dune sand, loess, and shale (Pettijohn, 1957) expressed as (a) weight % and (b) cumulative weight % curves.

as cumulative curves. The shale curve covers the size ranges of silt and clay. The loess consists of silt-sized rock flour, which is thought to have been produced by glacial grinding and transported by wind over long distances to dry depositional sites. The dune sand has a high degree of sorting through the winnowing action of waves and currents.

Based on the lithological and mineralogical compositions of particles, the clastic sedimentary rocks may be further classified into various subgroups. For example, Figure VI-8 provides one subdivision scheme for sandstone and mudstone, based on the relative abundances of sand-sized quartz, rock

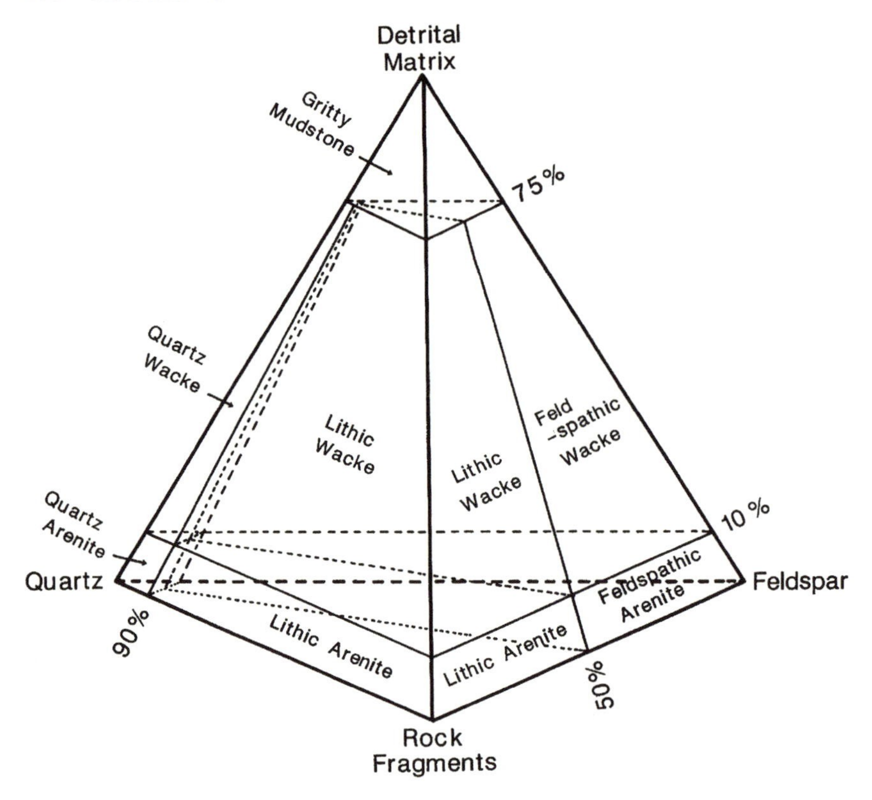

FIGURE VI-8. Classification scheme for sandstone and mudstone based on the relative abundances of sand-sized quartz, rock fragments, feldspar, and (silt + clay)-sized matrix (modified after Pettijohn et al., 1987; Dott, 1964).

fragments, feldspar, and (silt + clay)-sized matrix. Figure VI-8 is modified after Pettijohn et al. (1987) and Dott (1964). The **wackes** correspond roughly to Pettijohn's (1957) **graywackes**, **feldspathic arenite** to **arkose**, and **lithic arenite** to **subgraywacke**. The precursor for some wackes (or graywackes) is thought to be the marine turbidites (Pettijohn, 1957; Gorsline and Emery, 1959). **Turbidite deposits** were transported by **turbidity currents**, which are dense and turbid bottom flows of water plus suspended sediments moving rapidly downward on a continental slope. Turbidity currents are initiated by sudden sliding of slope sediments by gravity, most likely triggered by earthquakes. The particle size distribution of turbidite deposits from San Pedro Basin off southern California shows a wide size range, as shown in Figure VI-9 (Gorsline and Emery, 1959).

The most likely sources for the sand-sized feldspar in the **sandstone** + gritty **mudstone** were plutonic granites and gneisses that underwent rapid erosion and deposition of weathering products before total decomposition of

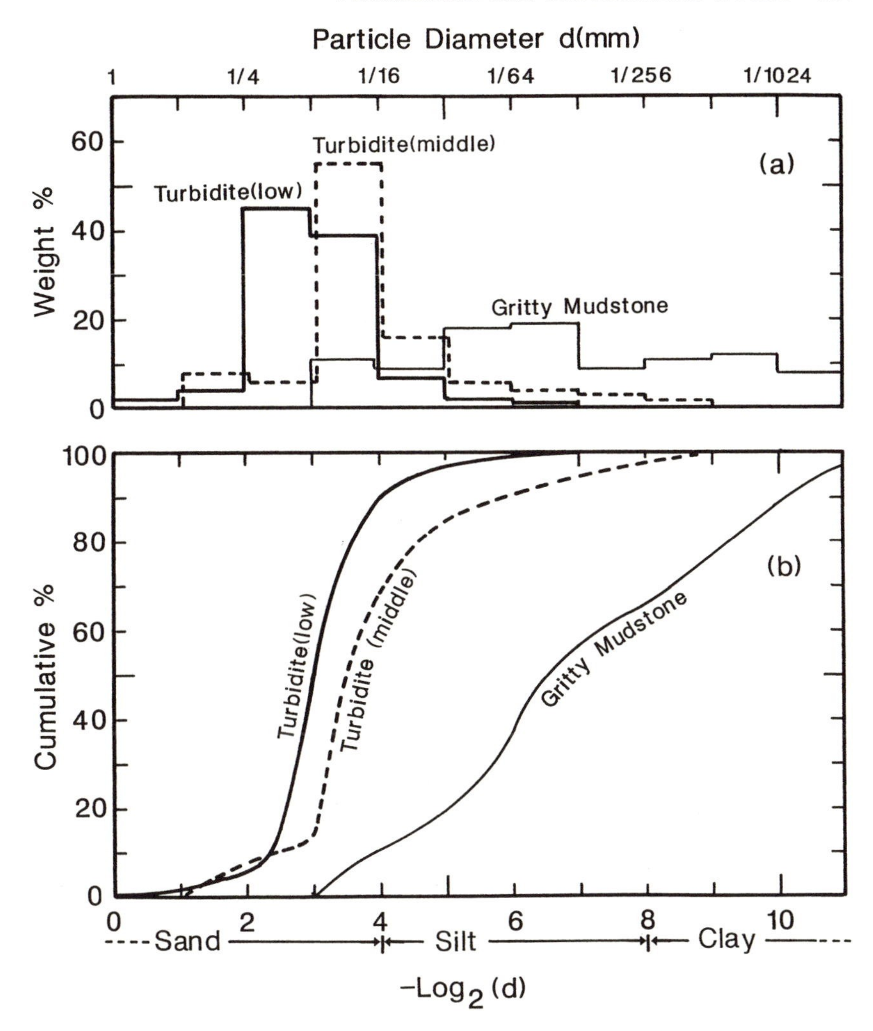

FIGURE VI-9. Grain size distribution patterns of the low and middle sections of a turbidite layer and mudstone deposits in San Pedro Basin off southern California (Gorsline and Emery, 1959) expressed as (a) weight % and (b) cumulative weight % curves.

feldspar, e.g., in continental rift zones and collision zones between two continental plates. The ultimate sources for quartz were the same as for feldspar, but most of the quartz grains in recent sandstones were probably derived from repeatedly recycled older sandstone and chert. The sources for lithic fragments could be volcanic, metamorphic, and sedimentary rocks, which again underwent rapid erosion and deposition, e.g., in active continental margins and island arcs. Violent volcanic eruptions and the grinding of bed rock by continental glaciation could also produce rock fragments covering a wide grain size range, e.g., **volcanic tuffs** and **glacial tills**. The (silt + clay)-sized matrix could be supplied by turbidity currents in tectonically unstable regions but also could be the end product of in situ chemical weathering of sand-sized lithic fragments and feldspar within the sandstone (Pettijohn et al., 1987). For more detailed classification schemes for sedimentary rocks, their origins and tectonic significance, one should refer to standard textbooks, e.g., Pettijohn et al. (1987) and Pettijohn (1957).

VI-4. Relative Abundances of Major Sedimentary Rock Types and Mass Balance

The most abundant sedimentary rock types are **shale**, **sandstone**, **carbonate**, and **evaporite** (excluding pure volcanogenic debris and sediments in the strata). Their relative abundances and average major element compositions have been estimated by Ronov and his coworkers (Ronov, 1982, and references therein), based on extensive stratigraphic sections and samples. The results are summarized in Table VI-4 along with other pertinent information. The relative abundances in weight percent of shale, sandstone, carbonate, and evaporites are roughly 51:23:25:1 in average continental sediments.

As shown in Figure VI-10a, the average sandstone is a mixture of average shale, pure quartz (SiO_2), and calcite ($CaCO_3$). Similarly, the average carbonate is a mixture of shale, pure calcite, magnesite ($MgCO_3$), quartz, and minor apatite. The slight excess Na over the shale contribution in both average sandstone and carbonate in Figures VI-10a and VI-10b is still doubtful, because the data given by Turekian and Wedepohl (1961) and Bowen (1979) do not show the same Na excess. The evaporites are essentially a mixture of 50% halite salt and 50% sulfates (gypsum and anhydrite) with minor impurities (Ronov, 1982). More quantitatively, one may set up mass balance equations for Al, Si, and Ca:

$$f_1 X_{i1} + f_2 X_{i2} + f_3 X_{i3} = X_{i0}, \qquad \text{(VI-3)}$$

where f_1, f_2, and f_3 are the fractions of the chosen end members, i.e., average shale, pure quartz, and pure calcite, respectively; X_{i1}, X_{i2}, and X_{i3} are the concentrations of the element i in the respective end members; and X_{i0} is the concentration of the element i in the chosen rock sample. Three elements give three mass balance equations that can be solved for f_1, f_2, and f_3. Also, $f_4 = 1-(f_1+f_2+f_3)$, where f_4 is the fraction of unspecified components that do not contribute much of the Al, Si, and Ca to the overall mass balances. For example, the average sandstone consists of shale, quartz, calcite, and other unspecified minerals in the weight ratio of 64:31:3:2. Similarly, the average carbonate consists of shale, quartz, calcite, and others (mainly magnesite and

TABLE VI-4

Average compositions of major sedimentary rock types and their mass-weighted average (all in weight %)

	Continent[a]				Continent	Oceanic	Global	Upper
	shale	*sandstone*	*carbonate*	*evaporite*	*ave. sed.*	*ave. sed.*	*ave. sed.*	*crust*
Mass (10^{24} g)	0.96	0.44	0.48	0.02	1.9	0.19	2.09	
wt. fraction	0.51	0.23	0.25	0.01				
density (g/cm^3)	2.49	2.28	2.52	2.3	2.45	1.03		2.8
SiO_2	55.84	66.71	13.5	0.81	47.2	36.28	46.21	64.2
TiO_2	0.78	0.54	0.14		0.56	0.52	0.55	0.55
Al_2O_3	16.38	10.54	2.79	0.083	11.48	10.49	11.39	14.7
Fe_2O_3	3.44	2.32	0.65	0.19	2.45	4.43	2.63	
FeO	3.31	2.36	0.64	0.026	2.39	0.67	2.23	
MnO	0.094	0.11	0.054		0.087	0.35	0.11	0.1
MgO	2.55	1.88	4.87	0.6	2.96	2.62	2.93	2.65
CaO	4.12	4.7	40.01	18.85	13.37	20.22	14	4.48
Na_2O	1.06	1.49	0.43	22.19	1.22[b]	1	1.20[b]	3.42
K_2O	3.02	1.96	0.6	0.32	2.14	1.87	2.12	3.13
P_2O_5	0.14	0.12	0.092		0.12	0.15	0.12	0.2
F	0.061	0.033	0.029	0.024	0.046	0.056	0.046	0.07
Corg	0.95	0.38	0.23		0.63	0.33	0.6	
CO_2	3.49	3.49	33.5	1.14	10.97	16.31	11.45	
SO_3	0.33	0.17	0.93	24.5	0.68	0.19	0.64	
S_{pyr}	0.28	0.2	0.12		0.22	0.001	0.2	
Cl	0.085	0.06	0.041	26	0.33		(> 0.30)	0.015
H_2O	4.5	2.5	1.47	5.39	3.29	4.72	3.42	
ΣFe_2O_3[c]	7.12	4.94	1.36	0.22	5.11	5.17	5.12	6.01

Sources: Ronov (1982) and Ronov and Yaroshevkiy (1976).

Notes: [a]The continental sediments include those from platforms, geosynclines, continental shelves, and slopes, and contain about 0.39×10^{24} g of additional volcanogenic debris and sediments. [b]If the Na in seawater is added to the continental or global average sediments (ave. sed.), the concentration of Na_2O becomes 2.0 wt %. [c]Total iron expresses as Fe_2O_3.

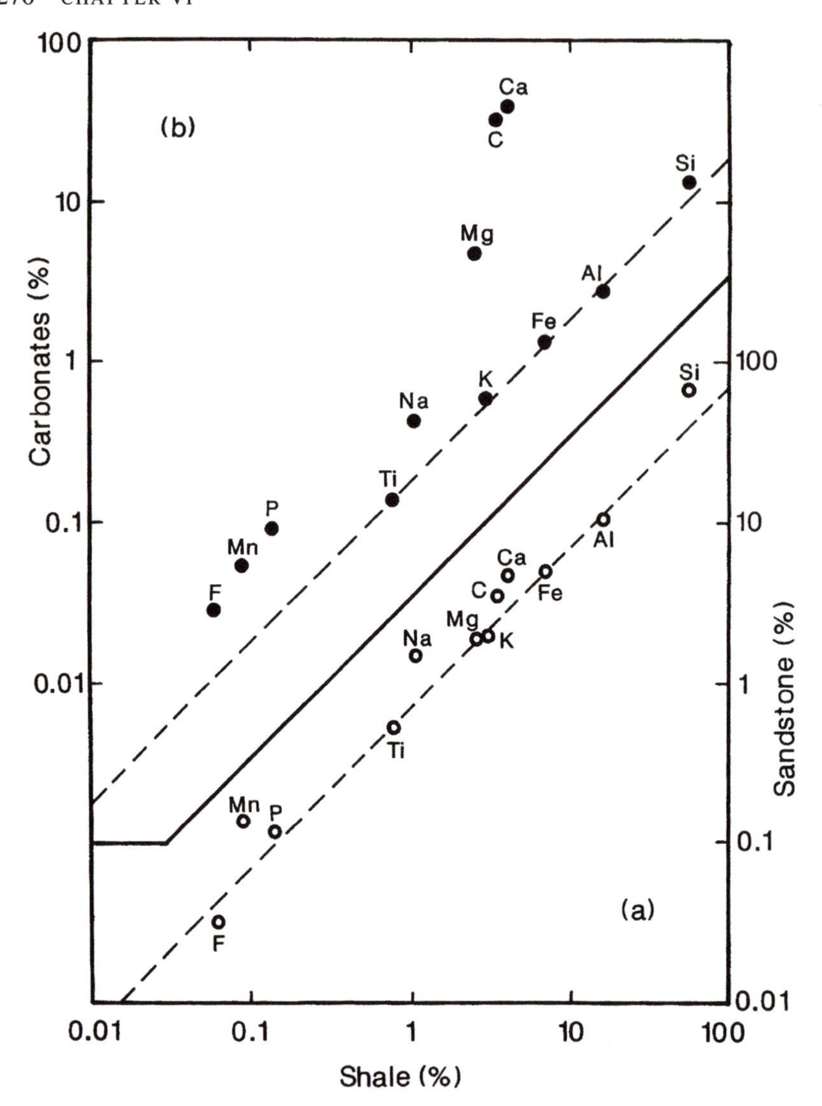

FIGURE VI-10. Average concentration (weight %) of major elements in shale plotted against that in (a) sandstone and (b) carbonate rocks, based on the data in Table VI-4.

minor apatite) in the ratios of 17:4:70:9. Therefore, mean continental sediments can also be expressed as a mixture of 70% average shale, 8% quartz, 20% calcite (including about 2% magnesite), 1% evaporite, and 1% others.

As will be discussed in Section VI-5, average shale itself also contains about 20% quartz; therefore, the calculated fractions of quartz given above are actually quartz in excess of that in shale fractions. As shown in Figure VI-11

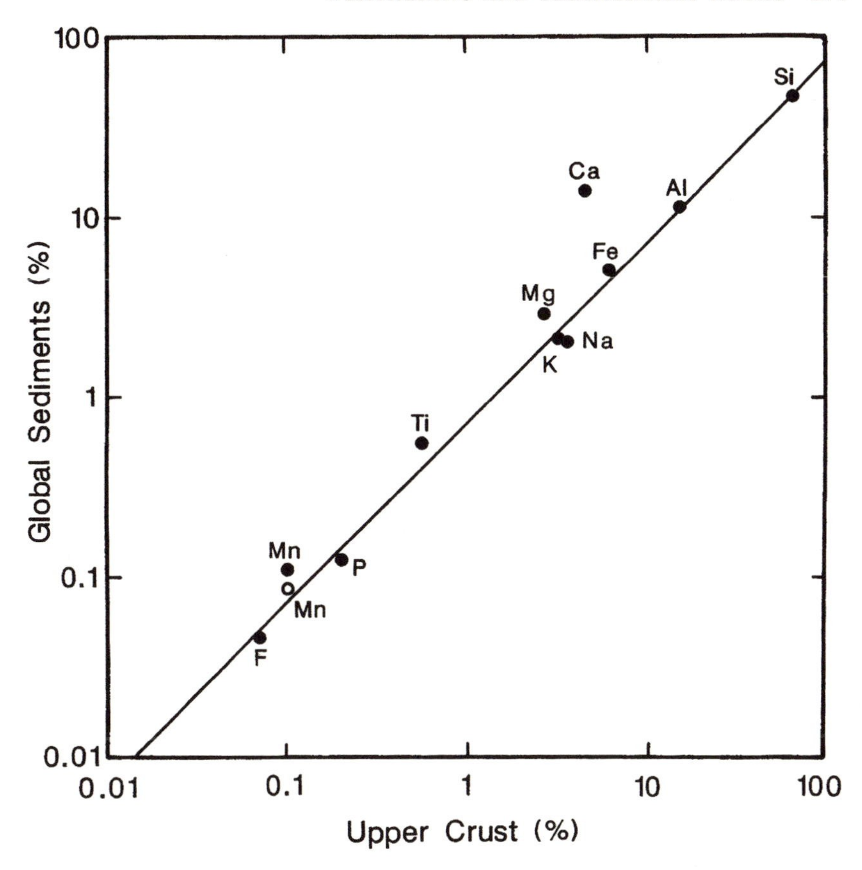

FIGURE VI-11. Average concentration of major elements in the upper crust plotted against that in mean global sedimentary rocks based on Table VI-4. The Mn value with the open circle is for the mean continental sedimentary rocks.

(based on Table VI-4), the mass-weighted mean composition of global sedimentary rock (including both the continental and the oceanic sediments which will be discussed in Chapter 7) is, as one would have expected, directly related to that of the upper crust. However, the obvious exception is Ca that is enriched in mean global sedimentary rock. The possible sources of the excess Ca will be discussed later. The compositions of mean continental and mean global sedimentary rocks are almost identical except for Mn, which is high in the latter due to the enrichment of Mn in oceanic sediments (Table VI-4).

The ratio of carbonate carbon and organic carbon in the global sedimentary rock is about 5:1 (Table VI-4), which is in good agreement with the value of about 4:1 obtained by a $^{13}C/^{12}C$ mass balance calculation (Li, 1972). In mean continental sediments and mean global sediments, the volatile species (C_{org},

CO_2, SO_3, S_{pyr}, Cl, and H_2O) account for about 17 weight %. Therefore, the mass ratio between weathered parent igneous rocks (M_{ig}) and sediments produced (M_{sed}) is about 0.83.

If one assumes that the sole source for the nonvolatile elements in various sedimentary reservoirs is the upper crust, then one can write the following mass balance equations:

$$M_{ig} X^i_{ig} = M_{sed} \sum f_j X^i_j + M_{sw} C^i_{sw}, \qquad \text{(VI-4a)}$$

and

$$\Sigma f_j = 1, \qquad \text{(VI-4b)}$$

where M_{sed} = total mass of sedimentary rocks today = 2.1×10^{24} g; M_{ig} = total mass of the upper crust weathered to produce the observed M_{sed}; M_{sw} = mass of seawater = 1.4×10^{24} g (Sverdrup et al., 1942); f_j = fraction of sedimentary rock type j (f_1 = shale, f_2 = sandstone, f_3 = carbonates, and f_4 = evaporite) in mean global sedimentary rock; and X^i_{ig}, X^i_j, and C^i_{sw} = average concentration of element i in the upper crust, sedimentary rock type j, and seawater, respectively. Equation VI-4a can be reduced to

$$\Sigma f_j X^i_j - \gamma \cdot X^i_{ig} = -0.67 C^i_{sw}, \qquad \text{(VI-5)}$$

where $\gamma = M_{ig}/M_{sed}$ is considered as an unknown. X^i_{ig}, X^i_j, and C^i_{sw} are all given (Table VI-4; for practical purposes, C^i_{sw} can be ignored except for Na). Therefore, from the four mass balance equations like VI-5 for Al, Si, Ca, and Na and the additional equation VI-4b, one can solve for f_1 to f_4 and γ. The results are $f_1 = 0.84$, $f_2 = 0.13$, $f_3 = 0.01$, $f_4 = 0.02$, and $\gamma = 0.93$. The value $f_3 = 0.01$ (carbonate fraction) is much lower than the "observed" 0.25 given by Ronov (1982; Table VI-4). Also, the M_{ig}/M_{sed} ratio of 0.93 is much higher than the "observed" value of 0.83 by Ronov (1982). If the X^i_j data of Ronov (1982) are replaced by another set of data given by Bowen (1979), one obtains $f_1 = 0.70$, $f_2 = 0.27$, $f_3 = 0.03$, and $\gamma = 0.94$. The fraction of carbonates (0.03) is still much too low as compared to the "observed" 0.25 given by Ronov (1982).

Sibley and Wilband (1977) argued that the total mass of the oceanic pelagic clays should be as large as that of shale, if one assumes that all the subducted pelagic sediments along the plate subduction zones before the mid-Mesozoic were mostly pelagic clays and have been preserved intact. Only after the mid-Mesozoic did the pelagic sediments start to contain a significant carbonate component due to the evolution of calcareous plankton in the open ocean. If this argument is correct, one may reduce the carbonate fraction in global average sediments to about one-half of the original value (from 0.25

down to 0.12). This is still too high. It is also doubtful that the subducted pelagic clays would be preserved intact without melting. When sediments were remelted or metamorphosed at high temperatures, the released volatiles would react with the crustal rock to produce new sediments on the Earth's surface. In addition, the pelagic clays are enriched in Mn, Co, Ni, Cu, Zn, Pb, Mo, Ba, Ir, Pt, and Pd relative to the average shale (by a factor of 5 to 10 as will be discussed in the next chapter). Therefore, the proposed high fraction of pelagic clays in the sedimentary reservoir would cause an excess of these elements in global mean sediments as compared to Earth's upper crust.

One possible solution to the "excess" calcium problem is that during the early history of Earth, the primordial volatiles may have preferentially leached Ca^{+2} from calcium-rich Archean crust (Garrels and Mackenzie, 1971) and a large mass of carbonate was thus deposited in the near-shore or inland environments. As argued by Sibley and Wilband (1977), carbonate rocks were not affected by the plate tectonic recycling of sediments before the mid-Mesozoic. Therefore, the total mass of carbonate rocks would not change much with time, maintaining a rough balance between weathering and deposition rates of carbonate rocks. Only after the mid-Mesozoic did carbonate rocks start to be subducted back to the mantle. One problem with this scenario is that no massive carbonate has been found in Archean sedimentary columns, but one may also argue for a preferential leaching and recycling of carbonates as well as evaporites relative to silicate sediments (Garrels and Mackenzie, 1971; Morse and Mackenzie, 1990).

VI-5. SHALES AND RELATED MATERIALS

As summarized in Table VI-5a and Figure VI-12 (see page 278), the composition of average shale is almost identical to the average upper crust except for Ca, Sr, and Na, which are depleted, and for some volatile elements, which are enriched in shale as compared to the upper crust. The depleted Ca and Sr are accumulated separately in carbonate rocks and the depleted Na in evaporites + seawater. The enriched volatiles in shale are C, N, S, Se, Te, Br, I, As, B, Bi, Cd, Hg, In, and Sb. These enriched volatiles are called **excess volatiles** or primary magmatic volatiles, and their sources should be related to magmatic hydrothermal solutions and volcanic vapor, and not to the direct weathering of the upper crust itself. These volatiles are the very same elements that are much enriched in the Kilauea volcanic vent emissions (E^i_{Al} values of greater than 500 relative to the associated basaltic rock; Olmez et al., 1986).

For comparison, the chemical data for the shale geostandard SCO-1 and the **hemipelagic mud** geostandard MAG-1 (as a precursor to shale) are also shown in Table VI-5 (Govindaraju, 1989; 1994). Surprisingly, the individually

TABLE VI-5a
Compositions of average shale and its related materials

	Upper crust	Soils		Loess	River mud	Shale		Hemipelagic mud	Micaschist
		Ave.	*SO-4*	*GSS-8*	*GSD-9*	*Ave.*	*SCO-1*	*MAG-1*	*SDC-1*
	(1)	(2)	(3)	(3)	(3)	(4)	(3)	(3)	(3)
Ag-47	0.06	0.13	+0.12	0.06	0.09	0.07	+0.13	0.08	0.04
Al-13 %	7.83	6.62	5.4	6.3	5.6	8.8 b	7.23	8.65	8.33
As-33	1.6	+11	+7.4	+13	+8.4	+13	+12.4	+9.2	0.22–
Au-79 ppb	2.3	*1 b	*0.6	1.4	1.3	2.5 w	2.1	2.4	1.2
B-5	12	+48	+43	+54	+53	+100	+72	+136	13–
Ba-56	570	470	700	480	430	580	570	480	630
Be-4	3.2	2	*1.3	1.9	1.8	3	1.8	3.2	3
Bi-83	0.054	+0.37	+0.14	+0.3	+0.42	+0.43 m	+0.37	+0.34	0.26–
Br-35	2.1	+5.4	+5.2	2.6	1.5	+20 w	1.0 g	+250	0.1–
C-6%	0.023	+2.0 b	+4.4	+1.6	+1.1	+1.2	+0.81	+2.2	0.027–
Ca-20 %	3.15	*1.54	*1.11	5.91	3.8	*1.6	1.87	*0.98	1
Cd-48	0.1	0.1	+0.34	0.13	+0.26	+0.3	0.14	+0.2	0.08–
Ce-58	58	68	54	66	78	82	62	88	93
Cl-17	150	100 b	*30	*70	*50	180	*51	+31000	32–
Co-27	17	13	10	13	14	19	11	20	18
Cr-24	69	61	64	68	85	90	68	97	64
Cs-55	3.7	+8.2	2.9	+7.5	5.1	5	+7.8	+8.6	4
Cu-29	39	23	21	24	32	45	29	30	30

Dy-66	3.5	4.1	3.5	4.8	5.1	4.7	4.2	5.2	6.7
Er-68	2	2.5	2.2	2.8	2.8	3	2.5	3	4.1
Eu-63	1.1	1	0.97	1.2	1.3	1.2	1.2	1.6	1.7
F-9	700	480	*300	580	490	740	770	770	600
Fe-26 %	4.17	2.94	2.37	3.1	3.4	4.72	3.59	4.75	5.36
Ga-31	18	18	11	15	14	19	15	20	21
Gd-64	3.9	4.6	3.9	5.4	5.5	5.1	4.6	5.8	7.2
Ge-32	1.5	1.7	3.5?	1.3	1.3	1.6	1		1.5
Hf-72	4	7.7	8	7	9.7	(5)	4.6	3.7	8.3
Hg-80	0.08	0.065	*0.032	*0.017	0.083	+0.18 m	0.05	*0.018?	0.023–
Ho-67	0.74	0.87	0.8	0.97	0.96	1.1	0.97	1	1.5
I-53	0.5	+3.8	+3	+1.6	0.61	+19 b		+380	
In-49	0.05	0.07	+0.1	0.043	0.056	+0.1	+0.11	+0.18	0.12
Ir-77 ppb	(0.05)					0.05 e			
K-19 %	2.56	1.86	1.73	2	1.7	2.66	2.28	2.95	2.72
La-57	30	(36)	28	36	40	43	30	43	42
Li-3	23	33	17	35	30	+66	45	+79	34
Lu-71	0.32	0.36	0.37	0.43	0.45	0.42	0.34	0.4	0.53
Mg-12 %	1.64	*0.78	*0.56	1.4	1.4	1.5	1.61	1.81	1.02
Mn-25	770	580	600	620	620	850	420	760	880
Mo-42	1.6	2	1	1.2	*0.64	2.6	1.4	1.6	0.25–
N-7	20	+2000 b	3800			+1000 v		+800	
Na-11 %	2.54	*1.02	*0.97	*1.3	*1.10	*0.59 b	*0.64	2.84	1.52
Nb-41	15	10 b	10	15	18	11	11	12	18

TABLE VI-5a *Continued*

		Soils		*Loess*	*River mud*	*Shale*		*Hemipelagic mud*	*Micaschist*
	Upper crust (1)	*Ave.* (2)	*SO-4* (3)	*GSS-8* (3)	*GSD-9* (3)	*Ave.* (4)	*SCO-1* (3)	*MAG-1* (3)	*SDC-1* (3)
Nd-60	26	26	25	32	34	33	26	38	40
Ni-28	55	27	24	32	32	(50)	27	53	38
Os-76 ppb	(0.05)					0.05 e			
P-15	860	800 b	920	790	660	700	1000	710	690
Pb-82	17	26	14	21	23	20	31	24	25
Pd-46 ppb	(1)						1 g	1.7	1.1
Pr-59	6.6	7.2	7.2	8.3	9.2	9.8	6.6	9.3	9.8
Pt-78 ppb	(1)						0.7	1	1
Rb-37	110	110	69	96	80	140	112	149	127
Re-75 ppb	(0.4)					0.4 e			
Rh-45 ppb									
Ru-44 ppb									
S-16	530	700 b	500	*120	*150	+2400	630	+3900	650–
Sb-51	0.2	+1.2	+0.71	+1	+0.81	+1.5	+2.5	+0.96	0.54–
Sc-21	14	11	8.4	12	11	13	11	17	17
Se-34	0.14	+0.29	+0.49	0.12	0.16	+0.6	+0.89	+1.2	0.032–
Si-14 %	30	33.0 b	32	27.4	30.3	27.5	29.2	23.5	30.8
Sm-62	4.5	5.2	4.7	5.9	6.3	6.2	5.3	7.5	8.2

Sn-50	3.3	2.6	2.5	2.8	2.6	(3)	3.7	3.6	3
Sr-38	350	167	*168	240	*170	*(170)	*174	*146	183
Ta-73	1.5	1.2	0.62	1.1	1.3	(1)	0.92	1.1	1.2
Tb-65	0.6	0.63	0.61	0.89	0.87	0.84	0.7	0.96	1.2
Te-52	0.003	+0.035	+0.03	+0.046	+0.04	+(0.07)	+0.077	+0.066	0.006−
Th-90	11	13.8	8.6	12	12	12	9.7	12	12
Ti-22	3300	3800	3400	3800	5500	4600	4000	4500	6050
Tl-81	0.53	0.62	*0.26	0.59	0.49	(0.7)	0.72	0.59	0.7
Tm-69	0.32	0.37	0.35	0.46	0.44	0.44	0.42	0.43	0.65
U-92	2.8	3	2.4	2.7	2.6	(2.7)	3.1	2.7	3.1
V-23	140	82	85	82	97	130	131	140	102
W-74	1.3	2.5	1	1.7	1.8	1.8	1.4	1.4	0.8
Y-39	22	23	22	26	27	26	26	28	40
Yb-70	2	2.4	2.1	2.8	2.8	2.8	2.3	2.6	4
Zn-30	67	74	94	68	78	95	103	130	103
Zr-40	170	260	270	230	+370	160	160	130	290

Sources: (1) Table V-4a, column $B/3 + 2G/3$. (2) National Environmental Monitoring Center in China (1990, 1994); (3) Govindaraju (1989); and (4) Turekian and Wedepohl (1961). REE from Nance and Taylor (1976); b: Bowen (1979); e: Esser (1991); g: Govindaraju (1994); m: Marowsky and Wedepohl (1971); v: Vinogradov (1959); w: Wedepohl (1969–1978).

Notes: The values in parentheses are educated guesses based on other related samples in the table. The "plus" and "asterisk" signs represent values higher and lower than those for the upper crust by more than a factor of two. The "negative" signs in the mica schist column represent values lower than those for the average shale. The geometric means of the following elements in soils are: Br = 3.4, Ca = 0.71%, Hg = 0.04, Mo = 1.2, and Sr = 120 ppm. All compositions are in ppm unless noted otherwise.

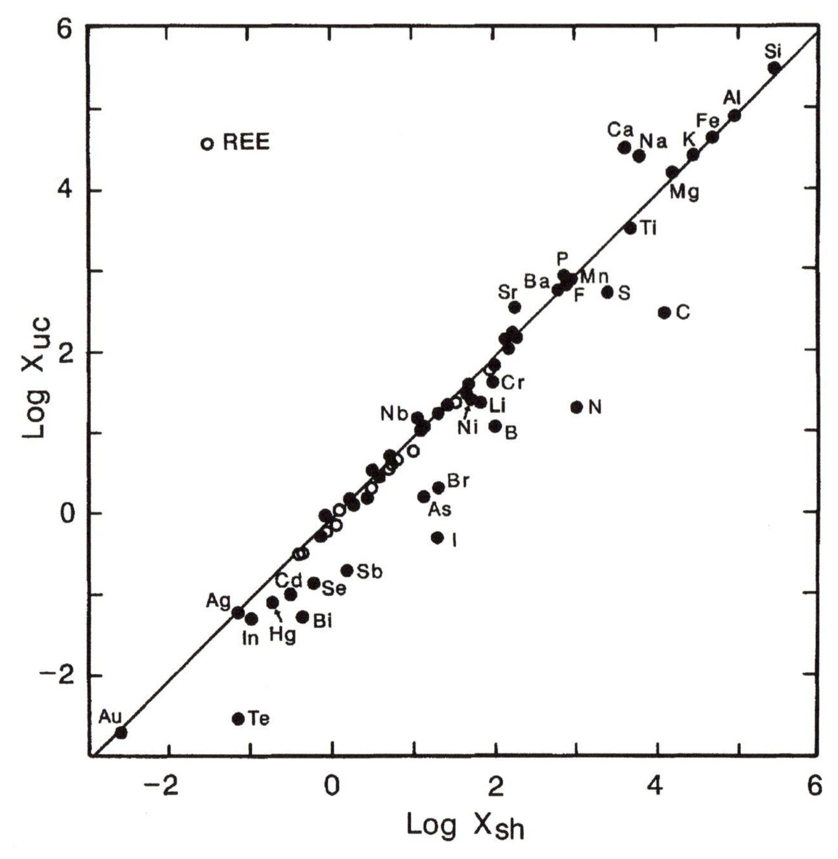

FIGURE VI-12. Concentrations of the elements in shale versus that in the upper crust (Table VI-5a). The data falling on the diagonal solid line have E^i_{Al} values of around 1. The elements falling off the solid line are all identified.

analyzed samples have compositions almost identical to average shale (within a factor of two), except for seasalt elements such as B, Br, Cl, I, Na, and S, which are enriched in the hemipelagic mud. This close agreement implies that the compositions of normal shales are relatively uniform on a global scale as a result of repeated cycles of erosion and deposition, and the average shale composition given in Table VI-5a is a good representation.

The **mica-schist** geostandard SDC-1 is a metamorphic rock of shale origin and has, again, a composition identical to that of average shale (Table Vl-5a). However, the originally enriched volatile elements in shale become depleted in mica-schist, probably resulting from volatilization processes through high-temperature and high-pressure metamorphism. Apparently metamorphism is another effective mechanism for mobilizing volatile elements in sediments.

The **loess** sample from northern China (geostandard GSS-8) and the river mud from the Yangtze River (geostandard GSD-9) represent samples integrated and homogenized over large surficial areas. Therefore, their compositions should be similar to those of average shale or to the upper crust when volatile elements are excluded, as shown in Table VI-5a. A few exceptions are Ca, Hf, and Zr, which are high in the loess and river mud, probably due to high content of carbonates and zircon in these samples as mentioned by Taylor et al. (1983).

Vinogradov (1959), Shacklette et al. (1971), Bowen (1979), and the National Environmental Monitoring Center of China (1990, 1995) provided the average composition of **soils**. Table VI-5a summarizes the arithmetic means of elements from China. Because most of the elements show a more or less normal distribution pattern in frequency plots of concentrations, the arithmetic means are only slightly larger than the geometric means. The exceptions are Ca, Sr, Br, I, Mo, and Hg which show logarithmic-normal distribution patterns. The geometric means of those elements are also given in the footnote of Table VI-5a. For comparison, the Canadian soil geostandard SO-4, which has a similar major element composition to average soil, is also given. The general agreement between the average soil and geostandard SO-4 compositions is again surprisingly good for other trace elements. The elements Na, K, Mg, Ca, Sr, and probably Au and F are depleted in the average soil relative to the upper crust, owing to their greater mobility during weathering processes as discussed earlier. The volatile elements are again enriched in soil as compared to the upper crust. In short, the average soil and shale are probably better representations of original source materials for inorganic suspended particles in the air and oceans than the upper crust. The enrichment factors E^i_{La} for REE in shales and related materials relative to the upper crust are mostly within the bounds of 1.0 ± 0.2 (Table VI-5b). Therefore the average upper crust, shale, soil, loess, river mud, river-suspended particles, and hemipelagic mud all have similar REE abundance patterns. The heavy REE depletions in river-suspended particles and near-shore hemipelagic sediments reported by Goldstein and Jacobsen (1987, 1988) and Sholkovitz (1988) may be caused by incomplete dissolution of heavy minerals during their sample preparation (Condie, 1991).

The mineralogical and chemical compositions of Pierre shale and equivalent rocks of Late Cretaceous age in the northern Great Plains Regions (Dakotas, Wyoming, and Montana), U.S.A., were characterized in detail by Schultz et al. (1980). Average compositions of normal, organic-rich, siliceous, and calcareous shales deposited in offshore marine environments are summarized in Table VI-6. The average composition of the normal Pierre shale in Table VI-6 is in good agreement with that of average shale in Table VI-5a. Besides clay minerals, quartz is an important mineral in all types of shale. In siliceous shale, cristobalite is abundant and was recrystallized from amorphous opal of radiolarian origin. The results of factor analysis from the

TABLE VI-5B
Enrichment factors E^i_{La} for rare earth elements relative to the average upper crust

	Soil		*Loess*	*River mud*	*River particles*[a]	*Shale*		*Hemipelagic mud*
	Ave	*SO-4*	*GSS-8*	*GSD-9*	*Ave*	*Ave*	*SCO-1*	*MAG-1*
La	1	1	1	1	1	1	1	1
Ce	0.98	1.00	0.95	1.01	1.09	0.99	1.07	1.06
Pr	0.91	1.31?	1.17	1.17	0.90	1.16	1.12	1.10
Nd	0.85	1.03	1.03	0.98	0.90	0.89	1.00	1.02
Sm	0.91	1.12	1.09	1.05	1.04	0.96	1.18	1.16
Eu	0.78?	0.94	0.91	0.89	0.91	0.76?	1.09	1.01
Gd	0.98	1.07	1.15	1.06	0.85	0.91	1.18	1.04
Tb	0.88	1.09	1.24	1.09	1.11	0.98	1.17	1.12
Dy	0.98	1.07	1.14	1.09		0.94	1.20	1.04
Ho	0.98	1.16	1.09	0.97	0.90	1.04	1.31	0.94
Er	1.06	1.18	1.17	1.05	1.00	1.05	1.25	1.05
Tm	0.96	1.17	1.20	1.03	0.83	0.96	1.31	0.94
Yb	1.02	1.13	1.17	1.05	1.17	0.98	1.15	0.91
Lu	0.94	1.24	1.12	1.05	1.04	0.92	1.06	0.87

Note: [a]Data for river-suspended particles are based on Tables VI-5a and VI-10.

original composition data for these samples are summarized in Figure VI-13a (factor loadings) and Figure VI-13b (factor scores). According to Figure VI-13a, factor 1 represents the association of elements Al, Ti, Ga, K, Ba, B, Zr, Sc, Cr, and Pb with clay minerals. Silica correlates only with Al and Ti, and represents both quartz and clay mineral components. Negative factor 1 (Ca, CO_2, Sr, and Mn) represents carbonate phases that may also contain apatite. Factor 2 represents the association of C, Mo, S, As, Se, Cd, U, V, Cu, and Fe(II) with organic matter and sulfide minerals. The close association between C and S is also common in anoxic lake and marine sediments (Urban, 1994; Berner et al., 1979). The concentrations of Ni, Co, and Zn are strongly correlated as factor 3 (not shown here) and correlate moderately with the factor 2 elements Fe(II), Cd, and S. In the factor score plot (Figure VI-13b), organic-rich shale samples are high in factor 2, and calcareous shale samples are high in negative factor 1 but a few are also high in factor 2. The ordinary and siliceous shale samples are not distinguishable in the factor score plot (Figure VI-13b). The only difference is that siliceous shales have high cristobalite content.

The average compositions of cratonic shales deposited during Archean and post-Archean times are significantly different (Condie, 1993). For example, **post-Archean shales** as compared to **Archean shales** tend to be low in elements enriched in basaltic rocks such as Co, Cr, Fe, Mg, Ni, Sc, and V; and tend to be high in elements enriched in granitic rocks such as Ba, REE,

TABLE VI-6

Average compositions of different types of shale in the Pierre Shale Members

	Normal (%)	Organic rich	Siliceous	Calcareous		Normal (ppm)	Organic rich	Siliceous	Calcareous
clays	68	58	61	51	As	13	113	12	18
quartz	23	20	34[a]	13	B	118	141	125	71
K-feldspar	0.8	2.8	0.6	0.7	Ba	730	600	460	440
plagioclase	3.9	1	1.1	1.3	Cd	0.54	2.2	0.71	2.2
calcite	0.4	0	0.3	28	Co	13	13	20	15
dolomite	1.4	0.4	0	1.9	Cr	90	108	70	78
organic	0.8	8.2	0.8	1.8	Cu	36	90	48	39
pyrite	0.4	3.9	0.2	1.6	Ga	18	14	9.2	12
SiO_2	61	53	67	39	La	30	19	30	22
TiO_2	0.65	0.61	0.45	0.45	Mn	500	150	6800	2700
Al_2O_3	16	13	12	11	Mo	1.4	86	1.3	5.9
Fe_2O_3	4.2	2.1	3.2	2.7	Ni	39	58	59	46
FeO	1.1	4.1	0.83	1.6	Pb	22	20	15	20
MgO	2.1	1.1	1.7	1.8	Sc	18	17	14	17
CaO	1.2	0.92	1.2	17	Se	1.4	60	1.2	11
Na_2O	1.1	0.65	0.69	0.63	Sn	1.3	1.4		0.7
K_2O	2.6	2.7	1.8	2.2	Sr	140	96	100	540
P_2O_5	0.15	0.15	0.07	0.22	U	4	13	2.6	5
F	0.07	0.079	0.082	0.073	V	170	490	210	180
C organic	0.63	5.9	0.58	1.3	Y	26	19	26	20
C carb	0.24	0.19	0.3	3.8	Yb	2.8	3	3	2
S total	0.29	3.9	0.22	0.95	Zn	130	140	180	120
Cl	0.013	0.05	0.01	0.02	Zr	170	180	130	140
H_2O^+	4.8	5.4	3.9	3					

Source: Schultz et al. (1980).

Note: [a]Value includes 23% of cristobalite.

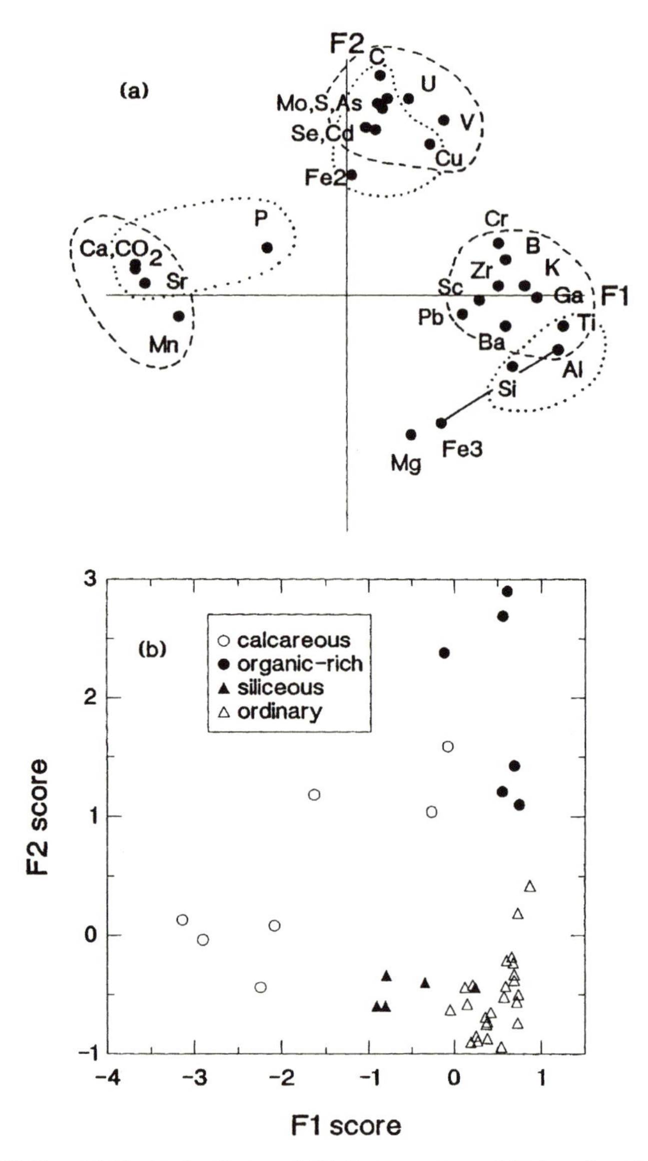

FIGURE VI-13. (a) Factor loading and (b) factor scores of factors 1 and 2 for different shale types from the Pirre shale (Schultz et al., 1980). The elements within each dotted enclosure or connected by solid lines have correlation coefficients greater than 0.5 between any pair. Fe2 and Fe3 are, respectively, di- and trivalent Fe. C and CO_2 are organic and carbonate carbon, respectively.

Hf, K, Li, Na, Nb, Pb, Rb, Si, Ta, Th, U, Y, and Zr (compare Tables VI-7 and V-3 excluding volatiles). However, Ca, P, Sr, and Ti do not fit into the above-mentioned pattern. Unfortunately Table VI-7 does not contain data for the "excess" volatile elements discussed earlier.

Archean and post-Archean shale data by Condie (1993) are all depleted in Ca, Sr, and Na (Table VI-7, $E^i_{\mathrm{Al}} \leq 0.4$), and are enriched in Sc-V-Cr-Ni and Cr-Ni ($E^i_{\mathrm{Al}} \geq 1.6$) as compared to juvenile upper continental crusts of the same ages (Table VI-7). Otherwise, the relative abundances of other elements are similar. Due to recycling, most sedimentary rocks are of Phanerozoic age (0–570 Ma; Garrels and Mackenzie, 1971), and the preserved mass of Archean sediments is negligibly small. The Phanerozoic shales integrate not only the weathering products of juvenile upper crust but also of upper crusts of all older ages. Therefore, when the Phanerozoic average shale is compared to the average present-day continental upper crust, the excess Cr and Ni disappear. The E^i_{Al} [Phanerozoic shale/average continental upper crust] values are mostly within the range of 1.0 ± 0.3 including REE data (Table VI-7), and E^i_{Al} values for Na, Ca, Sr are less than 0.4. A few exceptions ($E^{\mathrm{Rb}}_{\mathrm{Al}} = 1.65$, $E^{\mathrm{Ta}}_{\mathrm{Al}} = 1.49$) probably represent noisy data. As mentioned by Condie (1993), average compositions of various rock types have uncertainties (one standard deviation divided by the mean) of about ±10–40% for trace elements and ±20–30% for REE, whereas the overall uncertainties in average juvenile continental upper crusts and the average present-day continental upper crust are about ±30–50% for trace elements plus Nd and Gd, and about ±20–30% for other REE. The enrichment of Sc, V, Cr, and Ni in Archean shales relative to the Archean upper crust also suggests that Archean shale's source rocks contained more basaltic (including komatiite) components than the presently preserved Archean upper crust (Condie, 1993). Looking at the REE data in Figure VI-14 and Table VI-7 more closely, one can see that the distribution patterns of REE between juvenile upper crust and shale of the same age do not necessarily match exactly. Also, heavy REE in the Archean upper crust are quite low compared to others in Figure VI-14. The upper crusts and shales of Archean and post-Archean ages all show negative Eu anomalies in Figure VI-14, suggesting removal of Eu from upper continental crust to lower continental crust by plagioclase throughout geological time. However, no negative Eu anomaly was detected in Australian Archean sediments and turbidites by Taylor and McLennan (1985) and McLennan and Taylor (1991).

VI-6. Trace Elements in Sandstone and Limestone

The average trace element concentrations in sandstone and limestone were compiled by Turekian and Wedepohl (1961) and Bowen (1979). However, their original data sources were so diverse that their so-called averages are

TABLE VI-7

Average compositions of cratonic shales, the enrichment factors of elements (E^i_{Al}) for shales relative to juvenile upper continental crusts of the same age, and E^i_{Al} [phanerozoic shale/average present-day upper continental crust]

	Archean	*Proterozoic*	*Phanerozoic*	E^i_{Al} *[shale/juvenile crust]*			
	>3.5–2.5 Ga	*2.5–0.8 Ga*	*0.8–0 Ga*	*Archean*	*Proterozoic*	*Phanerozoic*	E^i_{Al} *[Phan. shale/ave.crust]*
SiO_2%	60.95	63.1	63.6	0.77	0.82	0.83	0.81
TiO_2	0.62	0.64	0.82	1.19	0.87	1.04	1.25
Al_2O_3	17.5	17.5	17.8	1	1	1	1
FeO	7.53	5.65	5.89	1.42	1.01	1.01	1.05
MgO	3.88	2.2	2.3	1.09	0.88	0.88	0.8
CaO	0.64	0.71	1.3	0.15	0.17	0.29	0.3
Na_2O	0.68	1.06	1.1	0.16	0.28	0.28	0.26
K_2O	3.07	3.62	3.84	1.03	1.03	1.16	1.18
P_2O_5	0.1	0.12	0.14	0.77	0.73	0.85	0.98
Rb ppm	111	165	163	1.29	1.52	1.51	1.65
Sr	61	108	136	0.18	0.32	0.43	0.4
Ba	456	642	551	0.68	0.79	0.68	0.73
Pb	13	27	22	0.61	1.36	1.17	1.09
Th	8.5	14.3	13.5	0.9	1.32	1.29	1.32
U	2.4	3.4	2.9	0.97	1.21	1.03	1.1
Zr	151	196	201	0.86	0.97	1.02	1.05
Hf	4.5	5.2	4.6	0.98	0.91	0.87	0.9

Nb	11.3	16.8	15.4	1.13	1.27	1.13	1.32
Ta	0.84	1.4	1.4	1.01	1.36	1.33	1.49
Y	28	35	33	1.32	1	0.94	1.16
La	30.7	38	38.8	0.88	1.14	1.3	1.15
Ce	60.9	81.7	82	0.91	1.18	1.24	1.2
Nd	27.7	37.5	32.3	0.94	1.2	1.09	1.06
Sm	4.85	6.68	5.75	1.03	1.09	0.98	1.05
Eu	1.12	1.32	1.14	0.96	0.97	0.95	0.91
Gd	4.55	5.6	5.22	1.09	0.98	0.94	1.04
Tb	0.71	0.9	0.81	1.11	1	0.93	1.03
Yb	2.43	2.86	2.95	1.34	1.09	1.09	1.3
Lu	0.39	0.48	0.47	1.26	1.14	0.98	1.23
Sc	21	17	16	1.63	0.92	0.88	1.01
V	154	100	117	1.81	0.88	0.92	1.14
Cr	507	115	104	2.31	1.61	1.74	0.84
Co	31	18	20	1.25	0.91	1.07	0.93
Ni	221	52	54	1.78	1.35	1.64	0.81

Source: Condie (1993).

Note: Oxides are in weight % and elements in ppm.

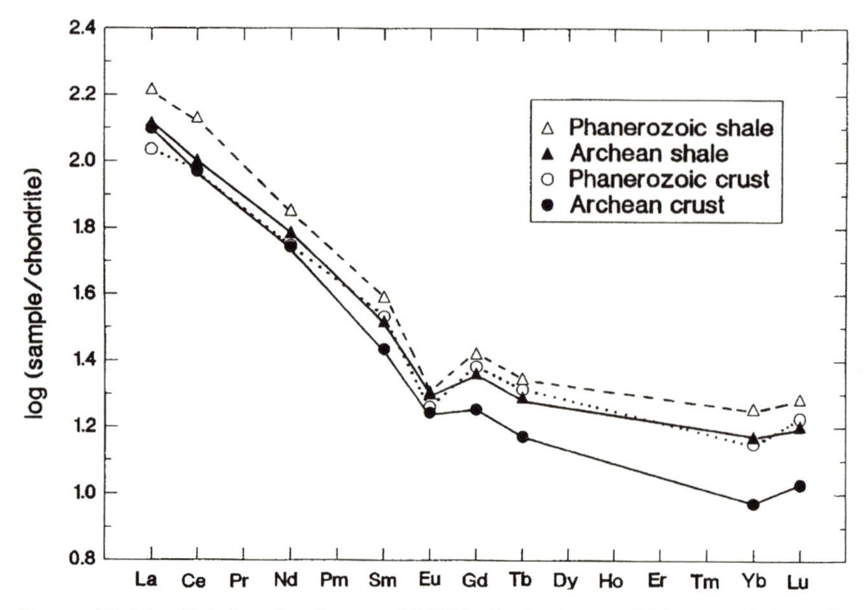

FIGURE VI-14. Relative abundances of REE in the Archean and Phanerozoic juvenile upper crusts and shales as normalized to that in the carbonaceous chondrites (data from Condie, 1993).

almost meaningless. It is difficult to use their average values to distinguish what phases control the concentrations of the elements in sandstone and limestone. Fortunately, one can utilize the compositional data of sandstone and limestone geostandard samples. For example, the compositions of sandstone geostandard GSR-4 and limestone geostandard GSR-6 from China are summarized in Table VI-8 (Govindaraju, 1989, 1994). The compositions of these geostandards are plotted against average shale composition in Figures VI-15a and VI-15b. Most of the elements fall on the solid diagonal line (slope of one) passing through Al (i.e., E^i_{Al} of one), or within a factor of two of the solid diagonal line (i.e., E^i_{Al} values between 0.5 and 2). Therefore, the content of these elements in sandstone and limestone geostandards is mainly contributed by the shale component. The obvious exceptions are Ca, Mg, Sr, and Pb in limestone, mainly contributed by carbonate phases; and Si, Ge, P, As, Zr, Hf, REE, Y, Nb, Th, U, W, Ag, and Au in sandstone, mainly contributed by weathering-resistant minerals such as quartz, apatite, zircon and monazite. The most recent compilation of the average compositions of cratonic sandstones as a function of age (Condie, 1993) also shows an enrichment of Si, Zr, Hf, Nb, Ta, Th, U, and Pb but not P, relative to cratonic shales of the same age (i.e., $E^i_{\mathrm{Al}} > 1.5$). It is still uncertain how important apatite, zircon, and monazite phases in sedimentary rocks are with regard to the global mass

TABLE VI-8
Compositions of sandstone and limestone geostandards

	Sandstone GSR-4	*Limestone GSR-6*		*Sandstone GSR-4*	*Limestone GSR-6*
Ag-47	0.062	0.043	Mn-25	150	460
Al-13%	1.86	2.66	Mo-42	0.76	0.38
As-33	9.1	4.7	Na-11%	0.044	0.060
Au-79 ppb	1.8	0.9	Nb-41	5.9	6.6
B-5	34	16	Nd-60	21	12
Ba-56	143	120	Ni-28	17	18
Be-4	0.97	0.8	P-15	960	230
Bi-83	0.18	0.16	Pb-82	7.6	18.3
C-6%	0.05	8.74	Pd-46 ppb		
Ca-20%	0.215	25.4	Pr-59	5.4	3.4
Cd-48	0.06	0.069	Pt-78 ppb		
Ce-58	48	25.4	Rb-37	29	32
Cl-17	42	80	S-16	860	370
Co-27	6.4	9	Sb-51	0.6	0.43
Cr-24	20	32	Sc-21	4.2	6
Cs-55	1.8	3.2	Se-34	0.098	0.099
Cu-29	19	23.4	Si-14%	42	7.29
Dy-66	4.1	1.6	Sm-62	4.7	2.4
Er-68	2	1.1	Sn-50	1.1	0.98
Eu-63	1	0.51	Sr-38	58	913
F-9	183	406	Ta-73	0.42	0.46
Fe-26%	2.25	1.76	Tb-65	0.79	0.35
Ga-31	5.3	7.1	Te-52	0.038	0.023
Gd-64	4.5	1.9	Th-90	7	4.1
Ge-32	1.2	0.67	Tl-22	1600	2000
Hf-72	6.6	1.8	Ti-81	0.36	0.36
Hg-80	0.008	0.016	Tm-69	0.32	0.17
Ho-67	0.75	0.33	U-92	2.1	1.9
In-49	0.026	0.042	V-23	33	36
K-19%	0.54	0.65	W-74	1.2	0.67
La-57	21	14.6	Y-39	22	9.1
Li-3	11	20	Yb-70	1.9	0.9
Lu-71	0.3	0.14	Zn-30	20	52
Mg-12%	0.049	3.1	Zr-40	214	62

Note: Units are ppm unless noted otherwise.

balance of trace elements. The close agreement between the upper crust and average shale compositions shown in Figure VI-12 (excluding volatiles) may suggest that these heavy minerals are quantitatively unimportant in the trace element mass balance budget.

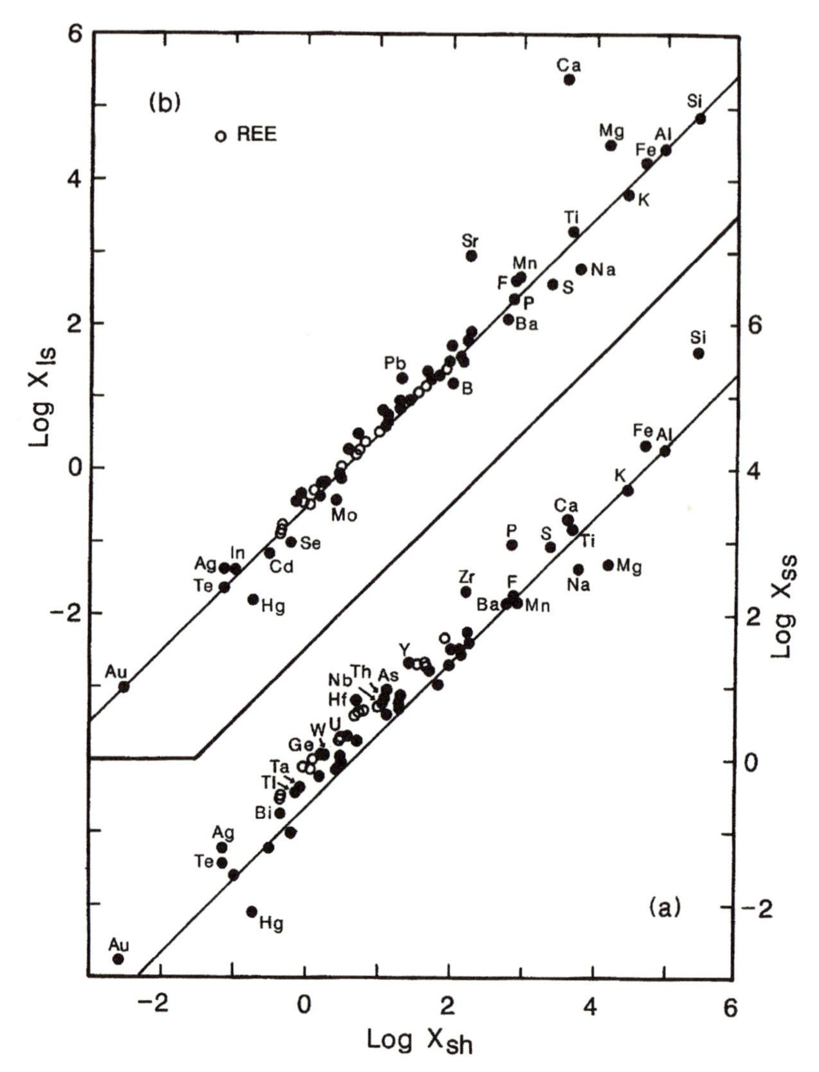

FIGURE VI-15. Concentrations of elements in shale (Table VI-5a) versus those in (a) sandstone geostandard GSR-4 and (b) limestone geostandard GSR-6 (Table VI-7). The solid diagonal lines represent an E^i_{Al} of one for sandstone and limestone relative to shale.

VI-7. IRON FORMATIONS

Iron formations (IF) are mostly pre-Cambrian ferruginous cherty sedimentary rocks and were deposited mainly in the middle Archean age (3500–

3000 m.y.; e.g., Guyana Shield, South Africa), late Archean age (2900–2600 m.y.; e.g., Block greenstone belts, West Australia), early Proterozoic age (2500–1900 m.y.; e.g., Lake Superior and Labrador Trough, North America; Transvaal-Griquatown belts, South Africa), and late Proterozoic to early Phanerozoic ages (750–450 m.y.; e.g., Maly Khingham-Uda areas of far eastern Russia) (James, 1983). Iron formations consist of four major iron mineral groups in variable proportions: i.e., oxides (hematite, magnetite,), carbonates (siderite, $FeCO_3$; ankerite, $[Ca, Mg, Fe, Mn] CO_3$), silicates (greenalite, $[Fe, Mg]_6SiO_{10}[OH]_8$; stipnomelane, $K_{0.6}[Mg, Fe]_6[Si, Al]_6[O, OH]_{22}$, minnesolaite, $Fe_3Si_4O_{10}[OH]_2$), and sulfides (pyrite, FeS_2). These iron minerals are usually interbedded with silica (as quartz, fibrous chaledony, and amorphous chert).

Gross (1980) subdivided the iron formations into Algoma type and Lake Superior type according to their deposition environments. Deposition of iron formations has been explained by the following processes. First, anoxic bottom water in a large basin became enriched in dissolved Fe^{+2} and silica, mainly from hydrothermal sources. When the anoxic bottom water moved up onto a shallow shelf, Fe and silica were deposited due to oxidation and evaporation, respectively. The Lake Superior type IF were deposited on large continental shelf environments, and the Algoma type IF were deposited in smaller shallow water basins around island arcs (Jacobsen and Pimentel-Klose, 1988)

Table VI-9 summarizes the average compositions of the Algoma and Lake Superior type IF (Gross, 1980), iron formation geostandard (IF-G) from the 3800-m.y.-old Isua Supracrustal Belt, West Greenland (Govindaraju, 1994), and Brockman IF from Joffre Member, Hamersley, Australia (Davy, 1983). Additional data for REE are from the iron formation geostandard (FeR-1) from Austin Brook, New Brunswick, Canada (Govindaraju, 1994); and Penge (PE-71) and Kuruman (KK-11) IF from Transvaal Supergroup, South Africa (Bau et al., 1997). As shown in Figures VI-16a and VI-16b, both the Algoma and Lake Superior type IF are enriched in many elements relative to the Archean bulk crust or post-Archean upper crust. Those enriched elements are also concentrated in the hydrothermal fluids of mid-ocean ridges relative to seawater (see Table VII-7, and some additional data from Von Damm, 1995). Some obvious exceptions are B, Ca, K, Sr, F, Mg, Mo, P, and S, especially F, Mg, Mo, P, and S, which are even very much depleted in hydrothermal solutions relative to seawater. Therefore, the main source for those elements in IF should be seawater.

The positive anomaly of Eu and the enrichment of the heavy over the light REE in the iron formations relative to the Earth's crust (Figure VI-17) again indicate a mixture of REE from hydrothermal fluids and seawater sources (Bau et al., 1997). The $^{143}Nd/^{144}Nd$ ratios in IF are higher than those for the bulk Earth at the time of IF deposition, suggesting a large mantle source of

TABLE VI-9
Compositions of average Algoma and Lake Superior type iron formations (IF), iron formation geostandard (IF-G), Brockman IF from Austrailia; and REE data from IF geostandard (FeR-1), and Penge (PE-71) and Kruman (KK-11) IF from South Africa

	Algoma IF-G (1)	*Superior Brock. (2)*	*Algoma ave. (3)*	*Superior ave. (3)*		*Algoma IF-G(1)*	*Superior Brock. (2)*	*Algoma ave. (3)*	*Superior ave. (3)*
Ag		0.2			S	700	640	15700	2000
Al	790	4710	19600	7940	Sb	0.63			
As	1.5	6			Sc	0.3		8	18
B		15	410	210	Si	192600	207000	228600	220200
Ba	1.5	90	190	16	Sn	0.3	5		
Be	4.7				Sr	3	55	116	37
Bi		3			Ta	0.2			
C	820	12600	6830	16400	Th	0.1	13		
Ca	11100	12700	13400	16000	Ti	84	300	1240	390
Cd		0.3			Tl	0.02			
Cl	25	170			U	0.02			
Co	29	2	41	28	V	2	5	109	42
Cr	4	10	118	112	W	220			
Cs	0.06				Y	9		54	47
Cu	13	90	149	14	Zn	20	50	330	40
F	50	110			Zr	1		98	81

Fe	390000	308000	263500	280300
Ga	0.7	1		
Ge	24	3		
Hf	0.04			
In	0.02			
K	100	498	5150	1660
Li	1	6		
Mg	11400	13900	12100	11600
Mn	325	1320	1900	4900
Mo	0.7	1		
Na	240	3900	3190	965
Nb	0.1			
Ni	23	10	103	37
P	280	790	1005	350
Pb	4	60		
Rb	0.4	50		

	Algoma (1)		*Superior (4)*	
	IF-G	*FeR-1*	*PE-71*	*KK-11*
La	2.8	9.8	3.14	1.9
Ce	4	7.5	3.78	3.2
Pr	0.4		0.42	0.39
Nd	1.8	7	1.78	1.71
Sm	0.4	1.7	0.35	0.35
Eu	0.39	3.1	0.2	0.14
Gd	0.74	1.5	0.61	0.49
Tb	0.11	0.2	0.1	0.08
Dy	0.8	1.8	0.76	0.55
Ho	0.2	0.4	0.21	0.14
Er	0.63	1	0.7	0.5
Tm	0.09	0.2	0.1	0.07
Yb	0.6	0.98	0.63	0.48
Lu	0.09	0.15	0.11	0.08

Source: (1) Govindaraju (1994); (2) Davy (1983); (3) Gross (1980); and (4) Bau et al. (1997).
Note: All compositions in ppm.

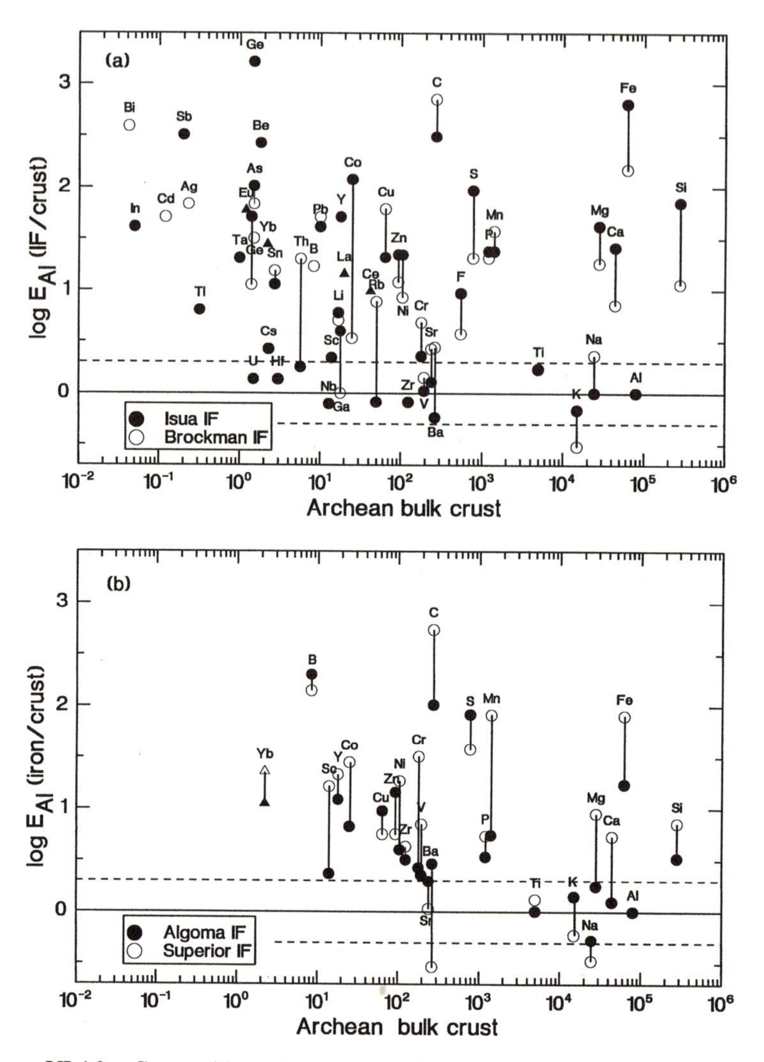

FIGURE VI-16. Composition of Archean bulk crust plotted against the enrichment factors (log E^i_{Al}) for (a) Isua iron formation (Table VI-9) relative to Archean bulk crust (Table V-5), and Brockman iron formation (Table VI-9) relative to post-Archean upper crust (Table V-4a), and for (b) average Algoma IF (Table VI-9) relative to Archean bulk crust, and average Lake Superior IF (Table VI-9) relative to post-Archean upper crust.

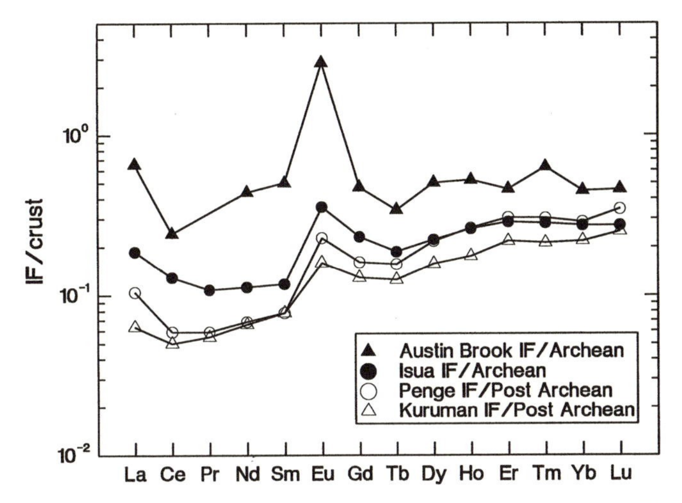

FIGURE VI-17. Rare earth compositions of various iron formations (Table VI-9) divided by those for Archean bulk crust and post-Archean upper crust (Tables V-3 and V-4).

REE in IF (Jacobsen and Pimentel-Klose, 1988; Bau et al., 1997). In short, the inputs from hydrothermal fluids and anoxic bottom seawater are two major sources of elements in IF. The anoxic bottom waters were probably formed in ocean basins where the organic fluxes from the surface ocean were remineralized, causing the $\delta^{13}C$ of the least metamorphosed carbonates in IF to become quite negative relative to seawater (−0.8 to −8 per mil versus 1 per mil in seawater; Perry et al., 1973). However, the carbon isotopic data may not be an ideal indicator for the carbon source in IF, because $\delta^{13}C$ of the hydrothermal fluids also ranges from −4 to −11 per mil (Shanks et al., 1995), and $\delta^{13}C$ can be changed in IF during metamorphic processes.

Hydrothermal solutions could be the major source of Fe in IF, but one cannot rule out the importance of Fe sources from the reduction of sedimentary iron into dissolved Fe^{+2} in anoxic basins as suggested by Holland (1973) and Drever (1974). Furthermore, the heavy $\delta^{32}S$ values (16 ± 3 per mil) of sulfide deposits Brunswick No. 6 IF and Brunswick No. 12 massive sulfides strongly suggest that nearly complete bacterial reduction of seawater sulfate to sulfide in closed or nearly closed back arc basins provided the S sources (Goodfellow and Peter, 1995). Hydrothermal solutions were not main S sources, because they have $\delta^{32}S$ of −5 to 8 per mil in the present mid-ocean ridge systems (Shanks et al., 1995). Further study on the Re-Os isotopes, $^6Li/^7Li$, and $^{11}B/^{10}B$ ratios in IF will provide additional insights about the sources of various elements in the iron formations.

TABLE VI-10

Concentrations of the elements in the world average river water and river-suspended particles

	River water $C(\mu g/l)$	River particles $X(\mu g/g)$	log X/C (cm^3/g)		River water $C(\mu g/l)$	River particles $X(\mu g/g)$	log X/C (cm^3/g)
Ag-47	0.3?	0.07*	2.37	N-7	375 m	1000*	3.4
Al-13	50	94000	6.27	Na-11	5300	7100	3.13
As-33	1.7	5	3.1	Nb-41			
Au-79	0.002?	0.0025*	3.1	Nd-60	0.04	35	5.94
B-5	18	70	3.59	Ni-28	0.5	90	5.26
Ba-56	60	600	4	Os-76			
Be-4	0.009 me	3*	5.52	P-15	25 m	1150	4.66
Bi-83				Pb-82	0.1	100	6
Br-35	20	5	2.4	Pd-46			
C-6	11500	12000*	3.02	Pr-59	0.007	8	6.06
Ca-20	13300	21500	3.21	Pt-78			
Cd-48	0.02	1	4.7	Ra-88	2.3E-8 mo		
Ce-58	0.08	95	6.07	Rb-37	1.5	100	4.82
Cl-17	7800 t	180*	1.36	Re-75			
Co-27	0.2	20	5	Rh-45			
Cr-24	1	100	5	Ru-44			
Cs-55	0.035	6	5.23	S-16	3700 t	2400*	2.81
Cu-29	1.5	100	4.82	Sb-51	0.07 a	2.5	4.55
Dy-66				Sc-21	0.004	18	6.65
Er-68	0.004	3	5.88	Se-34	0.06 me	0.6*	4
Eu-63	0.001	1.5	6.18	Si-14	5000	285000	4.76

F-9	100	740*	3.87	Sm-62	0.008	7	5.94
Fe-26	40	48000	6.08	Sn-50			
Ga-31	0.09	25	5.44	Sr-38	60	150	3.4
Gd-64	0.008	5	5.8	Ta-73	0.002	1.3	5.81
Ge-32	0.005 f	1.6*	5.51	Tb-65	0.001	1	6
Hf-72	0.01	6	5.78	Te-52			
Hg-80				Th-90	0.04	14	5.54
Ho-67	0.001	1	6	Ti-22	10	5600	5.75
I-53	7 t	19*	3.43	Tl-81			
In-49				Tm-69	0.001	0.4	5.6
Ir-77				U-92	0.24	3	5.34
K-19	1500	20000	4.12	V-23	0.76 s	170	5.34
La-57	0.05	45	5.95	W-74	0.03 t	1.8*	4.78
Li-3	12	25	3.32	Y-39		30	
Lu-71	0.001	0.5	5.7	Yb-70	0.004	3.5	5.94
Mg-12	3100	11800	3.58	Zn-30	0.7 s	350	5.7
Mn-25	8.2	1050	5.11	Zr-40			
Mo-42	0.5	3	3.78				

Sources: Martin and Whitfield (1981) unless indicated otherwise. a: Andreae et al. (1981), f: Froelich and Andreae (1980), m: Meybeck (1982), me: Measures and Edmond (1983), Measures and Burton (1980), mo: Moore and Scott (1986), s: Shiller and Boyle (1985, 1987), t: Turekian (1971); data with asterisks are from Table VI-5.

VI-8. Partition of Elements between River-Suspended Particles and River Water, and the Adsorption Model

The concentrations of trace elements in worldwide average river water and river-suspended particles were compiled by Turekian (1971), Martin and Meybeck (1979), Martin and Whitfield (1981), and Li (1982). Their results are selectively given in Table VI-10 (see pages 294–95) along with other, newer data sources. The recent measurements of REE in river waters (Goldstein and Jacobsen, 1987, 1988; Elderfield et al., 1990) are in general agreement with the data in Table VI-10. Average river-suspended particles are similar to average shale in composition (compare Table VI-5 and Table VI-10). Exceptions are Cd, Cu, Pb, and Zn, whose concentrations are more than a factor of two higher in river-suspended particles than in shale, probably indicating pollution inputs (Martin and Meybeck, 1979). By knowing the average total discharge rates of river water (37×10^{15} l/yr) and of river-suspended particles (14 to 18×10^{15} g/yr) to the ocean, one can in principle estimate the river flux of each element to the ocean.

The concentrations of elements in the river-suspended particles divided by those in the associated river water gives the distribution coefficient (K_d). These are also listed in Table VI-10, and plotted in Figure VI-18 as a function of the electron binding energies, I_z. The striking features of Figure VI-18 are as follows. There is a general positive correlation between log K_d and I_z for mono- to trivalent cations except for Cs, Rb, K, Ba, and Sr; and a negative correlation between log K_d and I_z for elements that have high oxidation states (greater than +3) and usually exist in oxygenated natural waters as oxyanions or hydroxyl complexes. However, B as H_3BO_3 does not follow this general trend.

In order to explain these features, we need to introduce the **surface complexation model** for the adsorption of elements onto oxide particles, as developed by Stumm et al. (1970) and Schindler (1975) and summarized nicely by Schindler and Stumm (1987) and Davis and Kent (1990).

Riverine particles are mainly a mixture of clays, silica, iron oxides, and manganese oxides. The hydrolysis of these particles produces hydrous oxide surface groups such as ≡Si-OH, ≡Mn-OH, and =Fe-OH. These surface hydroxyl groups have amphoteric properties. Protonation or deprotonation results in surface charge on these particles, e.g.,

$$\equiv\mathrm{MeOH}_2^+ \leftrightarrow \equiv\mathrm{MeOH} + \mathrm{H}^+,$$

$$K^s_{a1} = \frac{\{\equiv\mathrm{MeOH}\}[\mathrm{H}^+]}{\{\equiv\mathrm{MeOH}_2^+\}}, \qquad \text{(VI-6a)}$$

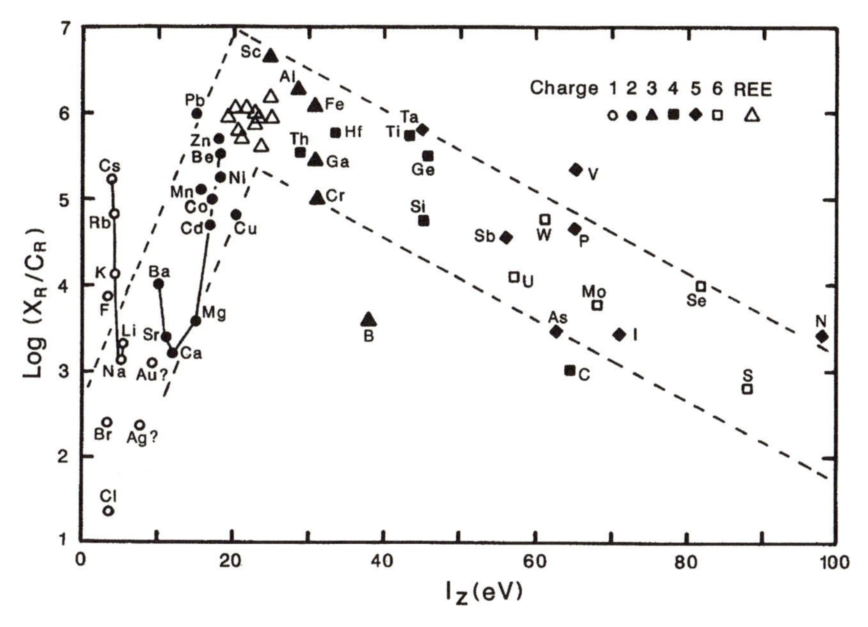

FIGURE VI-18. Elemental concentrations in river-suspended particles (X_R) divided by that in river water (C_R) based on Table VI-10 versus the electron binding energy to cations (I_z). The dotted lines delineate the general trend. The solid lines connect the alkali and alkaline earth elements.

$$\equiv\text{MeOH} \leftrightarrow \equiv\text{MeO}^- + \text{H}^+,$$

$$K^s_{a2} = \frac{\{\equiv\text{MeO}^-\}[\text{H}^+]}{\{\equiv\text{MeOH}\}}, \qquad \text{(VI-6b)}$$

where Me = metal (e.g., Fe, Mn, Si, Al) of solid oxide; [ ] = concentration of species in aqueous phases (mol/cm^3); { } = concentration of surface species on solid oxides (mol/g) = $[\,]/C_p$ (C_p = suspended solid particle concentration in g/cm^3); and K^s_{a1} and K^s_{a2} are **intrinsic acidity constants**.

The pH at which $\{\text{MeOH}_2^+\}$ is equal to $\{\text{MeO}^-\}$ is called the **pH of zero point of charge** (pH_{zpc}). At pH greater than pH_{zpc}, $\{\text{MeOH}_2^+\}$ is less than $\{\text{MeO}^-\}$ and the solid oxide is negatively charged. For example, the pH_{zpc} of δMnO_2 in NaCl solution is about 2.8 ± 0.3 (Morgan and Stumm, 1964). Therefore, in natural waters with pH of 7 to 8, δMnO_2 is negatively charged and the major surface species are $\{\text{MnO}^-\}$ and $\{\text{MnOH}\}$.

According to the surface complexation model, the adsorption of cation M^{+z} on the hydrous oxide surface can be visualized by the reaction

$$M^{+z} + \equiv\text{MeOH} \leftrightarrow \equiv\text{MeO} — M^{z-1} + \text{H}^+, \tag{VI-7a}$$

$$^*K_1^s = \frac{[\text{H}^+]\{\equiv\text{MeO} — M^{z-1}\}}{[M^{+z}]\{\equiv\text{MeOH}\}} = \frac{[\text{H}^+]}{\{\equiv\text{MeOH}\}} K_d, \tag{VI-7b}$$

where $^*K_1^s$ is called the **intrinsic stability constant** of the cation complex on a given solid oxide $= \exp(-\Delta G_{\text{chem}}/RT)$; ΔG_{chem} = the Gibbs free energy of the adsorption reaction; and $K_d = \{\equiv\text{MeO-}M^{z-1}\}/[M^{+z}]$ = **distribution coefficient** for cation M^{+z}.

The $-\Delta G_{\text{chem}}$ or $\log {}^*K_1^s$ involves essentially the Gibbs free energy change caused by the breakup of the O-H chemical bond (i.e., electron pair sharing) and the formation of the O-M chemical bond. Therefore $\log {}^*K_1^s$ or $-\Delta G_{\text{chem}}$ is a measure of the relative O-M bond strength for various cations. The higher the $\log {}^*K_1^s$ value, the stronger is the O-M bond, and the higher the K_d value, under constant [H$^+$] and particle composition conditions. An analogy to the above reaction is the hydrolysis of the cation, i.e.,

$$M^{+z} + \text{H}_2\text{O} \leftrightarrow \text{HO} — M^{z-1} + \text{H}^+, \tag{VI-8a}$$

$$^*K_1 = \frac{[\text{H}^+][\text{HO} — M^{z-1}]}{[M^{+z}][\text{H}_2\text{O}]}, \tag{VI-8b}$$

where $^*K_1$ is the **first hydrolysis constant**. The $\log {}^*K_1$ is again a measure of the relative O-M bond strength for various cations. Therefore, the $\log {}^*K_1^s$ and $\log {}^*K_1$ values for various cations should be linearly correlated, as proven by Dugger et al. (1964) and Schindler et al. (1976). Because the electron binding energies, I_z, are also linearly correlated to $\log {}^*K_1$ (Figure I-21b), one can predict that I_z should be proportional to $\log {}^*K_1^s$, and thus also to $\log K_d$, as proven by Figure VI-18 for mono- to trivalent cations.

Similarly, the adsorption of oxyanions (A) on a hydrous oxide surface can be visualized as

$$\equiv\text{MeOH} + \text{H}_{m-1}A^- \leftrightarrow \equiv\text{Me} — A\text{H}_{m-1} + \text{OH}^-, \tag{VI-9a}$$

$$K_1^s = \frac{[\text{OH}^-]\{\equiv\text{Me} — A\text{H}_{m-1}\}}{\{\equiv\text{MeOH}\}[\text{H}_{m-1}A^-]} = \frac{[\text{OH}^-]}{\{\equiv\text{MeOH}\}} K_d, \tag{VI-9b}$$

where K_1^s = the intrinsic stability constant of the anion complex on a given solid oxide; $K_d = \{\equiv\text{Me-}A\text{H}_{m-1}\}/[\text{H}_{m-1}A^-]$ = distribution coefficient for anion A. Log K_1^s is essentially the Gibbs free energy change caused by the breakup of the Me-OH bond and formation of the Me-A bond, and thus is a measure of the relative Me-A bond strength among various oxyanions on solids.

An analogy to the above reaction is

$$H_2O + H_{m-1}A^- \leftrightarrow H_mA + OH^-,$$
$$K' = [H_mA][OH^-]/[H_{m-1}A^-] = K_w/K_1, \qquad \text{(VI-10)}$$

where $K_1 = [H^+][H_{m-1}A^-]/[H_mA]$ = the **first dissociation constant of the oxyacid** H_mA, and $K_w = [H^+][OH^-]$ = the dissociation constant of water.

This reaction represents the formation of the H-A bond, and $\log K'$ or $-\log K_1$ is proportional to the H-A bond strength among various oxyanions. Therefore one would expect an inverse relationship between $\log K_1$ and $\log K_1^s$ (or $\log K_d$) for various oxyanions, as proven in Figure VI-19. A similar inverse relationship also applies to the second and third intrinsic stability constants of the oxyanion complex on a solid (K_2^s and K_3^s) against the second and third dissociation constants of oxyacids (K_2 and K_3) as also shown in Figure VI-19. As already shown in Figure I-23, $\log K_1$ and $\log K_2$ are linearly correlated to the I_z of the central metal of the oxyanion. Therefore one can predict a general inverse relationship between $\log K_d$ and I_z for oxyanions, as already demonstrated in Figure VI-18.

The exceptional negative correlation between $\log K_d$ and I_z for alkali and alkaline earth elements with larger ionic radii (i.e., Cs, Rb, K, Ba, and Sr) needs further explanation. According to the **adsorption model of James and Healy** (1972), the change in the Gibbs free energy during the adsorption of hydrated cations (such as alkali and alkaline earth cations) onto a solid surface (ΔG_{ads}) involves the following terms: **chemical bond energy** (ΔG_{chem}); **coulombic electrostatic energy** (ΔG_{coul}) if the solid particles are charged; and the so-called **secondary solvation energy** (ΔG_{solv}) if hydration waters around the adsorbing ion are partially removed at the solid-liquid interface, or

$$\Delta G_{ads} = \Delta G_{chem} + \Delta G_{coul} + \Delta G_{solv}. \qquad \text{(VI-11)}$$

A negative ΔG_{ads} value enhances the adsorption reaction.

As discussed already, ΔG_{chem} values are proportional to $-\log {}^*K_1^s$ and $-I_z$ values for cations; thus they are always negative as long as a chemical bond is formed. ΔG_{coul} is shown to be equal to $ze\Psi$, where z is the valence of the adsorbing ion; e is the charge of an electron; and Ψ is the potential at the adsorbing ion's equilibrium position on the solid-liquid interface. Ψ is, in turn, a function of $(pH_{zpc} - pH)RT/e$ and the ionic strength of the solution. Ψ is negative for clay minerals and δMnO_2 in a natural water pH of, e.g., 7, because pH_{zpc} values for these materials range from about 2 to 4.6 (Parks, 1967; Morgan and Stumm, 1964). The negative Ψ enhances the adsorption of positively charged ionic species ($\Delta G_{coul} < 0$) and hinders the adsorption

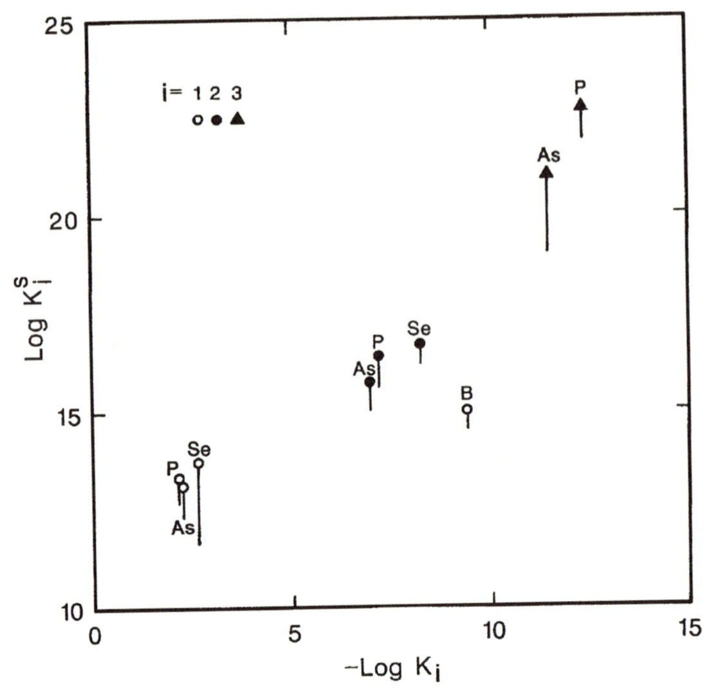

FIGURE VI-19. General inverse relationship between the first to third intrinsic stability constants (log K_i^s) of various oxyanions on goethite (Goldberg, 1985, 1986; Goldberg and Glaubig, 1985) and the first to third dissociation constant (log K_i) of oxyacids (H_3AsO_4, H_3PO_4, H_2SeO_4, and H_3BO_3).

of negatively charged ionic species ($\Delta G_{coul} > 0$). The ΔG_{solv} is related to the energy needed to remove some of the hydration water molecules around the hydrated cations during the adsorption process. The ΔG_{solv} term is always positive and hinders the adsorption of hydrated cations. However, the larger the radius of the cation or the lower the ionic potential, the less strongly the hydration waters are bonded to the cation, and the less positive is the ΔG_{solv} term, which thus hinders the adsorption less.

For the adsorption of alkaline earth cations onto clays, the ΔG_{coul} term is negative and almost constant for all alkaline earth cations. Because the ionic radii increase systematically from Be^{+2} (0.45 Å) to Ba^{+2} (1.35 Å), the positive ΔG_{solv} term should decrease from Be^{+2} to Ba^{+2}, whereas the I_z decreases from Be^{+2} (18.2 eV) to Ba^{+2} (10.0 eV); thus ΔG_{chem} increases from Be^{+2} to Ba^{+2}. Therefore, by combining the decreasing ΔG_{solv} and the increasing ΔG_{chem} trends, it is possible to obtain a least negative ΔG_{ads} (or minimum log K_d) around Ca^{+2}, as in Figure VI-18. A similar argument can be applied to the alkali cations. In short, the introduction of the ΔG_{solv} term is an important feature of the James and Healy adsorption model for explain-

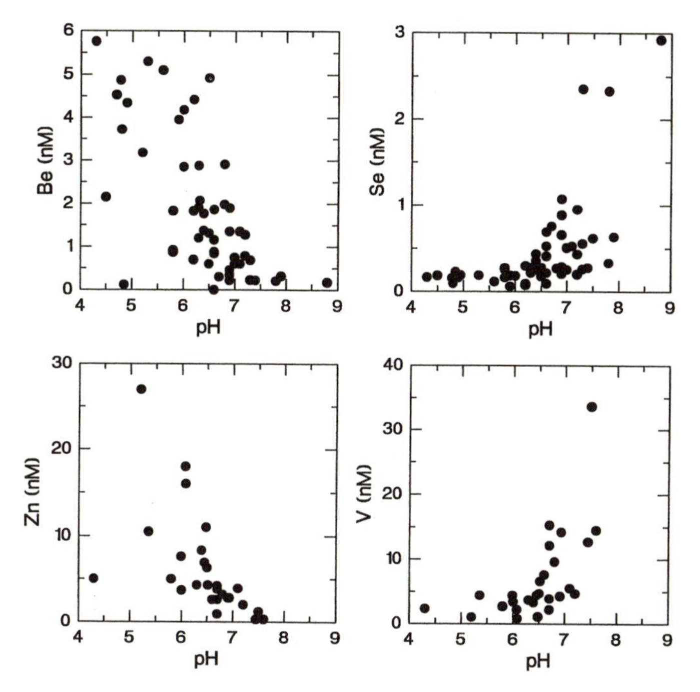

FIGURE VI-20. Dissolved concentrations of (a) Be, (b) Zn, (c) Se, and (d) total V versus pH in Amazon and Orinoco rivers.

ing the adsorption behavior of alkali and alkaline earth cations. The low K_d for borate at the given I_z (Figure VI-18) may imply the formation of yet unknown, more soluble borate complexes in natural water (Li, 1981). Certainly further studies are needed.

The observed decreasing concentrations of Be (Measure and Edmond, 1983; Brown et al., 1992) and Zn (Shiller and Boyle, 1985) with increasing pH in the least polluted Amazon and Orinoco rivers (Figures VI-20a and VI-20b) are consistent with the surface absorption model of cations depicted in equation VI-7. Normalization of Be data by Si concentration, and of Zn data by total dissolved salts, does not change the observed pH dependency (Measure and Edmond, 1983; Shiller and Boyle, 1985). Therefore the surface sorption processes do control the partition of cations between suspended particles and river water. Furthermore, the desorption of many trace mono- and divalent cations (e.g., Mn, Co, Ni, Cu, Zn, Ce, Ba, Ra, and Cs) during estuarine mixing as summarized by Li et al. (1984), also suggests that the adsorbed cations on riverine particles can react rather quickly to changes in solution chemistry.

The observed increasing concentrations of V (Shiller and Boyle, 1987) and Se (Yee et al., 1987) oxyanions with increasing pH in the Amazon and Orinoco rivers (Figures VI-20c and VI-20d) are also consistent with the adsorption model of oxyanions shown in equation VI-9. However, the normalization of Se data by sulfate concentration and of V data by total dissolved salts (Shiller and Boyle, 1987) causes the pH dependency to disappear. The implication is that the concentrations of Se and V in the Amazon and Orinoco rivers are mainly controlled by the weathering rates of evaporites and/or sulfides and carbonates in the source rock regions. Nevertheless, the amounts of oxyanion adsorbed on river-suspended particles are still regulated by adsorption isotherms.

VI-9. Concluding Remarks

Presently, the most important agent for chemical weathering of crustal rocks is carbonic acid, which is mainly produced by microbial oxidation of organic matter in soils. The end products of the weathering of the Earth's upper crust are mainly shale, limestone, sandstone, evaporite, and seasalts in the ocean. The mass balance of the elements between the end products and the upper crust shows that Ca and the volatile elements C, N, S, Se, Te, Cl, Br, I, Hg, Cd, B, As, Bi, In, and Sb are overabundant in the end products. Except for Ca, these excess volatile elements were introduced to the Earth's surface by magmatic hydrothermal solutions and vapors.

The observed partition pattern of elements between river-suspended particles and river water can be explained adequately, to the first approximation, by the adsorption model of James and Healy (1972). The model takes into account chemical bond formation, electrostatic attraction and repulsion, and partial removal of hydration water around cations during the adsorption processes.

Chapter VII

DISTRIBUTION OF ELEMENTS IN THE OCEAN

INTRODUCTION

THIS CHAPTER summarizes the concentration profiles of various elements in the ocean and discusses probable chemical speciations of the elements in oxygenated normal seawater at the standard state. It also illustrates the relationship between the compositions of seawater and marine solid phases, such as algae, plankton, zooplankton fecal pellets, sediment trap material, pelagic clays, calcareous ooze, siliceous ooze, manganese nodules, seamount manganese crusts, and marine phosphorites. The surface complexation model of adsorption is elaborated further in this chapter.

The last section explains how the compositions of hydrothermal solutions at mid-ocean ridges have evolved from the seawater-basalt interaction at high temperature. The importance of chemical fluxes from the hydrothermal solution of mid-ocean ridges, as compared to that from continental runoff, is also discussed.

VII-1. CONCENTRATIONS OF ELEMENTS IN THE OCEANS

The average concentrations of elements in surface and deep waters of the Atlantic and Pacific oceans, and in the oceans as a whole (the average of the deep Atlantic and Pacific ocean data) are summarized in Table VII-1 along with their data sources. The surface waters here represent mainly the well-mixed surface layer of the ocean with an average thickness of about 100 m. The rest of the oceans are designated as the deep waters.

The **mean residence time** of a given element in the ocean (τ) is defined by Barth (1952) as the total mass of the given element in the ocean divided by its dissolved input from rivers, i.e.,

$$\tau\,(\text{years}) = \frac{C_{sw} \cdot V_o}{C_R \cdot F_R} = 38000\frac{C_{sw}}{C_R}, \qquad \text{(VII-1)}$$

where C_{sw} and C_R are, respectively, the average concentrations of the given dissolved element in seawater and river water; V_o is the total volume of the ocean $= 1.4 \times 10^{21}$ liters; and F_R is the total river water flux to the

TABLE VII-1

Average concentrations of elements in the surface and deep waters of the Atlantic and Pacific oceans and in the whole ocean

		Atlantic		*Pacific*		*Whole ocean*
	Type	*Surface (nM)*	*Deep (nM)*	*Surface (nM)*	*Deep (nM)*	*Ave. (10^{-9} g/l)*
Ag-47	n	0.0007	0.007	0.001	0.023	2.5
Al-13	s	37	20	5	0.5	300
As-33 V	n	20	21	20	24	1700
III	s			0.3	0.07	5.2
Au-79	pc	0.053 E-3		0.055 E-3		0.03
B-5	c	0.42 E6				4.5 E6
Ba-56	n	35	70	35	150	15000
Be-4	n	0.010	0.020	0.004	0.025	0.21
Bi-83	s	0.25 E-3		0.2 E-3	0.02 E-3	0.004
Br-35	c	0.84 E6				67 E6
C-6	n	2 E6	2.2 E6	2 E6	2.4 E6	28 E6
Ca-20	c	10 E6	11 E6	10 E6	11.3E6	450 E6
Cd-48	n	0.010	0.35	0.010	1.0	76
Ce-58	s	0.066	0.019	0.011	0.004	1.6
Cl-17	c	530 E6				18.8 E9
Co-27	s			0.12	0.02	1.2
Cr-24 VI	n	3.5	4.5	3	5	250
III	s			0.2	0.05	2.6
Cs-55	c	2.3				310

Cu-29	n	1.3	2	1.3	4.5	210
Dy-66	n	0.005	0.0061			1.5
Er-68	n	0.0036	0.0053			1.3
Eu-63	n	0.0006	0.0010	0.0007	0.0018	0.21
F-9	c	0.068 E6				1.3 E6
Fe-26	m	2	7	0.2	2	250
Ga-31	m	0.0450	0.0300	0.0025	0.0150	1.7
Gd-64	n	0.0034	0.0061	0.0040	0.0100	1.3
Ge-32	n	0.001	0.020	0.005	0.100	4.3
Hf-72	s		0.020	0.022	0.019	3.4
Hg-80	m	0.0025	0.0025	0.0017	0.0017	0.42
Ho-67	n	0.0015	0.0018	0.0010	0.0036	0.45
I-53 V	n	410	450	350	460	58000
-I		0.035	~0	0.090	0.020	0.34
In-49	s	0.0005	0.0002	0.0015	0.0010	0.1
Ir-77	(n)			7.8 E-6		0.0015
K-19	c	10 E6				390 E6
La-57	n	0.013	0.028	0.019	0.051	5.6
Li-3	c	0.026 E6				0.18 E6
Lu-71	n	0.0008	0.0012	0.0004	0.0024	0.32
Mg-12	c	53 E6				1.3 E9
Mn-25	s	1.9	1.8	1.9	0.8	72
Mo-42	c	107				10300
N-7	n	5	20000	5	40000	0.42 E6

TABLE VII-1 *Continued*

		Atlantic		*Pacific*		*Whole ocean*
	Type	*Surface (nM)*	*Deep (nM)*	*Surface (nM)*	*Deep (nM)*	*Ave. (10^{-9} g/l)*
Na-11	c	470 E6				10.8 E9
Nb-41	(n)					10
Nd-60	n	0.013	0.023	0.013	0.034	4.2
Ni-28	n	2	7	2	10	530
Os-76	(n)					1.7 E-3
P-15	n	50	1400	50	2800	65000
Pb-82	s	0.150	0.020	0.050	0.005	2.7
Pd-46	n			0.18 E-3	0.66 E-3	0.07
Pr-59	n	0.0030	0.0050	0.0032	0.0073	0.87
Pt-78	n	0.0003	0.0003	0.0005	0.0014	0.1
Ra-88	n	16 E-6	41 E-6	20 E-6	71 E-6	130 E-6
Rb-37	c	1400				0.12 E6
Re-75	c			0.043		8
Rh-45	(n)					
Ru-44	(n)			< 0.05 E-3		≤ 0.005
S-16	c	28 E6				898 E6
Sb-51	c	1	1			150
Sc-21	m	0.014	0.020	0.008	0.018	0.86
Se-34 VI	n	0.5	1.0	0.5	1.3	90
IV	n	0.03	0.5	0.07	0.9	55
Si-14	n	1000	30000	1000	150000	2.5 E6

Sm-62	n	0.0027	0.0044	0.0027	0.0068	0.84
Sn-50	s	0.020	0.005			0.6
Sr-38	c	0.089 E6	0.090 E6	0.089 E6	0.090 E6	7.8 E6
Ta-73	(n)					≤ 2.5
Tb-65	n	0.0007	0.0010	0.0005	0.0016	0.21
Te-52 VI	s	0.0009	0.0004	0.0010	0.0004	0.05
IV	s	0.0004	0.0002	0.0005	0.0001	0.02
Th-90	s					0.05
Ti-22	m	0.060	0.300	0.005	0.200	10
Tl-81	pc	0.069				14
Tm-69	n	0.0008	0.0010	0.0004	0.0020	0.25
U-92	c	13.5				3200
V-23	n	35	35	40	49	2150
W-74	(n)					100
Y-39	(n)					13
Yb-70	n	0.0030	0.0045	0.0022	0.0130	1.5
Zn-30	n	0.8	1.6	0.8	8.2	320
Zr-40	m		0.200	0.080	0.185	17

Source: Mainly from Whitfield and Turner (1987) with additional data from Li (1991, and references therein); Colodner (1991) for Atlantic Pt; Hodge et al. (1985) for Pacific Pt; Flegal et al. (1995) for Ag; and Cutter and Cutter (1995) for Sb.

Note: Ex indicates 10^x.

ocean = 37×10^{15} liters/yr. Larger τ or greater C_{sw}/C_R values imply that the element stays in the ocean longer before it is removed by particles and settles in ocean bottom sediments. In addition, $1/\tau$ has units of yr^{-1} and can be considered as the removal rate constant of an element from the ocean to sediments. It is a measure of the **relative reactivity** of the element in the ocean (Goldberg et al., 1971; Li, 1982). The smaller the $1/\tau$ or C_R/C_{sw} value, the less particle-reactive is the element in the ocean, and vice versa.

Log C_R/C_{sw} (and thus $-\log \tau$) values calculated from Tables VII-1 and VI-10 are plotted against the electron binding energy, I_z, in Figure VII-1. Interestingly, the general trend in Figure VII-1 is similar to that shown in Figure VI-18 (log X_R/C_R versus I_z). The implication is that the relative reactivity of the dissolved elements in both rivers and oceans is related systematically to the electron binding energy I_z.

Elemental concentration profiles in the oceans can be classified into four major types (Li, 1991) according to Table VII-1:

1. **Conservative type** (c) where the ratio of element to chloride concentrations is constant (within 10%) throughout the ocean.

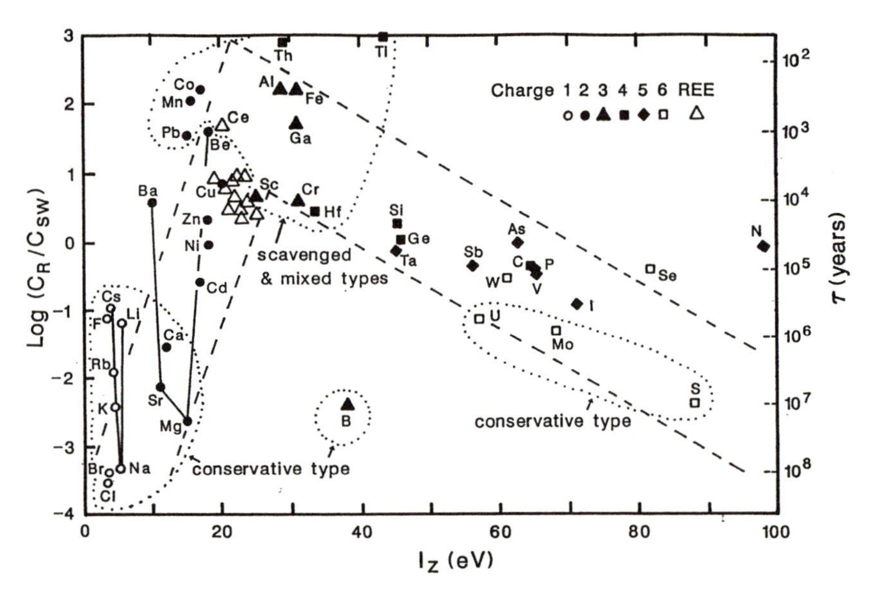

FIGURE VII-1. Ratio of elemental concentrations in rivers to those in seawater (log C_R/C_{sw}) plotted against the electron binding energy, I_z. The elements within the dotted enclosures with low mean ocean residence time (τ) are conservative type, and with high τ are scavenged and mixed types. The parallel dashed lines indicate the general trend between log C_R/C_{sw} and I_z. The solid lines connect alkali and alkaline earth group elements.

2. **Scavenged type** (s) where the concentration decreases from the surface to deep waters and from the deep Atlantic to the deep Pacific ocean (Figure VII-2a). Usually, the concentration in surface water is higher in the Atlantic than in the Pacific due to larger atmospheric inputs of aerosols.
3. **Nutrient type** (n) where the concentration increases from the surface to deep waters and from the deep Atlantic to the deep Pacific Ocean (Figure VII-2b).
4. **Mixed type** (m) where the concentration increases with depth as for n type but decreases from the deep Atlantic to the deep Pacific as for s type, (Figure VII-2c). The concentration in surface water is often higher in the Atlantic than in the Pacific.

One additional minor type is called the **pseudo-conservative type** (pc). For the pc type, the elemental concentration differences between the surface and deep waters, and between the deep Atlantic and deep Pacific, are relatively small as for c type, but the element has a shorter mean oceanic residence time. Possible pc-type elements are Tl (Figure VII-2d) and Au (Falkner and Edmond, 1990). The concentration of In decreases from surface to deep water like the s type, but increases from the deep Atlantic to the deep Pacific, like the n type, (Orians, pers. comm.). We may call In an abnormal scavenged-type element.

Table VII-1 summarizes the types of concentration profiles for various elements. As shown in Figure VII-1, the conservative-type elements have the longest τ, while the scavenged- and mixed-type elements have the shortest τ. The nutrient-type elements are in between.

How these different types of concentration profiles are produced can be best explained by a simple four-box model of the ocean as shown in Figure VII-3. Here I is the flux of a given element to the Atlantic surface ocean through rivers and wind-transported continental aerosols, I_b is the biogenic particle flux of the element from the surface to the deep Atlantic, and I_s is the net deposition flux of the element to the bottom sediments in the Atlantic. The symbol w is the advective water flux from the surface to the deep Atlantic, which continues on to the deep Pacific and eventually upwells back to the surface. w_m is the water-mixing flux between the surface and deep Atlantic waters. The concentrations of the element in the surface and the deep Atlantic are, respectively, C_o and C. Symbols with asterisks are for the Pacific ocean.

In a steady-state ocean, the input and output fluxes of a given element in each box should be balanced. For example, in the deep Atlantic Ocean box, the following equation holds:

$$(w + w_m)C_o + I_b = (w + w_m)C + I_s,$$

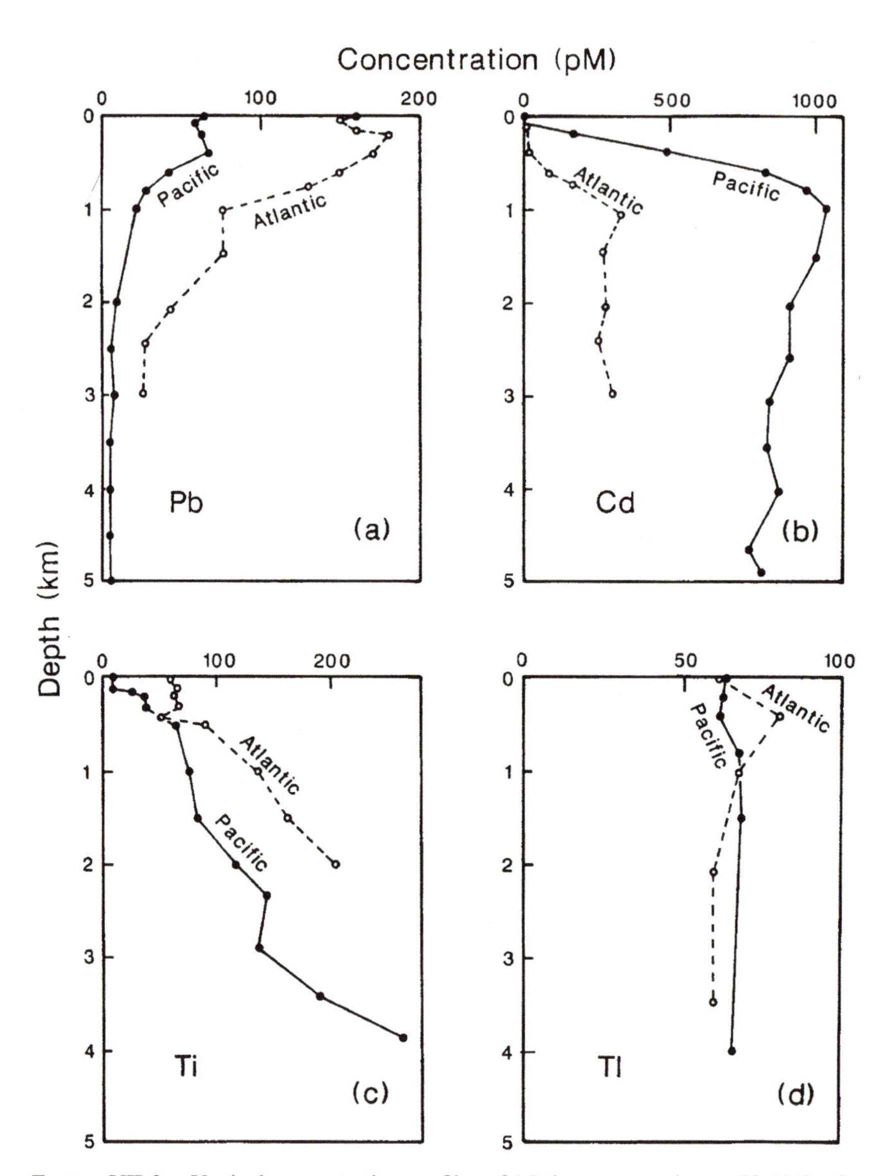

FIGURE VII-2. Vertical concentration profiles of (a) the scavenged type, Pb (Atlantic: Schaule and Patterson, 1983; Pacific: Schaule and Patterson, 1981), (b) the nutrient type, Cd (Atlantic: Bruland and Franks, 1983; Pacific: Bruland, 1980), (c) the mixed type, Ti (Orians et al., 1990), and (d) the pseudo-conservative type, Tl (Atlantic: Flegal and Patterson, 1985; Pacific: Murozumi and Nakamura, 1980), in the oceans.

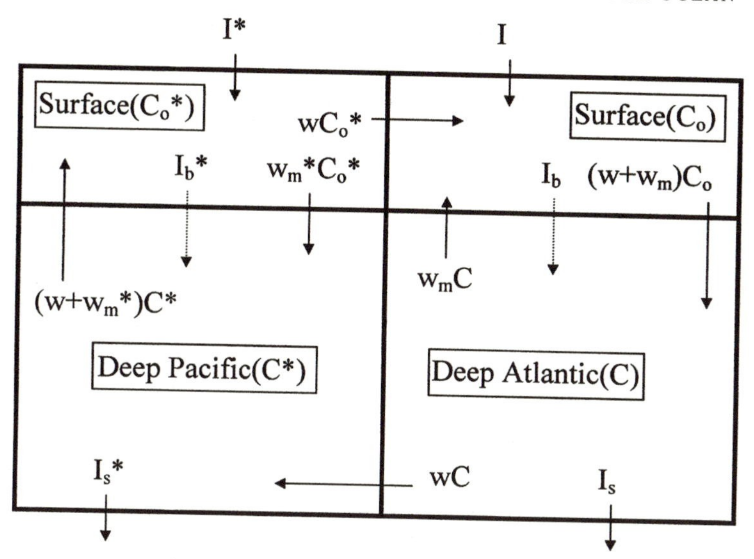

FIGURE VII-3. Four-box model of the ocean depicting various water and elemental fluxes (see text).

or, after rearrangement,

$$(w + w_m)(C - C_o) = I_b - I_s. \tag{VII-2a}$$

Similarly, for the deep Pacific box, one obtains

$$wC + w_m^* C_o^* + I_b^* = \left(w + w_m^*\right)C^* + I_s^*,$$

or, after rearrangement,

$$w\left(C^* - C\right) + w_m^*\left(C^* - C_o^*\right) = I_b^* - I_s^*. \tag{VII-2b}$$

Also, for the ocean as a whole, $I + I^* = I_s + I_s^*$, and $I \approx I_s$, $I^* \approx I_s^*$ for elements with short mean residence time.

By definition, $C - C_o$, $C^* - C_o^*$, and $C^* - C$ are all near zero for the conservative-type elements; therefore from equations VII-2a and VII-2b, $I_b - I_s$ and $I_b^* - I_s^*$ should be near zero, or $I_b = I_s$ and $I_b^* = I_s^*$. However, these fluxes are relatively small, as compared to the total inventory of each element in the ocean, because of the long τ's for the c-type elements. The conditions $I_b = I_s$ and $I_b^* = I_s^*$ also apply to the pseudo-conservative-type elements, but these values are high relative to the total inventory because of their shorter τ's.

For the scavenged-type elements, $C - C_o < 0$, $C^* - C_o^* < 0$, and $C^* - C < 0$ by definition. Thus, from equations VII-2a and VII-2b, one obtains $I_b - I_s < 0$

and $I_b^* - I_s^* < 0$. That means the output flux to the sediments $(I_s$ and $I_s^*)$ or the input flux from the continent (I and I^*) is larger than the internal biological flux $(I_b$ and $I_b^*)$. For the nutrient-type elements, $C - C_o > 0$, $C^* - C_o^* > 0$, and $C^* - C > 0$ by definition; thus, $I_b - I_s > 0$ and $I_b^* - I_s^* > 0$; namely, the internal biological fluxes are larger than the output fluxes to the sediments, indicating significant remineralization of biogenic particles through microbial oxidation processes in the deep oceans. For the mixed-type elements, $C - C_o > 0$, $C^* - C_o^* > 0$, and $C^* - C < 0$. If $C - C_o \approx C^* - C_o^*$ and $w_m \approx w_m^*$, then from equations VII-2a and VII-2b, one obtains $I_b - I_s > 0$ and $I_b - I_s > I_b^* - I_s^*$. This means that the difference between the internal biological flux and the output flux to sediments (or the continental input flux) is greater in the Atlantic than in the Pacific ocean.

In short, the most important factors for controlling elemental concentration profiles are the relative magnitudes of the biological particulate fluxes $(I_b$ and $I_b^*)$, the depositional fluxes to the sediments $(I_s$ and $I_s^*)$, and the input fluxes from the continents to the surface ocean (I and I^*) in the Atlantic and Pacific Oceans.

VII-2. Chemical Speciation of Elements in the Ocean

The dominant chemical speciation of each element in normal oxygenated seawater at the standard state (1 atmosphere pressure and 25°C; Li, 1991, and references therein) is summarized in Table VII-2. The most common chemical forms of the mono- to trivalent and some tetravalent cations are hydrated cations, chloride, carbonate, and hydroxyl complexes. Cations with charge equal to or greater than four exist mostly as hydroxo-, hydroxo-oxo, and oxo-anions, and oxo-carbonate complexes.

The relative abundances of hydrated cation, chloride, carbonate, and hydroxyl complexes for a given cation in seawater are mainly dependent on the complexation constants of the cation with these inorganic ligands and on the activities of the ligands in seawater. For example, the first complexation constants of the hydrated cation M^{+z} with these ligands are

$$\beta_1^{\mathrm{OH}} = \frac{[M\mathrm{OH}^{z-1}]}{[M^{+z}][\mathrm{OH}^-]} \tag{VII-3a}$$

$$\beta_1^{\mathrm{Cl}} = \frac{[M\mathrm{Cl}^{z-1}]}{[M^{+z}][\mathrm{Cl}^-]} \tag{VII-3b}$$

$$\beta_1^{\mathrm{CO}} = \frac{[M\mathrm{CO}_3^{z-2}]}{[M^{+z}][\mathrm{CO}_3^{-2}]}, \tag{VII-3c}$$

where [ ] represents the activity of a given species in this section.

TABLE VII-2
Types of concentration profiles and major chemical speciations of the elements in the ocean

Z	*Type*[a]	*Speciation*[b]	*Z*	*Type*[a]	*Speciation*[b]
Ia			**Ib**		
Li-3	c	Li^+	Cu-29	n	$CuCO_3{}^0$, Cu-org.
Na-11	c	Na^+	Ag-47	n	$AgCl_2{}^-$
K-19	c	K^+	Au-79	pc	$AuOH(H_2O)^0$
Rb-37	c	Rb^+	**IIb**		
Cs-55	c	Cs^+	Zn-30	n	Zn^{+2}, $ZnCl^+$
Fr-87	(c)	Fr^+	Cd-48	n	$CdCl_2{}^0$
IIa			Hg-80	m	$HgCl_4{}^{-2}$
Be-4	n	$BeOH^+$, $Be(OH)_2{}^0$	**IIIa**		
Mg-12	c	Mg^{+2}	B-5	c	$B(OH)_3{}^0$, polymer?
Ca-20	c	Ca^{+2}	Al-13	s	$Al(OH)_3{}^0$, $Al(OH)_4{}^-$
Sr-38	c	Sr^{+2}	Ga-31	m	$Ga(OH)_4{}^-$
Ba-56	n	Ba^{+2}	In-49	s	$In(OH)_3{}^0$
Ra-88	n	Ra^{+2}	Tl-81	pc	Tl^+, $TlCl^0$
IIIb			**IVa**		
Sc-21	m	$Sc(OH)_3^0$	C-6	n	$HCO_3{}^-$
Y-39	(n)	$YCO_3{}^+$, $Y(CO_3)_2{}^-$	Si-14	n	$H_4SiO_4{}^0$
La-57	n	$LaCO_3{}^+$, $La(CO_3)_2{}^-$	Ge-32	n	$H_4GeO_4{}^0$, $H_3GeO_4{}^-$
Ac-89		$AcCO_3{}^+$, Ac^{+3}	Sn-50	s	$SnO(OH)_3{}^-$
IVb			Pb-82	s	$PbCO_3{}^0$, $PbCl^+$
Ti-22	m	$Ti(OH)_4^0$	**Va**		
Zr-40	m	$Zr(OH)_5{}^-$	N-7	n	$NO_3{}^-$
Hf-72	s	$Hf(OH)_5{}^-$	P-15	n	$NaHPO_4{}^-$, $HPO_4{}^{-2}$
Vb			As-33	n	$HAsO_4{}^{-2}$
V-23	n	$NaHVO_4{}^-$, $HVO_4{}^{-2}$		s	$As(OH)_3{}^0$, $As(OH)_4{}^-$
Nb-41	(n)	$Nb(OH)_6{}^-$, $Nb(OH)_5{}^0$	Sb-51	(n)	$Sb(OH)_6{}^-$
Ta-73	(n)	$Ta(OH)_5{}^0$, $Ta(OH)_6{}^-$	Bi-83	s	$Bi(OH)_3{}^0$
VIb			**VIa**		
Cr-24	n	$CrO_4{}^{-2}$, $NaCrO_4{}^-$	S-16	c	$SO_4{}^{-2}$
	s	$Cr(OH)_3{}^0$, $Cr(OH)_2{}^+$	Se-34	n	$SeO_4{}^{-2}$
Mo-42	c	$MoO_4{}^{-2}$		n	$SeO_3{}^{-2}$
W-74	(n)	$WO_4{}^{-2}$	Te-52	s	$TeO(OH)_5{}^-$, $Te(OH)_6{}^0$
VIIb				s	$TeO(OH)_3{}^-$
Mn-25	s	Mn^{+2}, $MnCl^+$	Po-84		$PoO(OH)_3{}^-$
Tc-43	c	$TcO_4{}^-$	**VIIa**		
Re-75	c	$ReO_4{}^-$	F-9	c	F^-, MgF^+
			Cl-17	c	Cl^-
			Br-35	c	Br^-
			I-53	n	IO_3^-

TABLE VII-2 *Continued*

Z	*Type*[a]	*Speciation*[b]	*Z*	*Type*[a]	*Speciation*[b]
VIIIb			**VIIIb**		
Fe-26	m	$Fe(OH)_3{}^0$	Os-76	(n)	OsO_4^0, reduced forms
Co-27	s	Co^{+2}, $CoCl^+$	Ir-77	(n)	$Ir(OH)_3^0$, $IrCl_6^{-3}$?
Ni-28	n	Ni^{+2}, $NiCl^+$	Pt-78	pc	$PtCl_4{}^{-2}$, $Pt(OH)_2^0$?
Ru-44	(n)	$RuO_4{}^0$, reduced forms			
Rh-45	(n)	$Rh(OH)_3{}^0$, $RhCl_6{}^{-3}$?			
Pd-46	n	$PdCl_4{}^{-2}$, $Pd(OH)_2{}^0$?			
Lanthanides			**Actinides**		
La-57	n	$LaCO_3{}^+$, $La(CO_3)_2{}^-$	Ac-89		$AcCO_3{}^+$, Ac^{+3}
Ce-58	s	$CeCO_3{}^+$, $Ce(CO_3)_2{}^-$	Th-90	s	$Th(OH)_4{}^0$
Pr-59	n	$PrCO_3{}^+$, $Pr(CO_3)_2{}^-$	Pa-91	(s)	PaO_2OH^0
Nd-60	n	$NdCO_3{}^+$, $Nd(CO_3)_2{}^-$	U-92	c	$UO_2(CO3)_2{}^{-2}$, $UO_2(CO_3)_3{}^{-4}$
Pm-61	n	$PmCO_3{}^+$, $Pm(CO_3)_2{}^-$	Np-93		$NpO_2{}^+$, $NpO_2(CO_3)^-$
Sm-62	n	$SmCO_3{}^+$, $Sm(CO_3)_2{}^-$	Pu-94		$PuO_2(CO_3)(OH)^-$
Eu-63	n	$EuCO_3{}^+$, $Eu(CO_3)_2{}^-$		(s)	$Pu(OH)_3{}^+$, $Pu(OH)_4{}^0$
Gd-64	n	$GdCO_3{}^+$, $Gd(CO_3)_2{}^-$	Am-95		$AmCO_3{}^+$
Tb-65	n	$TbCO_3{}^+$, $Tb(CO_3)_2{}^-$	Cm-96		$CmCO_3{}^+$
Dy-66	n	$DyCO_3{}^+$, $Dy(CO_3)_2{}^-$	Bk-97		$BkCO_3{}^+$
Ho-67	n	$HoCO_3{}^+$, $Ho(CO_3)_2{}^-$	Cf-98		$CfCO_3{}^+$
Er-68	n	$ErCO_3{}^+$, $Er(CO_3)_2{}^-$			
Tm-69	n	$TmCO_3{}^+$, $Tm(CO_3)_2{}^-$			
Yb-70	n	$YbCO_3{}^+$, $Yb(CO_3)_2{}^-$			
Lu-71	n	$LuCO_3{}^+$, $Lu(CO_3)_2{}^-$			

[a]Types of the concentration profiles in the oceans: c = conservative, pc = pseudo-conservative, n = nutrient type, s = scavenged type, m = mixed type. The types in parentheses are only educated guesses.

[b]Li (1991, and references therein). The question marks are educated guesses for the speciations.

In seawater with pH = 8, $m_{\mathrm{Cl}^-} = 0.53\,\mathrm{M}$, $m_{\mathrm{CO}_3^{-2}} = 80\,\mu\mathrm{M}$ (representing the Antarctic deep water), assuming total activity coefficients of $\gamma_{\mathrm{Cl}^-} = 0.63$ and $\gamma_{\mathrm{CO}_3^{-2}} = 0.02$ at standard state (Whitfield, 1979), the conditions for $[M\mathrm{OH}^{z-1}]/[M^{+z}] \geq 1$, $[M\mathrm{Cl}^{z-1}]/[M^{+z}] \geq 1$, $\left[M\mathrm{CO}_3^{z-2}\right]/[M^{+z}] \geq 1$, $[M\mathrm{Cl}^{z-1}]/[M\mathrm{OH}^{z-1}] \geq 1$, and $\left[M\mathrm{CO}_3^{z-2}\right]/[M\mathrm{OH}^{z-1}] \geq 1$ are (from equations VII-3a–c)

$$\log \beta_1^{\mathrm{OH}} \geq -\log[\mathrm{OH}] = 14 - pH = 6 \qquad \text{(VII-4a)}$$

$$\log \beta_1^{\mathrm{Cl}} \geq -\log[\mathrm{Cl}^-] = 0.48 \qquad \text{(VII-4b)}$$

$$\log \beta_1^{\mathrm{CO}_3} \geq -\log\left[\mathrm{CO}_3^{-2}\right] = 5.8 \qquad \text{(VII-4c)}$$

$$\log \beta_1^{\mathrm{Cl}} - \log \beta_1^{\mathrm{OH}} \geq \log[\mathrm{OH}^-] - \log[\mathrm{Cl}^-] = -5.52, \text{ and} \qquad \text{(VII-4d)}$$

$$\log \beta_1^{\mathrm{CO}_3} - \log \beta_1^{\mathrm{OH}} \geq \log[\mathrm{OH}^-] - \log\left[\mathrm{CO}_3^{-2}\right] = -0.2. \qquad \text{(VII-4e)}$$

Equations VII-4a to VII-4e are plotted as solid lines in Figure VII-4 to define the stability field boundaries for $[M^{+z}]$ (hydrated cations), $[MCl^{z-1}]$, $[MOH^{z-1}]$, and $[MCO_3^{z-2}]$. The dashed lines represent the uncertainty of the boundary lines introduced by the uncertainty in the estimates of activity coefficients. Superimposed in Figure VII-4 are the first complexation constants for various cations and oxycations at standard state.

According to Figure VII-4, the alkalis and alkaline earths (minus Be) are mainly in the hydrated cation forms, whereas Ag, Cu, Cd, and Hg are mostly chloride complexes. For Tl, Mn, Ni, Co, and Zn, both hydrated cations and chloride complexes are important species. These speciations are all in good agreement with the compilations given in Table VII-2.

The β_1^{OH} values for Pd^{+2} and Pt^{+2} have a large uncertainty, and the β_1^{Cl} values for Rh^{+3} and Ir^{+3} are still unknown, but considering their high electric polarizability (see Section I-6), their chloride complexes are probably important. The β_1^{OH} values for OsO_4° and RuO_4° are small (Baes and Mesmer, 1976) and their β_1^{Cl} values are probably also small. Therefore, OsO_4° and RuO_4° should be the stable species in normal oxygenated seawater. However, Koide et al. (1986) suggested that a significant fraction of Os and Ru may exist in lower oxidation states in seawater due to **biologically mediated reduction**. This is also the case for As, Cr, Se, Te, I, and Pu (Table VII-2).

According to Figure VII-4, the rare earth cations plus Y^{+3}, trivalent actinides, Pb^{+2}, Cu^{+2}, PuO_2^{+2}, and UO_2^{+2} occur mostly as carbonate complexes, whereas Zr^{+4}, Hf^{+4}, Fe^{+3}, Ga^{+3}, In^{+3}, Be^{+2}, and probably Cr^{+3} and Ti^{+4} are hydroxyl complexes. These speciations are again in good agreement with Table VII-2, even though the complexes with higher coordination numbers of ligands are totally ignored in Figure VII-4. However, the importance of complexes with higher ligand coordination numbers (i.e., $[ML_i]/[ML_{i-1}] \geq 1$) in seawater with pH = 8, $m_{Cl^-} = 0.53\,M$, and $m_{CO_3^{-2}} = 80\,\mu M$ can easily be deduced from the relationships:

$$\log \beta_i^{OH} - \log \beta_{i-1}^{OH} \geq \log[OH] = 6, \quad \text{(VII-5a)}$$

$$\log \beta_i^{Cl} - \log \beta_{i-1}^{Cl} \geq -\log[Cl] = 0.48, \quad \text{(VII-5b)}$$

$$\log \beta_i^{CO_3} - \log \beta_{i-1}^{CO_3} \geq \log[CO_3^{-2}] = 5.8, \quad \text{(VII-5c)}$$

where

$$\beta_{i-1} = \frac{[ML_{i-1}]}{[M^{+z}][L]^{i-1}},$$

$$\beta_i = \frac{[ML_i]}{[M^{+z}][L]^i},$$

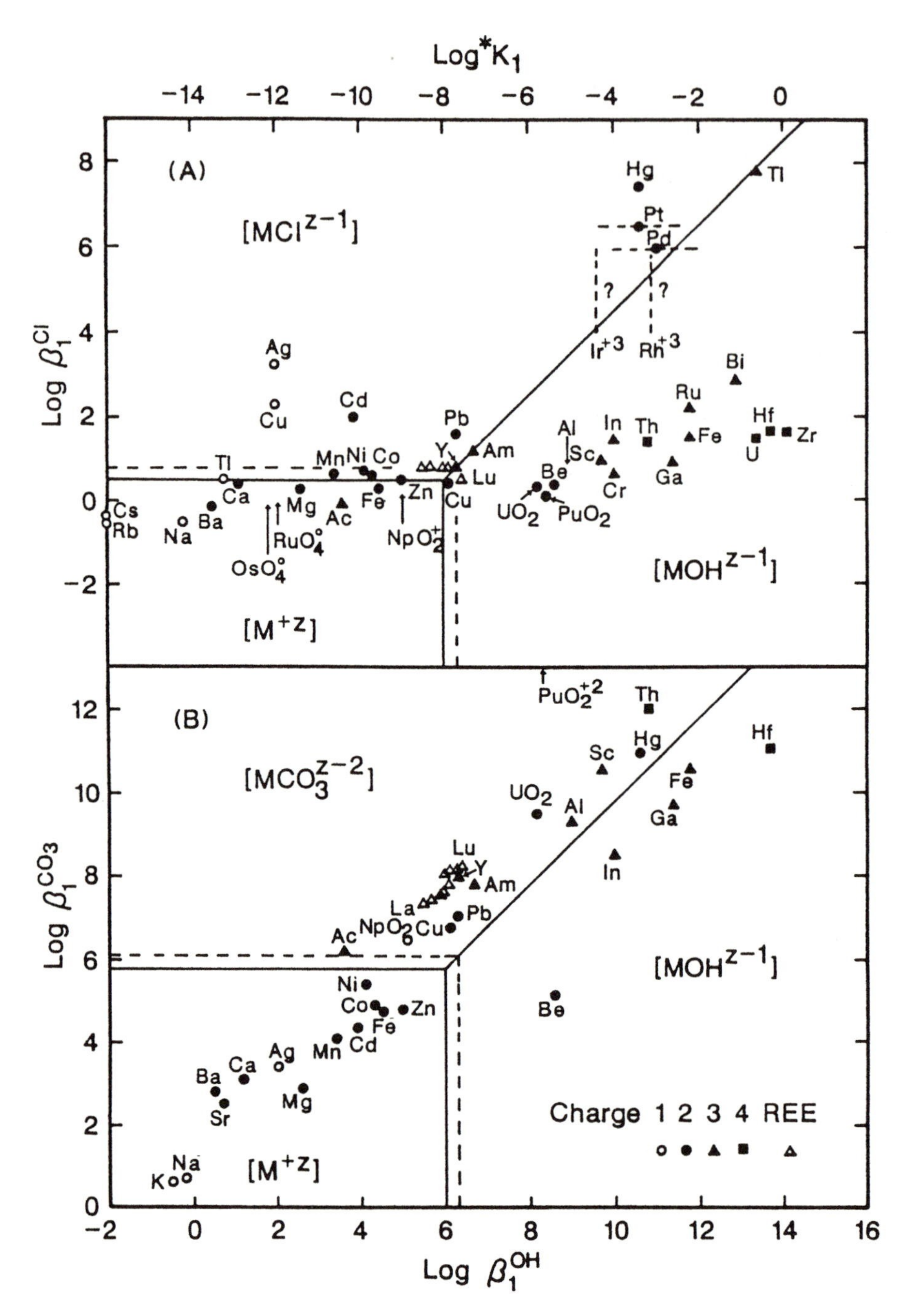

FIGURE VII-4. First hydrolysis constants of cations ($\log {}^*K_1$) or first complexation constants of cations with hydroxyl $\left(\log \beta_1^{OH} = 14 + \log {}^*K_1\right)$ plotted against (a) the first complexation constants of cations with chloride ion ($\log \beta_1^{Cl}$), and (b) the first complexation constants of cations with carbonate ion ($\log \beta_1^{CO_3}$). The stability fields for the major given species are delineated by solid and dashed lines (see text). The vertical arrows and question marks represent estimated data points (Li, 1991; with permission from Elsevier Science).

and L is a given ligand (Cl^-, CO_3^{-2}, OH^-, etc.). For example, the negative logarithms of complexation constants β_1^{Cl}, β_2^{Cl}, β_3^{Cl}, and β_4^{Cl} for Hg^{+2} are, respectively, 7.4, 14.3, 14.5, and 15.3. Therefore, one can easily show that $HgCl_4^{-2}$ is the most abundant species among the various chloride complexes.

In Figure VII-4, the $[MCO_3^{z-2}]/[MOH^{z-1}]$ ratios for Al, Sc, and Th are greater than one; thus the carbonate complexes should be dominant. However, these elements are given as hydroxyl complexes with high coordination numbers in Table VII-2. According to the known β_1^{OH} to β_4^{OH} values for Al^{+3}, Sc^{+3}, and Th^{+4}, the species $Al(OH)_3^{\circ}$, $Sc(OH)_3^{\circ}$, and $Th(OH)_4^{\circ}$ are the most abundant among the various hydroxyl species. However, the $\beta_2^{CO_3}$ to $\beta_4^{CO_3}$ values for Al and Sc and $\beta_4^{CO_3}$ values for Th^{+4} are still unknown. Therefore, one cannot yet ascertain the importance of the carbonate complexes for these elements.

As the data on the complexation constants of ions with various ligands increase and improve, the speciation results will change accordingly. One should also be fully aware that the speciation calculations are based on the assumption of thermodynamic equilibrium, which is not always true in the actual marine environment. For example, according to the thermodynamic calculations, the high-oxidation-state species of As, Cr, Se, Te, I, and Pu should be the stable forms. However, low-oxidation-state species of these elements are also found in the ocean (Table VII-1), probably through biologically mediated reduction processes. It is also still not feasible to quantify thermodynamically the extent of metal complexation with mostly yet-to-be-identified natural organic ligands, which cover a wide range of molecular weights and size (McCarthy et al., 1997; Benner et al., 1997). Using the separation scheme of Amberlite resin, Sugimura and Suzuki (1985) demonstrated the importance of metal-organic complexes for many transitional metals in the ocean.

VII-3. Marine Algae and Plankton

The average compositions of marine algae based on extensive studies by Yamamoto (1983) and Yamamoto et al. (1985), and of marine plankton by Martin and Knauer (1973) along with additional data compiled by Bowen (1966, 1979), are summarized in Table VII-3. Bowen's 1966 compilation is mainly based on data by Vinogradov (1953) and Black and Mitchell (1952). Bowen's 1979 compilation often gives only the range of concentrations. In this case, the geometric mean of the range is adopted here. The average composition of marine algae by Yamamoto is in good agreement with Bowen's brown algae (within a factor of 2 or better). The relatively high Al, Sc, Si, and Ti concentrations in Yamamoto's average may indicate a higher content of clay particles and colloids in Yamamoto's samples. Bowen's compilation is

TABLE VII-3
Average composition of marine algae, phytoplankton, and zooplankton

	Average algae (1)	*Brown algae* (2)	*Phytoplankton* (3)	*Zooplankton* (3)		*Average algae* (1)	*Brown algae* (2)	*Phytoplankton* (3)	*Zooplankton* (3)
Ag-47	0.26	0.28	0.6	0.1	N-7 %		1.8 b	5.7 d	8.8 b
Al-13	700	62	440	50	Na-11 %	0.96	3.3		
As-33	24	30	21 b		Nb-41				
Au-79		0.008 f			Nd-60				
B-5	100	120	33 b	64 b	Ni-28	2.7	3	8	3
Ba-56	30	31	20	21	Os-76				
Be-4	(0.03)	0.003 b			P-15	1100	2800	8000 d	4900 y
Bi-83					Pb-82	9.5	8.4	7	3
Br-35	180	740			Pd-46				
C-6 %	32 b	35	34 d	42 b	Pr-59				
Ca-20 %	2	1.2	0.53	0.84	Pt-78			0.0002 h	
Cd-48		0.4	1.5	3.5	Ra-88		E-7		
Ce-58	1.2				Rb-37	6.9	7.4		
Cl-17	3700	4700			Re-75		0.013 f		
Co-27	0.74	0.7	0.38 b	1 b	Rh-45				
Cr-24	1.7	1.3	4	1 b	Ru-44				
Cs-55	0.093	0.067	0.11 b	0.032 b	S-16 %	1.5	1.2	0.42 b	
Cu-29	14	11	7	13	Sb-51	0.36	0.16 b		
Dy-66					Sc-21	0.32	0.07 b		

Er-68					
Eu-63	0.018				
F-9	6	4.5			
Fe-26	540	690	1500		150
Ga-31	0.14				
Gd-64					
Ge-32	0.01 e				
Hf-72	0.18				
Hg-80	0.11	0.03	0.16		0.11
Ho-67					
I-53	430	1500	270 b		
In-49					
Ir-77			20E-6 w		
K-19 %	1.3	5.2?	1.3		1.2
La-57	0.76				
Li-3	6	5.4			
Lu-71	0.02				
Mg12 %	1.1	0.94 b	1.6		0.83
Mn-25	100	53	13		4
Mo-42	0.34	0.45	1 b		3 b
Se-34	0.14	0.06 b			
Si-14 %	0.48	0.15	4.8		0.37 y
Sm-62	0.14				
Sn-50	1.1	1.1	3.5 b		2.2 b
Sr-38	1200	1400	120		140
Ta-73	0.038				
Tb-65	0.027				
Te-52					
Th-90	0.18	0.1 b	0.42 b		0.32 b
Ti-22	35	12	27		
Tl-81			0.2 fl		0.2 fl
Tm-69					
U-92	1.1	0.8 b	0.7 b		0.5 b
V-23	3.7	2	4 b		13 b
W-74	0.035 f				
Y-39					
Yb-70	0.088				
Zn-30	150	150	120		91
Zr-40	0.7 e				

Sources: (1) Yamamoto (1983) and Yamamoto et al. (1985); (2) Bower (1966); (3) Martin and Knauer (1973), group II phytoplankton and Monterey Bay zooplankton; b: Bowen (1979); d: Duarte (1992); e: Eisler (1981); f: Fukai and Meinke (1962); fl: Flegal et al. (1986); h: Hodge et al. (1986); w: Wells et al. (1988); y: Yamamoto (1983).

Note: Values are in ppm (dry weight) unless indicated otherwise.

high in Na, K, Br, I and P. Redfield et al. (1963) also proposed that idealized plankton have P/N/C molar ratios of 1/16/106 (the so-called **Redfield ratios**). As discussed by Bowen (1979) and Li (1984, 1991), the average composition of marine algae is similar to that of marine plankton (Figures VII-5a and 5b, based on Table VII-3). Therefore, marine algae can represent the overall composition of the primary producers in the ocean.

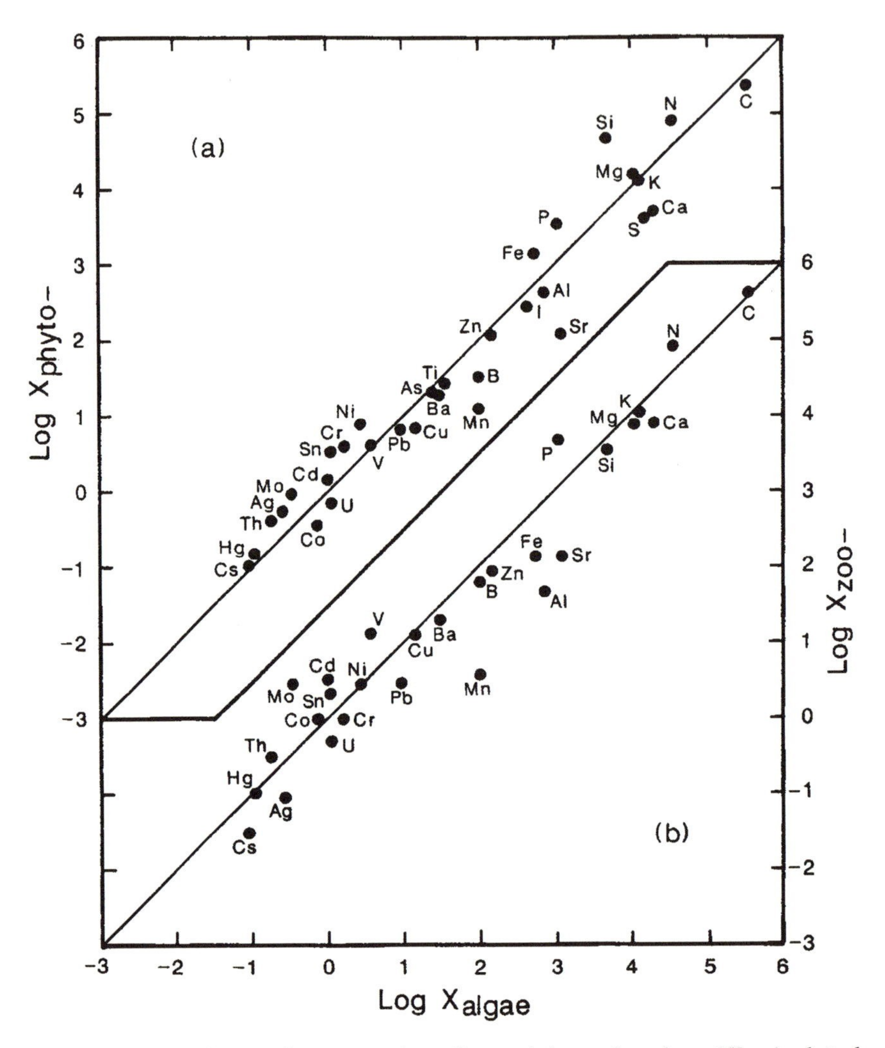

FIGURE VII-5. Elemental concentrations (in ppm) in marine algae (X_{algae}) plotted against those in marine (a) phytoplankton (X_{phyto}) and (b) zooplankton (X_{zoo}). Data are from Table VII-3.

A plot of average algae versus average shale compositions (Figure VII-6) shows that the relative abundances of Si, Al, Fe, Ti, F, V, Cr, REE, Sc, Th, Ga, Hf, Cs, and W in algae are essentially the same as in the average shale (solid circles and triangles in Figure VII-6). In other words, these elements all have E^i_{Al} values of about one to five in algae relative to shale. These elements are mainly A-type ions (with electron configuration of noble gases as well as [Xe]4f^{14}) and lanthanides; and they were named **"biophobe" elements** by Li (1984) regardless of their actual biological function. These elements are also the very same elements that represent factor 1 in the factor loading plots of Yamamoto's algae data (Figure VII-7a). These elements are most likely incorporated into the algal cell as colloidal or very fine-grained particles of shale material through the formation of so-called coated pits or vesicles on the cell membranes (Bretscher, 1985). This mechanism may also

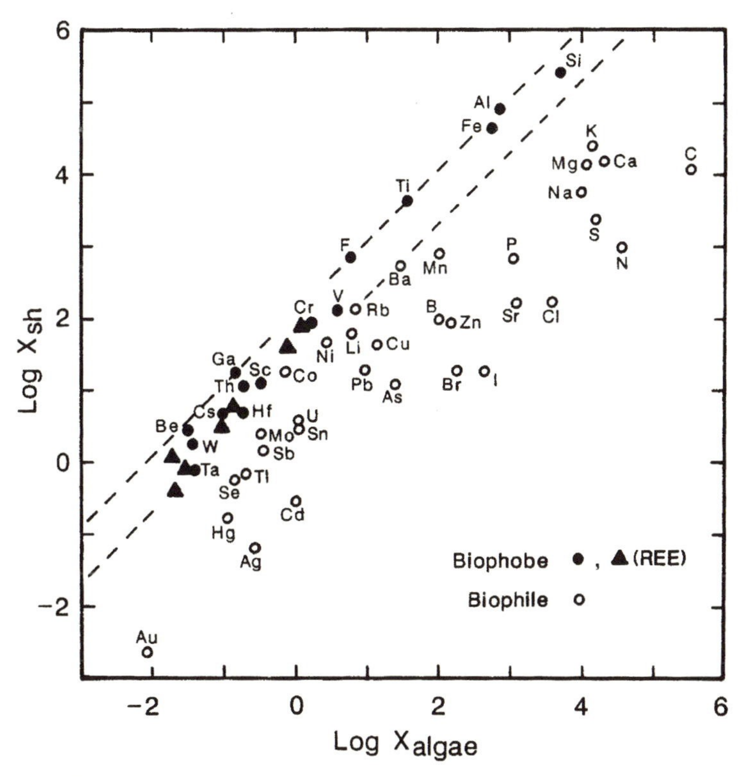

FIGURE VII-6. Elemental concentrations (in ppm) in marine algae (X_{algae}; Table VII-3) plotted versus those in average shale (X_{sh}; Table VI-5a). The elements within the two dashed lines have enrichment factors (E^i_{Al}) of 1 to 5 for algae relative to shale and are termed biophobic. The open circle elements have E^i_{Al} values of greater than 5 and are called biophilic.

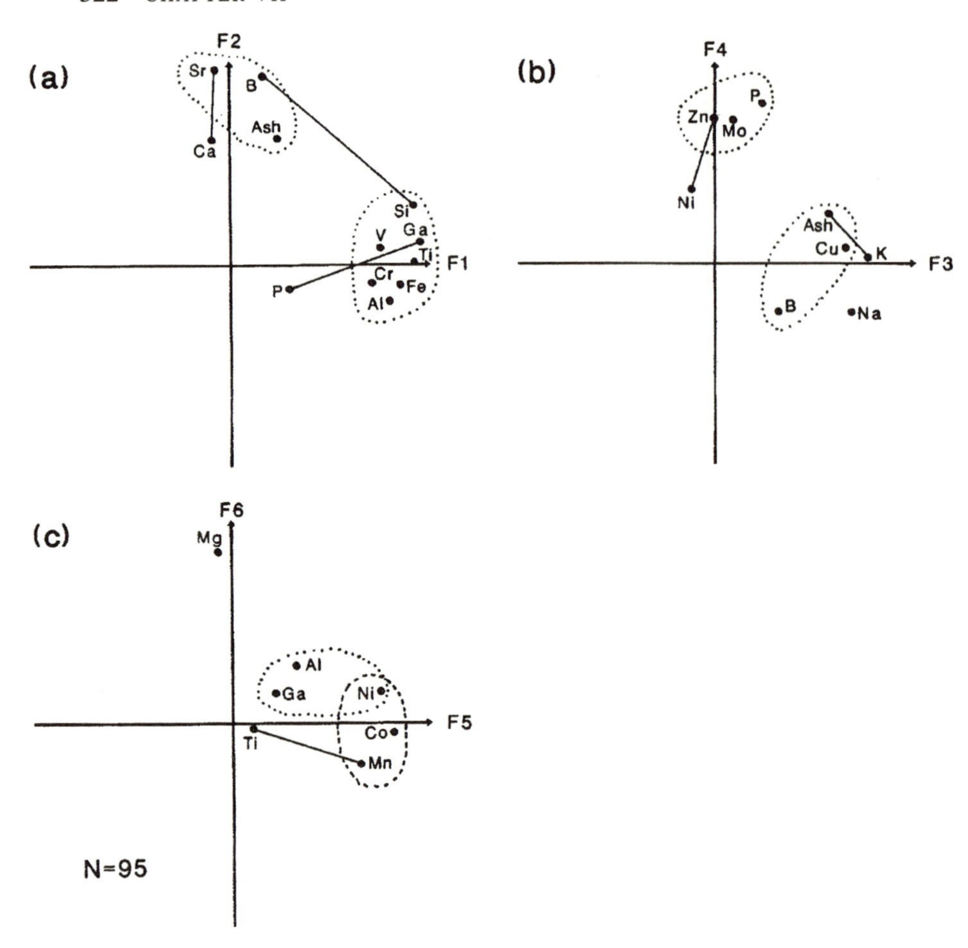

FIGURE VII-7. Plots of factor loadings obtained from compositional data of 95 marine algae (Yamamoto et al., 1985). The elements within the dotted enclosures have correlation coefficients (γ) greater than 0.5 between any pair. The elements connected by solid lines also have $\gamma \geq 0.5$.

explain the observed larger variability in the content of these elements in algae and plankton. Because the dissolved concentrations of these elements are extremely low in seawater (except for Si and F), it is hard to envision an effective uptake mechanism of these elements into the cell simply through adsorption processes of dissolved species on the cell membranes. As shown by Barbeau et al. (1996), protozoan grazers may digest colloidal iron in their acidic food vacuoles and convert the refractory iron into forms utilizable by marine phytoplankton.

Other elements have E^i_{Al} values of about 10 to 10^3 in algae relative to shale (Figure VII-6; open circles), and were called **"biophile" elements** by Li (1984, 1991), again regardless of their actual biological functions. The biophile elements are mainly A-type ions with long mean oceanic residence times, B-type ions [with electron configurations of nd^{10}, $nd^{10}(n+1)s^2$], and some transition metal cations. The relationship among the biophile elements is somewhat complicated as shown in the factor loading plots of Figures VII-7a–c. Factor 1 represents the colloidal aluminosilicates. The elements that directly or indirectly correlate with the ash content of algae are Ca, Sr, B, K, and Cu (as factors 2 and 3). Among these elements, Ca and K are the most abundant in algae besides combustible C and N. The close associations between Ca and Sr, among Mn-Co-Ni, and among P-Mo-Zn reflect their similar biogeochemical behaviors.

A plot of $\log X_{\mathrm{org}}/C_{\mathrm{sw}}$ against the electron binding energies, I_z (Figure VII-8), where X_{org} is the concentration of elements in algae, again resembles Figure VI-18 ($\log X_{\mathrm{R}}/C_{\mathrm{R}}$ vs. I_z). Apparently, the James and Healy adsorption model described in Section VI-8 is equally applicable to the marine organic

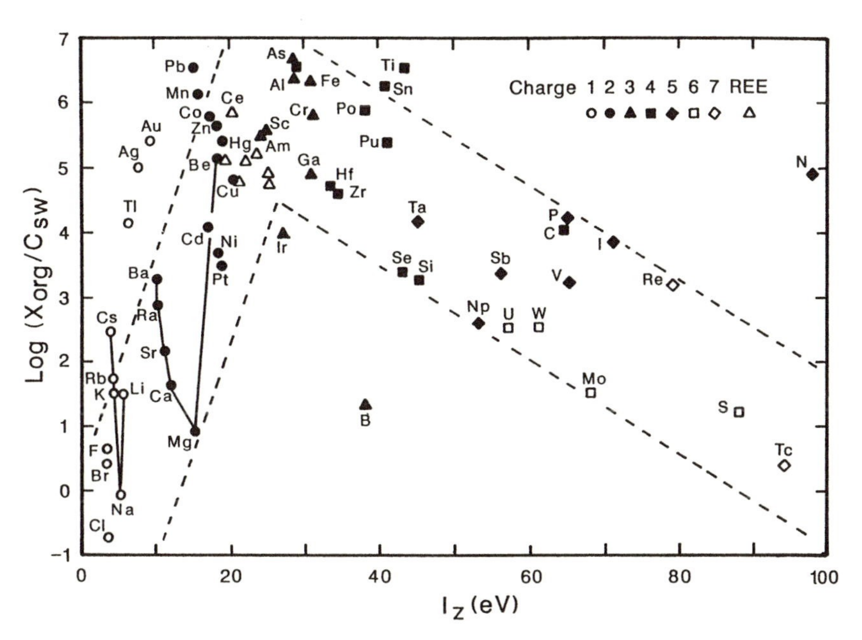

FIGURE VII-8. Plot of $\log X_{\mathrm{org}}/C_{\mathrm{sw}}$ (in units of $\mathrm{cm^3/g}$) versus the electron binding energy I_z. X_{org} is the elemental concentration in average marine organisms (assumed to be equal to that of marine algae or plankton) and C_{sw} the concentration in seawater (Li, 1991; with permission from Elsevier Science).

particle–seawater system. Apparently, the partition of elements between algae and seawater is mainly controlled by the relative bond strength between adsorbing ions and cell membranes (ΔG_{chem}), electrostatic attraction or repulsion (ΔG_{coul}), and partial removal of hydration water around cations (ΔG_{solv}) during the surface adsorption process. In other words, the incorporation of metals by marine organisms should be primarily dependent on the binding of metals to hydrophilic functional groups (such as —COOH, —NH_2, —SH, —OH, etc.) of proteins (which are embedded in the lipid bilayer of cell membranes), and on the subsequent transport of organometallic complexes into the cell interior through many biochemical mechanisms (Hughes, 1981; Ochiai, 1987). As demonstrated by Balistrieri et al. (1981), the complexation constants of various metals to a given organic ligand ($\log \beta^L$) do correlate linearly with $\log {}^*K_1$ (the first hydrolysis constant for cations; see Section I-9) and with I_z (Figure VII-9, using the EDTA ligand as an example), confirming the importance of the relative bond strengths. The causes for the separation of the "biophobe" and "biophile" groups will be discussed in Section VII-5.

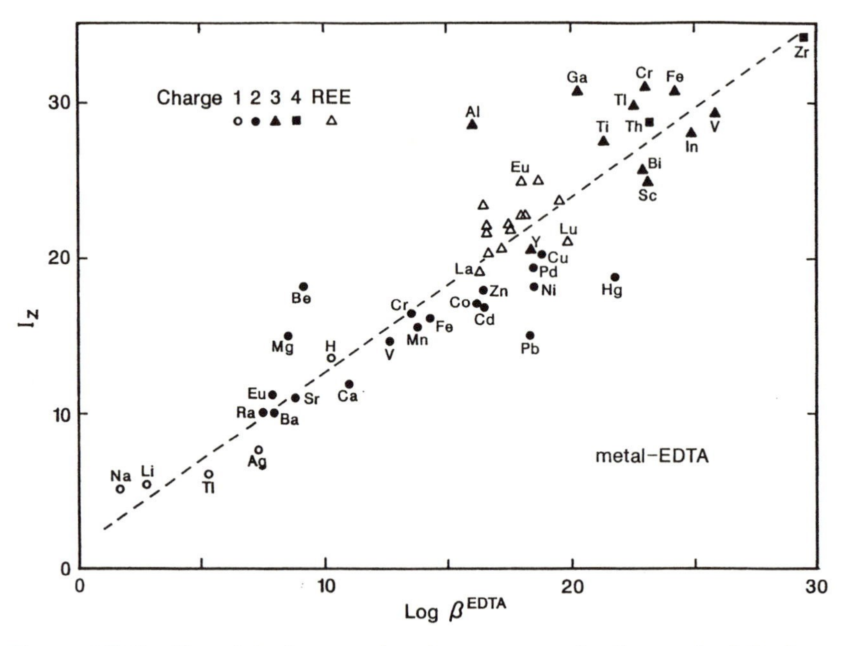

FIGURE VII-9. Plot of the first complexation constants of cations to the fully deprotonated EDTA ligand ($\log \beta^L$; Schwarzenbach, 1957; Dean, 1985) versus the electron binding energy I_z.

VII-4. Zooplankton Fecal Pellets and Sediment Trap Material

As discussed in Section VII-2, the gravitational settling of biogenic particles is the most important mechanism for transferring elements from the surface to the deep ocean. In order to measure and characterize the biogenic particle flux, a device called the **sediment trap** has been deployed extensively at different water depths (e.g., Berger and Soutar, 1967; Wiebe et al., 1976; Honjo, 1978, 1980; Deuser and Ross, 1980; Spencer et al., 1978; Brewer et al., 1980; Masuzawa et al., 1989, and references therein). The settling biogenic particles collected by sediment traps mainly consist of **zooplankton fecal pellets**, debris of planktonic foraminifera, radiolarians, diatoms, coccoliths, and other marine organisms.

Euphausiid zooplankton and their fecal pellets (Table VII-4a) are both enriched in biophile elements relative to shale (Figure VII-10a) but the degree of enrichment is much lower in fecal pellets than in zooplankton. This probably results from the ingestion of both phytoplankton and clay particles by zooplankton and preferential excretion of indigestible clay particles into fecal pellets. For example, "green" fecal pellets, which originate from the surface ocean, contain as much as 25% clay particles, along with 50% carbonate, 5% opal, 15% organic matter, and 3% organic nitrogen (Honjo, 1978).

The vertical fluxes of the major sediment trap components (i.e., organic matter, opal, carbonate, and clays) change drastically with depth, as exemplified by the Japan Sea data (Table VII-4b; Figure VII-11 (see page 328)). The dissolution of carbonate and opal and the oxidation of organic material within the water column and at the water-sediment interface, decrease the fluxes with depth for these components, whereas the flux of clay particles increases with depth below 2000 m, indicating additional horizontal inputs of clay materials from other areas. The bulk chemical compositions of sediment trap materials obtained at water depths of 890 m and 3240 m and of the bottom sediments in the Japan Sea (Masuzawa et al., 1989) are summarized in Table VII-4c. The enrichment factors $\left(E^{i}_{\text{Sc}}\right)$ for the trap material relative to the bottom sediments are also plotted in Figure VII-10b. The biophile elements are very much enriched in the shallow trap, but the degree of enrichment decreases drastically with increasing depth, a result of the preferential oxidation of organic matter pellets. The vertical particle fluxes to the deep Sargasso Sea also vary in a regular annual cycle (Bacon et al., 1985, and references therein), reflecting the annual cycle of phytoplankton bloom and zooplankton grazing in the surface ocean.

TABLE VII-4

A. Compositions (ppm) of zooplankton euphausiid, its excreted fecal pellets and shale

	Euphausiid	Fecal Pellets	Shale		Euphausiid	Fecal Pellets	Shale
Ag	0.71	2.1	0.07	Hg	0.35	0.34	0.18
Cd	0.74	9.6	0.3	Mn	4.2	240	850
Ce	0.21	200?	82	Ni	0.66	20	50
Co	0.18	3.5	19	Pb	1.1	34	20
Cr	0.85	38	90	Sb	0.071	71?	1.5
Cs	0.062	6	5	Sc	0.009	2.8	13
Cu	48	230	45	Se	4.4	6.6	0.6
Eu	0.0023	0.66	1.2	Sr	120	78	170
Fe	64	24000	47200	Zn	62	950	95

Sources: Fowler (1977) for euphausiid and fecal pellets; Table VI-5 for shale.

B. Vertical fluxes ($mg/m^2/day$) of various materials obtained from sediment traps at various depths in the Japan Sea

Depth (m)	*Organic matter*	*Opal*	*Carbonates*	*Clays*	*Total*
Sediment trap					
890	47.7	57	12.1	22.2	139
1100	38.5	47.6	9.5	20.4	116
1870	12	18.5	4.8	15	50.4
2720	4.6	18.5	4.1	22.1	49.4
3240	5.9	16.3	3.6	34.3	60
Bottom sediment					
3350	2.9	2.8	1	40.3	47

Source: Masuzawa et al. (1989).

C. Compositions (ppm) of sediment trap materials obtained at depths 890 m and 3240 m, and bottom sediments from the Japan Sea

	890 m	*3240 m*	*Bottom sediments*		*890 m*	*3240 m*	*Bottom sediments*
Ag	1.65	1.25	0.11	La	3.8	16.6	34.8
Al	8200	35500	75700	Mn	470	3150	3870
As	15.3	15.8	15.4	Rb	17	73	102
Ba	823	840	613	Sb	1.34	2.32	1.97
Br	1670	1500	65?	Sc	1.7	7.3	14
Ca	32300	22400	6100	Se	8.9	7.7	5.3
Co	6.1	21.5	24.2	Sr	564	231	154
Cs	1.2	4.81	9.24	Ta	0.15	0.66	1.13
Fe	6300	33000	40500	Th	1.38	6.46	14.6
Hf	0.49	2.25	5.14	V	22	71	139
I	502	219	366	Zn	113	192	172
K	2400	13400	23700				

Source: Masuzawa et al. (1989).

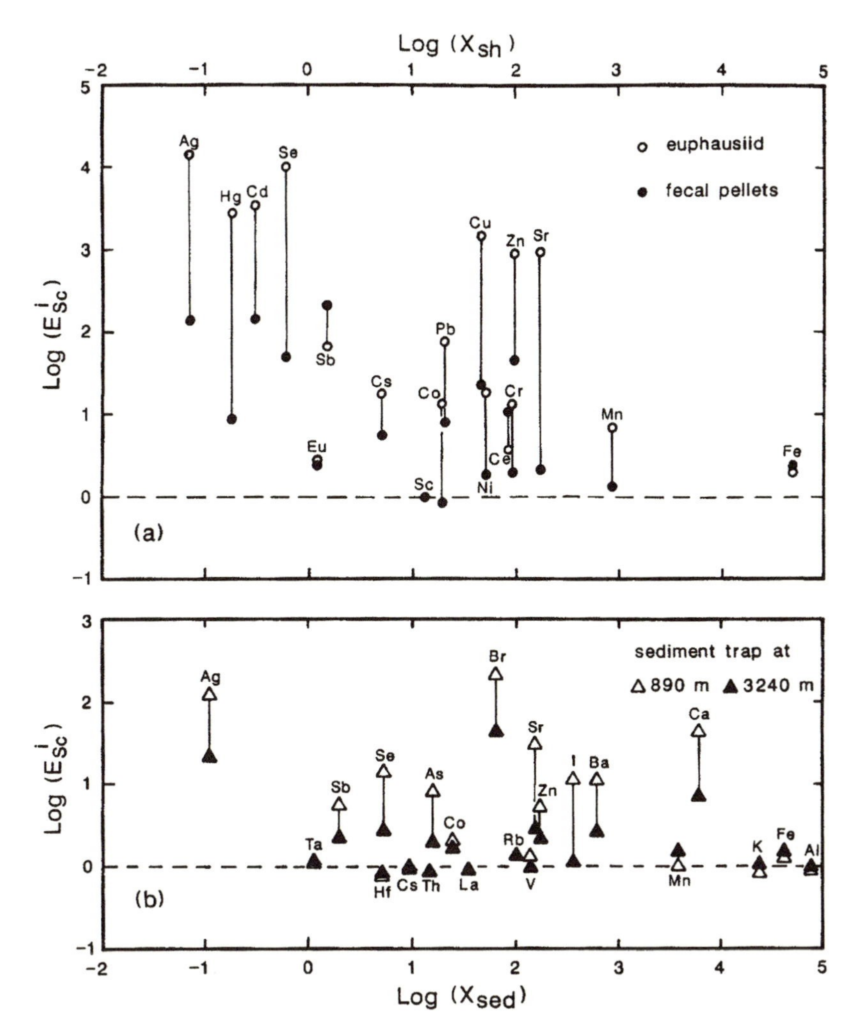

FIGURE VII-10. (a) Enrichment factors of elements, E^i_{Sc}, in the zooplankton euphausiid and its fecal pellets (Table VII-4a; Fowler, 1977) relative to the average shale, and (b) enrichment factors of elements, E^i_{Sc}, in sediment trap materials obtained at 890 m and 3240 m in the Japan Sea relative to the bottom sediments (Table VII-4b; Masuzawa et al., 1989).

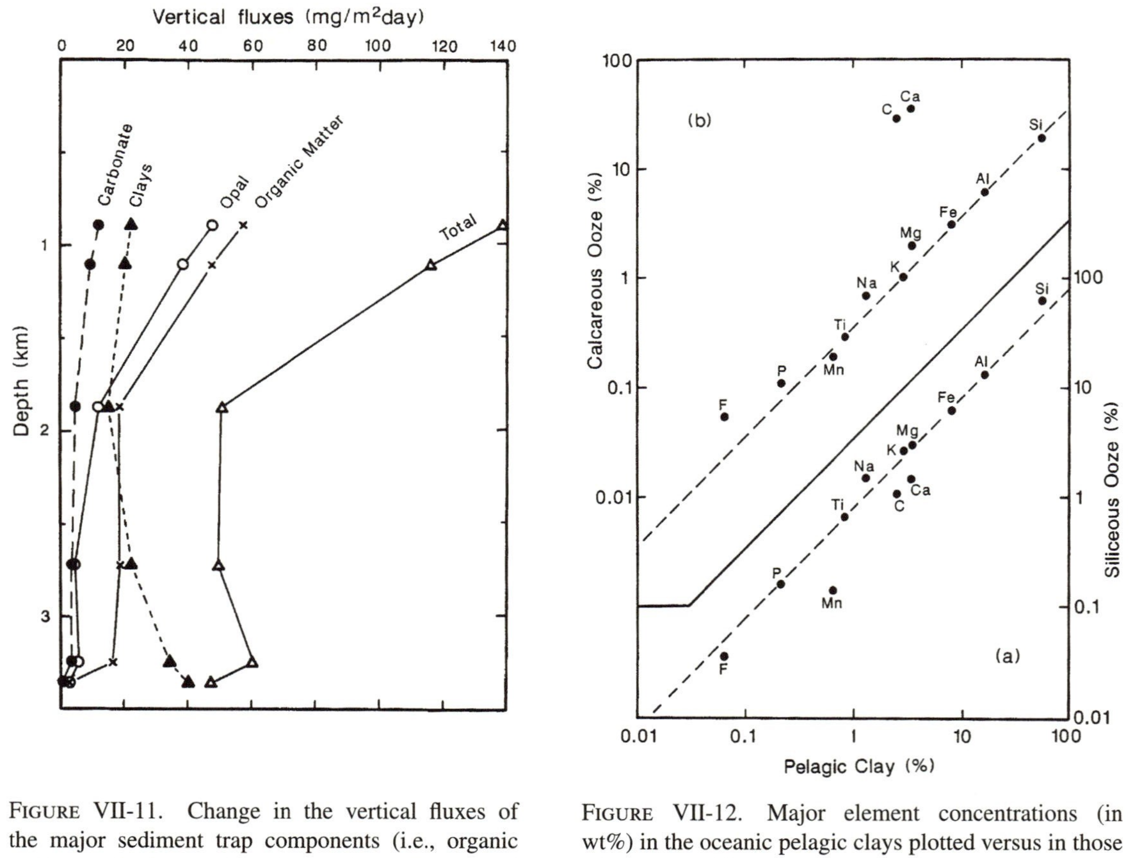

FIGURE VII-11. Change in the vertical fluxes of the major sediment trap components (i.e., organic matter, opal, clays, and carbonates) as a function of depth in the Japan Sea (Masuzawa et al., 1989).

FIGURE VII-12. Major element concentrations (in wt%) in the oceanic pelagic clays plotted versus in those (a) average siliceous ooze and (b) average calcareous ooze (Table VII-5; Ronov, 1982).

VII-5. Marine Sediments

The major sediments on the sea floor are **pelagic clays** (red clays), **calcareous ooze**, and **siliceous ooze**. Their geographical distribution on the sea floor is summarized and discussed by Broecker and Peng (1982). The total mass of marine sediments is about 0.19×10^{24} g. The value of 0.49×10^{24} g given by Ronov and Yaroshevskiy (1976) and Ronov (1982) is in error, because they assumed that the basaltic seismic Layer 2 consisted of normal marine sediments. The relative abundance and average composition for major elements of these three marine sediment types are summarized in Table VII-5 and plotted in Figures VII-12a and b. Similar to limestone on the continents, the average calcareous ooze is essentially a mixture of 69% $CaCO_3$

Table VII-5
Average compositions of major marine sediments in per cent

	Pelagic clay	*Calcareous ooze*	*Siliceous ooze*	*Ave. marine sediments*[b]
Weight fraction	0.37	0.53	0.1	
Density[a] (g/cm^3)	1.1	1.2	0.57	1.03
SiO_2 %	54.16	18.81	62.74	36.28
TiO_2	0.81	0.29	0.66	0.52
Al_2O_3	16.07	6.08	13.19	10.49
Fe_2O_3	6.98	2.55	4.92	4.43
FeO	0.85	0.46	1.15	0.67
MnO	0.63	0.19	0.14	0.35
MgO	3.42	1.98	3.02	2.62
CaO	3.33	35.55	1.47	20.22
Na_2O	1.28	0.7	1.52	1
K_2O	2.85	1.03	2.7	1.87
P_2O_5	0.21	0.11	0.16	0.15
F	0.063	0.054	0.036	0.056
C(organic)	0.27	0.38	0.3	0.33
CO_2	2.45	28.85	1.08	16.31
SO_3	(0.5)			(0.19)
S(pyrite)	0.001	0.001	0.001	0.001
H_2O	6.63	2.97	6.91	4.72
ΣFe_2O_3[c]	7.92	3.06	6.2	5.17

Source: Ronov and Yaroshevskiy (1976).
[a]The density of the surface sediments is calculated from equations given by Hamilton (1976).
[b]Marine sediments do not include coastal sediments, which are counted as continental sediments.
[c]The total iron content expressed as Fe_2O_3.

(with minor Mg, Sr), 38% red clay, and 3% apatite. The average siliceous ooze is a mixture of about 18% SiO_2, and 82% red clay (Figure VII-12a). Table VII-6 summarizes the abundances of the major and trace elements in the average oceanic pelagic clay (X_{op}); geostandards for the pelagic clay (OOPE-501), calcareous ooze (OOPE-401), and siliceous silt (OOPE-402); **manganese nodule** (X_{mn}); **seamount manganese crust** from the Marshall Islands (X_{mc}); and **marine phosphorites** (X_{phos}). The compositions of the average pelagic clay and the single pelagic clay geostandard from the Pacific Ocean (OOPE-501) are surprisingly similar, implying again the relative uniformity of their composition throughout the whole ocean floor (as is the case for shale). As shown in Table VII-6 and Figure VII-13 (see page 335), the average compositions of pelagic clay and shale are the same within a factor of two or better. The obvious exceptions are Mn, Na, Ba, Cu, Ni, Co, Pb, Mo, Te, Pd, Pt, Ir, and Os. Those elements are mostly enriched in ferromanganese oxide phases. In addition, P, Y, and REE are enriched in apatite phases of pelagic clay as inferred from the factor analysis of Pacific pelagic clay samples (Li, 1982). As shown in Figure VII-14a (see page 335), the elements in pelagic clays are split into three phase groups. They are aluminosilicates (Al, Si, K, Rb), ferromanganese oxides (Fe, Mn, Ce, Co, Cu, Ni, Mo, Pb, Sr, Ti, Zn, Zr, C_{org}), and apatites (P, Ca, Y).

The E^i_{Al} values for the various elements in siliceous silt (geostandard OOPE402) and calcareous ooze (geostandard OOPE401) relative to average pelagic clays are mostly around one. Therefore, most of the trace elements in these samples are again contributed by the red clay fraction, except for Ca, Sr, C, and Mg, which are directly related to carbonate phases.

In a steady-state ocean, another scale for the mean oceanic residence time (τ^*) of an element is defined as

$$\tau^* \text{ (years)} = \frac{C_{\mathrm{sw}} \cdot V_{\mathrm{o}}}{X_{\mathrm{op}} \cdot F_{\mathrm{s}}} = 1.3 \cdot 10^9 \frac{C_{\mathrm{sw}}}{X_{\mathrm{op}}}$$

(Goldberg and Arrhenius, 1958), where X_{op} is the concentration of the element in average oceanic pelagic clays, and F_{s} is the total sedimentation rate of oceanic pelagic clays ($\sim 1.1 \times 10^{15}$ g/yr). The deposition of elements by marine carbonate (1.2×10^{15} g/yr) is negligible except for Ca, C, Sr, and Mg. The $X_{\mathrm{op}}/C_{\mathrm{sw}}$ or $1/\tau^*$ values are again a measure of the relative reactivity or affinity of dissolved elements to the marine sediments. Therefore, one would expect a close relationship between $X_{\mathrm{op}}/C_{\mathrm{sw}}$ and $C_{\mathrm{R}}/C_{\mathrm{sw}}$ or between τ^* and τ, as proven in Figure VII-15 (see page 336). However, the slope in the log-log plot is not one; thus, their relationship is not a linear one. The nonlinearity is probably caused by modification of C_{R} by desorption as well as coagulation of various elements during estuarine mixing, and inclusion of detritus components of elements in X_{op} (Li, 1982, and references therein). The outliers Ag and Au probably represent excessively high C_{R} values caused

TABLE VII-6
Compositions of various marine sediments, manganese nodules, manganese crust, and phosphorite

	Pelagic clay		*Calcareous ooze*	*Siliceous silt*	*Manganese nodule*		*Mn crust*	*Phosphorite*
	Average (1)	*OOPE-501* (2)	*OOPE-401* (2)	*OOPE-402* (2)	*Average* (1)	*OOPE-601* (2)	*Marshall Island* (3)	*average* (4)
Ag-47	0.11		(0.3) g	(0.04) g	0.09	(0.2)		2
Al-13 %	8.4	8.45	1.9	4.74	2.7	2.76	0.41	0.91 gu
As-33	20	32	(30) g	20	140	110	230	23
Au-79 ppb	2	5	(2) g	4	2	8		1.4
B-5	230	70	(10) g	70	300	(90)		16
Ba-56	2300	1100	100	1500	2300	1900	1000	350
Be-4	2.6	2.1	1	1.6	2.5	(8)		2.6
Bi-83	0.53	(5)?			(7)			0.06
Br-35	~0				21			
C-6 %			org 0.3	0.34		0.18		2.1
	0.45	0.27	carb 8.8	0.74	0.1	0.11	0.12	0.6
Ca-20 %	1	2.17	28	4.57	2.3	1.98	2.2	31.4 gu
Cd-48	0.42	(3)?	(4) g	(2) g	10	9	3	18
Ce-58	106	100	(20) g	33	530	500	900	104
Cl-17	~0		(17000) g	(30000) g		8000		300 g
Co-27	74	160	12	30	2700	3100	8400	7
Cr-24	90	90	34	80	35	17	9.1	125
Cs-55	6	5	(2) g	3	1	(30)?		
Cu-29	250	320	30	140	4500	5100	380	75
Dy-66	12	(10)		(2) g	31	(30)		19.2
Er-68	7.2	(9)		(2) g	18	(20)	24	23.3

Table VII-6 *Continued*

	Pelagic clay		*Calcareous ooze*	*Siliceous silt*	*Manganese nodule*		*Mn crust*	*Phosphorite*
	Average (1)	*OOPE-501* (2)	*OOPE-401* (2)	*OOPE-402* (2)	*Average* (1)	*OOPE-601* (2)	*Marshall Island* (3)	*average* (4)
Eu-63	3.5	(5)	(0.6) g		9	(10)	8.1	6.5
F-9	1300	(1200)	(700) g		200	(300)		31000 gu
Fe-26 %	6.5	6.45	1.7	3.53	12.5	12	12.3	0.77 gu
Ga-31	20	14	5	11	10	(20)		4
Gd-64	13	(30)		(4) g	32	(50)	39	12.8
Ge-32	1.6	(2)	(4) g	(2) g	0.8			
Hf-72	4.1	(4)	(2) g	(2) g	8	(10)		
Hg-80	0.1				0.15			0.06
Ho-67	2.8	(2)		(0.3) g	7	(3)	8.2	4.2
I-53	28				400			24
In-49	0.08			(0.02) g	0.25			
Ir-77 ppb	0.4				7			
K-19 %	2.5	1.7	0.42	1.15	0.7	0.98	0.38	0.42 gu
La-57	42	80	7	15	157	150	190	133
Li-3	57	60	13	18	80	70		5
Lu-71	1.1	(2)	(0.3) g	(0.3) g	1.8	(2)	3.9	2.7
Mg-12 %	2.1	1.91	2.1	1.91	1.6	1.65	0.88	0.18 gu
Mn-25 %	0.67	1.37	0.16	0.28	18.6	23.2	20.4	0.12
Mo-42	27	38	4	2.8	400	430	370	9
N-7	600				200			100 m
Na-11 %	2.8	2.6	0.43	1.7	1.7	1.5	1.5	0.45 gu

Nb-41	14	12	(10) g	10	50	48		10 m
Nd-60	51	(80)	(50) g	13	158	150	150	98
Ni-28	230	370	38	100	6600	8400	3900	53
Os-76 ppb	0.14				2			
P-15	1500	3146	1000	520	2500	2840	3900	138000 gu
Pb-82	80	62	11	24	900	710	1400	50
Pd-46 ppb	6	(7)	(1) g	(3) g	6	(6)	1.1	
Pr-59	13	(20)			36		34	21
Pt-78 ppb	5	(20)		(7) g	200	190	350	
Ra-88 ppb	0.028							
Rb-37	110	90	11	46	17	16		
Re-75 ppb	0.3				1			
Rh-45 ppb	0.4			(3) g	13	(3)	14	
Ru-44 ppb	0.2				8			
S-16	2000	(1500)	1900 g	1700	4700	(1200)		7200 gu
Sb-51	1	(4)		(0.8) g	40	(20)		7 g
Sc-21	19	32	6	17	10	12		11
Se-34	0.2		(2) g	(6) g	(0.6)	(6)		4.6
Si-14 %	25	22.8	5.6	27.7	7.7	7.57	2.2	5.6 gu
Sm-62	13	20	(3) g	2.5	35	40	30	20

TABLE VII-6 *Continued*

	Pelagic clay		*Calcareous ooze*	*Siliceous silt*	*Manganese nodule*		*Mn crust*	*Phosphorite*
	Average (1)	*OOPE-501* (2)	*OOPE-401* (2)	*OOPE-402* (2)	*Average* (1)	*OOPE-601* (2)	*Marshall Island* (3)	*average* (4)
Sn-50	4	4	210	3.2	2	(3)		3
Sr-38	180	290	1200	340	830	900	1200	750
Ta-73	1	(1)	(1) g	(0.8) g	(10)	(5)		
Tb-65	1.9	(4)		(0.7) g	5.4	(8)	5	3.2
Te-52	1		(100) g?	(1) g	(10)	(1)		
Th-90	13	14	3	5	30	31		6.5
Ti-22	4600	5900	1800	3500	6700	8800	7700	640
Tl-81	1.8	(3)		(1) g	150	(100)		
Tm-69	1.1	(2)		(0.3) g	2.3	(2)	3.6	1.2
U-92	2.6	2.5	(2) g	1.5	5	5		120
V-23	120	150	57	85	500	400	500	100
W-74	4				100			
Y-39	60	150	9	16	150	160	190	260
Yb-70	7.3	15	(1) g	2.2	20	21	24	13
Zn-30	170	160	100	90	1200	770	540	200
Zr-40	150	190	80	100	560	600		70

Sources: (1) Li (1991) and references therein; (2) Berkovits et al. (1991); (3) Hein et al. (1998); (4) Altschuler (1980); g: Govindaraju (1994); gu: Gulbrandsen (1966); m: Manheim et al. (1980).

Notes: The values in parentheses for geostandards are uncertified, and thus probably less reliable, especially those with questions marks. Composition in ppm unless noted otherwise.

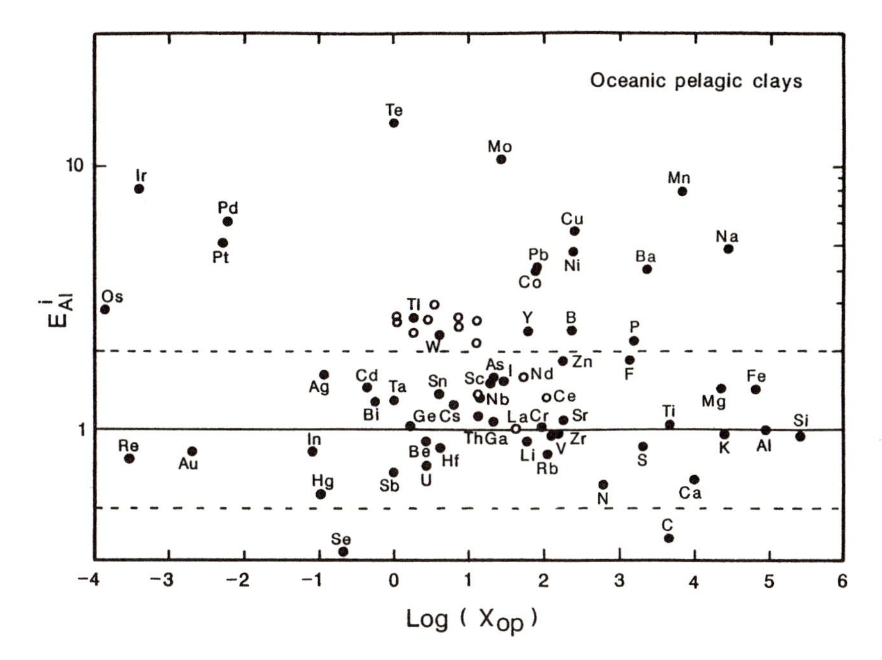

FIGURE VII-13. Enrichment factors of element, E^i_{Al}, in oceanic pelagic clays relative to average shale plotted against the logarithm of the element concentrations in oceanic pelagic clays (X_{op}, in ppm; Table VII-6). The open circles are REE data.

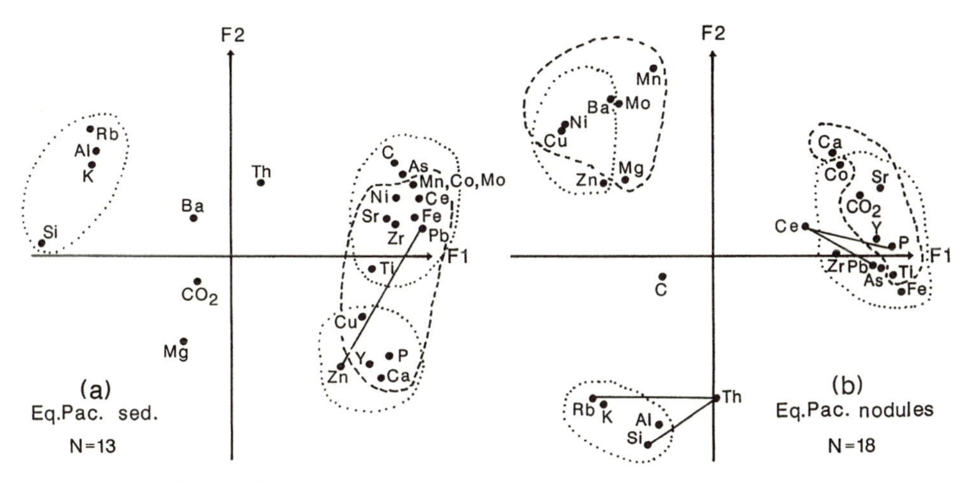

FIGURE VII-14. Factor loading 1 vs. 2 for pelagic sediments obtained from (a) Equatorial Pacific (28°N–13°S, 117°W–175°E), and (b) associated manganese nodules (Li, 1982).

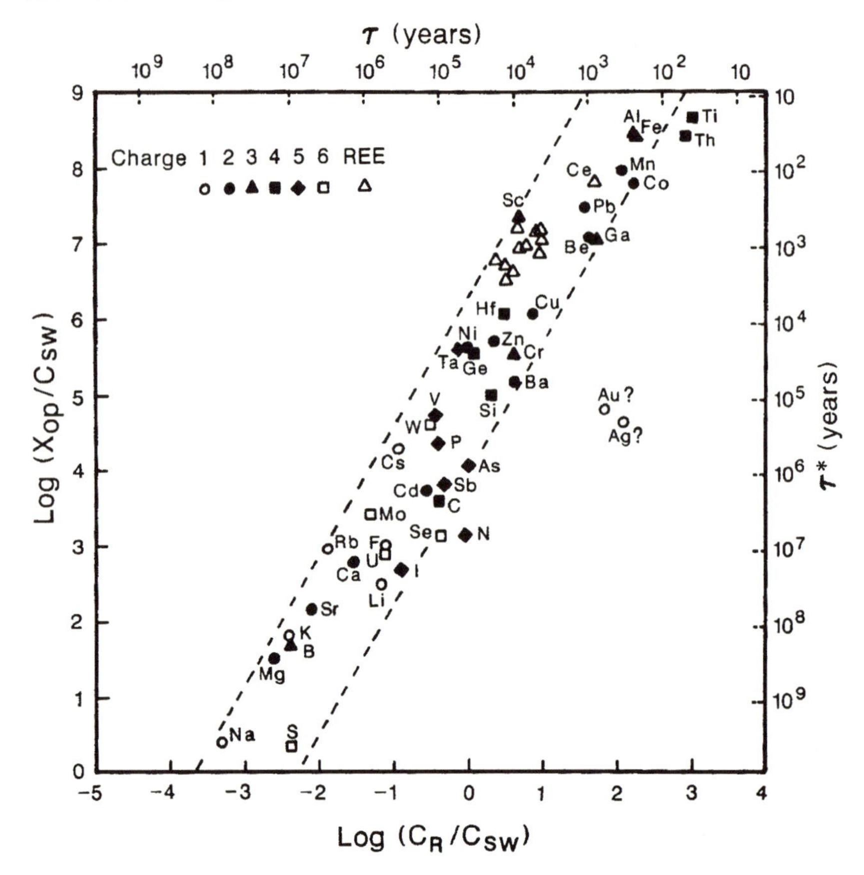

FIGURE VII-15. Plot of log C_R/C_{sw} and corresponding mean ocean residence time τ against log X_{op}/C_{sw} and its corresponding mean oceanic residence time τ^*. C_R, C_{sw}, and X_{op} are respectively the concentrations of element in river, seawater, and oceanic pelagic clays.

by pollution input or sample contamination. Similarly, the C_R values for Th and Ti are probably also too high. New measurements are certainly needed. Notice that in Figure VII-15, X_{op} for Ca, C, Sr, and Mg also includes a contribution from the marine carbonate reservoir.

The values of log X_{op}/C_{sw} and associated τ^* are plotted against I_z in Figure VII-16a. The elements with an asterisk are radionuclides, whose log X_{op}/C_{sw} values are calculated from their activities in filtered particles and seawater from the deep Atlantic ocean or from adsorption experiments (Li, 1991). For the calculation of log X_{op}/C_{sw} values for As, Cr, Pu, Se, and Te, the C_{sw} values for the reduced species were used (Table VII-1). This is because the species of these elements adsorbed by solid particles are predominantly the reduced species, as indicated by a 10^3-fold difference in the K_d (distribution

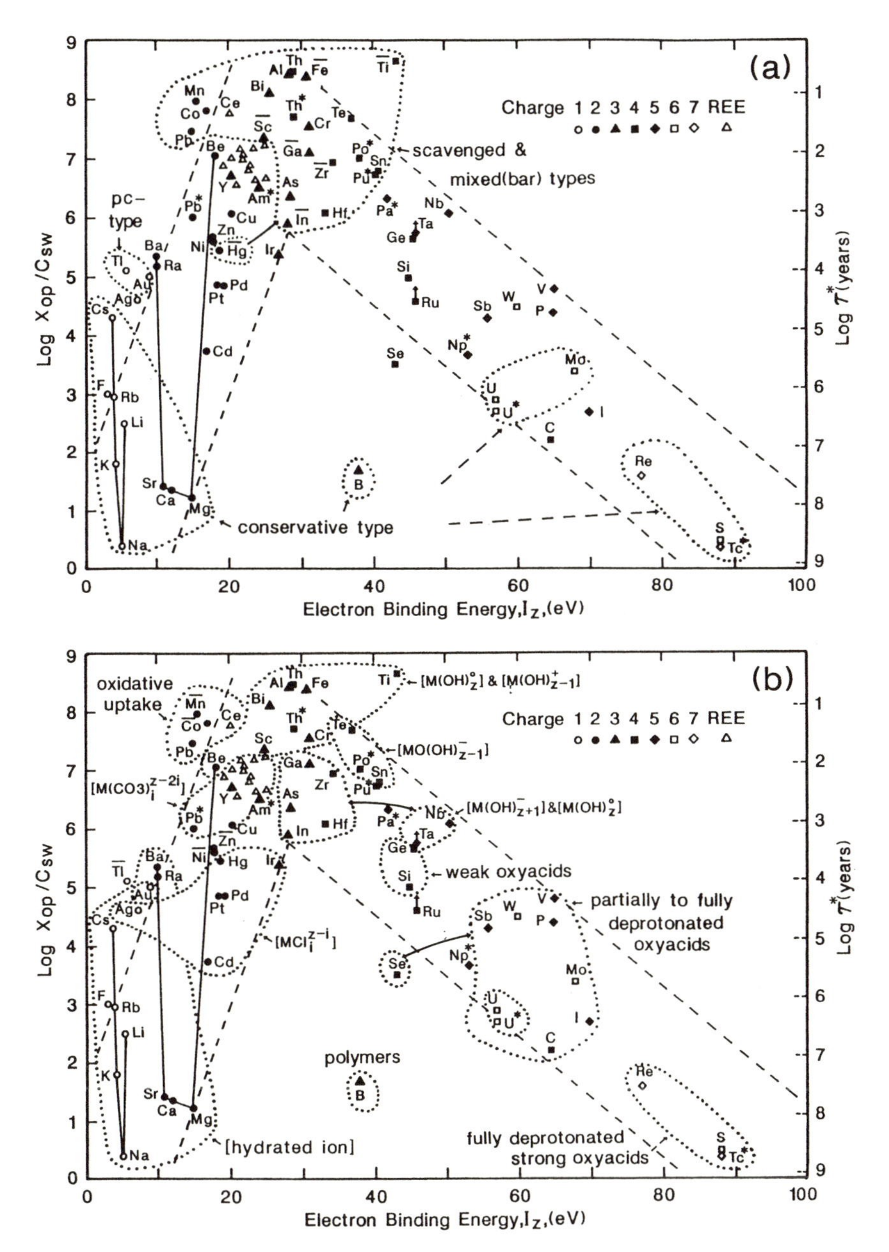

FIGURE VII-16. Log X_{op}/C_{sw} and its corresponding mean oceanic residence time τ^* are plotted against the electron binding energy I_z. In (a) the conservative type, pseudo-conservative type, and scavenged plus mixed type (with bar) elements are grouped by dotted enclosures. Ungrouped elements are nutrient type (n). In (b), the systematic change in the chemical speciations is shown. Elements with a bar indicates the dominance of both hydrated cationic and chloride complex forms. (Li, 1991; with permission from Elsevier Science.)

coefficient) values between reduced and oxidized forms (Li, 1991, and references therein). Figure VII-16a is again similar to Figure VII-8 ($\log X_{org}/C_{sw}$ vs. I_z) and Figure VI-18 ($\log X_R/C_R$ vs. I_z). The similarity includes features such as the general "inverted V" relationship between $\log X_{op}/C_{sw}$ and I_z, and the inverse correlation between $\log X_{op}/C_{sw}$ and I_z for alkali and alkaline earth elements with larger ionic radii. Therefore, one can conclude again that the partition of elements between oceanic pelagic clays and seawater is in accord with the adsorption model of James and Healy (1972). However, the exceptionally high $\log X_{op}/C_{sw}$ values for Mn, Co, Pb, Ce, and Tl compared to the neighboring elements in Figure VII-16 (as well as in Figure VII-8), may indicate **oxidative uptake** of these elements on the particle surface. For example, $Mn^{+2} \rightarrow Mn^{+4}$, $Co^{+2} \rightarrow Co^{+3}$, $Pb^{+2} \rightarrow Pb^{+4}$, $Ce^{+3} \rightarrow Ce^{+4}$, and $Tl^{+} \rightarrow Tl^{+3}$ (Li, 1991, and references therein).

For comparison, the types of concentration profiles for various elements in the oceans are also indicated in Figure VII-16a. Similar to Figure VII-8, the scavenged- and mixed-type elements are all clustered at the tip of the "inverted V" with high $\log X_{op}/C_{sw}$ or short τ^* values. The conservative-type elements all have low $\log X_{op}/C_{sw}$ and long τ^* values. The psuedo-conservative-type Au and Tl have short τ^*'s. The nutrient-type elements have intermediate $\log X_{op}/C_{sw}$ and τ^* values.

The systematic changes in the chemical speciation of elements in seawater along with $\log X_{op}/C_{sw}$ and I_z are also summarized in Figure VII-16b. In general, the speciation changes from hydrated cations → chloride complexes → carbonate complexes → fully hydrolyzed hydroxo complexes when both $\log X_{op}/C_{sw}$ and I_z increase systematically from Na^+ to Th^{+4}. Then, the speciation changes to negatively charged hydroxo complexes → weak oxyacids → partially to totally deprotonated oxyacids → totally deprotonated strong oxyacids when the I_z continues to increase but the $\log X_{op}/C_{sw}$ continues to decrease.

Because Figures VII-16a and VII-8 are so similar, one might expect $\log X_{op}/C_{sw}$ and $\log X_{org}/C_{sw}$ to be closely related to each other, as proven in Figure VII-17. The often observed inverse relationship between $\log C_{org}/C_{sw}$ and the logarithm of the mean oceanic residence time of elements (Yamamoto, 1972; Yamamoto et al., 1985; Masuzawa et al., 1989; Knauss and Ku, 1983) is as expected, because $\log \tau^*$ is inversely related to $\log X_{op}/C_{sw}$ as discussed earlier.

As in Figure VII-6, the "biophobe" elements all fall on the solid line with a slope of one in Figure VII-17. However, the original "biophile" elements such as Mn, Co, Ni, Ba, and Mo in Figure VII-6 also fall on the same "biophobe" solid line in Figure VII-17 due to the enrichment of these elements in the pelagic clays relative to shale (see Figure VII-13). The E^i_{Al} values for other biophile elements range between 5 and 10^3 relative to the pelagic clays. The biophile elements include the platinum group elements and actinides in addition to B-type cations and A-type ions with long τ^*.

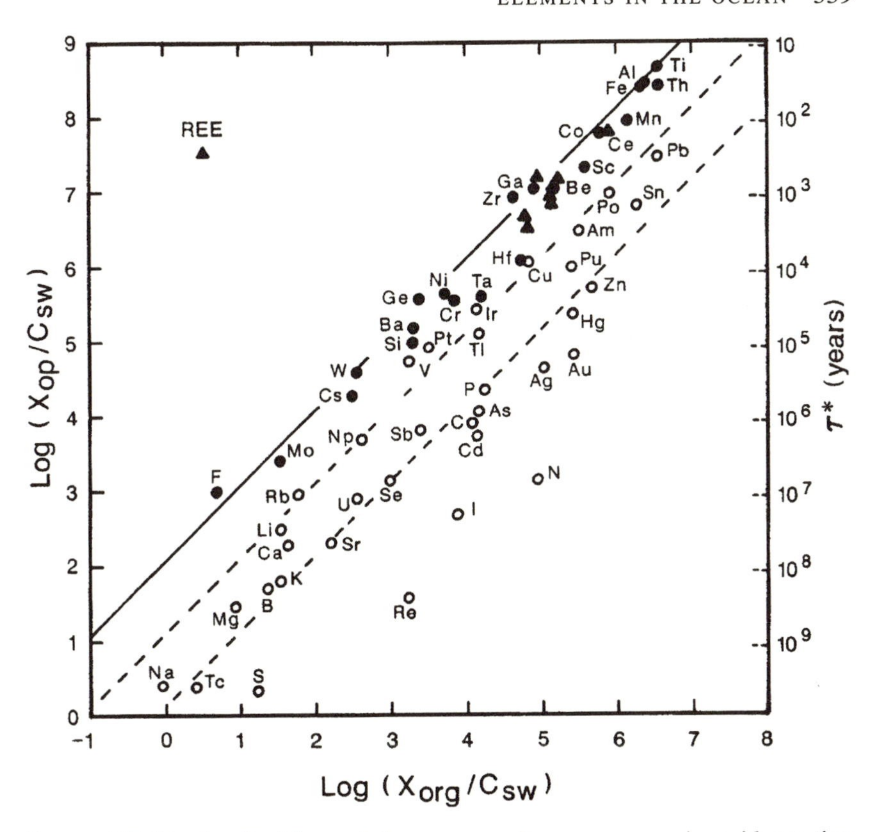

FIGURE VII-17. $\log X_{\mathrm{op}}/C_{\mathrm{sw}}$ and the corresponding mean oceanic residence time τ^* plotted versus $\log X_{\mathrm{org}}/C_{\mathrm{sw}}$. The dashed lines have a slope of one and are one order of magnitude apart. The solid circles and triangles are biophobe elements with enrichment factors E^i_{Al} of roughly 1 to 10 in marine organisms relative to pelagic clays. The open circles are biophile elements with E^i_{Al} values roughly greater than 10.

The electrons in the *d* subshell of the B-type and platinoid cations are highly polarizable, as discussed in Section I-6. Thus these cations form extra-strong chemical bonding with organic ligands, especially those containing donor atoms of sulfur and nitrogen (—SH and —NH_2), which are also highly polarizable compared to O^{-2} (Table I-6). The severe toxicity of highly polarizable cations such as Cu^+, Cu^{+2}, Ag^+, Pd^{+2}, Rh^{+3}, Ru^{+4}, Au^+, Au^{+3}, Hg^+, Hg^{+2}, Pt^{+2}, Pt^{+4}, Ir^{+3}, Ir^{+4}, and Os^{+4} (Table I-7) may be partly caused by their blockage of vital functional groups in biomolecules (Ochiai, 1987). The fact that actinides are biophile whereas lanthanides are biophobe can be explained as follows: The energy level difference between $5f$ and $6d$ subshells in actinides is small as compared to that between $4f$ and $5d$ subshells in lanthanides. Therefore, the electrons in the $5f$ subshell of actinides can easily move into the 6*d* subshell (thus becoming highly polarizable) to form strong chemical bonds with various organic ligands. In contrast, no 4*f* subshell electrons of lanthanides can move to the 5*d* subshell (Cotton and Wilkinson, 1988). A-type cations with long τ^* such as Na^+, K^+, Mg^{+2}, and Ca^{+2} can be effectively transported through the cell membrane by the so-called sodium pump (Na/K-ATPase) and calcium pump (Ca/Mg-ATPase), as well as by the so-called ionophore (ion carrier) and ion channels for these elements (Ochiai, 1987).

Finally, the distribution coefficients (K_d) for many radiotracers were determined in adsorption experiments using a pelagic clay–seawater system

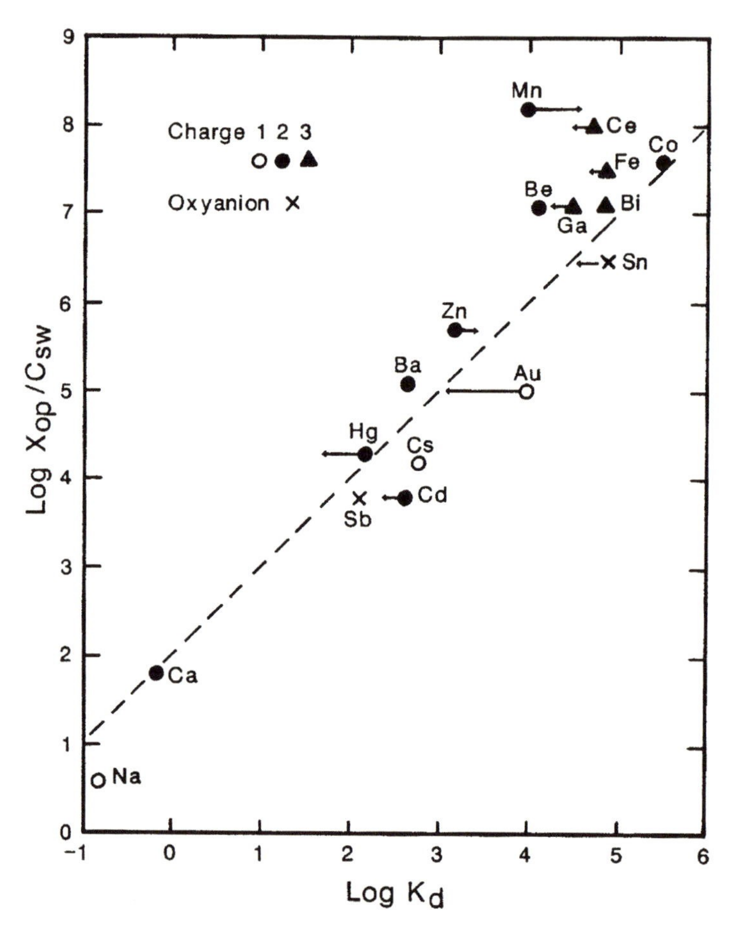

FIGURE VII-18. Distribution coefficients (K_d) for various radiotracers determined in a pelagic clay–seawater system with suspended particle concentrations between 0.4–1.7 g/l (Li et al., 1984) plotted against the corresponding natural log X_{op}/C_{sw} values. The dashed straight line has a slope of one.

(Nyffeler et al., 1984; Li et al., 1984). These measured K_d values are more or less linearly correlated to the corresponding natural X_{op}/C_{sw} as shown in Figure VII-18. The linearity supports the contention that adsorption is the major mechanism of uptake of elements by pelagic clays in the ocean. The difference between X_{op}/C_{sw} and the experimental K_d values by two orders of magnitude can be easily explained by the fact that there is ample time for pelagic clays in the ocean to be continuously coated by layers of new Fe and Mn oxides to provide additional adsorption sites for other trace elements.

VII-6. Marine Manganese Nodules and Seamount Manganese Crusts

Deep-sea manganese nodules and seamount manganese crusts nucleate on hard surfaces (such as shark teeth, volcanic fragments, and rocks) and grow mostly at extremely slow rates of a few millimeters per one million years (Broecker and Peng, 1982). Their occurrences, mineralogy, chemistry, and origins are discussed extensively by Glasby (1977), Bischoff and Piper (1979), and Baturin (1988). The focus in this section is on their chemical compositions.

The composition of the manganese nodules geostandard (OOPE601) from the Pacific ocean agrees well with that of the average manganese nodule, mostly within a factor of two or better (Table VII-6). The implication is that the compositions of manganese nodules are relatively uniform throughout the ocean floor, and the nodules were formed under similar conditions. The concentrations of elements in manganese nodules (X_{mn}) and their E^i_{Al} relative to pelagic clays are summarized in Figure VII-19a. In the figure, the elements are designated with different symbols to differentiate their preferred association with manganese oxide, iron oxide, phosphate, and aluminosilicate detritus phases. These associations are based on factor analysis of nodule data from the Pacific ocean (Figure VII-14b) and on the correlation coefficients among different elements (Li, 1982; Baturin, 1988). The main difference between pelagic clays and manganese nodules in the factor analysis is that the ferromanganese phase in pelagic clays is not separated into distinct manganese oxide and iron oxide phases as in the nodules (Figures VII-14).

The elements associated strongly with the manganese oxide phases are mono- and divalent cations with low to moderate $\log {}^*K_1$ values (Tl, Ag, Mg, Ba, Ni, Cu, and Zn) and oxyanions of Mo and Sb. The E^i_{Al} values for these elements (except Ag, Mg, Ba) are similar to the E^{Mn}_{Al} values (Figure VII-19a), suggesting that these elements are mainly incorporated into manganese oxide phases of both manganese nodules and pelagic clays. The low E^i_{Al} values for Ag, Mg, and Ba relative to E^{Mn}_{Al} suggest that the aluminosilicate phase is also a significant contributor of these elements in manganese nodules. The

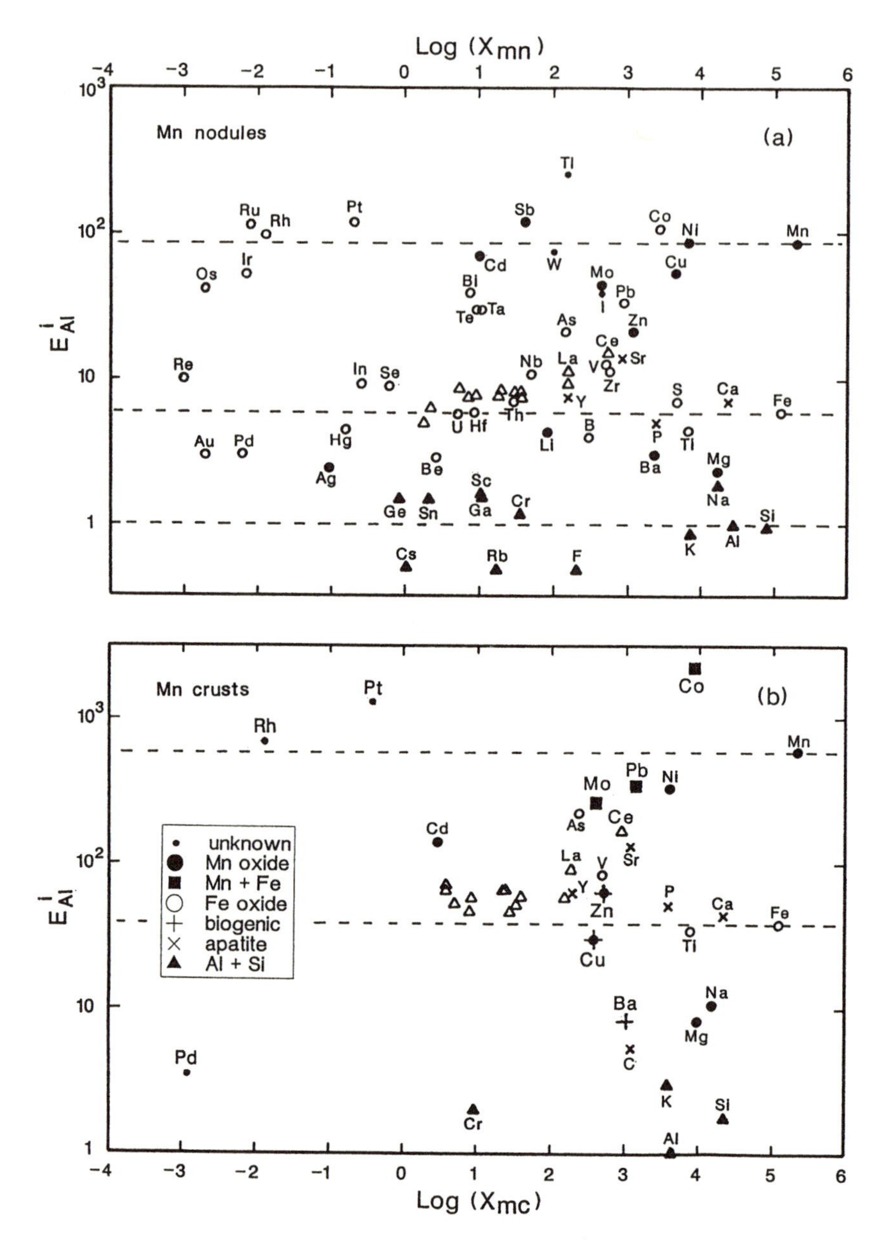

FIGURE VII-19. (a) Enrichment factors of elements, E^i_{Al}, in the average manganese nodule relative to the pelagic clays plotted against the elemental concentrations in manganese nodules (log X_{mn}). (b) E^i_{Al} values in the Marshall Island seamount manganese crusts relative to pelagic clays plotted versus the logarithmic concentrations of elements in the Marshall Island manganese crusts. The horizontal dashed lines pass through the Al, Fe, and Mn data points. The open triangles are rare earth elements, which are mostly associated with both iron oxide and apatite phases.

elements that correlate strongly with the iron oxide phase exist in seawater mainly as hydroxyl and carbonate complexes of tri- to pentavalent cations (As, B, Bi, In, Ir, Rh, REE, Y, Ti, Th, Zr, Hf, Nb, and Ta), oxyanions (I, P, Re, Ru, Os, S, Se, and Te), divalent cations with high $^*K_1$ values (Hg and Be). The E^i_{Al} values for B, In, REE, Y, Ti, Th, Zr, Hf, Nb, P, Se, Re, Pd, and Hg are similar to $E^{\text{Fe}}_{\text{Al}}$. Therefore, these elements are incorporated into the iron oxide phases of both manganese nodules and pelagic clays. The high E^i_{Al} values for Co, Pb, Pt, Rh, Ru, Ir, and Os relative to $E^{\text{Fe}}_{\text{Al}}$ suggest additional uptake mechanisms for these elements in manganese nodules. For example, the oxidation of Co^{+2}, Pb^{+2}, and Pt^{+2} into Co^{+3}, Pb^{+4}, and Pt^{+4} on the nodule surface, and the oxidation of Rh, Ir, Ru, and Os from di- or trivalent to tetravalent are suggested (Goldberg et al., 1986). The apatite phase is closely related to the iron oxide phase in manganese nodules (Figure VII-14b) and contributes significant amounts of Ca, Sr, P, C, REE, and Y in manganese nodules. The E^i_{Al} values of near one for Si, K, Na, F, Cr, Rb, Cs, Sc, Ga, Sn, and Ge (Figure VII-19a) suggest that these elements are contributed mainly by aluminosilicate detritus phases in both manganese nodules and pelagic clays.

The specific association of elements with manganese oxide, iron oxide, and aluminosilicate phases can be explained by the fundamental differences in physicochemical properties of these phases. Those include the pH of zero point of charge (pH_{zpc}), intrinsic acidity constants $\left({}^*K_1^s\right)$, and dielectric constant (ε) (Li, 1991). For example, the manganese oxide phases in the nodules are mainly todorokite with variable amounts of vernadite (δMnO_2). Todorokite is thought to form at the nodule-sediment interface through the oxidation of Mn^{+2} supplied by the interstitial water of the underlying sediments. Todorokite forms tunnel structures, using building blocks of octahedral $\left[(Mn^{+4}, Mn^{+2})O_6\right]$ units (Ostwald, 1988, and references therein). The tunnels accommodate mono- or divalent cations with low $\log {}^*K_1$ values (such as alkali and alkaline earth elements) and H_2O molecules, whereas the divalent transitional metal cations can easily substitute for Mn^{+2} in the $\left[(Mn^{+4}, Mn^{+2})O_6\right]$ units. The todorokite tunnels are not large enough to accommodate highly hydrolyzable cations such as Au^+, Be^{+2}, Pd^{+2}, and tri- and tetravalent cations with high $\log {}^*K_1$ values. Therefore, it is not surprising to find that only the mono- and divalent cations with relatively low to moderate $\log {}^*K_1$ values are covariant with Mn content in the nodules. The obvious exceptions are Co, Pb, and Pt, which have moderate $\log {}^*K_1$ values, but are associated mainly with the iron oxide phases of the nodules. One plausible explanation has been that Co^{+2}, Pb^{+2}, and Pt^{+2} are easily oxidized into highly hydrolyzable Co^{+3}, Pb^{+4}, and Pt^{+4} on the nodule surface. Thus, their incorporation into todorokite is precluded, but their adsorption onto hydrous iron oxide phases is enhanced (Li, 1991, and references therein).

Furthermore, the pH_{zpc} values for aluminosilicate clay minerals are in the range of 2 to 4.6, for δMnO_2 (and probably also todorokite) about 2.8, and for goethite about 7.5 to 8.3 (Stumm and Morgan, 1981). Therefore, both manganese oxide and aluminosilicate particles are negatively charged in seawater with a pH of 8, and can electrostatically attract positively charged ions (ΔG_{coul}) in addition to the chemical bond formation (ΔG_{chem}). In contrast, goethite may be positively charged and attract anions. The dielectric constants for aluminosilicate minerals range from 4.5 to 8 (Keller, 1966), and for δMnO_2 the constant is about 32 (Murray, 1975). Because ΔG_{solv} is inversely related to the dielectric constant (James and Healy, 1972), δMnO_2 (and probably also todorokite) has a much smaller positive ΔG_{solv} term than aluminosilicate and inhibits the adsorption of hydrated cations to a lesser degree. Also, there are some hints that the intrinsic acidity constants ($^*K_1^s$; see Section VI-8) for the manganese oxide phases are probably much smaller than those for aluminosilicate minerals (Schindler and Stumm, 1987). In other words, H^+ is more strongly held by hydrous manganese oxide than by aluminosilicate surfaces. If this is true, the chemical bond formed between the adsorbed metal cation and the hydrous oxide surface should be much stronger in manganese oxide than in aluminosilicate minerals. All these factors explain why the aluminosilicate detritus phases in both manganese nodules and pelagic clays are not important sinks for many cations as compared to manganese oxide phases. In contrast, goethite particles in seawater with a pH of 8 are only slightly charged or neutral. Therefore, adsorption of ions onto goethite occurs mostly through chemical bond formation.

The concentrations of elements in the Marshall Island seamount manganese crusts (X_{mc}, Table VII-6) and their E^i_{Al} values relative to the pelagic clays are summarized in Figure VII-19b. The elemental associations with major phases of seamount manganese crusts, based on factor analysis (Figure VII-20), are also shown. In Figure VII-20, the factor 1 (Mn, Mg, Ni, Zn, and Cd) represents the manganese oxide phases. Mo and Co may partially relate to factor 1. The association of Mo, As, and V as factor 2 may represent phases closely related to the iron oxide phase, but Fe also strongly correlates to aluminosilicate phases (Al, Si, and K as negative factor 3) through Si. Factor 3 (Ca, P, and Y) represents the apatite phase. As shown in the factor score diagram, two samples are very much enriched in the apatite component. Factor 4 (Cu, Ba, Zn, and partially Ni and Ce) may suggest a biogenic origin (Wen et al., 1997). Again, according to the factor score diagram, five samples are enriched in biogenic component. An interesting contrast is that Co and Pb are strongly associated with Fe in manganese nodules as mentioned before, but they do not show any preference for either Fe or Mn in manganese crust. A possible explanation is that the major manganese oxide phase in the seamount manganese crust is δMnO_2 (vernadite). Vernadite is a poorly ordered hydrated manganese oxide and may be represented roughly by the formula $(R_2O, RO, R_2O_3)_{0.5} \cdot MnO_2 \cdot (1\text{–}2)H_2O$, where R represents mono- to tri- or even tetravalent cations that are incorporated into vernadite

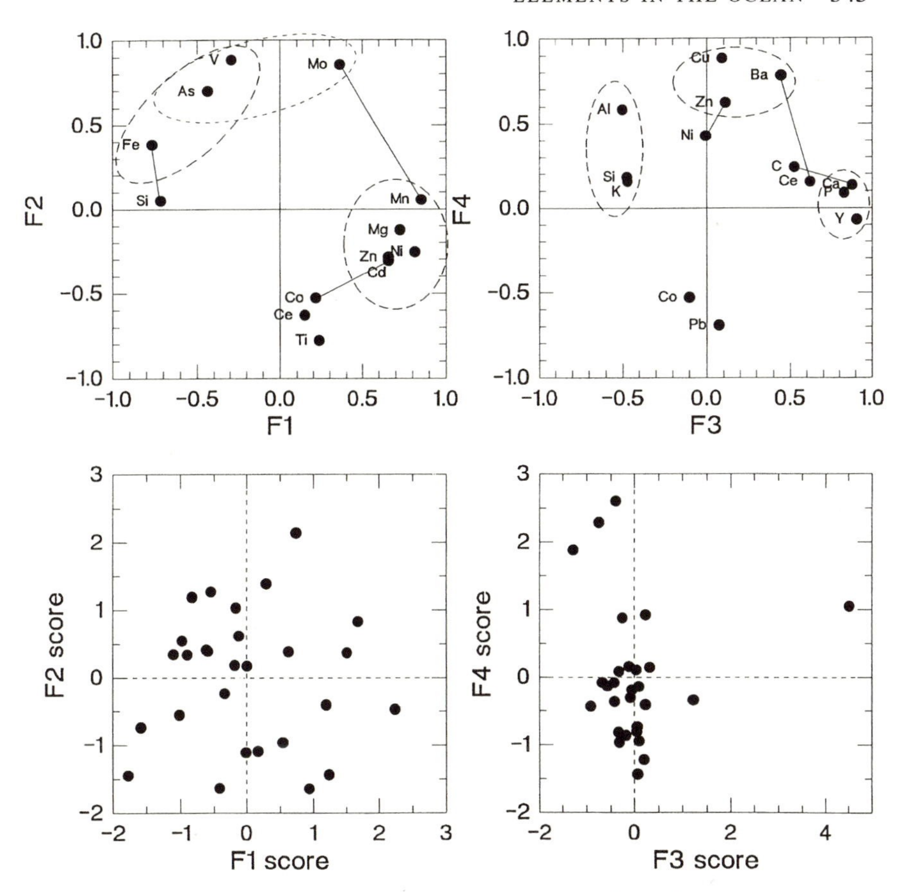

FIGURE VII-20. Factor loading and score results for Marshall Island manganese crust composition data (Hein et al., 1988).

through both surface adsorption and substitution for Mn. Therefore Co and Pb, in either low or high oxidation states, can be easily accommodated into vernadite as well as goethite. As shown by adsorption experiments in seawater (Li et al., 1984), vernadite and goethite are almost equally effective in adsorbing cations with high $\log {}^{*}K_1$ values. For both minerals, the strong chemical bond formation (ΔG_{chem}) overwhelms any electrostatic (ΔG_{coul}) and solvation (ΔG_{solv}) effects.

Finally, the plot of $\log X_{\text{mn}}/C_{\text{sw}}$ versus I_z (Figure VII-21) is similar to Figure VII-8 ($\log X_{\text{org}}/C_{\text{sw}}$ vs. I_z) and Figure VII-16 ($\log X_{\text{op}}/C_{\text{sw}}$ vs. I_z). The conclusion is that the partition pattern of elements between manganese nodules (as well as manganese crusts) and seawater is once more in accord with the adsorption model of James and Healy (1972).

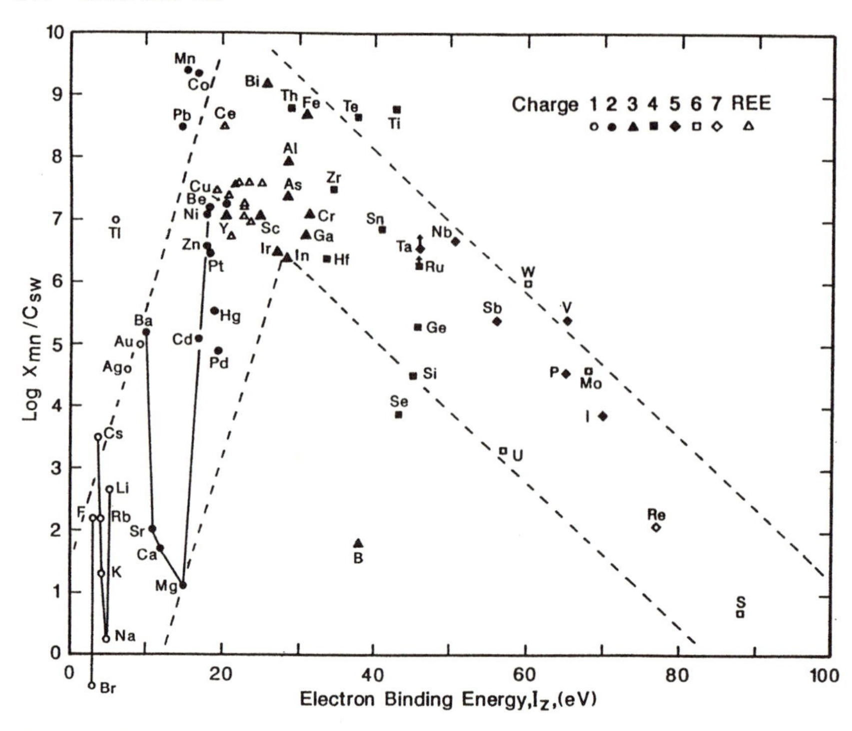

FIGURE VII-21. Elemental concentrations in average marine manganese nodules over those in seawater (log X_{mn}/C_{sw}) plotted versus the electron binding energy I_z (Li, 1991; with permission from Elsevier Science).

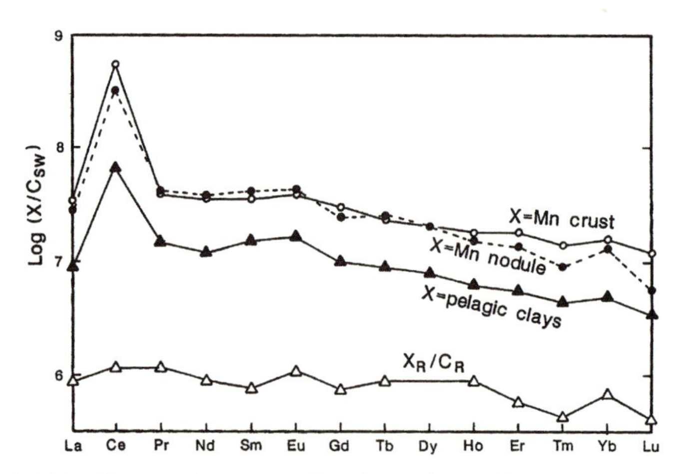

FIGURE VII-22. Rare earth concentrations in marine solid phases over that in seawater (log X/C_{sw}) where X is for seamount manganese crust from Marshall Island, average manganese nodules, and pelagic clays. For comparison, similar ratios for river-suspended particles and a river-water system are given (log X_R/C_R).

The natural partition coefficients (K_d) for REE in pelagic clays, manganese nodules, and seamount crusts relative to seawater, and in river-suspended particles relative to river water, are summarized in Figure VII-22. The systematic decrease of K_d from Eu to Lu may reflect their increasing tendency to form stronger carbonate complexes in solution; thus, they are less reactive to particles (Erel and Morgan, 1991). The exceptionally high K_d for Ce in marine systems indicates its oxidative uptake mechanism.

VII-7. Marine Phosphorite

Marine phosphorite is a sedimentary rock composed mainly of microcrystalline carbonate-fluoroapatite (15 to 20% P_2O_5), sulfide minerals, carbonates, organic matter, and some aluminosilicate detritus minerals. Apatite aggregates can be in the form of nodules, oolites, pellets, laminae, etc. Marine phosphorite is thought to be the end product of diagenetic processes within organic-rich sediments, especially in areas under strong upwellings, e.g., the Namibian shelf and the Peru-Chile continental slope (Bentor, 1980; Burnett and Froelich, 1988).

The average composition of marine phosphorites is summarized in Table VII-6 (Altschuler, 1980), and their E^i_{Al} values relative to pelagic clays are plotted in Figure VI-23. The E^i_{Al} for Si, Al, Fe, K, Na, Mg, Mn, Ti, Ba,

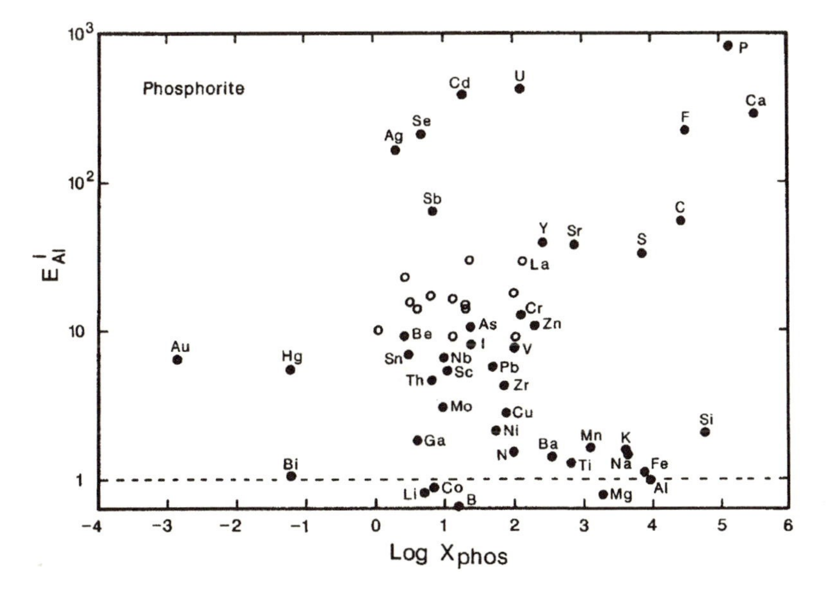

Figure VII-23. Enrichment factors of elements, E^i_{Al}, in average marine phosphorites relative to pelagic clays plotted versus the concentrations of elements in phosphorites (log X_{phos}; Table VII-6). The open circles are REE data.

B, Co, Li, Ga, and Bi are all near one (within a factor of 2); thus, these elements are most likely related to aluminosilicate detritus phases. The moderate enrichments of REE, Y, Zr, V, Cr, Th, Sc, and Nb, and high enrichments of U, Cd, and Ag are related to their displacement of Ca in apatite, facilitated by their similar ionic radii. However, the enrichments of some biophile elements (Se, S, Sb, Zn, Au, Hg, Sn, Mo, Cu, etc.) are probably also related to the sulfide and organic matter phases in the sample. Because phosphorite can accommodate so many trace metals, its usage as fertilizer often causes metal pollution of soils.

VII-8. Hydrothermal Vents of the Mid-Ocean Ridges

Many hot springs have been discovered along the mid-ocean ridge system since 1977 (Von Damm, 1990, and references therein). These **hydrothermal vents** are the manifestation of convective circulation of seawater through the oceanic crust, triggered by the hot magma emplaced in the mid-ocean ridge. The highest exit temperature of these hydrothermal solutions is about 350°C. The chemistry of these solutions is quite variable and is nicely summarized by Von Damm (1990, 1995). Table VII-7 provides one example from the vent named the Hanging Garden (HG) on the mid-Pacific Rise at 21°N. As compared to seawater, the HG vent solution is extremely depleted in Mg, S, P, F, and U; more or less similar in B, Br, Cl, Na, and Sr concentrations; and enriched in other elements from one to six orders of magnitude (Table VII-7). The solution is acidic (pH = 3.3) and anoxic (high sulfide concentration). The alteration of seawater composition to that of the observed vent solution can be explained in the following way.

As shown by Bischoff and Seyfried (1978), and Bischoff and Rosenbauer (1983), the heating of seawater (up to 500°C at 1000 bars) causes the precipitation of anhydrite ($CaSO_4$) and magnesium-hydroxysulfate-hydrate [$MgSO_4 \cdot (1/4)Mg(OH)_2 \cdot (1/2)H_2O$] and results in high H^+ and low Ca^{+2}, Mg^{+2}, and SO_4^{-2} in seawater, i.e.,

$$Ca^{+2} + SO_4^{-2} \leftrightarrow CaSO_4,$$

$$\frac{5}{4}Mg^{+2} + SO_4^{-2} + H_2O \leftrightarrow MgSO_4 \cdot \frac{1}{4}Mg(OH)_2 \cdot \frac{1}{2}H_2O + \frac{1}{2}H^+,$$

$$\frac{5}{4}Mg^{+2} + CaSO_4 + H_2O \leftrightarrow MgSO_4 \cdot \frac{1}{4}Mg(OH)_2 \cdot \frac{1}{2}H_2O + \frac{1}{2}H^+ + Ca^{+2}.$$

TABLE VII-7

Concentrations of elements in the Hanging Garden hydrothermal vent at 21°N (C_{HG}) and comparisons with those in seawater (C_{SW}, Table VII-1), the mid-oceanic ridge basalts (X_{MORB}, Table V-5), and river water (C_R, Table VI-9)

	C_{HG} (mg/l)	C_{HG}/C_{sw}	X_{MORB}/C_{HG}	$(C_{HG}-C_{sw})/C_R$		C_{HG} (mg/l)	C_{HG}/C_{sw}	X_{MORB}/C_{HG}	$(C_{HG}-C_{sw})/C_R$
Ag	0.004	1600	8	13	P	0.018	0.28		−2
Al	0.12	400	680000	2.4	Pb	0.075	27000	6.5	750
As	0.035	20		20	Rb	2.8	23	0.46	1800
B	6	1.3		83	S^{+6}	13	0.014		
Ba	>1.5	>100	<9.3	25	S^{-2}	270		3	−170
Be	120E-6	570	13000	13	Se	0.0048	33	42	23
Br	69	1		100	Si	450	180	530	90
C	71	2.5		3.7	Sr	5.8	0.74	19	−33
Ca	480	1	170	2.3	Th	0.3E-6	6	620000	0.006
Cd	0.02	260	7	1000	U	~0	~0		−13
Cl	18000	0.96		−100	Zn	6.9	21000	12	9900
Co	0.013	11000	3600	65	REE				
Cs	0.028	90	0.89	790	Ce	1640E-6	1000	7400	20
Cu	2.8	13000	26	1900	Nd	500E-6	120	22000	13
F	0.14	0.11		−12	Sm	137E-6	160	28000	17
Fe	139	540000	600	3500	Eu	275E-6	1300	4700	270
K	950	1.4	0.94	370	Gd	92E-6	70	56000	11
Li	9.4	52	0.96	770	Dy	69E-6	46	91000	8.4
Mg	0	0		−420	Er	35E-6	27	120000	
Mn	49	680000	28	6000	Yb	33E-6	22	120000	7.9
Na	10400	0.96	2	−75	pH	3.3			

Sources: HG vent, Von Damm et al. (1985), except for Th and U (Chen et al., 1986) and REE (Michard and Albarebe, 1986).

If the seawater is in direct contact with basaltic rock at high temperature, some additional reactions become possible. For example,

$$7Mg^{+2} + 10H_2O + 5CaMg(SiO_3)_2$$
$$\leftrightarrow 5Ca^{+2} + 4H^+ + \underset{\text{quartz}}{2SiO_2} + \underset{\text{hypothetical chlorite}}{Mg_6Si_8O_{20}(OH)_4 \cdot 6Mg(OH)_2}$$

in the case of high seawater/rock ratio; and

$$2Mg^{+2} + \underset{\text{diopside}}{3CaMg(SiO_3)_2} + 2SiO_2 + 2H_2O$$
$$\leftrightarrow Ca^{+2} + 2H^+ + \underset{\text{tremolite}}{CaMg_5SiO_{22}(OH)_2}$$

in the case of low seawater/rock ratio (Mottl, 1983). The Mg in the hypothetical chlorite can be partially substituted by Al and Fe, and Si by Al. The Mg in tremolite can be partially substituted by Fe and Mn, and tremolite becomes actinolite. Furthermore, some SO_4^{-2} in seawater can be reduced to S^{-2} by the oxidation of Fe^{+2} to Fe^{+3} in iron-containing minerals, e.g.,

$$SO_4^{-2} + 12FeSiO_3 \rightarrow S^{-2} + 4Fe_3O_4 + 12SiO_2.$$

The H^+ ions produced by the incorporation of Mg^{+2} into secondary minerals can react with basalt to leach out other trace metal cations (Mn^{+2}, Fe^{+2}, Zn^{+2}, Cu^+, etc.) into the solution. The end result is the observed acidic and anoxic hydrothermal vent solution depleted in Mg^{+2} and SO_4^{-2} and enriched in various elements.

The **extent of leaching** of cations from basaltic rock into the vent solution can be roughly represented by the ratio of the concentration of a given element in the mid-ocean ridge basalt (MORB) to that in the vent solution, as given in Table VII-7 for the HG vent. Note that the plot of $\log X_{MORB}/C_{HG}$ versus I_z shows a nice general linear relationship in Figure VII-24, where C_{HG} is the concentration of a given element in the Hanging Garden vent solution. Therefore, the extent of leaching of cations from the basalt is more or less controlled by how strongly various cations are bonded in the aluminosilicate framework of the basaltic rock. The relatively low $\log X_{MORB}/C_{HG}$ values for Pb, Zn, Cd, and Mn may suggest that the leaching of these elements is enhanced by the formation of polysulfide complexes in the solution (Jacobs and Emerson, 1982). The deviation of Eu from the correlation line is as expected, because Eu can exist in the system as both Eu^{+2} and Eu^{+3}. As compared to seawater and river water, the HG hydrothermal solution is very much enriched in Eu (probably as Eu^{+2}). In contrast, seawater is exceptionally depleted in Ce (Figure VII-25) because of the oxidative uptake of Ce by pelagic clays and manganese deposits.

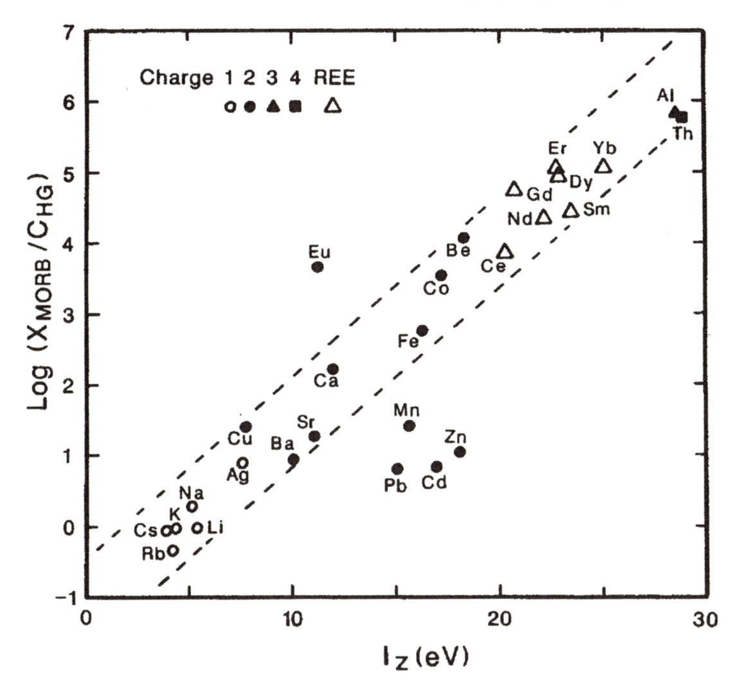

FIGURE VII-24. Elemental concentrations in average mid-ocean ridge basalts over those in the Hanging Garden hydrothermal vent ($\log X_{\mathrm{MORB}}/C_{\mathrm{HG}}$; Table VII-7) versus the electron binding energy I_z.

The relative importance of dissolved elemental fluxes from the mid-ocean ridge hydrothermal vents and from rivers to the ocean is still hotly debated (Elderfield and Schultz, 1996). The critical point is the magnitude of the water fluxes from these sources. Based on the ^{3}He/heat ratio of ridge hot springs (Edmond et al., 1979) and the Sr isotopic mass balance model in the ocean (Palmer and Edmond, 1989), a **global vent water flux** (F_V) of about 0.14×10^{18} cm^3/yr was deduced (as compared to the global runoff, F_{R}, of 37×10^{18} cm^3/yr). If one assumes that the composition of the Hanging Garden vent is representative of the average vent, then elements with a $(C_{\mathrm{HG}} - C_{\mathrm{sw}})/C_{\mathrm{R}}$ ratio of greater than 260 ($= F_{\mathrm{R}}/F_V$) have greater vent inputs than river inputs. Alkali elements (Li, K, Rb, Cs), Fe, Mn, Cu, Zn, Pb, Cd, and Eu fit this criterion (Table VII-7). Negative $(C_{\mathrm{HG}} - C_{\mathrm{sw}})/C_{\mathrm{R}}$ values (i.e., $C_{\mathrm{HG}} < C_{\mathrm{sw}}$) for Mg, F, P, and U indicate that the mid-ocean ridge basalts are a sink for these elements. The $(C_{\mathrm{HG}} - C_{\mathrm{sw}})/C_{\mathrm{R}}$ ratio of -420 for Mg means that the removal rate of Mg in the ridge hydrothermal systems is higher than its input rate from rivers, an unlikely event. Certainly the F_V of 0.14×10^{18} cm^3/yr is an over estimate. For example, the production rate of **new oceanic crust** (F_{MORB}) is about 44×10^{15} g/yr (= expansion rate of ridge area × thickness of

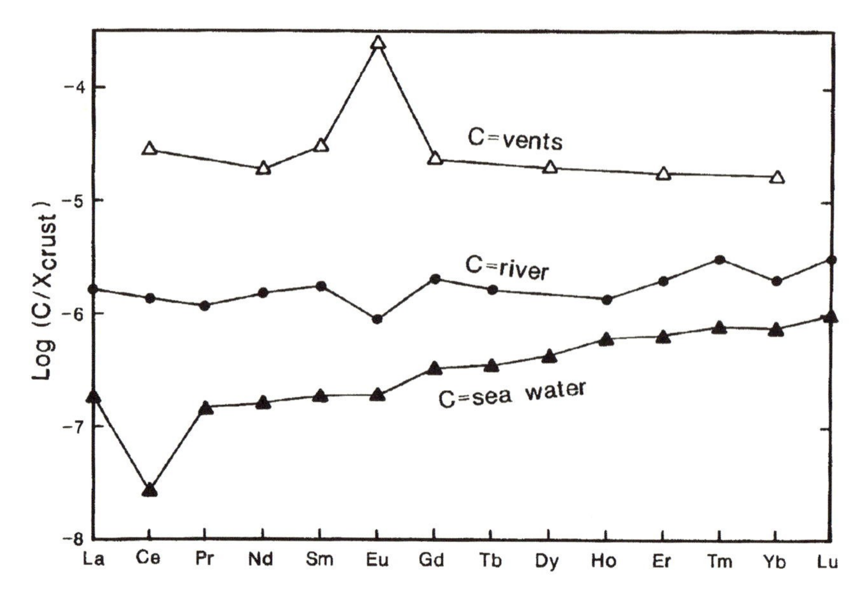

FIGURE VII-25. Concentrations of REE in the Hanging Garden hydrothermal vent, river water, and seawater, normalized to that in the average upper crust.

ocean crust × density of ocean crust = 3.1×10^{10} cm²/yr × 5 km × 2.9 g/cm³) or $F_V/F_{\mathrm{MORB}} = 3.2$. Therefore, the $X_{\mathrm{MORB}}/C_{\mathrm{HG}}$ ratio of less than 3.2 for alkali elements (Table VII-7) means that the vent waters leach out all the alkali elements from the newly formed oceanic crust and more from other sources. Again, this is not a reasonable conclusion. Based on geophysical constraints (Morton and Sleep, 1985) and on the Li isotopic mass balance model in the ocean (Chan et al., 1992), the vent water flux could be as low as 13×10^{15} cm³/yr (an F_{R}/F_V ratio of 2800). Only Fe, Mn, and Zn have $(C_{\mathrm{HG}} - C_{\mathrm{sw}})/C_{\mathrm{R}}$ ratios of greater than 2800 (Table VII-7). This means that the vent fluxes may be comparable to the river fluxes for Fe, Mn, and Zn. Further constraints on F_V values are badly needed. One also cannot ignore many element fluxes through the seafloor during the alteration of the upper oceanic crust (Staudigel et al., 1996).

VII-9. CONCLUDING REMARKS

The observed partitioning patterns of elements between a given marine solid phase (including pelagic clays, deep-sea manganese nodules, seamount manganese crusts, and marine algae and plankton) and seawater can be explained again by the adsorption model of James and Healy (1972). The exceptions are Ce, Co, Mn, Pb, and Tl, which are often concentrated in the solid phase through oxidative uptake processes. Except for elements that reside within the crystal lattice of aluminosilicate detritus phases or form their own authigenic minerals in the ocean (e.g., $CaCO_3$, apatite, barite, ferromanganese oxide coatings, authigenic clays, etc.), the most important removal mechanism of elements from the ocean is the adsorption of elements onto the surface of fine particles, as first suggested by Goldberg (1954) and Krauskopf (1956). In other words, the composition of seawater is mainly controlled by the compositions of source materials (the crustal igneous rocks and primary magmatic volatiles) and the physicochemical principles that regulate the partition of various chemical species between seawater and oceanic sediments.

Since post-Archean time, the composition of Earth's upper crust has probably been quite constant; therefore, one should expect that the average compositions of seawater and the sedimentary rocks also have been quite constant. Recent studies of brine inclusions in marine halite of Permian, Silurian, and Cambrian ages (Das et al., 1990; Horita et al., 1991), indeed, strongly suggest a remarkable constancy of seawater composition over these periods up to the modern time. However, the fugacities of O_2 and CO_2 gases in the atmosphere have changed considerably since post-Archean time (Berner, 1990); thus, the concentrations of dissolved O_2 and CO_2 in the ocean should not be constant.

Chapter VIII

BIOSPHERE AND HOMO SAPIENS

INTRODUCTION

HOW VARIABLE is the chemical composition of diverse life forms? How is the chemical composition of living organisms related to their living environments? This chapter provides some answers to these questions. Taking the human body as an example, this chapter also illustrates the relationships of chemical compositions among the various parts of the human body, human diets, urine, and milk.

In order to understand the effects of fossil fuel burning on the environments, this chapter summarizes the elemental compositions of fossil fuels (including coals, crude oils, oil shales, and black shales) and their major components (aluminosilicate impurities, sulfides, organic matter of different molecular weight, etc.). It also introduces the concept of relative volatility of elements during coal combustion and volcanic emissions to explain the observed enrichment of elements in aerosols from polar regions. Finally, the continental scale effects of fossil fuel burning and human activities on the chemistry of rainwater and river waters are discussed.

VIII-1. ARE ALL CREATURES CREATED EQUAL?

By compiling many elemental composition data for marine and terrestrial plants and animals, Bowen (1966, 1979) noticed a remarkable similarity among them. Figure VIII-1 gives some examples of the good linear relationships among the elemental concentrations in marine brown algae and several other organisms (Li, 1984), based on Bowen's compilations. An immediate question arises as to the quality of these compilations based on diverse data sources and analytical methods. The answer is provided by the so-called biological **standard reference materials** (SRM), which have been prepared by the National Bureau of Standards and carefully analyzed for many elemental concentrations by many laboratories using the best available analytical techniques (Gladney et al., 1987). Compositional data for some biological reference materials are summarized in Table VIII-1, and are

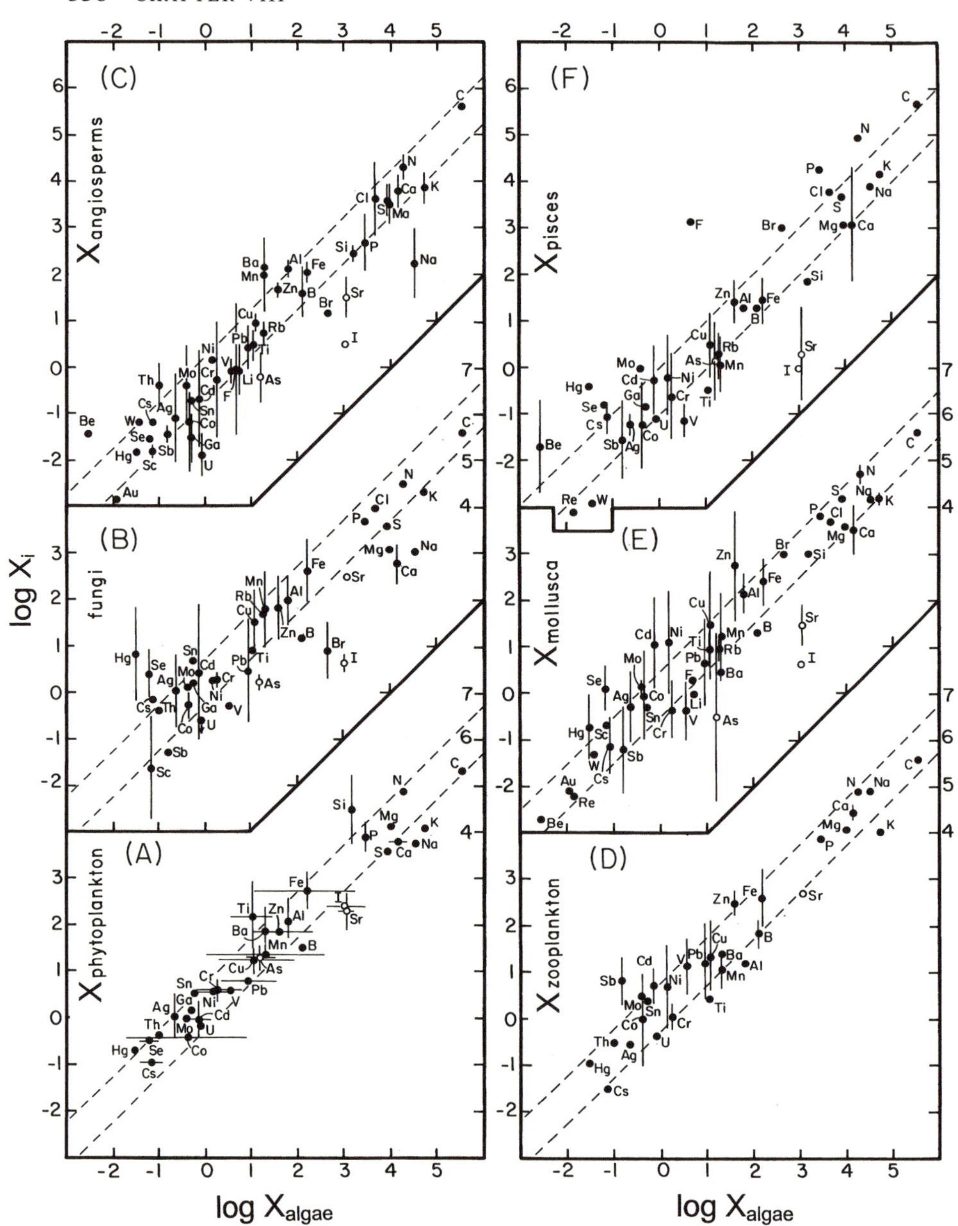

FIGURE VIII-1. Log-log plots of elemental concentration data for marine brown algae against those for (A) marine phytoplankton, (B) fungi, (C) woody angiosperms, (D) marine zooplankton, (E) marine mollusca, and (F) marine fishes (Li, 1984). Horizontal and vertical lines represent the ranges of concentrations given by Bowen (1979).

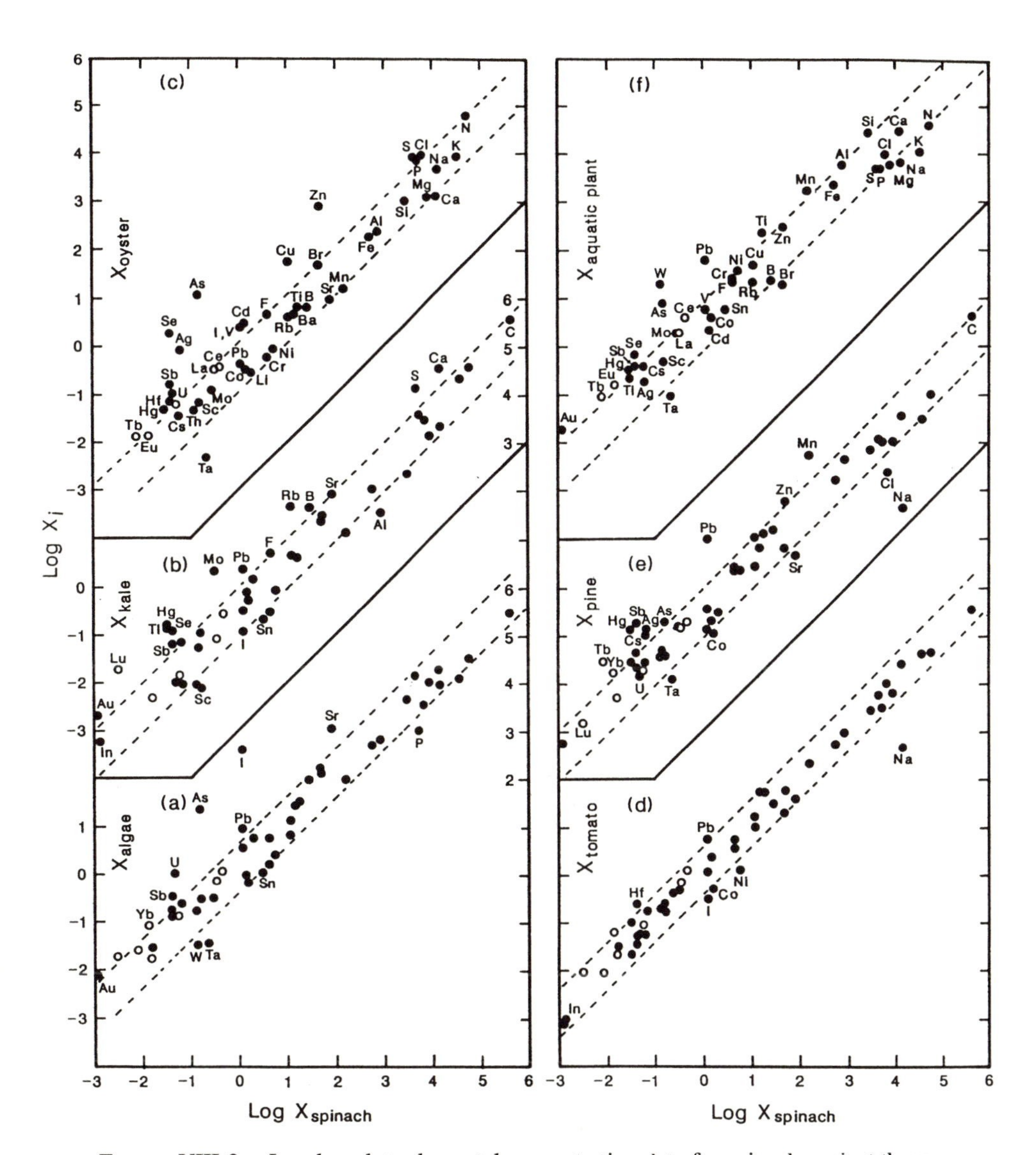

FIGURE VIII-2. Log-log plots elemental concentration data for spinach against those for (a) marine algae, (b) kale, (c) oyster, (d) tomato leaves, (e) pine needle, and (f) aquatic plant, based on Table VIII-1. Open circles are REE data.

TABLE VIII-1

Elemental compositions of various biological standard reference materials

	Kale	*Oyster SRM1566*	*Spinach SRM1570*	*Citrus leaves SRM1572*	*Tomato leaves SRM1573*	*Pine needle SRM1575*	*Aquatic plant SRM8030*	*Bovine liver SRM1577*
Ag-47	0.33	0.94	0.065		0.18	0.15	0.2	0.062
Al-13	40	255	810	76.5	1000	510	6140	16
As-33	0.13	13	0.15	3	0.25	0.21	8	0.055
Au-79 ppb	2.1		1.2	0.11	0.8	0.6	20	2.8
B-5	49	7	28	67	33	17	25	2.9
Ba-56	4.9	5.2	14.9	23.5	57	7.2		0.94
Be-4			0.016	0.007	0.032			0.004
Bi-82								
Br-35	25	53	48	8.4	21	6.9	20	9.1
C-6 %	44.8		40.8		37.8	50.5		50.6
Ca-20	41000	1400	13300	31300	28300	4200	31000	122
Cd-48	0.89	3.43	1.43	0.046	2.5	0.22	2.2	0.28
Ce-58	0.21	0.42	0.46	0.45	1.3	0.21	4	0.02
Cl-17	3600	9900	6600	400	10700	280	10000	2700
Co-27	0.063	0.37	1.56	0.016	0.53	0.12	4	0.23
Cr-24	0.37	0.65	4.3	1	4	2.6	26	0.12
Cs-55	0.076	0.041	0.061	0.093	0.057	0.11	0.4	0.017
Cu-29	4.9	63	12	16	11	3	51	190
Dy-66	<0.024			0.043	0.068			0.0029

Er-68				0.022	0.051			0.0005
Eu-63	0.0067	0.016	0.015	0.014	0.022	0.0055	0.17	0.00033
F-9	5.9	5.2	4.4	4	5.5	2.8	24	0.08
Fe-26	120	195	540	100	580	190	2400	265
Ga-31	0.027				0.076			0.004
Gd-64			0.06	0.039	0.075	0.028		0.0021
Ge-32								
Hf-72	0.013	0.08	0.04		0.25	0.023		0.0042
Hg-80	0.17	0.056	0.03	0.081	0.1	0.14	0.34	0.016
Ho-67			0.008	0.013				0.00025
I-53	0.14	2.8	1.2	1.5	0.32	0.15		0.23
In-49	0.00062		0.0013		0.00096			0.00007
Ir-77								
K-19	24400	9300	35600	18300	44400	3670	11400	9800
La-57	0.087	0.37	0.34	0.2	0.71	0.16	2	0.016
Li-3	1.6	0.32	2	0.23		0.34		0.16
Lu-71	0.02		0.003	0.0016	0.0093	0.0016		0.00004
Mg-12	1600	1330	8650	5600	6850	1220	6030	608
Mn-25	15	17	154	23	224	650	1760	10.2
Mo-42	2.3	0.14	0.3	0.15	0.53	0.15	2	3.2
N-7 %	4.28	6.62	5.6	3.6	4.93	1.2	4.1	1.05

TABLE VIII-1 *Continued*

	Kale	*Oyster SRM1566*	*Spinach SRM1570*	*Citrus leaves SRM1572*	*Tomato leaves SRM1573*	*Pine needle SRM1575*	*Aquatic plant SRM8030*	*Bovine liver SRM1577*
Na-11	2370	4950	14200	160	470	50	6700	2400
Nb-41								
Nd-60			0.31	0.32	0.62	0.16		0.014
Ni-28	0.9	1	5.6	0.72	1.3	2.5	40	0.16
Os-76								
P-15	4480	7600	5240	1310	3370	1190	5140	11300
Pb-82	2.5	0.48	1.2	13.4	5.9	10.7	64	0.35
Pd-46	.002–.026							
Pr-59					0.187			
Pt-78 ppb	0.2?			0.06				
Rb-37	52	4.5	11.5	4.8	17.3	11.7	23	18.4
Re-75								
Rh-45								
Ru-44	0.0045							
S-16	15700	8700	4350	4080	6200	1320	5200	7900
Sb-51	0.069	0.19	0.04	0.034	0.036	0.2	0.4	0.0096
Sc-21	0.0095	0.076	0.17	0.01	0.17	0.041	0.5	0.0009
Se-34	0.13	2.1	0.04		0.053	0.047	0.7	1.1
Si-14	250	1100	2900	1900	3000	814	28500	17.6
Sm-62	0.067	0.07	0.056	0.05	0.092	0.02		0.0016

Sn-50	0.22		3.1	0.24			6	0.018
Sr-38	96	10	80	98	42	5		0.17
Ta-73		0.0055?	0.23		0.43	0.013	0.1	0.003
Tb-65		0.015	0.008	0.009	0.009	0.031	0.1	0.0008
Te-52								0.09
Th-90	0.01	0.052	0.13		0.21	0.04		0.0049
Ti-22	3.3	7.3	18	22	56	14	240	2.7
Tl-81	.15–.53		0.031		0.022	0.029	0.24	0.002
Tm-69	0.012							0.00012
U-92	0.39	0.12	0.046	0.04	0.059	0.016	0.3	0.001
V-23	0.061	2.7	1.2	0.24	1.2	0.39	6	0.058
W-74			0.14	0.0081	<0.04	0.05	20	0.008
Y-39								
Yb-70			0.013	0.012	0.063	0.018		0.00035
Zn-30	32	850	50	30	61	67	310	130
Zr-40								2.3

Sources: All from Gladney (1989) except Kale from Bowen (1979, 1985)

Notes: ? indicates doubtful data. Compositions in ppm unless noted otherwise.

plotted in Figure VIII-2 (see page 357), along with Yamamoto's (1983) marine algae data in Table VII-3. The two parallel dashed lines in each log-log plot of Figures VIII-1 and VIII-2 (and many correlation plots in this chapter) are one order of magnitude apart, enclose as many data points as possible, and have a slope of one. The striking similarity of chemical compositions among **spinach**, **algae**, **kale**, **oyster**, **tomato leaves**, **pine needles**, and **aquatic plants** (with a few exceptional elements) confirms Bowen's original observation. Figure VIII-2a also confirms that the concentrations of Sr, I, As, and probably U in marine algae are relatively high as compared to other organisms (see Figure VIII-1) as noticed earlier by Bowen (1979). Oysters tend to concentrate trace elements such as Zn, Cu, As, Se, and Ag (Figure VIII-2c).

Similar to Figure VII-6 (X_{shale} vs. X_{algae}), the plot of elemental compositions of the average soil against that of spinach (as representative of land plants) also nicely separates out the **biophobe** (solid circles and triangles) and **biophile** (open circles) group elements (Figure VIII-3). However, considering the uncertainty inherent in the data and possible specific enrichment or depletion of certain elements in different organisms (Peterson, 1971), the biophobe and biophile groupings in Figures VIII-3 and VII-6 need not be exactly the same. It is also apparent from Figure VIII-3 that the relative abundances of rare earth elements in spinach are almost identical to those in the average soil.

As discussed in Sections VII-3 and VII-5, the partitioning of elements between algae and seawater and between marine pelagic clays and seawater is mainly controlled by the relative strengths of bonds formed between adsorbing ions and particle surfaces. Therefore, the compositions of pelagic clays and algae are closely coupled through seawater. Similarly, one can argue that the compositions of soils and spinach are also coupled through soil pore waters. Because the average elemental compositions of soil and pelagic clays are similar (Tables VI-5a and VII-6, and Figure VII-13), one can expect the compositions of marine algae and spinach to be similar, as already proven by Figure VIII-2a. Kabata-Pendias and Pendias (1984) have nicely summarized the positive correlation between the concentrations of a given element in plants and in associated soil solution or soil for many elements under well-controlled laboratory conditions, confirming the importance of adsorption during the uptake processes of elements by the growing plants.

Erdman et al. (1976) specifically measured the chemical compositions of many native plants and associated soils to investigate their relationship. Figures VIII-4a and VIII-4b (Li, 1984) are examples of the good linear relationship between the average chemical compositions of trees (buckbush and hickory) and associated soils for the biophile and biophobe group elements separately, as was seen in the case for spinach and soil (Figure VIII-3). However, based on raw data, the correlation coefficients between the concentrations of various elements in sumac stems and those of associated

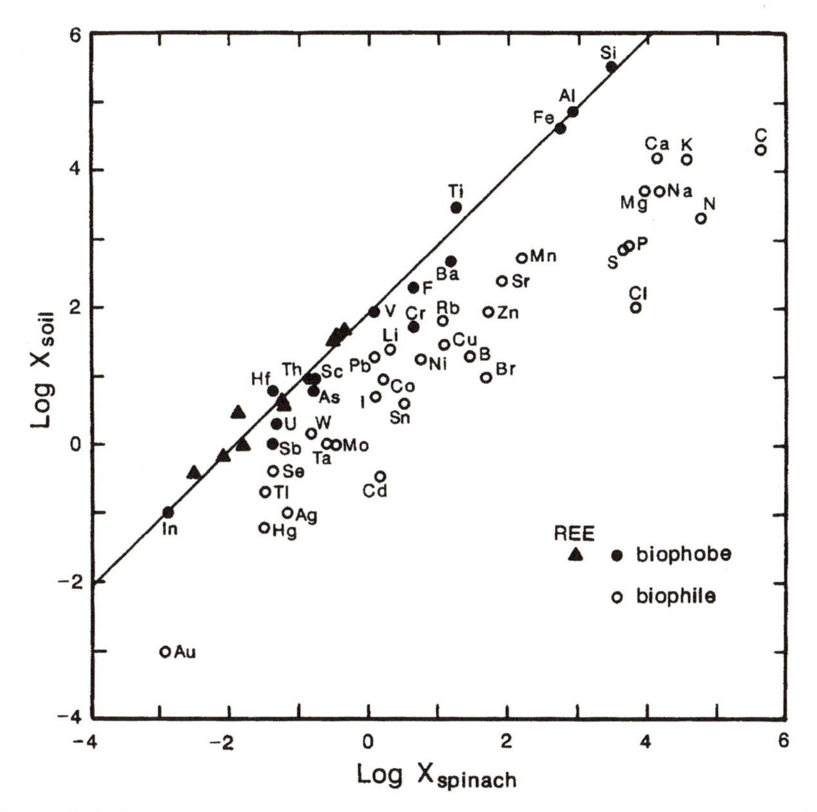

FIGURE VIII-3. Log-log plot of compositional data for the spinach reference material (SRM1570) versus that for average soil (Table VI-5a).

soils are usually low in the studied areas (Erdman et al., 1976). A possible explanation is that the fraction of each element available to an individual plant in the soil is not necessarily proportional to the total concentration in the soil; it is also dependent on pH, redox condition, and organic ligand concentrations in the soil porewater around the individual plant.

Biological functions of individual elements in living organisms are important research subjects in the expanding field of inorganic biochemistry. The details are beyond the scope of this book. One may refer to recent textbooks by Frausto da Silva and Williams (1991), Cowen (1993), and Frieden (1984). In short, six basic structural elements, H, C, N, O, P, and S, are the building blocks of biological macromolecules and polymers. For example, proteins (polymers of twenty different amino acids) are mostly made from H, C, N, O, and S. Nucleic acids, including deoxyribonucleic acid (DNA) and ribonucleic acid (RNA), are made from H, C, N, O, and P. Polysaccha-

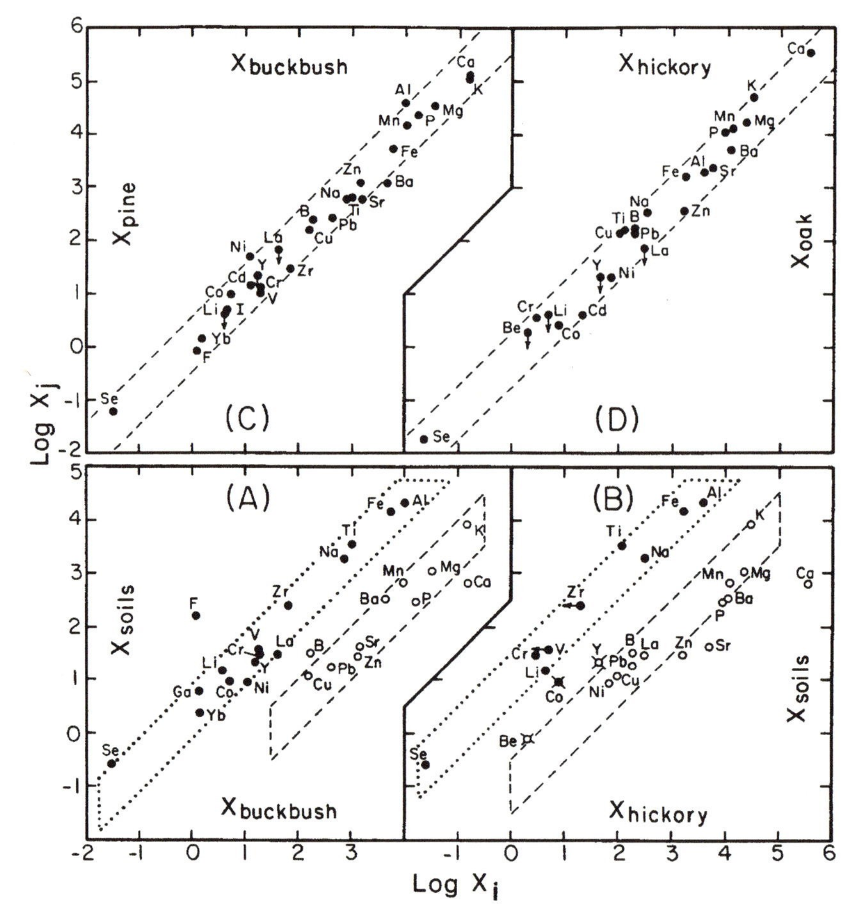

FIGURE VIII-4. Correlation plots of average elemental composition data for (A) buckbush vs. soils, (B) hickory vs. soils, (C) buckbush vs. pine, and (D) hickory vs. oak (Li, 1984). Data points with arrows represent maximum values.

rides (including cellulose, chitin, and starch) and membrane lipids are made from H, C, and O. The elements Mg, Ca, Sr, Ba, Si, and F in addition to H, C, O, P, and S are involved in the formation of the exoskeleton (Mg-rich calcite and aragonite; opal), skeleton, teeth [apatite $Ca_5(PO_4)_3(OH,F)$], and biogenic minerals such as celestite ($SrSO_4$), barite ($BaSO_4$), magnetite (Fe_3O_4), weddellite [$Ca(COO)_2 \cdot 2H_2O$], whewellite [$Ca(COO)_2 \cdot H_2O$] and ferritin [$Fe(OH)_3$], etc.

Na^+, K^+, and Cl^- are major electrolytes in the cytoplasm of cells and circulating fluids (e.g., blood). They are central to the control of osmotic and electrolytic balance across cell membranes, and to the transmission of current for the nervous system. The alkaline earth elements Ca and Mg perform many essential functions for transmitting nerve impulses, triggering muscular contraction, and activating various enzyme systems. The most famous example is the association of Mg with chlorophylls for converting light energy into free-radical energy.

The first-row transition metals (V, Cr, Mn, Fe, Co, Ni, Cu, Zn), As, Se, and the second-row transition metal Mo are essential in acid-base, electron transfer, and redox catalysis reactions in living organisms. Those metals combine with specific proteins to form so-called **metalloenzymes** (strong bonding) or **metal-ion activated enzymes** (weak bonding). Each metalloenzyme has a unique catalytic function. For example, reversible binding of O_2 to Fe^{+2} in hemoglobin and myoglobin (the oxygen binding proteins of blood and muscle), or to Cu^{+} in hemocyanin, is essential for respiratory and metabolic processes. Co is an essential metal in vitamin B_{12}, which affects growth and red blood cell formation. Mo forms so-called molybdoenzymes, which are essential for nitrogen fixation. Furthermore, iodine is essential for the synthesis of thyroid hormones in thyroid glands of vertebrates. Thyroid hormones affect normal development and metabolism in vertebrates.

Biological functions for the ultra-trace second- and third-row transition metals, B-type cations, lanthanides, and actinides are not yet apparent, but some of those elements may some day also prove to perform essential biological functions.

Finally, Table VIII-2 summarizes the **major carbon reservoirs** on the Earth's surface. The living biomass is only a small fraction (10^{-5}) of the total surface carbon reservoir. Moreover, 99% of the living biomass is plant material, with other species making up only 1%, which includes 0.004% for human beings.

TABLE VIII-2
Major carbon reservoirs on the Earth's surface (in unit of 10^{15} gC)

Global sediments		*Ocean*		*Continent*		*Atmosphere (1980)*	
organic	12,500,000	dissolved inorganic	37,400	plants	700	CO_2	712
				animals	1 to 2		
carbonate	65,300,000	dissolved organic	1,000	human	0.03	CH_4	3
				bacteria + fungi	3		
		particulate organic	30	standing dead organic	30	CO	0.2
		biota	3	litter	60		
				peat	160		
				soil organic	1,500		
Subtotal	77,800,000		38,400		2,400		715

Sources: Sediment data are from Table VI-4, and the rest from Bolin (1983) and references therein.

VIII-2. Human Body

Human beings are a part of the biological system. The chemical compositions of various parts of the human body have been studied extensively and the data have been published in many medical journals. Study of the human body provides additional insights into what factors control the chemical compositions of living organisms. The chemical compositions of various parts of the human body have been compiled by Snyder et al. (1975) and Iyengar et al. (1978). However, their data sources are so diverse, it is better to discuss compositional relationships based on internally consistent data sets provided by one or two laboratories such as the Radiological Protection Service (RPS) at Sutton, England, and Oak Ridge National Laboratory (ORNL), United States. Table VIII-3a summarizes the fresh weight of major human organs in a so-called **reference man**, who has a total body weight of 70 kg, along with percentages of dry weights and ash weights in each organ (Snyder et al., 1975). Table VIII-3b gives the fresh weight compositions of selected human organs from the Sutton area near London (Hamilton, 1979, and references therein). The composition of the total soft tissue in the last column was calculated from the relative weights given in Table VIII-3a, assuming that the trace element compositions of adipose tissue (fat), the gastro-intestinal tract, and skin are essentially similar to that of muscle. The data for C, N, and Na are from Snyder et al. (1975).

Table VIII-3a
Fresh weight of various human organs in reference man; and percentages of dry and ash weight on each organ

	Fresh weight (g)	*% of dry weight*	*% of ash weight*
Total body	70000	40	5.3
Total soft tissue	60000	36	0.67
Adipose tissue	15000	85	0.2
Blood	5500	20	1
Brain	1400	21	1.5
GI tract	1200	21	0.83
Kidney	310	23	1.1
Liver	1800	28	1.3
Lung	1000	22	1.1
Muscle	28000	21	1.2
Skin	2600	38	0.69
Others	3190		
Skeleton	10000	67	28
Bone	5000	83	54
Marrow, etc.	5000		

Source: Condensed from Snyder et al. (1975).

TABLE VIII-3B

Elemental compositions of various human organs and the calculated total soft tissue

	Blood	*Brain*	*Kidney*	*Liver*	*Lung*	*Lymph*	*Muscle*	*Total soft tissue*
Ag	0.008	0.004	0.002	0.006	0.002	0.001	0.002	0.027
Al	0.39	0.5	0.4	2.6	18	33	0.5	0.83
As		0.1?	0.3	0.005	0.02	<0.2	0.002	0.0036
B	0.13	0.06	0.6	0.2	0.6	0.6	0.1	0.11
Ba	0.1	0.006	0.01	0.01	0.03	0.8	0.02	0.026
Be		0.00075	0.0002	0.0016				
Bi		0.01	0.4	0.004	0.01	0.02	0.007	0.0083
Br	4.7	1.7	6.5	4	7.5	0.9	4	4.1
C %	9.8 s	12 s	13 s	14 s	10 s		11 s	
Ca	62	57	170	54	120	140	41	45
Cd	0.0052	0.3	14	2	0.48	0.06	0.03	0.17
Ce	0.002		0.013	0.29		0.4		
Cl	3000	1400	330	1400	270	2200	720	940
Co	0.00033							
Cr	0.003	0.01	0.03	0.008	0.5	2.2	0.005	0.013
Cs	0.005	0.007	0.009	0.012		0.02		
Cu	1.2	5.6	2.1	7.8	1.1	0.8	0.7	1.1
F	0.07	0.03	0.01	0.06	0.04	0.09	0.01	0.018?
Fe	490	57	90	210	290	110	31	81
Ga		0.0006	0.0009	0.0007	0.005	0.007	0.0003	0.00036
Hg	0.0078			0.077				
I	0.04	0.02	0.04	0.2	0.07	0.03	0.01	0.019
K	1860	2500	2400	2400	2000	690	2850	2700
La		0.0008	0.003		0.01	0.07		
Li	0.006	0.004	0.01	0.007	0.06	0.2	0.005	0.006
Mg	46	190	210	170	140	180	230	210

TABLE VIII-3B *Continued*

	Blood	*Brain*	*Kidney*	*Liver*	*Lung*	*Lymph*	*Muscle*	*Total soft tissue*
Mn	0.05	0.2	1.3	0.5	0.08	1.1	0.04	0.064
Mo	0.001		0.4	0.4	0.12		0.01	0.024
N %	2.9 s	1.3 s	2.7 s	2.8 s	2.8 s		2.75 s	
Na	1800 s	1800 s	2000 s	1000 s	1800 s		750 s	
Nb	0.005		0.01	0.04	0.02	0.06	0.03	0.027
Ni	0.03	0.4	1	0.2	0.2	0.3	0.2	0.19
P	330	2800	1700	2000	1000	1400	1400	1400
Pb	0.3	0.3	1.4	2.3	0.4	0.4	0.1	0.19
Rb	2.7	4	4	7	3.5	5.5	5	4.8
S	1800	1500	1500	2000	1200	1100	1100	1200
Sb	0.005	0.007	0.006	0.01	0.06	0.2	0.009	0.0094
Se	0.06	0.09	0.1	0.3	0.1	0.05	0.11	0.11
Si	3.9	23	11	15	43	490	4.1	5.5
Sn	0.009	0.06	0.2	0.4	0.8	1.5	0.07	0.086
Sr	0.021	0.08	0.1	0.1	0.2	0.3	0.05	0.053
Th	0.002				0.01	0.2		<0.003
Ti		0.8	0.4	0.4	3.7	8 s	0.2	0.26
Tl	0.0005	<0.001	<0.003	0.009				
U	0.0008	0.0008		0.0008	0.001	0.01	0.0002	0.00029
V		0.03	0.03	0.04	0.1	0.4	0.01	0.012
W	0.001							
Y	0.005	0.004	0.006	0.01	0.02	0.06	0.004	0.0045
Zn	6.7	13	37	57	10	14	39	36
Zr	0.02	0.02	0.02	0.03	0.06	0.3	0.02	0.021

Sources: Hamilton (1979); s: Snyder et al. (1975).

Note: Compositions in ppm unless noted otherwise.

The correlation plots for various organs against muscle compositions (Figure VIII-5) again show their general similarity in composition. The compositions of **brain** and **muscle** are almost identical within a factor of five or better, except for high As in the brain. Compared to muscle, the **blood** is enriched in Fe, Cl, F, and I; the **kidney** in As, Bi, Cd, Mo, Mn, and Pb; and the **liver** in Cd, Cu, Fe, I, Mo, Mn, and Pb. The enrichment of Fe, Si, Al, Ti, Sn, Cr, Sb, V, Mo, Y, Li, Ga, U, Cd, and As in the **lung** indicates the retention of fine aerosol particles in lung tissue. Similarly, the enrichment of Si, Al, Sn, Cr, Mn, Ba, Sb, V, Y, Li, Ga, U, Zr, B, F, and As in

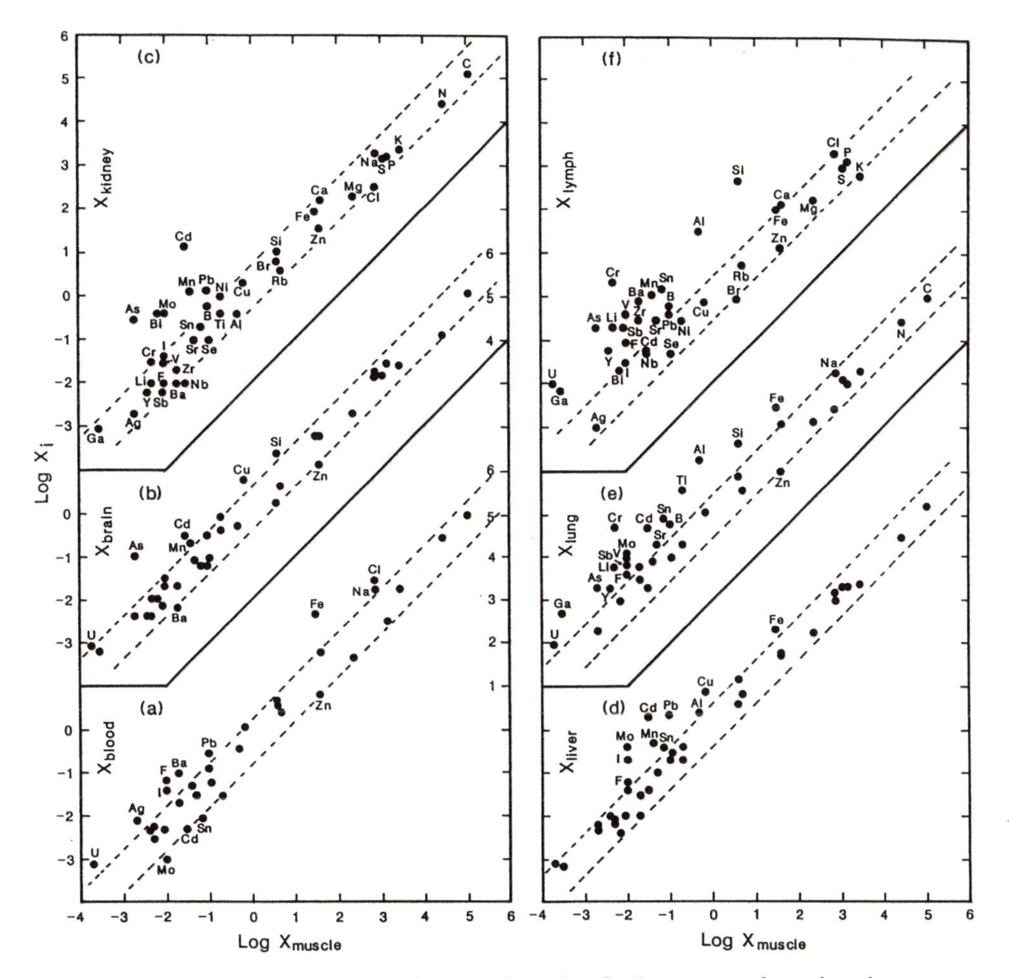

FIGURE VIII-5. Log-log plots of composition data for human muscle against those for (a) blood, (b) brain, (c) kidney, (d) liver, (e) lung, and (f) lymph, based on Table VIII-3b.

lymph nodes indicates the filtration and engulfing of fine alien particles by lymphatic cells and tissue (Figure VIII-5). Also, muscle is enriched in Zn as compared to other organs (Figure VIII-5). It is well known that iodine is mostly concentrated in the **thyroid** (Snyder et al., 1975).

Table VIII-4 summarizes the average compositional data for the total soft tissue, the skeleton, and the reference man given by Snyder et al. (1975), along with data for the average diet of people in the United Kingdom (Hamilton, 1979), **human urine** (Bowen, 1982), and **human milk** from Sweden (WHO/IAEA, 1989). The original data for diet and urine are given in units of g/day and mg/day, which are converted into $\mu g/g$ and $\mu g/cm^3$ by assigning the daily intake of food and discharge of urine to be abut 1.4 kg/d and 1.4 l/d, respectively (Hamilton, 1979). A plot of the total soft tissue data by Snyder et al. (1975) against those by Hamilton (1979) shows their close agreement (Figure VIII-6, see page 373). However, the Zr datum by Snyder et al. (1975) is too high, and the F datum by Hamilton (1979) is too low. The mineral composition of **skeletal bone** is mainly hydroxyapatite, $Ca_5(PO_4)_3OH$. Besides Ca, P, and OH, bone is also enriched in Ag, Al, B, Ba, Be, Cd, Co, Cr, F, Mg, Mo, Ni, Pb, Rb, Si, Sn, Sr, U, Y, and probably REE as compared to soft tissues (Tables VIII-3 and VIII-4). Radiotracer experiments performed on laboratory rats showed similar trends (Durbin, 1960).

As shown in Figure VIII-7a (see page 374), the compositions of the average British diet and spinach (RSM1570) are also similar, except that the diet may be enriched in Se and depleted in Al, Mn, Mg, Sr, and U. The composition of the reference man is also closely related to that of diet (Figure VIII-7b), except that F, Ca, P, Sr, Pb, and Cd are enriched in the human body due to the formation of bone. The fair resemblance among spinach, the average diet, and human body compositions attests to the overall similarity among all organisms, as discussed earlier.

Urine is the waste product of metabolism in the human body. The partition of elements between the human body and urine (X_{man}/C_{urine}) again relates well to the electron binding energies I_z (Figure VIII-8, see page 375), and shows similar features as in Figure VII-8 (X_{algae}/C_{sw} vs. I_z). Apparently, the relative bond strengths of cations and anions to organic matter are an important factor controlling the chemistry of urine.

The mean residence time (τ) of a given element in the human body can be defined as

$$\tau = \frac{M \cdot X_{man}}{F_{urine} \cdot C_{urine}} = 50\frac{X_{man}}{C_{urine}}, \qquad \text{(VIII-1)}$$

where M = body weight of the reference man = 7×10^4g; F_{urine} = discharge rate of urine by the reference man = $1.4 \times 10^3 cm^3/day$; X_{man} and C_{urine} = concentrations of a given element in the reference man (g/g) and urine (g/cm^3), respectively.

TABLE VIII-4
Elemental compositions of the reference man and related organic samples

	Total soft tissue (1)	*Skeleton (1)*	*Reference man (1)*	*Average diet (2)*	*Urine (3)*	*Milk (4)*	*Residence time years (days)*
Ag	0.013	(0.59 h)	0.04 h	0.02	0.0007		7.8
Al	0.67	(36 h)	2.6 h	1.7	0.07	0.33 s	5.1
As	0.3?	0.01?	0.26?	0.036	0.035	0.0005 s	1
B	0.23	0.74	0.3	2	0.71		(21)
Ba	0.03	2	0.31	0.44	0.018		2.4
Be	0.00045	0.001	0.0005	<0.011	<0.00007		
Bi				0.0036			
Br	2.8	2.8	2.9	6.1	2.8		(52)
C %	23	25	23	21 s	0.34 s	0.61 s	(260)
Ca	230	1.0E+5	14000	990	150	240	13
Cd	0.63	1.2	0.71	0.046	0.0007?		
Cl	1400	1400	1400	3900	3100	400	(23)
Co	<0.02	0.28	0.021		0.0007	0.00027	4.2
Cr	0.03	0.48	0.094	0.23	0.0007	0.0015	18
Cs	0.023	0.016	0.021	0.0094	0.007		(150)
Cu	1.1	0.72	1	2.2	0.035	0.19	3.9
F	0.48	250	37	0.36	1.2	0.017	4.2
Fe	55	81	60	17	0.14	0.45	59
Ge				0.27	1		
Hg	0.22?			<0.012	0.003	0.0033	
I	0.22	(0.16 h)	0.19	0.16	0.12	0.056	(79)
K	2000	1500	2000	2000	2000	550	(50)
Li	0.01	(0.024)		0.08	0.5?		
Mg	130	1100	270	180	81	34	(170)

TABLE VIII-4 *Continued*

	Total soft tissue (1)	*Skeleton (1)*	*Reference man (1)*	*Average diet (2)*	*Urine (3)*	*Milk (4)*	*Residence time years (days)*
Mn	0.12	0.52	0.17	2	0.0003?	0.0032	
Mo	<0.075	<0.48	<0.13	0.093	0.043	0.0004	(150)
N %	2.5	3	2.6	1 s	1 s	0.52 s	(130)
Na	1100	3200	1400	3400	2300	88	(30)
Nb		(<0.04 h)		0.014	0.3		
Ni	0.088	<0.5	0.14	<0.22	0.007	0.011	2.7
P	1300	70000	11000	1400	640	140	2.4
Pb	0.18	11	1.7	0.23	0.033	0.017	7.1
Rb	7.8	21	9.7	3.2	1.1		1.2
S	2000	1700	2000	680	570	140 s	(175)
Sb		0.2	0.03	0.025	0.0014	0.003	2.9
Se	0.22		0.11 h	0.15	0.035	0.013	(160)
Si			260		7	0.34 s	5.1
Sn	0.1	<1.2	0.24	0.14	0.02		1.6
Sr	0.055	32	4.6	0.62	0.17		2.6
Te	0.14						
Th		0.016 h	0.005 h		0.0001		6.8
Ti	0.15			0.58			
Tl				<0.0014			
U	0.00051	0.0059	0.0013	0.00071	0.0002		(330)
V	<0.3				0.00013		
W					0.005		
Y		(0.038 h)		0.012			
Zn	30	48	33	10	0.31	0.7	15
Zr	7?	(<0.05 h)		0.038			

Sources: (1) Snyder et al. (1975), (2) Hamilton (1979), (3) Bowen (1982), (4) WHO/IAEA (1989), h: Hamilton (1979), s: Snyder et al. (1975).

Notes: ? indicates questionable data. The parentheses in the skeleton column represent data for bone. The residence time for carbon is calculated from the respiration rate of 270 gC/day for the reference man. Units are μg/g for solids and $\mu g/cm^3$ for liquids.

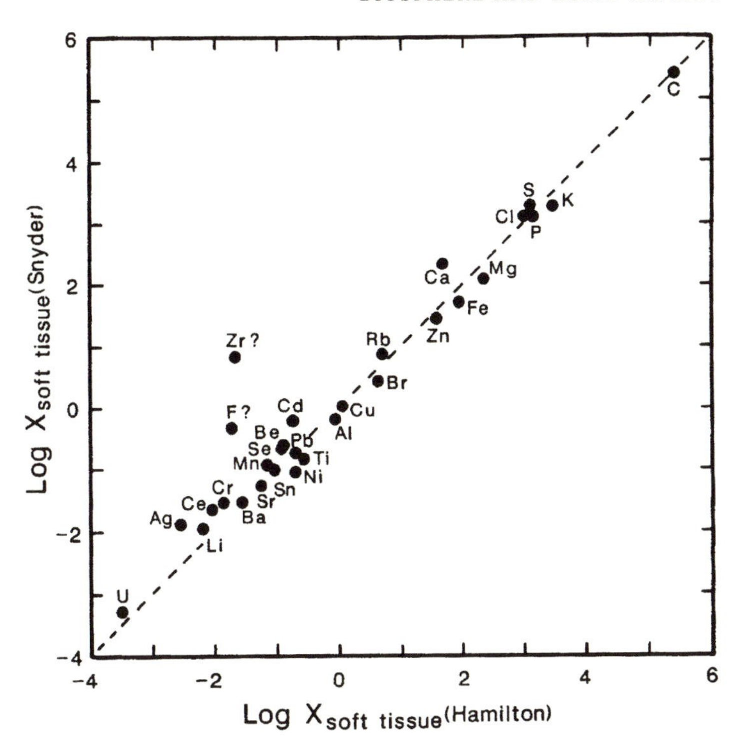

FIGURE VIII-6. Log-log plot of the total soft tissue composition of reference man based on data by Hamilton (1979) and Snyder et al. (1975) (Table VIII-3b and Table VIII-4).

Therefore, the τ's are directly proportional to X_{man}/C_{urine} ratios and their values are summarized in Table VIII-4 and Figure VIII-8. The exceptionally high τ and X_{man}/C_{urine} values for Ca, F, Sr, Pb, Ag, and Rb (which are enriched in bone) reflect the slow metabolism rate of bone tissue relative to the soft tissue. The residence times given here are also maximum, because the other outputs of elements from the human body through sweat and digestive fluid excretions are ignored. For the calculation of τ for carbon, one also should take into account CO_2 exhalation through the lung (270 g C/day, Snyder et al., 1975). In principle, the residence time can also be derived from the diet intake rate. However, a large fraction of ingested food simply passes through the digestive system without becoming a part of the human body. Daily mass balances between intakes of various elements through food + fluids + air, and outputs through urine + feces + air + sweat + hair are summarized by Snyder et al. (1975, pp. 365–367).

Human breast milk from several countries was studied by the World Health Organization and the International Atomic Energy Agency (WHO/IAEA,

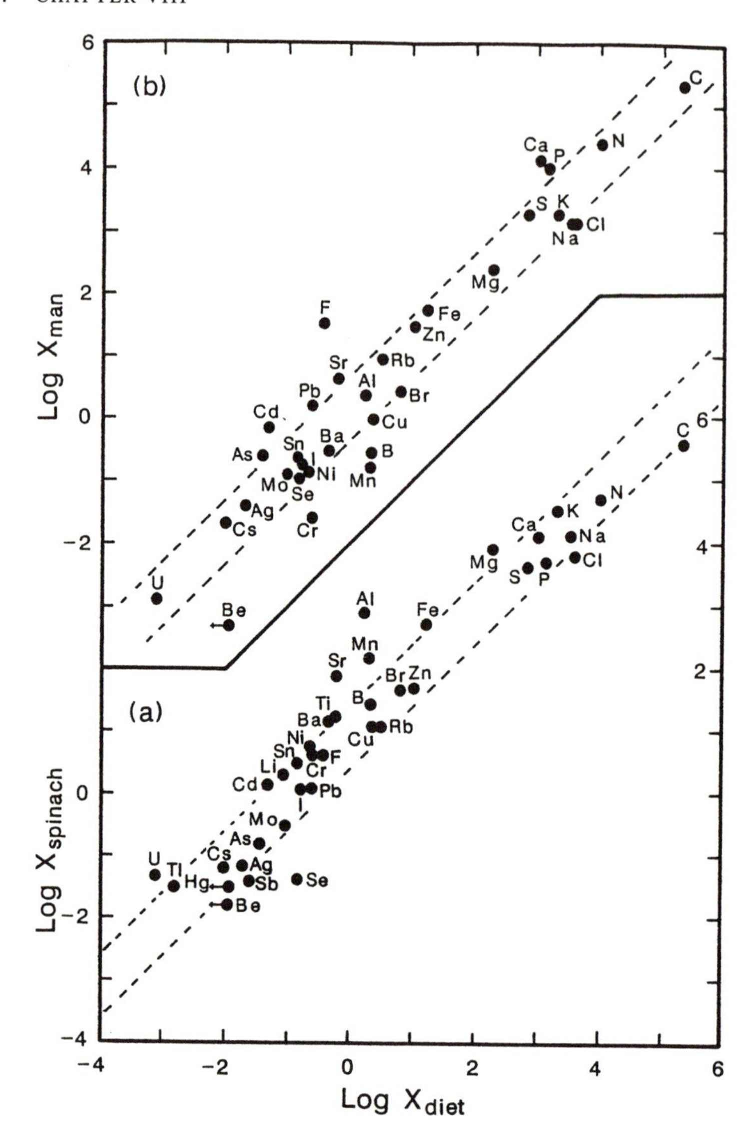

FIGURE VIII-7. Log-log plots of the compositions of average British diet versus those for (a) spinach (SRM1570) and (b) reference man (Table VIII-4). The data points with arrows indicate maxima.

1989). A few trace elements do show some regional variation, but the overall compositions are similar. Table VIII-4 provides the compositional data from Sweden as an example. As shown in Figure VIII-9, the milk composition is similar to that of human muscle, except that Ca, F, and I are much higher in milk.

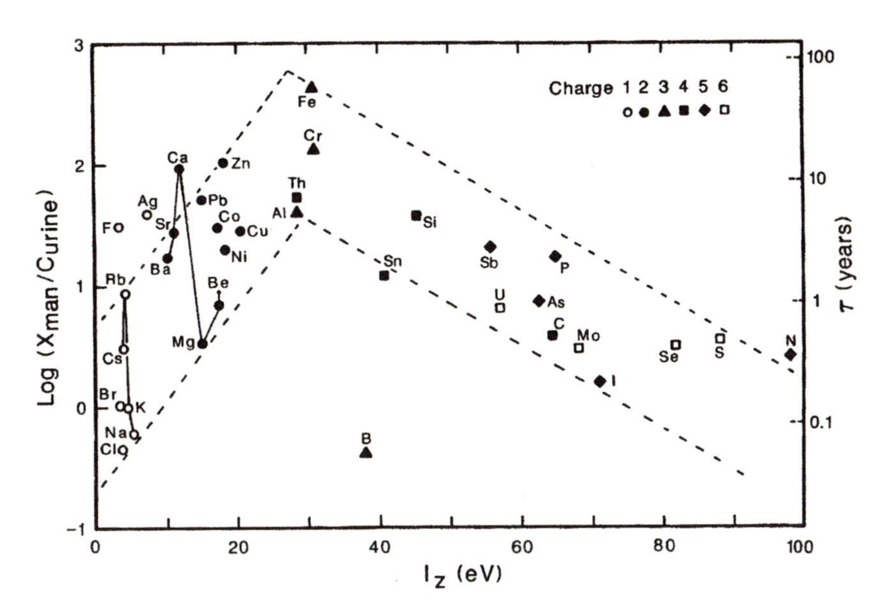

FIGURE VIII-8. Plot of log X_{man}/C_{urine} or the residence times of elements (τ) in the human body versus electron binding energy I, where X_{max} and C_{urine} are the concentrations of elements in the reference man and urine (Table VIII-4).

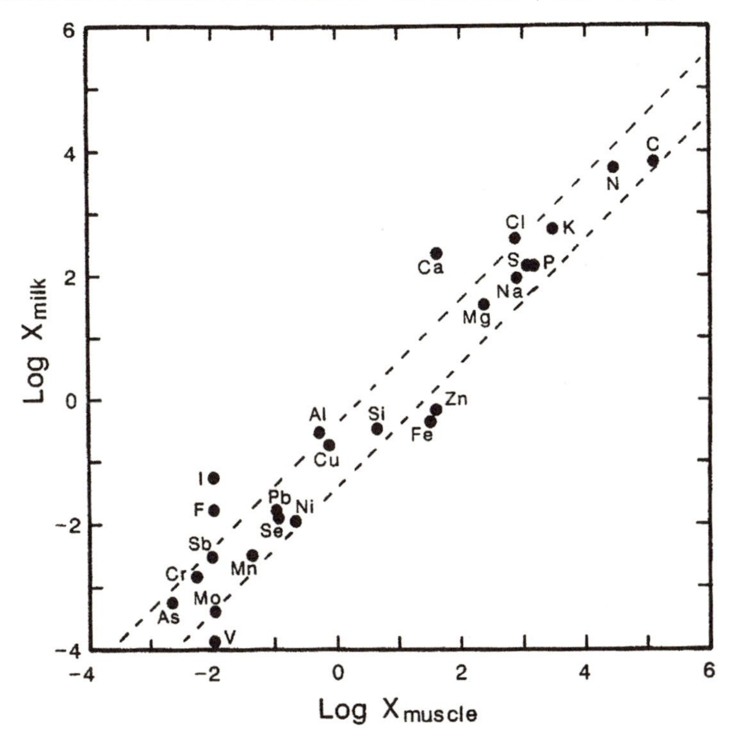

FIGURE VIII-9. Log-log plot of elemental concentrations for human muscle and those for human breast milk.

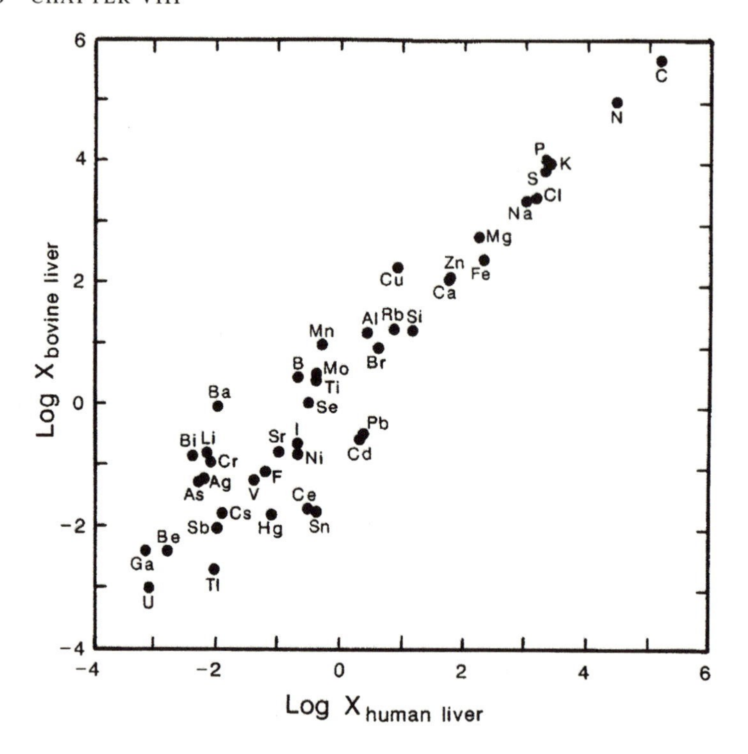

FIGURE VIII-10. Log-log plot of human liver and bovine liver compositions.

Figure VIII-10 shows the similarity in composition between human liver and bovine liver reference samples (SRM1577, Table VIII-1), confirming the regularity in the composition of all living matter.

VIII-3. COALS, CRUDE OILS, AND ORGANIC-RICH SHALES

Coals

Accumulation of terrestrial plant material in swamps and subsequent burial and alteration through pressure and heat over geological time has resulted in coal deposits. The compositions of average coal over the whole world, Illinois coal, and the standard reference coal (SRM1632) are similar (Table VIII-5).

As shown in Figure VIII-11 (see page 380), the elements falling on or near the straight line have E^{i}_{Al} values of near 1 in Illinois coal relative to

TABLE VIII-5

Compositions of various coals, crude oils, and organic rich shales

	Average coal (1)	*Illinois coal* (2)	*Coal SRM1632* (3)	*Average petroleum* (4)	*Alberta crude oil* (5)	*Black shale SDO-1* (3)	*Oil shale SGR-1* (3)
Ag-47	0.05	0.03	0.063	0.0001			
Al-13	10000	12000	17300	0.5		65000	34500
As-33	5	7.4	5.8	0.01	0.11	69	67 g
Au-79	0.0004		0.00092	0.001 b	0.00044		(0.0089)
B-5		98	41	0.002		(130)	54
Ba-56	200	75	330	0.1		400	290
Be-4	1	1.6	1.6	0.0004		3.3	1.1
Bi-82	0.05		1.1				0.94
Br-35	5	10	18		0.49		
C-6 %	80	70	71	85 n		10 a	6.4 a
Ca-20	1500	5100	4200	5		7500	60000
Cd-48	0.2	0.59	0.21	0.01	0.01		(0.93)
Ce-58	12	12	21	0.01		79	36
Cl-17	500	800	880		39		(32)
Co-27	4	6	5.6	0.2	0.054	47	12
Cr-24	10	16	20	0.3	0.093	(66)	30
Cs-55	0.3	1.2	1.5	0.004 b	0.0043	(6.9)	5.2
Cu-29	15	13	17	0.14		(60)	66
Dy-66	2.5	1	1.2			6.0 g	(1.9)
Er-68	0.5		0.7	0.001		3.6 g	1.1
Eu-63	0.5	0.25	0.36	0.005 b	0.00094	(1.6)	0.56
F-9	80	63	80				2000
Fe-26	8000	19000	8500	2.5	11	65000	21000
Ga-31	5	3	5.9	0.01		17	(11)
Gd-64	1.2		1.3			7.4 g	(2)

TABLE VIII-5 *Continued*

	Average coal (1)	*Illinois coal* (2)	*Coal SRM1632* (3)	*Average petroleum* (4)	*Alberta crude oil* (5)	*Black shale SDO-1* (3)	*Oil shale SGR-1* (3)
Ge-32	4	4.8	2.6	0.001			(1.6)
Hf-72	0.9	0.49	0.98			(4.7)	1.4
Hg-80	3?	0.16	0.12	0.05 b	0.051	(0.19)	(0.31)
Ho-67	0.19		0.25			(1.2) g	(0.38)
I-53	1	1.2	3.2		0.72		
In-49	0.02	0.13	0.035				(0.096)
Ir-77	0.005		0.028				
K-19	3000	1600	2800	3 b		28000	14000
La-57	5	6.4	10	0.005		39	20
Li-3	10		26	1 b		29	(150)
Lu-71	0.07	0.08	0.13			0.54 g	(0.14)
Mg-12	2000	500	1600	0.1		9300	27000
Mn-25	50	40	41	0.1	0.1	330	260
Mo-42	3	6.2	3.8	10		130	35
N-7 %	1.5	1.3	1.2				
Na-11	400	300	380	2	3.6	2800	22000
Nb-41	10		5			11	(5.2)
Nd-60	9		9	0.01 b		37	16
Ni-28	10	19	15	10	9.4	100	(29)
Os-76							
P-15	130	45	140			480	1400
Pb-82	10	15	28	0.3		28	38
Pd-46	0.005						(0.0052)
Pr-59	1.8		3.8			8.9 g	3.9
Pt-78			0.23				(0.003)
Rb-37	20	17	21	0.015 b	0.15	(130)	83
Re-75				<0.2 b			
Rh-45							
Ru-44	0.02		0.018				
S-16 %	1.5	3.4	1.3	0.34	0.83	5.4	1.5

Sb-51	1	0.81	3.4	1 b	0.0062	13	3.4
Sc-21	5	2.5	3.8	0.001	0.0078		4.6
Se-34	3	2	3	0.17 b	0.052		(3.5)
Si-14 %	3.0	2.3	3.1			23	13
Sm-62	1.2	1.1	1.6			7.7	2.7
Sn-50	2	0.94	9.3	0.01		(2.9)	(1.9)
Sr-38	150	30	150	0.1		75	420
Ta-73	0.2	0.14	0.25			(1.1)	(0.42)
Tb-65	0.22	0.18	0.28			(1.2)	0.36
Te-52	0.1		0.71				(0.25)
Th-90	2	1.9	3.2			11	4.8
Ti-22	500	600	940	0.1		4300	1600
Tl-81	0.2	0.59	0.55				(0.33)
Tm-69	0.06		0.3			(0.45) g	(0.17)
U-92	1	1.3	1.4	0.01		49	5.4
V-23	20	29	34	50	14	160	130
W-74	0.5	0.63	0.74				2.6
Y-39	7		7.5	0.001		41	(13)
Yb-70	0.37	0.53	0.79			3.4	(0.94)
Zn-30	50	87	37	5 b	0.46	64	74
Zr-40	50	41	34	0.1 b		170	(53)

Sources: (1) Bowen (1979), (2) Gluskoter et al. (1977), (3) Govindaraju (1989), (4) Bertine and Goldberg (1971), (5) Hitchon et al. (1975), a: carbon values include 0.28% and 3.2% carbonate carbon in black shale and oil shale, respectively. b: Barwise and Whitehead (1983) g: Govindaraju (1994), n: Neumann et al. (1981).

Notes: Values in parentheses for geostandards are uncertified and thus may be less reliable. Compositions are in ppm unless noted otherwise.

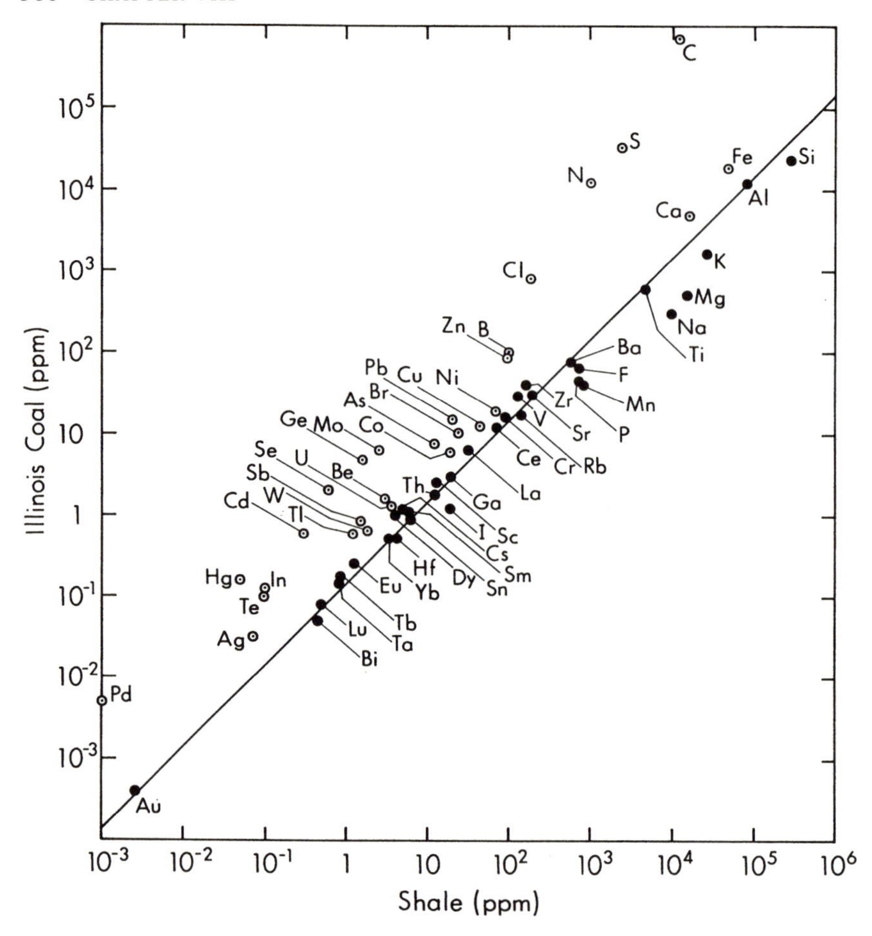

FIGURE VIII-11. Plot of average scale and Illinois coal compositions. The open circles are elements with enrichment factor values of greater than 2 relative to shale.

shale; thus, these elements in Illinois coals are mainly contributed by shale components. Most biophile elements plus Be, Fe, Ge, and U are enriched in coal ($E^i_{Al} > 1$) and are associated mainly with sulfide and organic phases, as will be discussed later.

Ruch et al. (1974) separated bulk coal into different specific gravity fractions and analyzed their elemental concentrations. For example, Table VIII-6 summarizes the ratios of elemental concentrations in different specific gravity fractions to those in bulk coal. A ratio greater than one indicates the enrichment of a given element in a given fraction. The elements B, Br, C, Ge, H,

TABLE VIII-6
Elemental concentrations in different specific gravity fractions of Illinois coal divided by those of bulk coal

	Cleaned coal $\rho < 1.25$	*Alumino-silicates* $\rho = 1.6–2.89$	*Sulfides* $\rho > 2.89$		*Cleaned coal* $\rho < 1.25$	*Alumino-silicates* $\rho = 1.6–2.89$	*Sulfides* $\rho > 2.89$
Ag	0.5			Mo	0.4	1.1	*17
Al	0.1	*3.7	0.6	N	*1.4		
As	1	2.1	*22	Na	0.3		0.7
B	*1.3	*1.6	0.1	Ni	0.3	*2.6	*3.4
Ba	0.5			P	0.3		
Be	0.9	0.6	*1.9	Pb	0.1	2.1	*22
Br	~ *3			Rb	0.4		
C	*1.3			S-organic	*1.0	0.2	0.006
Ca	0.08	*5.4	0.1	S-pyrite	0.2	0.7	*20
Cd	0.06	1.5	*47	S-sulfate	0.5	*7.8	3
Ce	0.5			Sb	0.6	1.5	*6.3
Co	0.4	*4	*5.8	Sc	0.3		
Cr	0.3	*2.4	*1.2	Se	0.3	1.8	*5.7
Cs	0.6			Si	0.1	*5.2	0.6
Cu	0.2	*2.4	*3.6	Sm	0.5		
Dy	0.7			Sn	0.2		
Eu	0.5			Sr	0.4		
Fe	0.4	3.6	*24	Ta	0.3		
Ga	0.4	*3.1	0.5	Tb	0.8		
Ge	*1.2	0.08	0.08	Th	0.4		
H	*1.2			Ti	0.3	*5.9	0.8
Hf	0.6			U	*1.6		
Hg	0.4	4.5	*22	V	0.4	*1.5	*2.2
K	0.2	*5.5	0.3	W	0.4		
La	0.2			Yb	0.3		
Lu	0.2			Zn	0.02	1.8	*48
Mg	0.3			Zr	0.1	*2	*2.1
Mn	0.05	*3.5	0.6				

Source: Ruch et al. (1974),
Note: Asterisk represents the highest value among different fractions.

organic S (S_{org}), and U are enriched in the fraction with specific gravity less than 1.25, representing relatively pure coal organic matter. The elements As, Be, Cd, Co, Cr, Cu, Fe, Hg, Mo, Ni, Pb, Sb, Se, V, Zn, and pyrite sulfur (S_{py}) are enriched in the fraction with specific gravity greater than 2.87, representing sulfide minerals (e.g., pyrite, arsenopyrite, sphalerite, galena chalcopyrite, etc.). The 1.6–2.89 specific gravity fraction is mainly composed of aluminosilicate minerals, and probably some minor minerals such as barite, gypsum, rutile, apatite, and carbonates (Harvey and Ruch, 1986).

Factor analysis of coal composition data for the Herrin No. 6 Coal Member in Illinois (Gluskoter et al., 1977) reveals the following relationships

(Figure VIII-12). The elements Al, Si, REE, K, Rb, Cs, Zr, Hf, Sc, Ga, Ti, Ta, and partly Cr, B, F, and V are related to factor 1, representing aluminosilicate detritus phases. Fe, S, and As as factor 2 are associated with pyrite and arsenopyrite, and the amount of ash in the coal samples is mainly contributed by both aluminosilicate and pyrite phases. The grouping of C, H, N, I, and a part of the total volatile matter (vol) as negative factor 2 represents relatively pure coal organic matter; Cl, Br, and Sb are also related to C and N (see factor 5). The close association of I, Be, and Ge in factor 6 may indicate that Be and Ge are associated with organic matter through the formation of iodine complexes (independent confirmations are needed). The association of organic sulfur (S_{org}), part of B, and total volatile (vol) (see negative factor 5) may suggest that B forms volatile organic sulfur compounds. The grouping

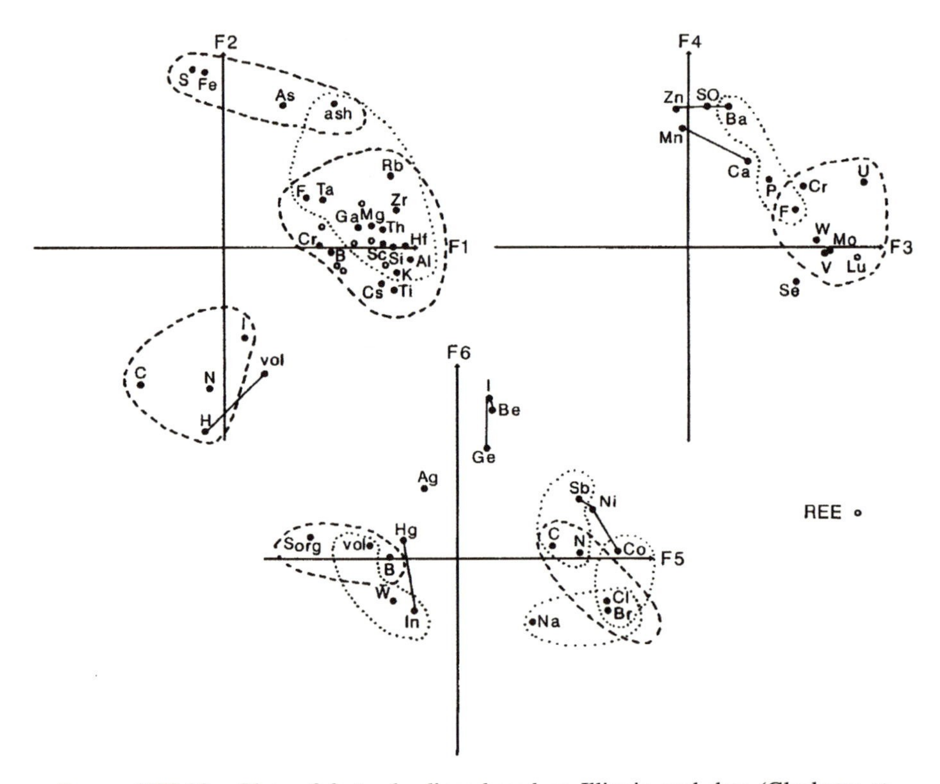

FIGURE VIII-12. Plots of factor loadings based on Illinois coal data (Gluskoter et al., 1977). Elements within each dotted loop or connected by a straight line have correlation coefficients of greater than 0.5 between any pair of them. S, SO, and S_{org} represent, respectively, pyrite, sulfate, and organic sulfur; vol is the total volatile matter. Open circles are rare earth elements.

of vol, W, In, and Hg may represent volatile sulfides; other sulfide phases are represented by Co-Ni-Sb and Mo-W-V-U-Cr-Se associations. The Ba-SO_4-Zn and Ba-P-F associations represent the barite and apatite phases, respectively. The Ca-Mn association may represent carbonate phases.

Crude Oils

Petroleum is a complex liquid hydrocarbon with various impurities. Petroleum is thought to originate mainly from lipidic fractions of organisms and to have undergone various maturation and migration processes through geological time (Eisma and Jurg, 1969). The compositions of average crude oil (Bertine and Goldberg, 1971; Barwise and Whitehead, 1983) and crude oils from Alberta, Canada (Hitchon et al., 1975) are summarized in Table VIII-5 and are plotted against the shale composition in Figures VIII-13a and 13b. The elements nearest the dashed line (solid circles) in Figures VIII-13a and 13b have E^i_{Sc} values of about 1 relative to shale. Therefore, these elements in crude oils are mostly contributed by colloidal clay mineral particles (Yen, 1975). The other enriched elements (open circles with E^i_{Sc} of about 10 to 10^5) in crude oils probably exist as metallo-organic complexes.

Filby (1975) operationally separated crude oil into three components: a methanol-soluble component, a methanol-insoluble but n-pentane-soluble component (resin), and an n-pentane-insoluble component (asphaltene). Each component was further separated into different molecular weight fractions. Neutron activation analysis data for nine elements in various components and fractions are summarized in Table VIII-7 along with other pertinent data. The asphaltene component contains the highest concentrations of all selected elements among the three major components (Table VIII-7). Except for As, the asphaltene and resin components can account for most of the trace elements in crude oil. In contrast, about one-half of the As is in the methanol-soluble component. Except for Ni and Sb, the highest molecular weight fractions of each component have the highest concentrations of other elements among different molecular weight fractions. These elements are probably incorporated into various molecular weight fractions through complexing with functional groups of organic ligands, such as ≡COOH, —NH_2, —SH (Stumm and Morgan, 1981; Filby, 1975). The high Ni concentrations in the low molecular weight fractions of resins and asphaltenes are mostly contributed by Ni-porphyrins, which are essentially the chlorophyll-*a* molecule with the central Mg replaced by a Ni cation. It is well known that VO^{+2} ions also form highly stable vanadyl-porphyrins in crude oils (Filby, 1975). As shown by Hitchon et al. (1975), Ni-V-S-Se concentrations in Alberta crude

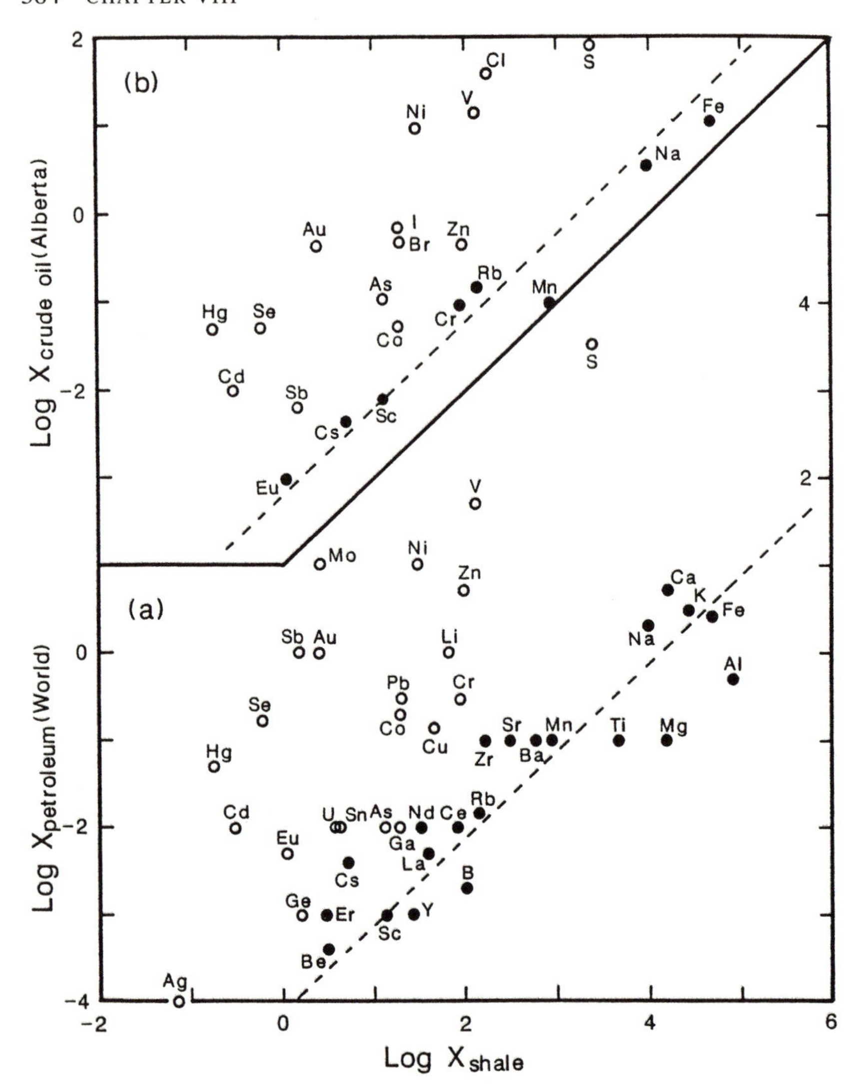

FIGURE VIII-13. Log-log plots of average shale against (a) average petroleum and (b) average Alberta crude oil compositions (Table VIII-5). The solid circles are elements with enrichment factor values around one relative to shale.

oils are highly correlated. The implication is that large portions of Ni and V form strong bonds with ligands (other than porphyrins) containing the functional groups —SH and —SeH in the crude oils. The other highly enriched elements in Figures VIII-13a and b are Au, Br, Cd, I, Mo, and Sb, and their chemical associations in crude oil need further study. The complexation of trace elements with various organic ligands in crude oils is further supported

TABLE VIII-7

Concentrations of trace elements in three major components of Crude Oil C-1 and in various molecular weight fractions of each major component (1975)

Major component	*Molecular wt. fraction (%)*	*Ni*	*Co*	*Fe*	*Hg*	*Cr*	*Zn*	*Cu*	*Sb*	*As*
Crude oil		94	13	73	21	0.63	9.3	< 0.3	0.052	0.66
Methanol soluble	0.3k–1k (93.8)	11	0.73	<1	0.41		< 1		0.0046	0.34
57.5%	> 1.6k(6.2)	<1	2.6	9.9	7.1		3.5		< 0.002	5
	total (100)	7.2	0.8	2	0.89	< 0.3	0.74	< 0.5	0.0033	0.55
Resins	0.3k–1k (29.4)	210	4.4	30	22	0.31	3.3	< 0.2	0.043	0.41
37.5%	1k–4k (21.2)	110	10	24	44	0.8	11	< 0.5	0.0026	0.2
	4k–8k (49.4)	80	25	240	72	3	27	1.3	0.0054	0.2
	total (100)	150	11	66	30	0.89	8.9	0.32	0.013	0.29
Asphaltenes	0.3k–1k (11.0)	1300	2.7	480	72	0.77	110	0.34	11	0.85
5.0%	1k–4k (23.2)	190	30	370	21	4.8	52	1.5	0.91	0.62
	4k–8k (50.6)	980	170	870	90	9.1	100	4	0.35	1.9
	8k–22k (15.2)	1100	180	1900	350	20	230	7.2	0.1	6.6
	total (100)	850	120	900	140	7.5	110	3	1.2	2.3

Sources: Filby (1975),

Notes: Concentrations in $\mu g/g$. k represents a factor of 1000.

by the fact that many trace metals in crude oils can be removed by acid treatment (Yen, 1975).

Organic Rich Shales

The elements C, S, As, Cd, Cu, Fe, Mo, Se, U, and V are enriched in the organic-rich shale of the Pierre Shale member (Section VI-5). Two additional good examples of organic-rich shales in the United States are the Devonian-Mississippian **black shales** of the eastern United States (from New York

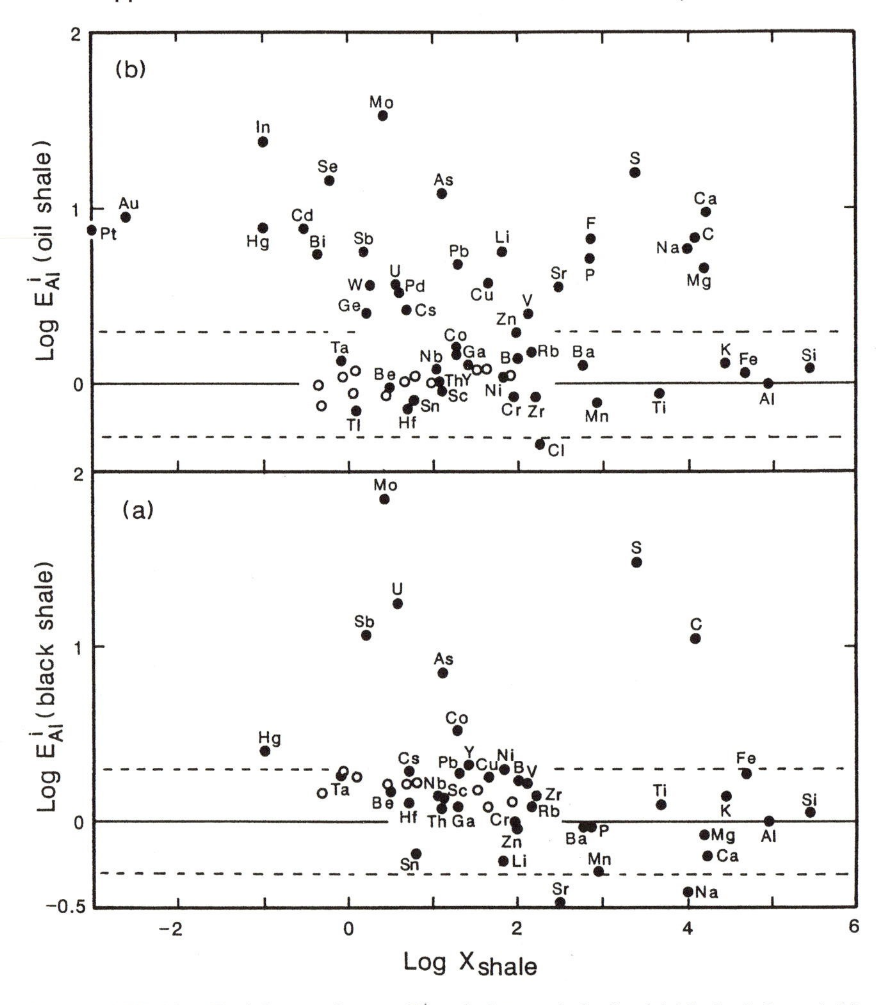

FIGURE VIII-14. Enrichment factors $E^i_{\rm Al}$ relative to shale for (a) black shale and (b) oil shale as plotted against the shale composition. The open circles are REE data.

to Oklahoma and from Iowa to Alabama) and the Eocene **oil shales** of the Green River Formation in Colorado, Utah, and Wyoming, which were formed in lacustrine to saline lake environments (Miknis and McKay, 1983). Their compositions are represented by the black shale geostandard SDO1 of the Huron Member of the Devonian Ohio Shale, and the oil shale geostandard SGR1 of the Green River Formation (Table VIII-5). As shown in Figures VIII-14a and VIII-14b, the elements enriched in the average coal are also concentrated in black shales and oil shales as compared to average shale. The enrichment of Na, Ca, Mg, Sr, Li, F, and P in oil shale represents authigenic sodium carbonate minerals (nahcolite, $NaHCO_3$, shortite, $Na_2CO_3 \cdot 2CaCO_3$, etc.) and apatite formed in a chemically stratified ancient saline lake (Smith, 1983). These organic-rich shales could be important sources for liquid hydrocarbons, uranium, and some trace elements in the future.

VIII-4. Relative Volatility of Elements and Compositions of Aerosol Particles

Coal Burning

It is well known that the introduction of CO_2 into the atmosphere by fossil fuel burning has caused global warming through the greenhouse effect of CO_2 (Broecker and Peng, 1982). Less well known is the effect on the geochemical cycles of elements of introducing various volatile elements into the atmosphere by fossil fuel burning (Lantzy and Mackenzie, 1979). For illustration purposes, the following provides an example of what happened to the elements in coals in a coal-fired electric power plant.

Table VIII-8 summarizes the compositional data of coal from Illinois, boiler slag, and fly ashes collected before and after electrostatic precipitation (i.e., inlet and outlet fly ashes) at the coal-fired Thomas Allen steam plant in Memphis, Tennessee (Klein et al., 1975; Lyon, 1977). In this plant, one gram of coal produced an average of 0.0573 g slag, 0.0503 g inlet fly ash, and 0.00025 g outlet fly ash after complete combustion at high temperature (about 1300 to 1500°C). The following mass balance relationship should hold for element i:

$$\begin{aligned}(X_i)_{\text{coal}} &= 0.0573(X_i)_{\text{slag}} + 0.0503(X_i)_{\text{inlet ash}} \\ &\quad + 0.00025(X_i)_{\text{outlet ash}} + a(X_i)_{\text{flu gas}} \\ &= (X_i)_{\text{sum}} + a(X_i)_{\text{flu gas}}, \qquad \text{(VIII-2)}\end{aligned}$$

where X_i is the concentration of element i in the various given materials, $(X_i)_{\text{sum}} = 0.0573(X_i)_{\text{slag}} + 0.0503(X_i)_{\text{inlet ash}} + 0.00025(X_i)_{\text{outlet ash}}$, and a

TABLE VIII-8

Compositions of coal, slag, inlet fly ash (in), and outlet fly ashes (out) from Thomas Allen steam plant, and logarithms of various ratios

	coal 1g	*slag 0.0573g*	*in 0.0503g*	*out 0.00025g*	*sum*[a]	*(sum/ coal)*	log(*vapor /slag*)[b]	log Ex *(in/slag)*	log Ex *(out/slag)*
As	5	18	110	440	6.7	1.33	0.59	1.01	1.53
Al	10440	102300	90900	76000	10450	1.01	−0.70	−0.21	−1.17
Ba	65	500	744	750	66	1.02	−0.25	0.23	0.06
Br	3.7	2	4		0.3	0.09	1.32	0.42	
Ca	4340	46000	25200	32000	3910	0.90	*c*	*c*	*c*
Cd	0.47	1.1	8	51	0.5	1.02	0.61	1.09	1.82
Ce	8.2	84	84	120	9.1	1.11	−0.57	−0.08	0.02
Cl	914	100	200	170	16	0.02	2.03	0.42	
Co	2.9	20.8	39	65	3.2	1.09	−0.10	0.38	0.54
Cr	18	152	300	900	24	1.33	−0.07	0.42	0.88
Cs	1.1	7.7	13	27	1	1.00	−0.17	0.32	0.61
Cu	8.3	20	140	170	8.2	0.99	0.58	1.07	1.05
Eu	0.1	1.1	1.3	1.3	0.13	1.29	−0.42	0.06	−0.16
Fe	10850	112000	121000	150000	12540	1.15	−0.50	−0.01	−0.03
Ga	4.5	5	81		4.4	0.97	0.97	1.46	
Hf	0.4	4.6	4.1	5	0.5	1.18	−0.69	−0.20	−0.25
Hg	0.122	0.028	0.05		0.004	0.03	1.70	0.35	
K	1540	15800	20000	24000	1920	1.25	−0.37	0.12	0.07
La	3.8	42	40	42	4.4	1.17	−0.62	−0.13	−0.36

Mg	1210	12400	10600	11000	1250	1.03	−0.74	−0.25	−0.56
Mn	33.8	295	298	430	32	0.95	−0.56	−0.07	0.04
Mo	9	21	210	450	12	1.32	0.75	1.24	1.47
Na	696	5000	10100	11300	797	1.15	−0.06	0.43	0.35
Ni	16	85	210	220	15	0.97	0.06	0.55	0.43
Pb	4.9	6.2	80	650	4.5	0.93	0.88	1.35	2.18
Rb	15.5	102	155	190	14	0.88	−0.24	0.25	0.22
Sb	0.5	0.64	12	55	0.7	1.30	1.04	1.52	2.09
Sc	2.2	20.8	26	36	2.5	1.14	−0.38	0.11	0.17
Se	2.2	0.08	25	88	1.3	0.58	2.51	2.76	3.20
Si	23100	229000	196000		22980	1.00	−0.74	−0.25	
Sm	1	8.2	10.5	9	1	1.00	−0.36	0.13	−0.24
Sr	23	170	250	429	22	0.97	−0.26	0.23	0.42
Ta	0.11	0.95	1.4	1.8	0.13	1.14	−0.26	0.23	0.24
Th	2.1	15	20	26	1.9	0.90	−0.33	0.16	0.17
Ti	506	4100	5980	10000	538	1.06	−0.27	0.22	0.40
U	2.18	14.9	30.1	95	2.4	1.10	−0.05	0.43	0.91
V	28.5	260	440	1180	37	1.31	−0.16	0.32	0.74
Zn	46	100	740	5900	44	0.97	0.63	1.10	1.92

Source: Klein et al., (1975).

Notes: [a]sum = equation VIII-2 in the text; [b]vapor/slag = equation VIII-4a for Br, Cl, Hg, and Se, and equation VIII-4b for the rest; [c]Ca is used for normalization. All compositions in ppm.

is the unknown amount of flu gas produced. For most of the elements, $(X_i)_{sum}/(X_i)_{coal}$ ratios are within the range of 1.0 ± 0.2 (Table VIII-8); therefore, the amount of those elements in the flu gas fraction should be negligibly small. For some elements (As, Cr, Eu, Mo, Sb, and V), this ratio is 1.3, probably indicating about 30% extra input from corrosion of boiler machinery, and/or reflecting analytical uncertainty (Klein et al., 1975). The $(X_i)_{sum}/(X_i)_{coal}$ ratio is 0.56 for Se and much less for Br, Cl, and Hg (probably also for I, F, S, Te, and N). Therefore, a large fraction of these volatile elements must be lost as flu gas through the chimney.

The fly ash may consist of two components: one is a refractory component that has a chemical composition similar to that of slag, and the other is a volatile component that condenses from the gas phase onto particles as the gas temperature cools. One may define the excess concentration of an element i in fly ash, $(X_i^{ex})_{ash}$, as the volatile component, i.e.,

$$(X_i^{ex})_{ash} = (X_i)_{ash} - (X_j)_{ash}(X_i/X_j)_{slag}, \qquad \text{(VIII-3a)}$$

where $(X_j)_{ash}$ is the concentration of the most refractory element j in the ash, i.e., the element with the lowest $(X_i)_{ash}/(X_i)_{slag}$ value, assuming that the most refractory element in the ash has no or negligible volatile component.

Dividing equation VIII-3a by $(X_j)_{ash}$ and by $(X_i/X_j)_{slag}$, one obtains

$$(X_i^{ex}/X_j)_{ash}/(X_i/X_j)_{slag} = (X_i/X_j)_{ash}/(X_i/X_j)_{slag} - 1, \qquad \text{(VIII-3b)}$$

or

$$\mathrm{Ex}_j^i(\mathrm{ash/slag}) = E_j^i(\mathrm{ash/slag}) - 1, \qquad \text{(VIII-3c)}$$

where Ex_j^i(ash/slag) and E_j^i(ash/slag) are, respectively, the enrichment factor of excess element i and that of element i in ash as normalized by element j and slag.

For abbreviation, we simply use Ex and E, whenever it is self-evident which are the reference and normalizing elements. For volatile elements, Ex is always much larger than one; therefore, Ex is almost equal to E. However, for less volatile elements, the difference between Ex and E values can be significant.

For elements with $(X_i)_{sum}/(X_i)_{coal}$ about 1 ± 0.2, the partition of element i between vapor phase (i.e., the volatile component in fly ash) and slag after burning 1 g of coal can be defined as

$$\begin{aligned}\mathrm{vapor/slag} = \{&0.0503[(X_i)_{in} - (X_i/X_j)_{slag}(X_j)_{in}] \\ &+ 0.00025[(X_i)_{out} - (X_i/X_j)_{slag}(X_j)_{out}]\}/D, \qquad \text{(VIII-4a)}\end{aligned}$$

where $D = [0.0573 + 0.0503(X_j)_{in}/(X_j)_{slag} + 0.00025(X_j)_{out}/(X_j)_{slag}]\,(X_i)_{slag}$, and the subscripts "in" and "out" represent inlet fly ash and outlet fly ash, respectively. For volatile elements such as Br, Cl, Hg, and Se with $(X_i)_{sum}/$

$(X_i)_{\text{coal}}$ less than 0.6, one needs to use the equation

$$\text{vapor/slag} = [(X_i)_{\text{coal}} - D]/D. \qquad \text{(VIII-4b)}$$

The higher the Ex, E, and vapor/slag values, the higher the volatility of elements during coal burning. The logarithmic values for Ex^i_{Ca} (inlet fly ash/slag), Ex^i_{Ca} (outlet fly ash/slag), and vapor/slag are summarized in Table VIII-8. As shown in Figure VIII-15, there are nice linear relationships

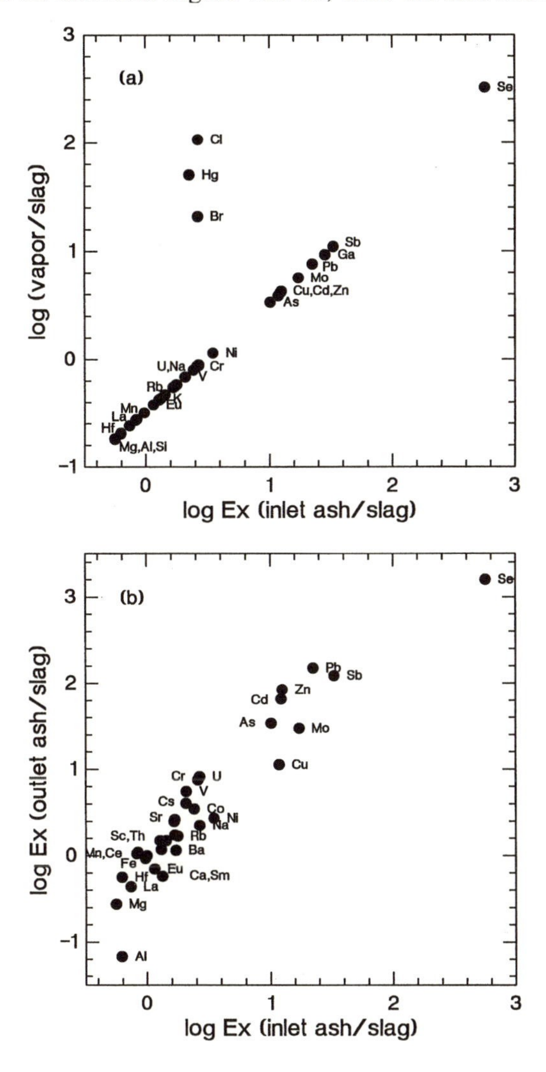

FIGURE VIII-15. Log Ex(inlet fly ash/slag) plotted versus (a) log (vapor/slag) and (b) log Ex(outlet fly ash/slag) for data from the coal-fired Thomas Allen steam plant.

among log(vapor/slag), logEx^i_{Ca} (inlet fly ash/slag), and log Ex^i_{Ca} (outlet fly ash/slag). The obvious exceptions are Br, Cl, and Hg. For consistency, we adopt the log(vapor/slag) values to represent the relative volatility scale of elements during coal burning processes. From the relative volatility scale, one can tell which elements will tend to escape into the atmosphere during coal burning.

Thermodynamic Equilibrium Model of Coal Combustion

The relative volatility scale obtained in the last section is purely empirical. The next question is does this scale make sense thermodynamically?

Let us consider the case that one ton of coal with the composition of average Illinois coal (Table VIII-5; Gluskoter et al., 1977) is burned at pressure of one bar and at a temperature of 1500°C in an isolated system. Enough oxygen is provided to convert all C, H, and S in the coal to CO_2, H_2O, and SO_2 gases and other major elements into solid oxides. The question is how these major elements from coal partition between gas and slag (assumed to be glass or ideal solid solution) phases at equilibrium. To answer this question, one may adopt the *HSC Chemistry PC* program (Roine, 1994), which can calculate the equilibrium composition of a given system, using the Gibbs energy-minimization method and the thermochemical database provided (including more than 7600 compounds).

The first input is the molar amounts of major elements in one ton of Illinois coal in oxide form (i.e., CO_2, H_2O, SO_2, CaO, FeO, Al_2O_3, K_2O, Na_2O, MgO, and SiO_2), Cl in HCl form, and F in HF. The output result is the molar amounts of major species of each major element in gas and glass phases. The obtained partial pressure of oxygen is about 10^{-3} bar. Next, along with major species of major elements, we add the molar amount of a chosen minor element in one ton of coal to the system, and obtain the molar amount of major species of this minor element in gas and glass phases. One then replaces the minor element with another minor element and repeats the calculation. The major species of each element in gas and glass phases, and the logarithm of the ratio of the total molar amount of each element in the gas phase to that in the glass phase, log (gas/glass), are summarized in Table VIII-9 (see page 394). Most of the elements in the glass phase exist as oxide and silicate components. The exceptions are halides (as halogenites of Na and K), and Cs, Rb, Ag, and Tl (as chlorides). In the gas phase, the major components are oxide, chloride, and metal species. Only Sb exists as sulfide gas. Thermodynamic data for noble metal compounds are usually not sufficient to obtain meaningful results. Therefore, the results for noble metals other than Rh are not listed in Table VIII-9. The relative volatility of noble metal elements Re and Os is probably comparable to those for halide group elements (Finnegan et al., 1990). It is well known that platinum group elements at high temperatures can form gaseous halide compounds (Cotton and Wilkinson, 1988). In short, the parameter log(gas/glass) is another useful thermodynamic scale representing the relative volatility of elements during coal combustion at high temperature. Because the elements in the glass phase exist mostly as oxide and silicate components, one expects a close relationship among log(gas/glass), the sublimation heats of oxides, and the boiling points of oxides for most of the elements, as proven in Figure VIII-16.

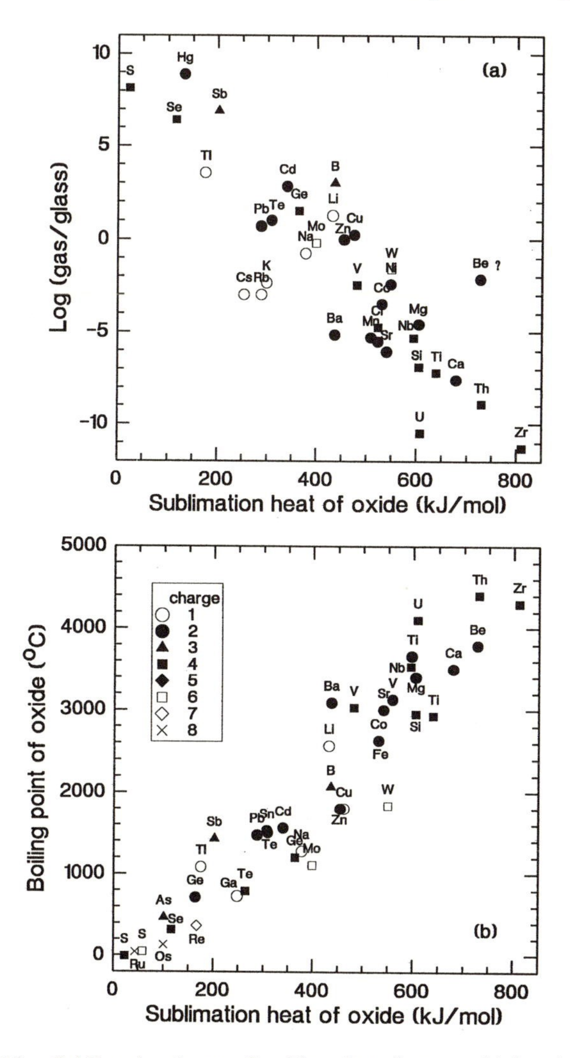

FIGURE VIII-16. Sublimation heats of oxides plotted versus (a) log (gas/glass) and (b) the boiling points of oxides for various elements.

TABLE VIII-9

Partition of elements between gas and glass phases [log(gas/glass)], and major species in gas and glass phases during the burning of average coal at 1500°C and one bar pressure

	log *(gas/glass)*	*Major species of elements in the gas and glass phases*		log *(gas/glass)*	*Major species of elements in the gas and glass phases*
Ag	3.14	Ag(g), AgCl(g); AgCl	Mn	−5.30	MnO(g), $MnCl_2$(g); $MnSiO_3$
Al	−5.60	$AlOF_2$(g); Al_2O_3, feldspars	Mo	−0.2	H_2MoO_4(g); $CaMoO_4$, Na_2MoO_4
As	1.28	AsO(g); $AlAsO_4$	Na	−0.76	NaCl(g); $NaAlSi_3O_8$
B	2.99	HBO_2(g), $NaBO_2$(g); CaB_2O_4	Nb	−5.33	NbO_2(g); NbO_2, Nb_2O_5
Ba	−5.16	$Ba(OH)_2$(g); $BaSiO_3$	Ni	−2.43	$Ni(OH)_2$(g); NiO
Be	−2.15	$Be(OH)_2$(g); BeO, $BeAl_2O_4$	P	−2.05	PO_2(g); $Ca_3(PO_4)_2$
Bi	2.25	Bi(g); BiO	Pb	0.69	PbO(g); $PbSiO_3$, PbO
Br	2.26	Br(g), NaBr(g); KBr, NaBr	Rb	3.94	RbCl(g); RbCl
Ca	−7.59	$Ca(OH)_2$(g); $CaSiO_3$, $CaAl_2SiO_6$	Rh	7.8	RhO_2(g); RhO
Cd	2.81	Cd(g); $CdSiO_3$, CdO	S	8.13	SO_2(g); FeS
Cl	1.9	KCl(g), NaCl(g); KCl, NaCl	Sb	6.9	SbS(g), SbO(g); SbO_2, Sb_2O_3
Co	−3.51	$CoCl_2$(g); Co_2SiO_4	Sc	−10.2	ScO(g); Sc_2O_3
Cr	−4.75	CrO_2(g); CrO_2	Se	6.41	SeO(g), SeO_2(g); $FeSe_{1.33}$
Cs	2.18	CsCl(g); CsCl	Si	−6.89	SiO_2(g); SiO_2

Cu	0.23	Cu(g), CuCl(g); CuO	Sn	-0.65	SnO(g); SnO_2
Eu	−7.99	$EuCl_3$(g); Eu_2O_3	Sr	−6.06	$Sr(OH)_2$(g); $SrSiO_3$
F	1.16	KF(g), NaF(g); NaF, KF	Ta	−6.18	TaO_2(g); Ta_2O_5
Fe	−5.52	$Fe(OH)_2$(g); FeO, $FeSiO_3$	Te	1.15	TeO(g); TeO, Te
Ga	−1.12	GaCl(g); Ga_2O_3	Th	−8.93	ThO_2(g); ThO_2
Ge	1.50	GeO(g); GeO_2	Ti	−7.2	TiO_2(g); TiO_2, $FeTiO_3$, $CaTiO_3$
Hg	8.88	Hg(g); HgO	Tl	3.53	TlCl(g), Tl(g); TlCl
I	3.82	I(g); I, KI, NaI	U	−10.5	UO_2(g); U_3O_5
In	3.9	InCl(g); In_2O_3	V	−2.48	VO_2(g); VO_2
K	−2.36	KCl(g); $KAlSi_3O_8$	W	−1.62	H_2WO_4(g); $CaWO_4$
La	−9.55	$LaCl_3$(g); $La_2Al_2O_6$	Zn	−0.01	Zn(g); ZnO, $ZnSiO_3$
Li	1.26	Li(OH)(g), LiCl(g); $LiAlSiO_4$	Zr	−11.3	ZrO2(g); ZrO_2, $ZrSiO_4$
Mg	−4.58	$Mg(OH)_2$(g); $MgSiO_3$			

Note: The interpolated log(gas/glass) value for Cs and Rb is about −3, if both exist as aluminosilicate components in the glass phase.

With a few exceptions, log(vapor/slag) values from Table VIII-8 are positively related to log(gas/glass) values from the thermodynamic modeling of coal combustion (Figure VIII-17). The exceptionally high log(gas/glass) values for Cs and Rb may be explained by the lack of good thermodynamic data for $CsAlSi_3O_8$ and $RbAlSi_3O_8$ solids in the database of the *HSC Chemistry PC* program, such that Cs and Rb form volatile chlorides in the model calculation (Table VIII-9), instead of more stable aluminosilicate species. The proper interpolated log(gas/glass) values for Cs and Rb from Figure VIII-17 should be about −3.

Volcanic Emissions

Volcanic eruption also represents a high-temperature process. Many volatile elements are injected into the atmosphere as gas and condensed phases on volcanic ashes. Is the relative volatility scale of the elements during volcanic emission similar to that during coal burning? This section presents an answer.

Logarithms of the enrichment factors of excess elements, log Ex, in volcanic plume or gas relative to volcanic rocks for Kilauea, Hawaii (Olmez et al., 1986); Mount Etna, Sicily (Buat-Menard and Arnold, 1978); Merapi Volcano, Indonesia (Symonds et al., 1987); and Kudryavy Volcano, Kuril Islands (Taran et al., 1995) are summarized in Table VIII-10.

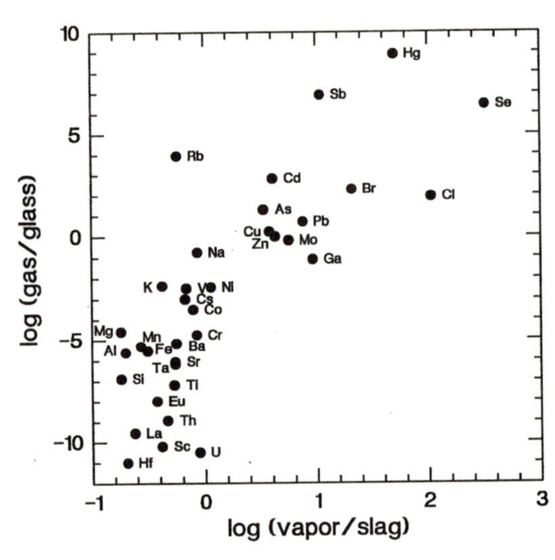

FIGURE VIII-17. Log (vapor/slag) from the coal-fired Thomas Allen steam plant plotted versus log (gas/glass) values from the thermodynamic modeling of coal combustion.

TABLE VIII-10
Logarithm of enrichment factors of excess elements, log Ex, in volcanic gas relative to volcanic rocks

	Kilauea /basalt	*Etna /crust*	*Merapi /andesite*	*Kudryavy /andesite*		*Kilauea /basalt*	*Etna /crust*	*Merapi /andesite*	*Kudryavy /andesite*
Ag	3.94	3.80	5.53		Li	(2.2)			2.41
As	4.94	3.77	5.13	4.21	Mg	(0.3)		2.30	−0.46
Al	−1.02	0.47	0.47	−0.54	Mn	0.28	1.16	2.34	
Au	5.24	4.65			Mo	3.14		6.24	3.71
B	(6.1)			4.71	Na	1.53	1.79	3.66	1.29
Ba	1.11		1.28	1.29	Ni	1.83	1.71		2.91
Bi	(6.9)		7.99	5.11	P	(1.0)		1.89	1.39
Br	5.64	5.10	6.82	4.31	Pb	(3.4)	3.40	6.53	4.11
Ca	(0.8)	1.32	*	*	Rb	1.93		4.60	1.70
Cd	5.34	4.19	7.48	4.51	Re	(6.9)		8.23	5.41
Ce	0.75			0.86	S	6.54	4.48	5.68	
Cl	6.14	5.18	5.68	6.41	Sb	2.74	3.43	3.38	3.61
Co	0.42	1.16	2.08	0.74	Sc	(−1.0)*	*	0.00	2.31
Cr	1.32	0.70	2.14	1.50	Se	6.94	5.85	9.06	5.01
Cs	2.24	2.37	6.12	1.80	Si	(0.3)		1.77	1.07
Cu	2.24	2.85	3.93	2.21	Sm	0.09			
Eu	0.41				Sn	5.14		5.76	3.51
F	5.34			4.41	Sr	(0.3)			0.02
Fe	0.29	0.13	2.60	0.35	Tb	0.54			
Ge	(2.0)			2.71	Te	(6.9)			5.51
Hf	(−1.0)			2.11	Th	0.82			
Hg	4.14	4.72		3.91	Ti	−0.11		2.30	−1.63
I	(6.9)			5.51	Tl	(4.9)			4.21
In	3.90		6.62		V	1.21	0.67	2.90	
Ir	4.74				W	2.34		6.34	3.51
K	1.73	2.23	3.86	1.70	Zn	2.24	3.32	5.21	2.91
La	0.33				Zr	(−1.0)			2.61

Source: Kilauea, Olmez et al. (1986); Etna, Buet-Menard et al. (1978); Merapi, Symonds et al. (1987); Kudryavy, Taran et al. (1995).

Note: Values in parentheses are interpolated estimates and are less reliable. The element (Ca or Sc) with asterisk is used for normalization.

Plots among log (gas/glass), log Ex(Etna/crust), log Ex(Kilauea/basalt), log Ex(Merapi/andesite), and log Ex(Kudryavy/andesite), as shown in Figures VIII-18 and 19, again all show a general linear relationship. In these figures, the elemental labels without symbols along one axis represent the known values for that axis but are unknown for the other axis. The linear relationship between log (gas/glass), as obtained from thermodynamic modeling of coal combustion, and log Ex from various volcanic systems is not totally unexpected, because the volcanic lava also consists mainly of oxide and silicate melt at a temperature around 1100°C (Symonds et al., 1987). The

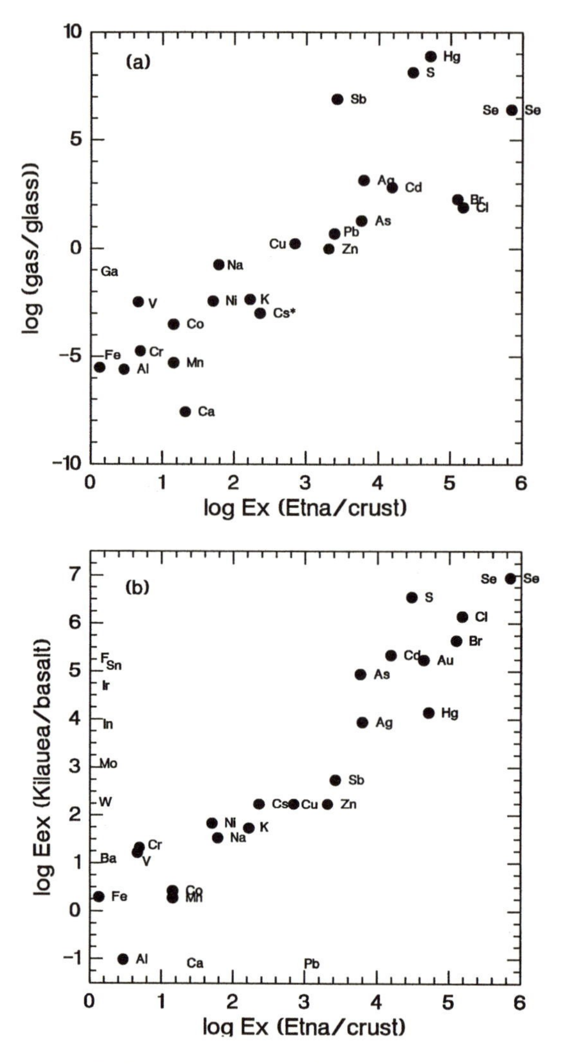

FIGURE VIII-18. Log Ex(Etna/crust) plotted versus (a) log (gas/glass) and (b) log Ex(Kilauea/basalt). The elemental labels without symbols along an axis represent values known for that axis but unknown for the other axis.

important conclusion is that the relative volatility scales of elements during coal combustion and volcanic emission are similar.

The elements Hf, Zr, and Sc are the least volatile, based on log (vapor/slag) and log (gas/glass) values. However, the log Ex values for these elements are relatively high in the Kudryavy volcano as compared to other volcanic systems (see Figure VIII-19 and Table VIII-10). One possible explanation is

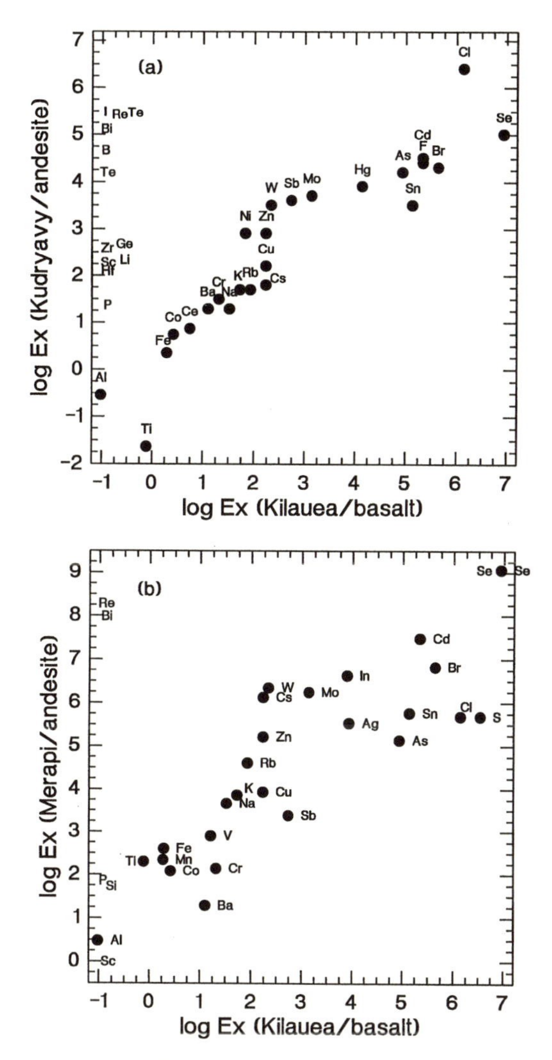

FIGURE VIII-19. Log Ex(Kilauea/basalt) plotted versus (a) log Ex(Kudryavy/andesite) and (b) log Ex(Merapi/andesite).

that the magmatic vapor of Kudryavy is altered seawater and is rich in Cl and F (Taran et al., 1995). Thus, the formation of gaseous halide species of Hf, Zr, and Sc may enhance their volatility. Finally, some unknown log Ex values for Kilauea are roughly interpolated from the known log Ex values from Kudryavy (Figure VIII-19; except for Hf, Zr, and Sc) and Merapi volcanoes (Figures VIII-19), and are listed in parentheses in Table VIII-10. The log Ex values of various elements from Kilauea present another useful set of the relative volatility scale.

Compositions of Aerosols from Polar Regions

The seasalt-corrected average compositions of aerosols and their log Ex values obtained for the south pole atmosphere (Tuncel et al., 1989) and the Norwegian Arctic atmosphere (Maenhaut et al., 1989) during winter and summer seasons in the respective hemispheres are summarized in Table VIII-11. The seasalt fractions of Na, K, Mg, Ca, Sr, Br, and I in the aerosol are subtracted by assuming that all the Cl is of seawater origin (i.e., $X_i - X_{Cl}(X_i/X_{Cl})_{seawater}$). To calculate log Ex, aerosol compositions are normalized by Sc and the average crust (Taylor and McLennan, 1985).

Plots among log Ex(Antarctic aerosol/crust), log Ex(Arctic aerosol/crust), and log (gas/glass) in Figure VIII-20 (see page 403) again demonstrate the general linear relationship. The linear relationship in Figure VIII-20 still holds if log (gas/glass) is replaced by log (Kilauea/basalt). The unusually low log Ex (Antarctic aerosol/crust) value for V in Figure VIII-20a is puzzling. One possible explanation is that the concentration of V in the Antarctic aerosol was underestimated. In short, the sources of volatile elements in polar aerosol particles should be inputs from natural and anthropogenic processes at high temperature, such as volcanic activities, forest fires, fossil fuel burning, ore smelting, municipal waste incineration, etc. However, there is no easy way to assign the relative contribution of natural and anthropogenic sources from the chemical compositions of aerosols alone without other independent information (Lantzy and Mackenzie, 1979; Nriagu, 1989). For example, in Arctic Greenland ice cores, the concentrations of sulfate, nitrate, and Pb have been increasing steadily for the last one hundred years due to anthropogenic inputs (Neftel et al., 1985; Mayewsky et al., 1986; Murozumi et al., 1969; Wolff and Peel, 1985; Legrand and Mayewski, 1997). However, the concentration of Pb in the Greenland snow has been decreasing since 1970, most likely reflecting the reduced usage of Pb as a gasoline additive during the last two decades (Boutron et al., 1994; Legrand and Mayewski, 1997). Therefore, the enriched volatile elements in the present Arctic aerosol must be mostly related to pollution inputs. A trajectory analysis of Arctic air masses also strongly supports the contention that industrial emissions from Eurasia (probably also from North America) are the major cause for the Arctic haze (Barrie, 1990). A factor analysis of 12 daily Arctic aerosol samples collected during March 1985 at Alert, Canada (Landsberger et al., 1990) shows four distinctive factors (Figure VIII-21, see page 404). Factor 1 includes Al, Ca,

TABLE VIII-11

Seasalt-corrected compositions of Antarctic and Arctic atmospheres, and logarithm of enrichment factors of excess elements normalized by Sc and average crust

	Antarctic		*Arctic*			*log Ex(Antarctic)*		*log Ex(Arctic)*	
	winter	*summer*	*winter*	*summer*	*average crust*	*winter*	*summer*	*winter*	*summer*
Ag	0.51	0.86	16	10	0.08	3.68	3.55	3.30	4.04
Al	320	730	39000	19400	84000	0.27	0.28	0.56	1.29
As	11	11	970	99	1	3.92	3.56	3.99	3.94
Au	0.03	0.04	0.5	0.12	0.003	3.88	3.65	3.22	3.55
Ba	50	40	2500	1290	250	2.17	1.72	2.00	2.66
Br		580	6040	390	2.8		4.84	4.33	4.09
Ca	610	670	25000	8000	53000	0.88	0.51	0.57	1.09
Cd	50	110	119	150	0.1	5.57	5.56	4.08	5.12
Ce	0.65	0.88	90	19	33	1.14	0.90	1.42	1.70
Co	0.45	0.77	32	6.9	29	1.03	0.89	1.00	1.30
Cr	20	29	650	121	185	1.90	1.71	1.53	1.75
Cs	0.04	0.06	12.6	2.3	1	1.46	1.28	2.10	2.31
Cu	130	190	1090	200	75	3.11	2.93	2.16	2.37
Eu	0.007	0.02	0.89	0.9	1.1	0.58	0.70	0.85	1.85
Fe	280	660	27000	5000	71000	0.29	0.32	0.45	0.72
Hf	0.022	0.04			3	0.65	0.54		
I	130	260	1530	1930	0.5	5.29	5.24	4.49	5.53
In	0.17	0.33	2.3	2.7	0.05	3.41	3.34	2.66	3.68
K	485	430	34000	5000	9100	1.59	1.17	1.56	1.68
La	0.43	0.56	12.7	4.4	16	1.28	1.03	0.84	1.37
Mg	872	550	5100		32000	1.29	0.67	−0.23	

TABLE VIII-11 *Continued*

	Antarctic		*Arctic*			*log Ex(Antarctic)*		*log Ex(Arctic)*	
	winter	*summer*	*winter*	*summer*	*average crust*	*winter*	*summer*	*winter*	*summer*
Mn	4.2	8.9	970	121	1400	0.10	0.05	0.77	0.82
Mo	4.8	4.3	200	50	1	3.56	3.16	3.30	3.64
Na	9872	3362	367680	50840	23000	2.51	1.68	2.20	2.29
Ni			1200	350	105			2.05	2.47
Pb			5600	700	8			3.85	3.89
Rb		2.4	144	80	32		1.38	1.64	2.34
S	7400	44000	1E + 06	350000	770	3.86	4.28	4.11	4.60
Sb	3.1	3.7	120	12.6	0.2	4.07	3.79	3.78	3.74
Sc	0.04	0.09	3	0.34	30				
Se	4.8	8.4	290	166	0.05	4.86	4.75	4.76	5.47
Si			107000	20000	265000			0.48	0.75
Sm	0.05	0.05	3.4	0.55	3.5	0.99	0.58	0.94	1.11
Sr			640	220	260			1.37	1.87
Ta	0.02	0.03			1	1.15	0.95		
Th	0.06	0.11	3.2	1.2	3.5	1.07	0.98	0.91	1.47
Ti	160	290	820	400	5400	1.33	1.23		0.74
V	0.42	1.1	1680	260	230	−0.43	−0.23	1.86	1.99
W	0.87	0.95	8	4.1	1	2.81	2.50	1.90	2.56
Yb	0.04	0.05			2.2	1.10	0.82		
Zn	170	250	6100	1280	80	3.20	3.02	2.88	3.15

Source: Antarctic, Tuncel et al. (1989); Arctic, Maenhaut et al. (1989).
Note: Concentrations are in pg/m^3.

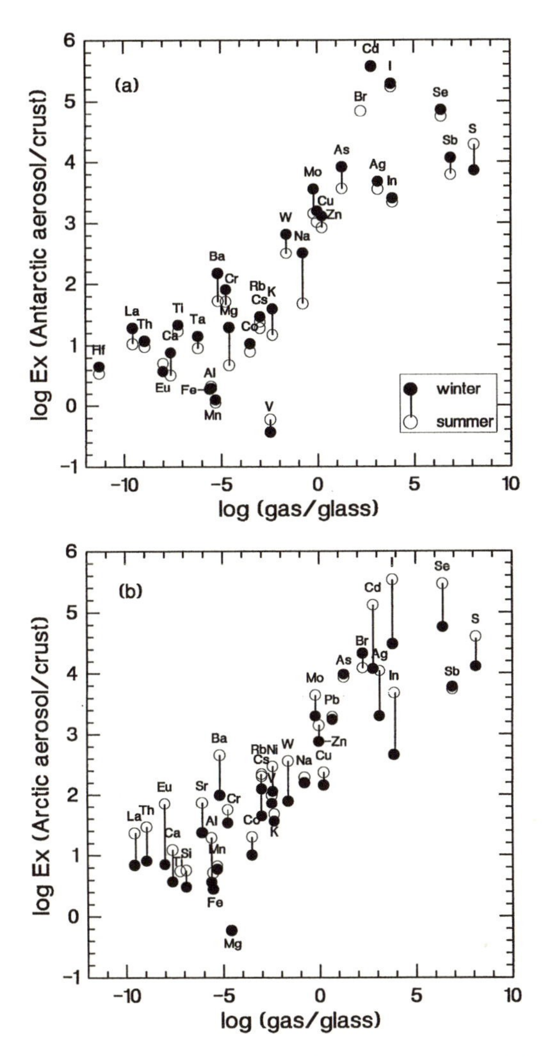

FIGURE VIII-20. Log (gas/glass) from the thermodynamic modeling of coal combustion plotted versus (a) log Ex(Antarctic aerosol/crust) and (b) log Ex(Arctic aerosol/crust) during winter and summer seasons.

Co, Fe, Mn, REE, Sc, Ti, Th, Zn, and partly Sb and K, representing crustal origin. Factor 2 includes As, Br, In, NH_4, NO_3, Se, SO_4, and partly I and Sb, representing pollution inputs. Factor 3 (Na, Cl, and partly I), is seasalts, and factor 4 is K, probably indicating another independent K pollution source such as fertilizer.

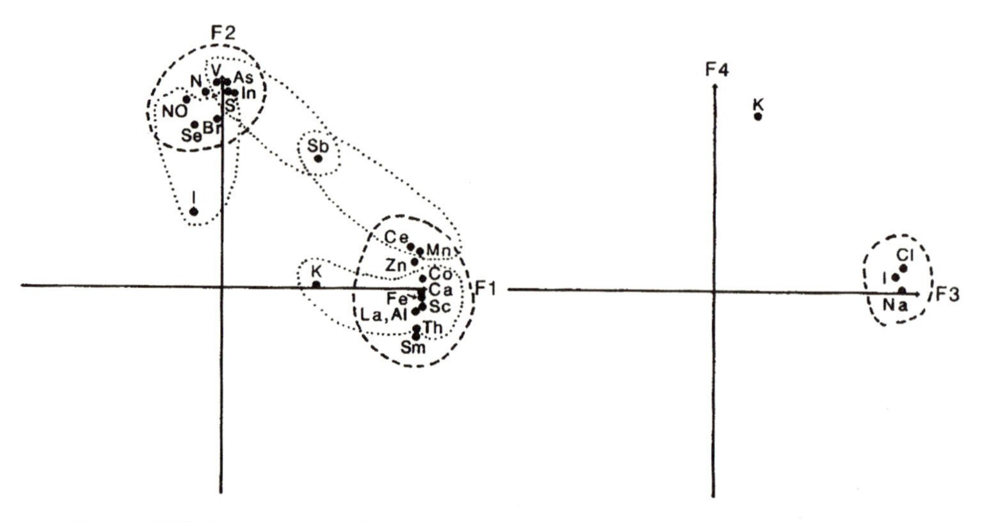

FIGURE VIII-21. Factor loading based on arctic aerosol (raw data from Landsberger et al., 1990). N and NO are ammonia and nitrate nitrogen, respectively.

In Antarctic ice cores, no increasing trend of sulfate with time has been observed for the last 220 years (Legrand and Delmas, 1987; Legrand and Mayewski, 1997), and many sulfate peaks in the ice cores are closely related to known historical volcanic eruptions. Also, the content of metals per unit volume of air is much lower in the Antarctic atmosphere than in the Arctic (Table VIII-11). Therefore, volcanic emission inputs are probably more important than pollution inputs in the Antarctic aerosol, as also suggested by Zreda-Gostynska et al. (1997). Zreda-Gostynska et al. (1997) showed that Pb isotopes in Antarctic ice and snow can be reproduced by two-component mixing of Pb isotopes from Antarctic seawater and Mount Erebus volcanic emissions (77°33′, 167°10′; the southernmost active volcano in the world).

Biogenic emissions of elements (including natural forest fires, aerosols and organic gases from the sea surface and surfaces of plant leaves) may account for more than 55% of the total natural fluxes to the atmosphere (including soil particles, seasalt spray, and volcanic emissions in addition to biogenic sources) for Hg and Se (as well as S), and 25 to 40% for As, Cd, Cu, Mo, Pb, and Zn (Nriagu, 1989). However, according to Lantzy and Mackenzie (1979) and Nriagu (1989), the fluxes of those elements into the atmosphere from fossil fuel burning are presently three to ten times higher than the natural fluxes. Biogenic emissions for relatively less volatile elements are probably negligible. Further studies are needed.

VIII-5. Effects of Fossil Fuel Burning on the Chemistry of Rain and River Waters: A Case Study

The burning of fossil fuels introduces volatile elements into the atmosphere, and these volatile elements eventually fall back onto the Earth's surface through **rain (wet)** and **aerosol (dry) precipitation**. How the burning of fossil fuels affects the chemistry of rain and river waters is the subject of this section.

Figure VIII-22 summarizes the equal-concentration contour maps for the precipitation-weighted mean concentrations (mg/l) of major ions in rainwater (Li, 1992), based on the National Atmospheric Deposition Program/National Trend Network (NADP/NTN) Annual Data Summary (1987). The coincidence of the lowest pH and the highest SO_4^{-2} and NO_3^- concentrations in the northeastern United States reflects industrial inputs of SO_2 and NO_x gases from fossil fuel burning and the smelting of sulfide ores. The NH_4^+ concentration is also high in this area, indicating some input of NH_3 gas through fossil fuel burning (Warneck, 1988). However, the concentration of NH_4^+ is highest in the central United States, where the pH is also highest. The sources of NH_3 gas may be animal waste from cattle ranches and/or ammonia fertilizer used in the area. The high pH is in part caused by the neutralization of H^+ by NH_3 gas through the overall reaction $NH_3 + H^+ \rightarrow NH_4^+$.

The rapid decrease in the concentrations of Cl^-, Na^+, and Mg^{+2} from the coasts inland indicates the dominant seasalt inputs for these ions in coastal areas. However, Mg^{+2} has another maximum over the central northern United States, where Ordovician, Devonian, and Silurian dolomitic carbonate rocks are dominant underlying rock types (*The National Atlas of the United States of America*, 1970). Therefore the major source of Mg^{+2} in the area is the dissolution of dolomitic dust particles in rainwater. The highest Ca^{+2} and K^+ rainwater concentrations are again in the central United States, indicating crustal dust and/or fertilizer particle inputs, which may also neutralize some H^+ ions in the rainwater through dissolution or ion exchange, e.g.,

$$CaCO_3 + 2H^+ \rightarrow Ca^{+2} + CO_2 + H_2O,$$

$$KAlSi_3O_8 + H^+ + (1/2)H_2O \rightarrow K^+ + 2SiO_2 + (1/2)Al_2Si_2O_5(OH)_4.$$

The precipitation-weighted mean concentrations of major ions multiplied by the annual wet precipitation rate gives the annual deposition rates of major ions per unit area (mg/m^2/yr or meq/m^2/yr) by rainfall. Figure VIII-23 shows, as an example, the contour maps for the annual deposition rates

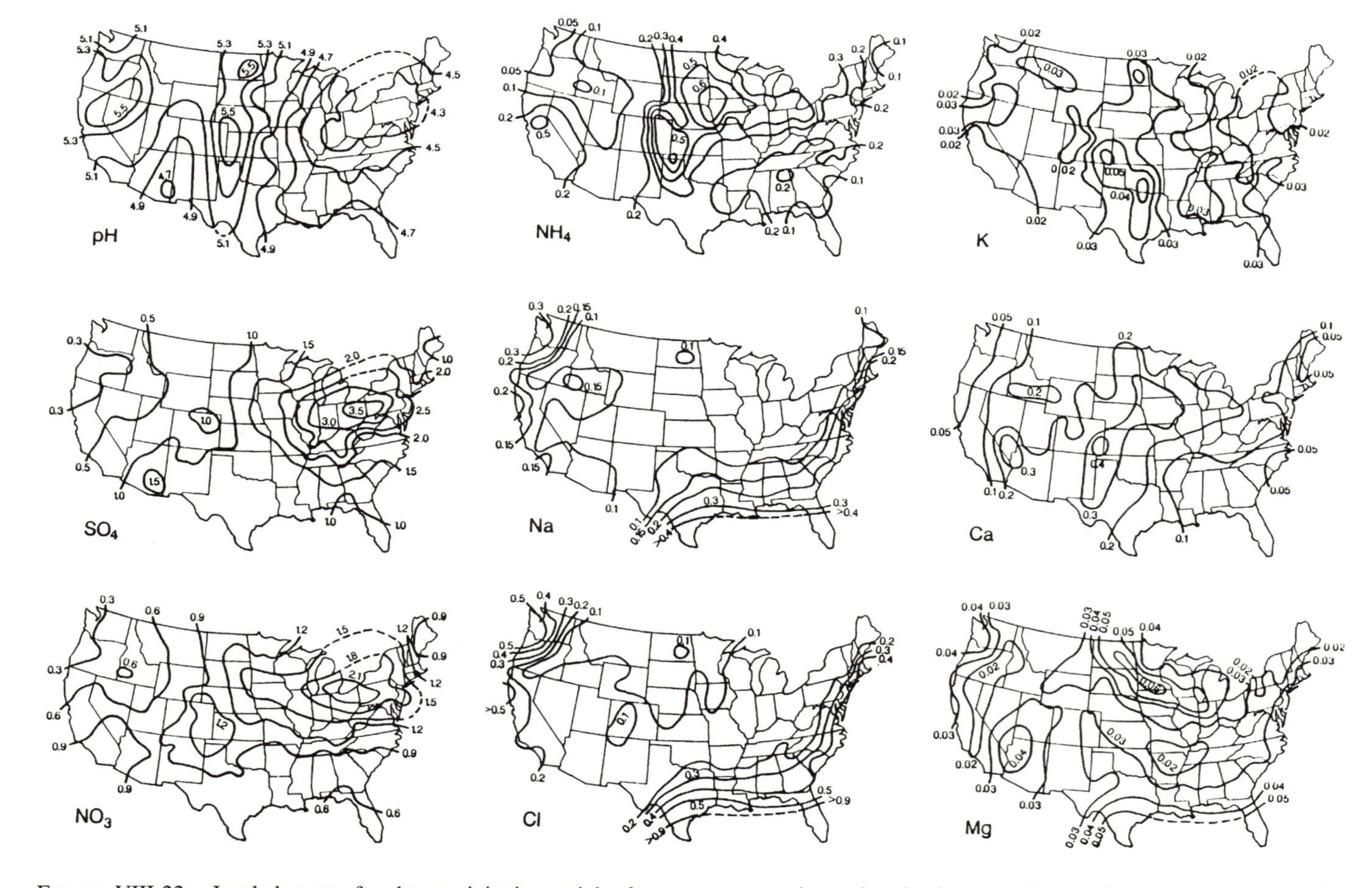

FIGURE VIII-22. Isopleth maps for the precipitation-weighted mean concentrations of major ions (mg/l) and pH in rainwaters over the continental United States (Li, 1992; with kind permission from Kluwer Academic Publishers).

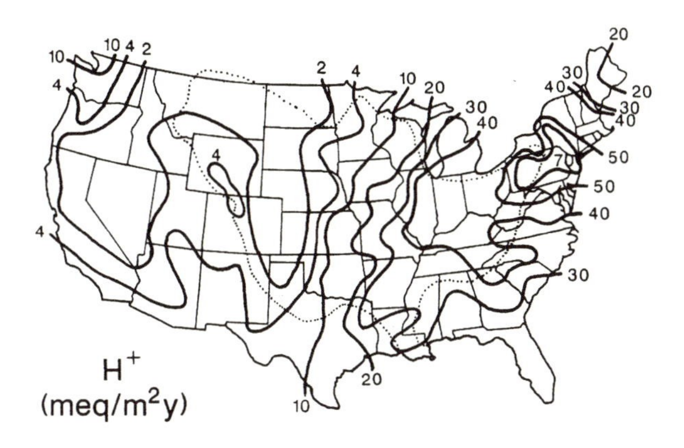

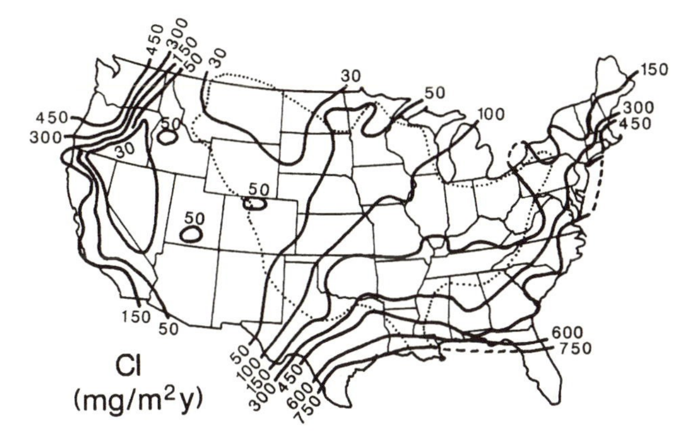

FIGURE VIII-23. Isopleth maps for the annual deposition rates of H^+ and Cl^- ions per unit area over the continental United States (Li, 1992; with kind permission from Kluwer Academic Publishers). The area enclosed by the dotted line is the Mississippi River drainage basin.

of Cl^- and H^+ during the year 1987. For complete data, one should refer to the NADP/NTN Annual Data Summary (1987). The dotted line in Figure VIII-23 encloses the Mississippi River drainage basin. From figures like Figure VIII-23, the area-weighted mean deposition rates of major ions by rainfall can be estimated for the whole Mississippi River drainage basin. The results are summarized in Table VIII-12 in units of meq/m^2/yr. If one assumes that the entire Cl^- burden in rainwater is of seasalt origin, then the percent contribution of seasalt in rainwater for each major ion can be estimated (Table VIII-12, parentheses). The average seasalt contributions of Ca^{+2} and SO_4^{-2} in rainwater over the Mississippi River basin are less than

2%, whereas Na^+, Mg^{+2}, and K^+ are, respectively, about 100, 38, and 11%. The output fluxes of major ions from the Mississippi River to the ocean during the years 1987 and 1905 are also summarized in Table VIII-12 in units of meq/m^2/yr. These fluxes are calculated from the discharge-weighted mean ion concentrations in the Mississippi River (data from USGS Water Resource Data for Louisiana-1987, station near St. Francisville; and Clarke, 1924, for the 1905 data) multiplied by the long-term average discharge rate of the river (Q = 580 km^3/yr) and divided by the river's drainage area (A = 3.27×10^6 km^2) (Milliman and Meade, 1983).

As shown in Table VIII-12 (values in parentheses), the seasalt contribution of major ions through rainwater to the Mississippi River are all minor. For example, in the year 1905 when Mississippi River water was least polluted, the seasalt Cl^- delivered from rainwater accounted for only 8% of the total Cl^- in the river. If one assumes the dry Cl^- deposition rate is roughly equal to the wet Cl^- deposition rate, the total seasalt Cl^- contribution to the pollution-corrected Mississippi river would still be only about 16%. This value is comparable to the value of 18% estimated for the Amazon River at peak water discharge (Stallard and Edmond, 1981). The overestimation of cyclic seasalt in world average river water by earlier works (for example, Conway (1942) 44%, Livingston (1963) 46%, Garrels and Mackenzie (1971) and Holland (1978) $\geq$ 27%, Meybeck (1979) 33%), was probably in part biased by many smaller coastal river basins that contain large seasalt inputs as compared to the continental-scale river basins [a suggestion made also by Berner and Berner (1987)].

The increases in the ion concentrations or fluxes for the Mississippi River from the year 1905 to 1987 (Table VIII-12) can be mostly explained as being due to anthropogenic inputs. For example, the increment of 74 meq SO_4^{-2} is most probably due to input from fossil fuel burning ($SO_2 + H_2O \rightarrow 2H^+ + SO_4^{-2}$; ignoring many intermediate steps). The coal and crude oil consumption for the United States ($\approx$ production + net import) were, respectively, 0.83×10^{15} g/yr and 1.1×10^{15} g/yr in 1987 and negligible in 1905 as compared to 1987 (*Statistical Abstract of the United States*, 1941, 1989). If the sulfur contents of coal and crude oil are on the order of 1% and 0.1%, respectively, then the burning of fossil fuel releases on the order of 100 meq SO_4^{-2} /m^2/yr over the continental United States ($8 \times 10^{12} m^2$, excluding Alaska). This is in good agreement with the observed 74 meq SO_4^{-2}/m^2/yr in the Mississippi River basin, considering some atmospheric export of anthropogenic sulfur to the Atlantic Ocean and Canada (Bischoff et al., 1984). Introduction of 74 meq SO_4^{-2}/m^2/yr through fossil fuel burning should also result in the release of about 1 meq Cl^-/m^2/yr (Table VIII-5) and 75 meq H^+/m^2/yr (= 74 + 1). Most NO_x and NH_3 gases released by fossil fuel burning (about 90 mmol/m^2/yr if all nitrogen in coals is converted to NO_x and NH_3) did not show up in the Mississippi River as dissolved NO_3^- (Table

TABLE VIII-12

Average deposition rates of ions into the Mississippi river basin from rainwater during the year 1987; river fluxes of ions from the Mississippi river basin (1987 and 1905); and various pollution inputs

	Rain	*Mississippi River*		*Net* Δ	*Fossil*	*Common*	$CaCO_3$	*Silicate*	*Rest*
	1987(a)	*1987(b)*	*1905(c)*	$= (b) - (c)$	*fuel input*	*salt input*	*input*	*input*	$\Delta - \Sigma$ *input*
H^+	17.0				75		−58	−17	0
NH_4^+	12.4								
Na^+	3.0 (100)	199 (1.8)	85 (4.1)	114		91		17	6
K^+	0.6 (11)	15 (.5)	13 (.6)	2					2
Ca^{+2}	7.2 (1.8)	358 (.04)	300 (.05)	58			58		0
Mg^{+2}	1.8 (38)	188 (.4)	130 (.6)	58			58		0
Cl^-	3.5 (100)	143 (2.9)	51 (8)	92	1	91			0
SO_4^{-2}	25 (1.4)	206 (.2)	132 (.6)[a]	74	74				0
HCO_3^-	~0	398	337	61			58		3
NO_3^-	13.5	13	8	5	90[b]				−85
SiO_2		20	22						
H_3PO_4		1.8							
$\Sigma^{\pm}$	42	760	528	232					

Notes: All rates in meq/m^2/yr, except SiO_2 and H_3PO_4 in mmol/m^2/yr. Values in parentheses are percent contribution from seasalt input (modified after Li, 1992).

[a] Sulfate is obtained by charge balance.

[b] As $NO_x + NH_3$. $\Sigma^{\pm}$ = the total charge of cations or anions.

VIII-12). Thus, they probably end up as NH_3 in the living organic matter of forests, grassland, and aquatic environments, e.g.,

$$NO_2 + H_2O \rightarrow HNO_3 + (1/2)H_2,$$

$$HNO_3 + H_2O \rightarrow (NH_3)_{org} + 2O_2,$$

$$NH_3 + H_2O \rightarrow NH_4^+ + OH^-,$$

$$NH_4^+ \rightarrow (NH_3)_{org} + H^+.$$

However, the increase of HNO_3 and NH_4^+ in the environment may also enhance denitrification and nitrification by bacteria in soils and convert some nitrogen back to the atmosphere as N_2 gas, e.g.,

$$4HNO_3 + 5CH_2O \rightarrow 2N_2 + 5CO_2 + 7H_2O,$$

$$4NH_4^+ + 3O_2 \rightarrow 2N_2 + 6H_2O + 4H^+.$$

The increment of 92 meq/m^2/yr for Cl^- and associated Na^+ is probably due to inputs from common salt (NaCl) consumption. The consumption rate ($\approx$production rate) of common salt in the United States was about 3.3×10^{12} g/yr and 36×10^{12} g/yr during 1905 and 1987, respectively (*Statistical Abstracts of the United States*, 1941, 1989). Therefore, from 1905 to 1987, the increment of common salt input in the continental United States was on average about 72 meq NaCl/m^2/yr, which is again in close agreement with the observed value (92 meq/m^2/yr) for the Mississippi river basin.

The increments of 58 meq/m^2/yr for Ca^{+2}, Mg^{+2}, and associated HCO_3^- can easily be explained by the dissolution of dolomite through the reaction

$$CaMg(CO_3)_2 + 2H^+ \rightarrow Ca^{+2} + Mg^{+2} + 2HCO_3^-,$$

where H^+ comes from fossil fuel burning. The rest of the H^+ (17 meq/m^2 yr) can replace Na^+ in silicate mineral particles by ion exchange and dissolution processes. The small leftover HCO_3^-, Na^+, and K^+ in Table VIII-12 (the column "Rest") can be attributed to a slight increase in the natural chemical weathering rate and fertilizer inputs or simply data noise.

In short, rainwater chemistry over the continental United States is currently mainly controlled by gas emission inputs from fossil fuel burning (SO_2, NO_x, NH_3, HCl) and from cattle ranches (NH_3) and ammonia fertilizers; dissolution of carbonate and silicate dust by anthropogenic acids; and ion exchange with silicate dust. Seasalt inputs decrease dramatically from coastal to inland regions. The significant increase of major ion concentrations in the Mississippi River from 1905 to 1987 can be semiquantitatively explained by anthropogenic inputs such as common salt consumption, fossil fuel burning, and dissolution of carbonate and silicate rocks by acids derived from fossil fuel burning. Introduction of 75 meq $H^+/m^2/yr$ through fossil fuel burning corresponds to an enhancement of the natural chemical weathering rate (528 meq ions/m^2/yr; Table VIII-12) by about 14%.

The consumption rates of N, P, and K (mainly as fertilizers and some other products over the continental United States) were, respectively, 130, 35, and 16 mmol/m^2/yr in 1987, and negligible in 1905 (*Statistical Abstract of the United States*, 1989). Again, these anthropogenic inputs did not show up in the Mississippi River as dissolved species (Table VIII-12). Some of these inputs might end up as organic matter in soils of farmlands and aquatic environments and as dissolved species in groundwater. Some of the P and K might also accumulate in the inorganic fraction of soils and as river-suspended particles. How much of the anthropogenic input of C, N, P, and K to the surface environment has been effectively converted into new organic matter reservoirs since the early century is still hotly debated (Bolin and Cook, 1983; Bolin, 1983).

VIII-6. Concluding Remarks

The relative abundances of elements in marine and terrestrial living organisms, including the human body as a whole, are shown to be remarkably similar (Figures VIII-1, 2, and 7). The E^i_{Al} values of 10 to 10^3 for biophile trace elements relative to soil or pelagic clays (Figure VIII-3) mainly reflect the formation of extra strong bonds between highly polarizable biophile trace elements and organic ligands containing —NH_2, —SH, and —I functional groups.

The relative abundances of elements in human muscle, brain, blood, kidney, and milk are again similar, except for a few specific elements (Figures VIII-5 and 9). The partition pattern of elements between the human body and urine, and the mean residence times of elements in the human body (Figure VIII-8) again can be explained by the concept of the relative bond strength between ions and organic matter.

Average coal is enriched in biophile trace elements and some biophobe elements such as B, Fe, Be, and U relative to average shale (Figure VIII-11). These enriched elements are associated mainly with the sulfide and organic phases of coal. The enriched biophile elements and V in crude oils (Figure VIII-13) are mainly concentrated in the resin and asphaltene fractions. Ni and V are also known to form very stable porphyrin complexes.

The relative volatility of elements during coal combustion and volcanic emission is similar. The enriched volatile elements in the Arctic aerosols relative to average crust are mostly of anthropogenic origin (e.g., fossil fuel burning, smelting of sulfide ores). In contrast, volcanic emissions are probable sources for the enriched volatile elements in the Antarctic aerosols.

The major ion chemistry of rainwater over the continental United States is currently controlled by inputs from seasalt aerosols, fossil fuel burning (SO_2, NO_x, NH_3, HCl), cattle ranching and agricultural activities (NH_3, K), and dissolution of carbonate and silicate dusts by anthropogenic acids. The chemistry of the Mississippi River is greatly affected by pollution inputs such as common salt consumption, fossil fuel burning, fertilizers, etc. Acids produced by fossil fuel burning have already enhanced the chemical weathering rate of the land by at least 14%.

APPENDIX

RAYLEIGH CONDENSATION AND EVAPORATION MODELS FOR SILICON ISOTOPES

Let R_s and R_v equal the $^{29}Si/^{28}Si$ atomic ratio in solid and vapor phases, respectively; $\alpha = R_s/R_v$= fractionation factor, and M is the total mass of ^{28}Si atoms in the vapor phase. For the Rayleigh condensation model, the decreasing rate of ^{29}Si atoms in the vapor phase should be equal to the removal rate of ^{29}Si atoms from the vapor by condensation, i.e.,

$$-\frac{d}{dt}(R_v \cdot M) = -R_s \frac{dM}{dt}. \qquad \text{(A-1)}$$

The above equation can be written as

$$M \cdot dR_v + R_v \cdot dM = \alpha \cdot R_v \cdot dM.$$

By rearranging and integrating, one obtains

$$\int_{M_0}^{M} \frac{\alpha - 1}{M} dM = \int_{R_v^0}^{R_v} \frac{dR_v}{R_v}$$

i.e.,

$$R_v = R_v^0 \left(\frac{M}{M_0}\right)^{\alpha-1} = R_v^0 \cdot f^{\alpha-1}, \qquad \text{(A-2a)}$$

or

$$R_s = \alpha \cdot R_v^0 \cdot f^{\alpha-1} \qquad \text{(A-2b)}$$

if α is constant; where M_0 and R_v^0 are, respectively, the initial total mass of ^{28}Si and the $^{29}Si/^{28}Si$ ratio in the vapor phase; and $f(= M/M^0)$ is the fraction of ^{28}Si vapor remaining. By definition (see Section I-2),

$$\delta_v = \left(\frac{R_v}{R_{\text{STD}}} - 1\right) \cdot 1000‰,$$

so

$$R_v = \left(\frac{\delta_v}{1000} + 1\right) \cdot R_{\text{STD}}. \qquad \text{(A-3)}$$

Substituting equation (A-3) in equation (A-2) one obtain

$$\delta_v = 10^3\left[\left(\frac{\delta_v^0}{1000}+1\right)\cdot f^{(\alpha-1)} - 1\right]‰. \qquad \text{(A-4)}$$

Similarly, for the Rayleigh evaporation model, one can obtain

$$R_s = R_s^0 \cdot F^{(1/\alpha-1)}, \qquad \text{(A-5a)}$$

$$R_v = R_s^0 \cdot F^{(1/\alpha-1)}/\alpha, \qquad \text{(A-5b)}$$

$$\delta_s = 10^3\left[\left(\frac{\delta_s^0}{1000}+1\right)\cdot F^{1/\alpha-1} - 1\right]‰, \qquad \text{(A-5c)}$$

where R_s^0 and F are, respectively, the initial $^{29}Si/^{28}Si$ ratio of the solid and the fraction of solid ^{28}Si remaining, and $\delta_s = (R_s/R_{STD} - 1)\cdot 1000‰$. δ_s and δ_v can be related by

$$\alpha = \frac{\delta_s/1000+1}{\delta_v/1000+1}$$

or $\delta_s \approx \delta_v + 10^3 \ln\alpha$, by using the approximation $\ln(1+x) \approx x$ when $x < 1$.

RAYLEIGH FRACTIONAL CRYSTALLIZATION MODEL OF MAGMA, AND RAYLEIGH PARTIAL MELTING MODEL OF SOLID ROCKS FOR TRACE ELEMENTS

For these models, one need only substitute R_v and R_v^0 with C and C^0; R_s and R_s^0 with X and X^0; and α with D in all equations given above, e.g.,

$$D = X/C, \qquad \text{(A-6a)}$$

$$X = D\cdot C^0 \cdot f^{(D-1)} \qquad \text{(A-6b)}$$

from equation (A-2b) for the Rayleigh fractional crystallization, and

$$C = (X^0/D)F^{(1/D-1)}$$

from equation (A-5b) for Rayleigh partial melting, where D is the bulk distribution coefficient of a tracer; X and C are, respectively, the concentration of the tracer in the solid and magma phases; f and F are, respectively, the fractions of molten magma and solid rock remaining. Further discussions are given in Section V-3.

Appendix Table A-1
Ionization energies (I_z) and electron affinity (I_{-1}) of the elements (in eV/atom) at 0° K

Z		I_{-1}	I_1	I_2	I_3	I_4	I_5	I_6	I_7
1	H	0.75	13.6						
2	He	0	24.59	54.4					
3	Li	0.62	5.39	75.6	122.5				
4	Be	<0	9.32	18.2	153.9	217.7			
5	B	0.28	8.3	25.2	37.9	259.4	340.2		
6	C	1.26	11.26	24.4	47.9	64.5	392.1	490	
7	N	<0	14.53	29.6	47.5	77.5	97.9	552.1	667
8	O	1.46	13.62	35.1	54.9	77.4	113.9	138.1	739.3
9	F	3.4	17.42	35	62.7	87.1	114.2	157.2	185.2
10	Ne	<0	21.56	41	63.5	97.1	126.2	157.9	207.3
11	Na	0.55	5.14	47.3	71.6	98.9	138.4	172.2	208.5
12	Mg	<0	7.65	15	80.1	109.2	141.3	186.5	224.9
13	Al	0.44	5.99	18.8	28.5	120	153.7	190.5	241.4
14	Si	1.39	8.15	16.4	33.5	45.1	166.8	205.1	246.5
15	P	0.75	10.49	19.7	30.2	51.4	65	220.4	263.2
16	S	2.08	10.36	23.3	34.8	47.3	72.7	88.1	280.9
17	Cl	3.62	12.97	23.8	39.6	53.5	67.8	97	114.2
18	Ar	<0	15.76	27.6	40.7	59.8	75	91	124.3
19	K	0.5	4.34	31.6	45.7	60.9	82.7	100	117.6
20	Ca	<0	6.11	11.9	50.9	67.1	84.4	108.8	127.7
21	Sc	0.19	6.54	12.8	24.8	73.5	91.7	111.1	138
22	Ti	0.08	6.82	13.6	27.5	43.3	99.2	119.4	140.8
23	V	0.53	6.74	14.7	29.3	46.7	65.2	128.1	150.2
24	Cr	0.67	6.77	16.5	31	49.1	69.3	90.6	161.1
25	Mn	<0	7.44	15.6	33.7	51.2	72.4	95	119.3
26	Fe	0.16	7.87	16.2	30.7	54.8	75	99	125
27	Co	0.66	7.86	17.1	33.5	51.3	79.5	102	129
28	Ni	1.16	7.64	18.2	35.2	54.9	75.5	108	133
29	Cu	1.23	7.73	20.3	36.8	55.2	79.9	103	139
30	Zn	<0	9.39	18	39.7	59.4	82.6	108	134
31	Ga	0.3	6	20.5	30.7	64			
32	Ge	1.2	7.9	15.9	34.2	45.7	93.5		
33	As	0.81	9.81	18.6	28.4	50.1	62.6	127.6	
34	Se	2.02	9.75	21.2	30.8	42.9	68.3	81.7	155.4
35	Br	3.37	11.81	21.8	36	47.3	59.7	88.6	103
36	Kr	<0	14	24.4	37	52.5	64.7	78.5	111
37	Rb	0.49	4.18	27.3	40	52.6	71	84.4	99.2
38	Sr	<0	5.7	11	43.6	57	71.6	90.8	106
39	Y	0.3	6.38	12.2	20.5	61.8	77	93	116
40	Zr	0.43	6.84	13.1	23	34.3	81.5		

Appendix Table A-1 *Continued*

Z		I_{-1}	I_1	I_2	I_3	I_4	I_5	I_6	I_7
41	Nb	0.89	6.88	14.3	25	38.3	50.6	102.6	125
42	Mo	0.75	7.1	16.2	27.2	46.4	61.2	68	126.8
43	Tc	0.55	7.28	15.3	29.5	(43)	(59)	(76)	(94)
44	Ru	1.05	7.37	16.8	28.5	(46)	(63)	(81)	(100)
45	Rh	1.14	7.46	18.1	31.1	(46)	(67)	(85)	(105)
46	Pd	0.56	8.34	19.4	32.9	(49)	(66)	(90)	(111)
47	Ag	1.3	7.58	21.5	34.8	(52)	(70)	(89)	(116)
48	Cd	<0	8.99	16.9	37.5	(55)	(73)	(94)	(115)
49	In	0.3	5.79	18.9	28	54			
50	Sn	1.2	7.34	14.6	30.5	40.7	72.3		
51	Sb	1.07	8.64	16.5	25.3	44.2	56	108	
52	Te	1.97	9.01	18.6	28	37.4	58.8	70.7	137
53	I	3.06	10.45	19.1	33	(42)	(71)	(83)	(104)
54	Xe	<0	12.13	21.2	32.1				
55	Cs	0.47	3.89	25.1					
56	Ba	<0	5.21	10					
57	La	0.5	5.58	11.1	19.2	50			
58	Ce	0.5	5.47	10.9	20.2	36.8			
59	Pr	0	5.42	10.6	21.6	39	57.5		
60	Nd	<0	5.49	10.7	22.1	40.4			
61	Pm	<0	5.55	10.9	22.3	41.1			
62	Sm	0.3	5.63	11.1	23.4	41.4			
63	Eu	<0	5.67	11.2	24.9	42.6			
64	Gd	0.5	6.14	12.1	20.6	44			
65	Tb	0.5	5.85	11.5	21.9	39.8			
66	Dy	<0	5.93	11.7	22.8	41.5			
67	Ho	<0	6.02	11.8	22.8	42.5			
68	Er	<0	6.1	11.9	22.7	42.7			
69	Tm	0.3	6.18	12.1	23.7	42.7			
70	Yb	<0	6.25	12.2	25	43.7			
71	Lu	0.5	5.43	13.9	21	45.2			
72	Hf	~ 0	6.25	14.9	23.3	33.3			
73	Ta	0.32	7.89	(16)	(22)	(33)	(45)		
74	W	0.82	7.98	(18)	(24)	(35)	(48)	(61)	
75	Re	0.15	7.88	13.1	26	37.7	(51)	(65)	(79)
76	Os	1.1	8.7	(17)	(25)	(40)	(54)	(68)	(89)
77	Ir	1.57	9.1	(17)	(27)	(39)	(57)	(72)	(88)
78	Pt	2.13	9	18.6	(29)	(41)	(55)	(75)	(92)
79	Au	2.31	9.23	20.5	(30)	(44)	(58)	(73)	(96)
80	Hg	<0	10.44	18.8	34.2	(46)	(61)	(77)	(94)
81	Tl	0.2	6.11	20.4	29.8	(50)	(64)	(81)	(98)
82	Pb	0.36	7.42	15	31.9	42.3	68.8	(84)	(103)

APPENDIX TABLE A-1 *Continued*

Z		I_{-1}	I_1	I_2	I_3	I_4	I_5	I_6	I_7
83	Bi	0.95	7.29	16.7	25.6	45.3	56	88.3	(107)
84	Po	1.9	8.42	(19)	(27)	(38)	(61)	(73)	(112)
85	At	2.8	9.2	(20)	(29)	(41)	(51)	(78)	(91)
86	Rn	<0	10.75	(21)	(29)	(44)	(55)	(67)	(97)
87	Fr		3.98	(23)	(34)				
88	Ra		5.28	10.2	(34)				
89	Ac		5.17	12.1	(19)				
90	Th		6.08	11.5	20	28.8	(65)	(80)	(94)
91	Pa		5.89						
92	U		6.05	(15)	(25)				
93	Np		6.19						
94	Pu		6.06						
95	Am		5.99						
96	Cm		6.02						
97	Bk		6.23						
98	Cf		6.3						
99	Es		6.42						
100	Fm		6.5						
101	Md		6.58						
102	No		6.65						
103	Lr								

Source: Ionization energies (I_z) are from Dean (1985). Electron affinities (I_{-1}) are from Hotop and Lineberger (1985), except REE which are from Bratsch (1983).

Notes: The ionization energy I_z at 298 K is

$$\begin{aligned} I_z(298\,\mathrm{K}) &= I_z\ (0\,\mathrm{K}) + \int_0^{298} [C_p(M^z) + C_p(e^-) - C_p(M^{z-1})]\,dT \\ &= I_z(0\,\mathrm{K}) + \int_0^{298} \frac{5}{2}k\,dT = I_z(0\,\mathrm{K}) + 0.064\ \mathrm{eV} \end{aligned}$$

if M^z (g), e^-(g), and M^{z-1}(g) are assumed to be ideal gases and their heat capacities, C_p, equal to zero at 0 K, and $\frac{5}{2}\,k$ at other temperatures (k = Bolzmann constant).

Appendix Table A-2
Abundance of the nuclides (atoms/10^6 Si)

Z	*A*	*Atom %*	*Process*	*Abundance*	*Z*	*A*	*Atom %*	*Process*	*Abundance*
1 H	1	99.9966		2.79E10	22 Ti	47	7.3	Ex	175
	2	0.0034	U	9.49E5		48	73.8	Ex	1771
2 He	3	0.0142	U, h?	3.86E5		49	5.5	Ex	132
	4	99.9858	U, h	2.72E9		50	5.4	Eq	130
3 Li	6	7.5	X	4.28	23 V	50	0.25	Ex, Eq	0.732
	7	92.5	U, x, h	52.82		51	99.75	Ex	292
4 Be	9	100	X	0.73	24 Cr	50	4.345	Ex	587
5 B	10	19.9	X	4.22		52	83.789	Ex	1.131E4
	11	80.1	X	16.98		53	9.501	Ex	1283
6 C	12	98.9	He	9.99E6		54	2.365	Eq	319
	13	1.1	H, N	1.11E5	25 Mn	55	100	Ex, Eq	9550
7 N	14	99.634	H	3.12E6	26 Fe	54	5.8	Ex	5.22E4
	15	0.366	H, N	1.15E4		56	91.72	Ex, Eq	8.25E5
8 O	16	99.762	He	2.37E7		57	2.2	Eq, Ex	1.98E4
	17	0.038	N, H	9.04E3		58	0.28	He, Eq, C	2.52E3
	18	0.2	He, N	4.76E4	27 Co	59	100	Eq, C	2250
9 F	19	100	N	843	28 Ni	58	68.27	Eq, Ex	3.37E4
10 Ne	20	92.99	C	3.2E6		60	26.1	Eq	1.29E4
	21	0.226	C, Ex	7.77E3		61	1.13	Eq, Ex, C	557
	22	6.79	He, N	2.34E5		62	3.59	Eq, Ex, O	1770
11 Na	23	100	C, Ne, Ex	5.74E4		64	0.91	Ex	449
12 Mg	24	78.99	N, Ex	8.48E5	29 Cu	63	69.17	Ex, C	361
	25	10	Ne, Ex, C	1.07E5		65	30.83	Ex	161
	26	11.01	Ne, Ex, C	1.18E5	30 Zn	64	48.63	Ex, Eq	613
13 Al	27	100	Ne, Ex	8.49E4		66	27.9	Eq	352

14 Si	28	92.23	O, Ex	9.22E5		67	4.1	Eq, S	51.7
	29	4.67	Ne, Ex	4.67E4		68	18.75	Eq, S	236
	30	3.1	Ne, Ex	3.1E4		70	0.62	Eq, S	7.8
15 P	31	100	Ne, Ex	1.04E4	31 Ga	69	60.108	S, eq, r	22.7
16 S	32	95.02	O, Ex	4.89E5		71	39.892	S, eq, r	15.1
	33	0.75	Ex	3.86E3	32 Ge	70	20.5	S, eq	24.4
	34	4.21	O, Ex	2.17E4		72	27.4	S, eq, r	32.6
	36	0.02	Ex, Ne, S	1.03E2		73	7.8	eq, s, r	9.28
17 Cl	35	75.77	Ex	2860		74	36.5	eq, s, r	43.4
	37	24.23	Ex, C, S	913		76	7.8	Eq	9.28
18 Ar	36	84.2	Ex	8.5E4	33 As	75	100	R, s	6.56
	38	15.8	O, Ex	1.6E4	34 Se	74	0.88	P	0.55
	40		S, Ne	26		76	9	S, p	5.6
	40			**25 ± 14**		77	7.6	R, s	4.7
19 K	39	93.2581	Ex	3516		78	23.6	R, s	14.7
	40	0.01167	S, Ex, Ne	0.44		80	49.7	R, s	30.9
	40			**5.48**		82	9.2	R	5.7
	41	6.7302	Ex	253.7	35 Br	79	50.69	R, s	5.98
20 Ca	40	96.941	Ex	5.92E4		81	49.31	R, s	5.82
	42	0.647	Ex, O	395	36 Kr	78	0.339	P	0.153
	43	0.135	Ex, C, S	82.5		80	2.22	S, p	0.999
	44	2.086	Ex, S	1275		82	11.45	S	5.15
	46	0.004	Ex, C, Ne	2.4		83	11.47	R, s	5.16
	48	0.187	Eq, Ex	114		84	57.11	R, S	25.7
21 Sc	45	100	Ex, Ne, Eq	34.2		86	17.42	S, r	7.84
22 Ti	46	8	Ex	192	37 Rb	85	72.165	R, s	5.12

Appendix Table A-2 *Continued*

Z	*A*	*Atom %*	*Process*	*Abundance*	*Z*	*A*	*Atom %*	*Process*	*Abundance*
37 Rb	87	27.835	S	1.97	50 Sn	117	7.672	R, S	0.293
	87			**2.11**		118	24.217	S, r	0.925
38 Sr	84	0.56	P	0.132		119	8.587	S, R	0.328
	86	9.86	S	2.32		120	32.596	S, R	1.245
	87	7	S	1.64		122	4.632	R	0.177
	87			**1.51**		124	5.787	R	0.221
	88	82.58	S, r	19.41	51 Sb	121	57.362	R, s	0.177
39 Y	89	100	S	4.64		123	42.638	R	0.132
40 Zr	90	51.45	S	5.87	52 Te	120	0.09	P	0.0043
	91	11.22	S	1.28		122	2.57	S	0.124
	92	17.15	S	1.96		123	0.89	S	0.0428
	94	17.38	S	1.98		124	4.76	S	0.229
	96	2.8	R	0.32		125	7.1	R, s	0.342
41 Nb	93	100	S	0.698		126	18.89	R, S	0.909
42 Mo	92	14.84	P	0.378		128	31.73	R	1.526
	94	9.25	P	0.236		130	33.97	R	1.634
	95	15.92	R, s	0.406	53 I	127	100	R	0.9
	96	16.68	S	0.425	54 Xe	124	0.121	P	0.00571
	97	9.55	R, s	0.244		126	0.108	P	0.00509
	98	24.13	R, s	0.615		128	2.19	S	0.103
	100	9.63	R	0.246		129	27.34	R	1.28
44 Ru	96	5.52	P	0.103		130	4.35	S	0.205
	98	1.88	P	0.035		131	21.69	R	1.02
	99	12.7	R, s	0.236		132	26.5	R, s	1.24
	100	12.6	S	0.234		134	9.76	R	0.459
	101	17	R, s	0.316		136	7.94	R	0.373

	102	31.6	R, S	0.588
	104	18.7	R	0.348
45 Rh	103	100	R, s	0.344
46 Pd	102	1.02	P	0.0142
	104	11.14	S	0.155
	105	22.33	R, s	0.31
	106	27.33	R, S	0.38
	108	26.46	R, S	0.368
	110	11.72	R	0.163
47 Ag	107	51.839	R, s	0.252
	109	48.161	R, s	0.234
48 Cd	106	1.25	P	0.0201
	108	0.89	P	0.0143
	110	12.49	S	0.201
	111	12.8	R, S	0.206
	112	24.13	S, R	0.388
	113	12.22	R, S	0.197
	114	28.73	S, R	0.463
	116	7.49	R	0.121
49 In	113	4.3	p, s, r	0.0079
	115	95.7	R, S	0.176
50 Sn	112	0.973	P	0.0372
	114	0.659	P, s	0.0252
	115	0.339	p, r, s	0.0129
	116	14.538	S, r	0.555

55 Cs	133	100	R, s	0.372
56 Ba	130	0.106	P	0.00476
	132	0.101	P	0.00453
	134	2.417	S	0.109
	135	6.592	R, s	0.296
	136	7.854	S	0.353
	137	11.23	S, r	0.504
	138	71.7	S	3.22
57 La	138	0.089	P	0.000397
	138			**0.000409**
	139	99.911	S, r	0.446
58 Ce	136	0.19	P	0.00216
	138	0.25	P	0.00284
	138			**0.00283**
	140	88.48	S, r	1.005
	142	11.08	R	0.126
59 Pr	141	100	R, S	0.167
60 Nd	142	27.13	S	0.225
	143	12.18	R, S	0.101
	143			**0.1**
	144	23.8	S, R	0.197
	145	8.3	R, s	0.0687
	146	17.19	R, S	0.142
	148	5.76	R	0.0477
	150	5.64	R	0.0467

APPENDIX TABLE A-2 *Continued*

Z	A	Atom %	Process	Abundance	Z	A	Atom %	Process	Abundance
62 Sm	144	3.1	P	0.008	73 Ta	180	0.012	p, s, r	2.48E−6
	147	15	R, s	0.0387		181	99.988	R, S	0.0207
	147			**0.0399**	74 W	180	0.13	P	0.000173
	148	11.3	S	0.0292		182	26.3	R, s	0.035
	149	13.8	R, S	0.0356		183	14.3	R, s	0.019
	150	7.4	S	0.0191		184	30.67	R, s	0.0408
	152	26.7	R, S	0.0689		186	28.6	R	0.038
	154	22.7	R	0.0586	75 Re	185	37.4	R, s	0.0193
63 Eu	151	47.8	R, s	0.0465		187	62.6	R	0.0324
	153	52.2	R, s	0.0508		**187**			**0.0351**
64 Gd	152	0.2	P, s	0.00066	76 Os	184	0.018	P	0.000122
	154	2.18	S	0.00719		186	1.58	S	0.0107
	155	14.8	R, s	0.0488		187	1.6	S	0.0108
	156	20.47	R, s	0.0676		**187**			**0.00807**
	157	15.65	R, s	0.0516		188	13.3	R, s	0.0898
	158	24.84	R, s	0.082		189	16.1	R	0.109
	160	21.86	R	0.0721		190	26.4	R	0.178
65 Tb	159	100	R	0.0603		192	41	R	0.277
66 Dy	156	0.056	P	0.000221	77 Ir	191	37.3	R	0.247
	158	0.096	P	0.000378		193	62.7	R	0.414
	160	2.34	S	0.00922	78 Pt	190	0.0127	P	0.00017
	161	18.91	R	0.0745		192	0.78	S	0.0105
	162	25.51	R, s	0.101		194	32.9	R	0.441
	163	24.9	R	0.0982		195	33.8	R	0.453
	164	28.19	R, S	0.111		196	25.2	R	0.338
67 Ho	165	100	R	0.0889		198	7.19	R	0.0963

68 Er	162	0.14	P	0.000351	79 Au	197	100	R	0.187
	164	1.61	P, S	0.00404	80 Hg	196	0.1534	P	0.00052
	166	33.6	R, s	0.0843		198	9.968	S	0.0339
	167	22.95	R	0.0576		199	16.873	R, S	0.0574
	168	26.8	R, S	0.0672		200	23.096	S, r	0.0785
	170	14.9	R	0.0374		201	13.181	S, r	0.0448
69 Tm	169	100	R, s	0.0378		202	29.863	S, r	0.1015
70 Yb	168	0.13	P	0.000322		204	6.865	R	0.0233
	170	3.05	S	0.00756	81 Tl	203	29.524	R, S	0.0543
	171	14.3	R, s	0.0354		205	70.476	S, R	0.1297
	172	21.9	R, S	0.0543	82 Pb	204	1.94	S	0.0611
	173	16.12	R, s	0.04		206	19.12	R, S	0.602
	174	31.8	S, R	0.0788		**206**			**0.593**
	176	12.7	R	0.0315		207	20.62	R, S	0.65
71 Lu	175	97.41	R, s	0.0357		**207**			**0.644**
	176	2.59	S	0.000951		208	53.31	R, s	1.837
	176			**0.001035**		**208**			**1.828**
72 Hf	174	0.162	P	0.000249	83 Bi	209	100	R, s	0.144
	176	5.206	S	0.00802	90 Th	232	100	R	0.0335
	176			**0.00793**		**232**			**0.042**
	177	18.606	R, s	0.0287	92 U	235	0.72	R	6.48E-5
	178	24.297	R, S	0.042		**235**			**0.00573**
	179	13.629	R, s	0.021		238	99.2745	R	0.00893
	180	35.1	S, R	0.0541		**238**			**0.0181**

Source: Anders and Grevesse (1989).

Notes: Abundance in atoms/10^6 Si. Abbreviations for various processes: U = cosmological nucleosynthesis, H = hydrogen burning, N = hot or explosive hydrogen burning, He = helium burning, C = carbon burning, O = oxygen burning, Ne = neon burning, Ex = explosive nucleosynthesis, Eq = nuclear statistical equilibrium, S = slow (s) process, R = rapid (r) process, P = proton (p) process, and X = cosmic ray spallation and others. Processes are listed in the order of importance, with minor processes (10–30% for r- and s- processes) shown in lower case.

Bold values are abundances of radionuclides 4.55 billion years ago.

APPENDIX TABLE A-3
Some minerals found in meteorites

Silicates, phosphates, and oxides
Akermanite (Ak) $Ca_2MgSi_2O_7$
Albite (Ab) $NaAlSi_3O_8$
Andradite $Ca_3Fe_2Si_3O_{12}$
Anorthite (An) $CaAl_2Si_2O_8$
Apatite $Ca_3(PO_4)_2$
Armalcolite $FeMgTi_2O_5$
Augite $Mg(Fe, Ca)Si_2O_6$
Baddeleyite ZrO_2
Bloedite $Na_2Mg(SO_4)_2 \cdot 4H_2O$
Brianite $CaNa_2Mg(PO_4)_2$
Buchwaldite $NaCaPO_4$
Calcite $CaCO_3$
Clinopyroxene $(Ca, Mg, Fe)SiO_3$
Chlorapitite $Ca_5(PO_4)_3Cl$
Chlorite $(Mg, Fe)_6Si_4O_{10}(OH)_8$
Chromite $FeCr_2O_4$
Cordierite $Mg_2Al_4Si_5O_{18}$
Corundum Al_2O_3
Cristobalite SiO_2
Diopside $CaMgSi_2O_6$
Dolomite $CaMg(CO_3)_2$
Enstatite (En) $MgSiO_3$
Epsomite $MgSO_4 \cdot 7H_2O$
Farringtonite $Mg_3(PO_4)_2$
Fassaite $Ca(Mg, Ti, Al)(Al, Si)_2O_6$
Fayalite (Fa) Fe_2SiO_4
Feldspar solid solution: Or-Ab-An
Ferrosilite (Fs) $FeSiO_3$
Forsterite (Fo) Mg_2SiO_4
Gehlenite (Geh) $Ca_2Al_2SiO_7$
Graftonite $(Fe, Mn)_3(PO_4)_2$
Grossular $Ca_3Al_2Si_3O_{12}$
Gypsum $CaSO_4 \cdot 2H_2O$
Hedenbergite $CaFeSi_2O_6$
Hematite Fe_2O_3
Hercynite (Her) $(Fe, Mg)Al_2O_4$
Hibonite $CaAl_{12}O_{19}$
Ilmenite $FeTiO_3$
Magnesiochromite $MgCrO_4$
Magnetite Fe_3O_4
Majorite $Mg_3(Mg, Si)Si_3O_{12}$
Melilite solid solution: Ak-Geh
Mica $X_2Y_{4\text{-}6}Z_8O_{20}(OH, F)_4$ $X = K, Na, Ca$; $Y = Al, Mg, Fe$; $Z = Si, Al, Fe, Ti$
Merrihueite $(K, Na)_2Fe_5Si_{12}O_{30}$
Monticellite $Ca(Mg, Fe)SiO_4$
Montmorillonite $(Ca_{0.5}, Na)_{0.7}Al_4[(Si, Al)_8O_{20}](OH)_4 \cdot nH_2O$
Na-phlogopite $NaMg_3AlSi_3O_{10}(OH)_2$
Nepheline $NaAlSiO_4$
Olivine solid solution: Fo-Fa
Orthoclase (Or) $KAlSi_3O_8$
Orthopyroxene $(Mg, Fe)SiO_3$
Panethite $(Ca, Na)_2(Mg, Fe)_2(PO_4)_2$
Perovskite $CaTiO_3$
Phyllosilicates: hydrous silicates
Pigeonite $(Fe, Mg, Ca)SiO_3$
Plagioclase solid solution: Ab-An
Potash feldspar $(K, Na)AlSi_3O_8$
Pyroxene solid solution: En-Fs-Wo
Quartz SiO_2
Rhoenite $Ca_4(Mg, Al, Ti)_{12}(Si, Al)_{12}O_{40}$
Richterite $Na_2CaMg_5Si_8O_{22}F_2$
Ringwoodite $(Mg, Fe)_2SiO_4$
Roedderite $(K, Na)_2Mg_5Si_{12}O_{30}$
Sanidine $KAlSi_3O_8$
Sarcopside $(Fe, Mn)_3(PO_4)_2$
Scheelite $CaWO_4$
Septechlorite $(Fe, Mg)_6(Si, Al, Fe)_4O_{10}(OH)_8$
Serpentine $(Mg, Fe)_6Si_4O_{10}(OH)_8$
Sodalite $Na_4Al_3Si_3O_{12}Cl$
Spinel (Sp) $MgAl_2O_4$
Spinel solid solution: Sp-Her
Stanfieldite $Ca_4(Mg, Fe)_5(PO_4)$
Thorianite ThO_2
Tridymite SiO_2
Ureyite $NaCrSi_2O_6$
V-rich magnite $(Fe, Mg)(Al, V)_2O_4$
Wollastonite (Wo) $CaSiO_3$
Yagiite $(K, Na)_2(Mg, Al)_5(Si, Al)_{12}O_{30}$
Zircon $ZrSiO_4$

Appendix Table A-3 *Continued*

Sulfides, nitrides, and others	
Alabandite (Mn, Fe)S,	Mackinawite FeS_{1-x}
Awaruite Ni_3Fe	Marcasite FeS_2
Barringerite $(Fe, Ni)_2P$	Molybdenite MoS_2
Brezinaite Cr_3S_4	Niningerite (Mg, Fe)S
Carlsbergite CrN	Oldhamite CaS
Caswellsilverite $NaCrS_2$	Osbornite TiN
Chalcopyrite $CuFeS_2$	Pentlandite $(Fe, Ni)_9S_8$
Cohenite $(Fe, Ni)_3C$	Perryite $(Ni, Fe)_5(Si, P)_2$
Cubanite $CuFe_2S_3$	Pyrite FeS_2
Daubreelite $FeCr_2S_4$	Pyrrhotite $Fe_{1-x}S$
Diamond C	Roaldite $(Fe, Ni)_4N$
Djerfisherite $K_3CuFe_{12}S_{14}$	Schreibersite $(Fe, Ni)_3P$
Graphite C	Sinoite Si_2N_2O
Greigite Fe_3S_4	Smythite Fe_9S_{11}
Haxonite $Fe_{23}C_6$	Sphalerite (Zn, Fe)S
Heazlewoodite Ni_3S_2	Suessite Fe_3Si
Heideite $(Fe, Cr)_{1+x}(Ti, Fe)_2S_4$	Taenite (Fe, Ni)
Kamacite (Fe, Ni)	Tetrataenite FeNi
Krinovite $NaMg_2CrSi_3O_{10}$	Troilite FeS
Lawrencite $(Fe, Ni)Cl_2$	Valleriite $CuFeS_2$
Lonsdaleite C	

Sources: Mason (1979) and Kerridge and Mathews (1988).

APPENDIX TABLE A-4

Common minerals in igneous rocks and possible association of trace elements with major cations of minerals

Minerals (chemical composition)	Minerals (chemical composition)
Albite $NaAlSi_3O_8$	Plagioclase $NaAlSi_2O_8$-$CaAl_2Si_3O_8$
Alkali or Potash feldspar $(K, Na)AlSi_3O_8$	Phlogopite $K_2(Mg, Fe)_6Si_6Al_2O_{20}(OH, F)_4$
Allanite $(Ca, Ce)_2(Al, Fe)_3Si_3O_{12}OH$	Pyroxene group $X_{1-p}Y_{1+p}Z_2O_6$
Amphibole group $X_{2\text{-}3}Y_5Z_8O_{22}(OH)_2$	X = Ca, Na; Y = Mg, Fe, Mn, Li, Ni, Al, Cr, Ti;
X = K, Na, Ca; Y = Al, Mg, Fe, Li;	Z = Si, Al
Z = Si, Al, Fe, Ti	Quartz SiO_2
Anorthite $CaAl_2Si_2O_8$	Rutile TiO_2
Apatite $Ca_5(PO_4)_3(OH, F, Cl)$	Scheelite $CaWO_4$
Beryl $Be_3Al_2(Si_6O_{18})$	Sodalite $Na_8(Al_6Si_6O_{24})Cl_2$
Biotite:	Sphene $CaTi(SiO_4)(O, OH, F)$
$K_2(Mg, Fe)_{6\text{-}4}(Fe, Al, Ti)_{0\text{-}2}[Si_{6\text{-}5}Al_{2\text{-}3}O_{20}](OH, F)_4$	Spinel $MgAl_2O_4$
Cassiterite SnO_2	Spodamene $LiAlSi_2O_6$
Chlorite $(Mg, Al, Fe)_{12}(Si, Al)_8O_{20}(OH)_{16}$	Stishovite SiO_2
Chromite $FeO \cdot Cr_2O_3$	Topaz $Al_2SiO_4(OH, F)$
Corumdum Al_2O_3	Tourmaline $NaMg_3Al_6(Si_6O_{18})(BO_3)_3(OH, F)_4$
Cristobalite SiO_2	Tridymite SiO_2
Diopside $CaMg(SiO_3)_2$	Wolframite $(Mn, Fe)WO_4$
Fluorite CaF_2	Xenotime YPO_4
Garnet $Mg_3Al_2Si_3O_{12}$	Zircon $ZrSiO_4$

	Trace element association with major cations
Hornblende	$Na^{+} \rightarrow$ K, Ca, Sr, REE, Y, Cu
$(Na, K)_{0\text{-}1}Ca_2(Mg, Fe, Al)_5Si_{6\text{-}7}Al_{2\text{-}1}O_{22}(OH, F)_2$	$K^{+} \rightarrow$ Rb, Tl, Cs, NH_4, Ba, Ra, Sr, Pb, Eu, REE
Hypersthene $(Mg, Fe)SiO_3$	$Mg^{+2} \rightarrow$ Fe, Co, Ni, Cu, Zn, Mn, (LiAl), Cr, Sc, Sb
Ilmenite $FeO \cdot TiO_2$	$Ca^{+2} \rightarrow$ Na, Cd, Sr, Pb, REE, Y, U, Th
Lepidolite $K_2(Li, Al)_{5\text{-}6}[Si_{6\text{-}7}Al_{2\text{-}1}O_{20}](OH, F)_4$	$Be^{+2} \rightarrow$ (Li, Na, K, Cs)Al
Leucite $KAlSi_2O_6$	$Fe^{+2} \rightarrow$ Mg, Co, Ni, Cu, Zn, Mn, Sb
Magnetite $FeO \cdot Fe_2O_3$	$Fe^{+3} \rightarrow$ Cr, As, Al, Ga, Ge, V, Ti, Sn, In, Sb
Mica group $X_2Y_{4\text{-}6}Z_8O_{20}(OH, F)_4$	$Cr^{+3} \rightarrow$ Ge, As, Ti, PGE(Ru, Os, Rh, Ir, Pt, Pd), Fe
X = K, Na, Ca; Y = Al, Mg, Fe, Li;	$Al^{+3} \rightarrow$ As, Cr, Fe, Ti, Ga, Ge, Si
Z = Si, Al, Fe, Ti	$Si^{+4} \rightarrow$ (AlLi), (AlNa), Ge, Fe
Monazite $(Ce, La, Th)PO_4$	$Ti^{+4} \rightarrow$ PGE, V, Hf-Zr, Nb-Ta, Mo-W, Sn, Sb, Ge, In, Al, As, Cr
Muscovite $K_2Al_4(Si_6Al_2O_{20})(OH, F)_4$	$Zr^{+4} \rightarrow$ Hf, Nb-Ta, Mo-W, Sn, Th, U, Fe
Nepheline $NaAlSiO_4$	$Sn^{+4} \rightarrow$ Nb-Ta, Mo-W, Hf-Zr, Fe, Mn, Ti, Sc, In
Olivine $(Mg, Fe)_2SiO_4$	$W^{+6} \rightarrow$ Mo, Nb-Ta, Hf-Zr
Orthoclase $KAlSi_3O_8$	$F^{-} \rightarrow$ Cl, Br, I, OH
Perovskite $CaTiO_3$	

Source: Mainly from Deer et al. (1992).

REFERENCES

Abbey, S. (1983) Studies in "Standard samples" of silicate rocks and minerals. 1969–1982. Geol. Surv. of Canada paper 83–15.

Adams, J. A. S., and Richardson, K. A. (1960) Thorium, uranium, and zirconium concentrations in bauxite. Econ. Geol. 55, 1653–1675.

Adams, J. B. (1975) Interpretation of visible and near-infrared diffuse reflectance spectra of pyroxenes and other rock forming mineral. In *Infrared and Raman Spectroscopy of Lunar and Terrestrial Minerals*, ed. Karr, C. Academic Press, San Diego.

Agee, C. B. (1990) A new look at differentiation of the earth from melting experiments on the Allende meteorite. Nature 346, 834–837.

Ahrens, L. H. (1952, 1953) The use of ionization potentials. Part 1, Ionic radii of the elements; Part 2, Anion affinity and geochemistry. Geochim. Cosmochim. Acta 2, 155–169; 3, 1–29.

———. (1954) Shielding efficiency of cations. Nature 174, 644–645.

———. (1983) *Ionization Potentials—Some Variations, Implications and Applications*. Pergamon, Oxford.

Ahrens, T. J. (1979) Equations of state of iron sulfide and constraints on the sulfur content of the earth. J. Geophys. Res. 84, 985–1008.

Albritton, C. C., Jr. (1989) *Catastrophic Episodes in Earth History*. Chapman and Hall, New York.

Alexander, C. M. O'D. (1994) Trace element distributions within ordinary chondrite chondrules: Implications for chondrule formation conditions and precursors. Geochim. Cosmochim. Acta 58, 3451–3467.

Allen, L. H. (1987) Chemical abundances. In *Spectroscopy of Astrophysical Plasmas*, eds. Dalgarno, A., and Layzer, D. Cambridge University Press, Cambridge, England, pp. 89–124.

Altschuler, Z. S. (1980) The geochemistry of trace elements in marine phosphorites. In *Marine Phosphorites*, ed. Bentor, Y. K. Society of Economic Paleontologists and Mineralogists. E. Brothers, Inc. Ann Arbor, Mich., pp. 19–30.

Alvarez, L. W., Alvarez, W., Asaro, F., and Michael, H. W. (1980) Extraterrestrial cause for the Cretaceous-Teritary extinction. Science 208, 1095–1108.

Amari, S., Hoppe, P., Zinner, E., and Lewis, R. S. (1993) The isotopic compositions and stellar sources of meteoritic graphite grains. Nature 365, 806–809.

Anders, E. (1964) Origin, age and composition of meteorites. Space Sci. Rev. 3, 583–714.

———. (1991) Interstellar grains in meteorites: diamond, graphite, SiC and TiC. Proc. Geol. Soc. China 34, 283–292.

Anders, E., and Ebihara, M. (1982) Solar system abundances of the elements. Geochim. Cosmochim. Acta 46, 2363–2380.

Anders, E., and Grevesse, N. (1989) Abundances of the elements: Meteoritic and solar. Geochim. Cosmochim. Acta 53, 197–214.

Anderson, D. L. (1989) *Theory of the Earth*. Blackwell Science Publications, Boston.

Anderson, D. L., Svendsen, B., and Ahrens, T. J. (1989) Phase relations in iron-rich systems and implications for the Earth's core. Phys. Earth Planet. Inter. 55, 208–220.

Andreae, M. O., Asmode, J. F., Foster, P., and Van't dack, L. (1981) Determination of antimony (III), antimony (V), and methylantimony species in natural waters by atomic absorption spectrometry with hydride generation. Anal. Chem. 53, 1766–1771.

Armstrong, J. T., and El Goresy, A. (1985) Willy: A prize noble Ir-Fremdling—Its history and implications for the formation of Fremdlings and CAI. Geochim. Cosmochim. Acta 49, 1001–1022.

Armstrong, J. T., Hutcheon, I. D., and Wasserburg, G. J. (1987) Zelda and Company: Petro-genesis of sulfide rich Fremdlinge and constraints on solar nebula processes. Geochim. Cosmochim. Acta 51, 3155–3173.

Arnett, D. (1996) *Supernovae and Nucleosynthesis: An Investigation of the History of Matter from the Big Bang to the Present.* Princeton University Press, Princeton, N.J.

Arnett, D., and Bazan, G. (1997) Nucleosynthesis in stars: Recent developments. Science 276, 1359–1362.

Arnett, D., and Truran, J. W., eds. (1985) *Nucleosynthesis—Challenges and New Developments.* University of Chicago Press, Chicago.

Arnett, D., Bahcall, J. N., Kirshner, R. P., and Woosley, S. E. (1989) Supernova 1987A. Annu. Rev. Astron. Astrophys. 27, 629–700.

Arth, J. G. (1976) Behavior of trace elements during magmatic processes—A summary of theoretical models and their applications. J. Res. U.S.G.S. 4, 41–47.

Audouze, J., and Mathieu, N., eds. (1985) *Nucleosynthesis and its Implications on Nuclear and Particle Physics.* NATO ASI Series C, vol. 163. Reidel, Boston.

Azbel, I. Y., and Tolstikhin, I. N. (1990) Geodynamics, magmatism, and degassing of the Earth. Geochim. Cosmochim. Acta 54, 139–154.

Bacon, M. P., Huh, C. A., Fleer, A. P., and Deuser, W. G. (1985) Seasonality in the flux of natural radionuclides and plutonium in the deep Sargasso Sea. Deep Sea Res. 32, 273–286.

Baertschi, P. (1976) Absolute ^{18}O content of Standard Mean Ocean Water. Earth Planet. Sci. Lett. 31, 341–344.

Baes, C. F., Jr., and Mesmer, R. E. (1976) *The Hydrolysis of Cations.* Wiley, New York.

———. (1981) The thermodynamics of cation hydrolysis. Am. J. Sci. 281, 935–962.

Balistrieri, L., Brewer, P. G., and Murray, J. W. (1981) Scavenging residence times of trace metals and surface chemistry of sinking particles in the deep ocean. Deep Sea Res. 28A, 101–121.

Barbeau, K., Moffett, J. W., Caron, D. A., Croot, P. L., and Erdner, D. L. (1996) Role of protozoan grazing in relieving iron limitation of phytoplankton. Nature 380, 61–64.

Barnes, C. A., Clayton, D. D., and Schramm, D. N., eds. (1982) *Essays in Nuclear Astrophysics.* Cambridge University Press, Cambridge, England.

Barrie, L. A. (1990) Arctic air pollution: A case study of continent-to-ocean-to-continent transport. In *The Long-Range Atmospheric Transport of Natural and Contaminant Substances*, ed. Knap, A. H. Kluwer Dordrecht, pp. 137–148.

Barshad, I. (1966) The effect of variation in precipitation on the nature of clay mineral formation in soils from acid and basic igneous rocks. Proc. Int. Clay Conf. Terus. 1, 167–173.

Barth, T. W. (1952) *Theoretical Petrology*. Wiley, New York.

Barwise, A. J. G., and Whitehead, E. V. (1983) Fossil Fuel Metals. In *Trace Elements in Solving Petrogenesis Problems and Controversies*, ed. Augustithis, S. S. Theophrastus Athens, Greece, pp. 599–643.

Basaltic Volcanism Study Project (1981) *Basaltic Volcanism on the Terrestrial Planets*. Pergamon Publ., New York.

Baturin, G. N. (1988) *The Geochemistry of Manganese and Manganese Nodules in the Ocean*. Kluwer, Dordrecht.

Bau, M., Hoehndorf, A., Dulski, P., and Beukes, N. J. (1997) Sources of rare earth elements and iron in Paleoproterozoic iron-formations from the Transvaal Supergroup, South Africa: evidence from neodymium isotopes. J. Geol. 105, 121–129.

Bell, J. F., Hawke, B. R., Owensby, P. D., and Gaffey, M. J. (1987) *Atlas of Asteroid Infrared Reflection Spectra (0.8–2.5 Microns)*. Internal report, Planetary Geosciences Division, Hawaii Institute of Geophysics, University of Hawaii, Honolulu.

Benner, R., Biddanda, B., Black, B., and McCarthy, M. (1997) Abundance, size distribution, stable carbon and nitrogen isotopic compositions of marine organic matter isolated by tangential-flow ultrafiltration. Mar. Chem. 57, 243–263.

Bentor, Y. K., ed. (1980) *Marine Phosphorites—Geochemistry, Occurrence, Genesis*. Society of Economic Paleontologists and Mineralogists. E. Brothers, Inc., Ann Arbor., Mich.

Benz, W., and Cameron, A. G. W. (1990) Terrestrial effects of the giant impact. In *Origin of the Earth*, ed. Newson, H. E., and Jones, J. H. Oxford University Press, Oxford, England.

Berger, W. H., and Soutar, A. (1967) Planktonic foraminifera field experiment on production rate. Science 156, 1495–1497.

Berkley, J. L., Brown IV, H. G., and Keil, K. (1976) The Kenna ureilite: An ultramafic rock with evidence for igneous, metamorphic, and shock origin. Geochim. Cosmochim. Acta 40, 1429–1437.

Berkley, J. L., Taylor, G. J., and Keil, K. (1980) The nature and origin of ureilites. Geochim. Cosmochim. Acta 44, 1579–1597.

Berkovits, L. A., Obolyaninova, V. G., Parshin, A. K., and Romanovskaya, A. R. (1991) A system of sediment reference samples: OO. Geostandards Newsletter 15, 85–109.

Berner, E. K., and Berner, R. A. (1987) *The Global Water Cycle: Geochemistry and Environment*. Prentice-Hall, Englewood Cliffs, N.J.

Berner, R. A. (1990) Atmospheric carbon dioxide levels over Phanerozoic time. Science 249, 1382–1386.

Berner, R. A., Baldwin, T., and Holden, G. R. (1979) Authigenic iron sulfides as paleosalinity indicators. J. Sediment. Petrol. 49, 1345–1350.

Bertine, K. K., and Goldberg, E. D. (1971) Fossil fuel combustion and the major sedimentary cycle. Science 173, 233–235.

Bievre, P. D., Gallet, M., Holden, N. E., and Barnes, I. L. (1984) Isotopic abundances and atomic weights of the elements. J. Phys. Chem. Ref. Data 13, 809–891.

Bild, R. W., and Wasson, J. T. (1976) The Lodran meteorite and its relationship to the ureilites. Mineral Mag. 40, 721–735.

———. (1977) Netschaevo: A new class of chondritic meteorite. Science 197, 58–60.

Binns, W. F., Fixsen, D. J., Garrard, T. L., Israel, M. H., Klarmann, J., Stone, E. C., and Waddington, C. J. (1984) Elemental abundance of ultra heavy cosmic rays. In *Particle Acceleration Processes, Shockwaves, Nucleosynthesis and Cosmic Rays*, eds. Koch-Miramond, L., and Lee, M. A. Adv. Space Res. 4, 25.

Bionta, R. M., and many others (1987) Observation of a neutrino burst in coincidence with supernova 1987A in the large magellanic cloud. Phys. Rev. Lett. 58, 1494–96.

Bischoff, A., and Keil, K. (1983) Ca-Al-rich chondrules and inclusions in ordinary chondrites. Nature 303, 588–592.

Bischoff, A., and Palme, H. (1987) Composition and mineralogy of refractory-metal-rich assemblages from a Ca-Al-rich inclusion in the Allende meteorite. Geochim. Cosmochim. Acta 51, 2733–2748.

Bischoff, J. L., and Piper, D. Z., eds. (1979) *Marine Geology and Oceanography of the Pacific Manganese Nodule Province*. Plenum, New York.

Bischoff, J. L., and Rosenbauer, R. J. (1983) A note on the chemistry of seawater in the range 350°–500°C. Geochim. Cosmochim. Acta 47, 139–144.

Bischoff, J. L., and Seyfried, W. E. (1978) Hydrothermal chemistry of seawater from 25 to 350°C. Am. J. Sci. 278, 838–860.

Bischoff, W. D., Paterson, V. L., and Mackenzie, F. T. (1984) Geochemical mass balance for sulfur- and nitrogen-bearing acid components: eastern United States. In *Geological Aspects of Acid Deposition*, ed. Bricker, O. P. Butterworth, Boston.

Black, W. A. P., and Mitchell, R. L. (1952) Trace elements in the common brown algae and in seawater. J. Mar. Biol. Ass. U. K. 30, 575–584.

Blum, J. D., Wasserburg, G. J., Hutcheon, I. D., Beckett, J. R., and Stolper, E. M. (1988) "Domestic" origin of opaque assemblages in refractory inclusions in meteorites. Nature 331, 405–409.

———. (1989) Origin of opaque assemblages in C3V meteorites: implications for nebular and planetary processes. Geochim. Cosmochim. Acta 53, 543–556.

Bodansky, D., Clayton, D. D., and Fowler, W. A. (1968) Nuclear quasi-equilibrium during silicon burning. Astrophys. J. Suppl. 16, 299–371.

Bolin B. (1983) C, N, P and S cycles: major reservoirs and fluxes. In *The Major Biogeochemical Cycles and Their Interactions*, eds. Bolin, B., and Cook, R. B. Wiley, New York.

Bolin, B., and Cook, R. B., eds. (1983) *The Major Biogeochemical Cycles and Their Interactions*, "Scientific Committee on Problems of the Environment" Scope 21. Wiley, New York

Born, M. (1919) A thermo-chemical application of the lattice theory. Verhandl. Deut. Physik Ges. 21, 13–24.

Born, M., and Lande, A. (1918) The crystal lattice and Bohr's atomic model. Verhandl. Deut. Physik Ges. 20, 202–209.

Boss, A. P. (1986) The origin of the moon. Science 231, 341–345.

———. (1996) A concise guide to chondrule formation models. In *Chondrules and the Protoplanetary Disk*, eds. Hewins, R. H., Jones, R. H., and Scott, E. R. D. Cambridge University Press, Cambridge, England.

Boutron, C., Candelone, J. P., and Hong, S. (1994) Past and recent changes in the large-scale tropospheric cycles of lead and other heavy metals as documented in

Antarctic and Greenland snow and ice: a review. Geochim. Cosmochim. Acta 58, 3217–3225.

Bowen, H. J. M. (1966) *Trace elements in Biochemistry*. Academic Press, New York.

———. (1979) *Environmental Chemistry of the Elements*. Academic Press, New York.

———. (1982) The elemental content of human diets and excreta. In *Environmental Chemistry*, vol. 2. Royal Society for Chemistry, Burlington House, London.

———. (1985) Kale as a reference material. In *Biological Reference Materials*, ed. Wolf, W. R. Wiley, New York, pp. 3–17.

Bowen, N. L. (1928) *The Evolution of the Igneous Rocks*. Dover, New York.

Boynton, W. V. (1975) Fractionation in the solar nebula: Condensation of yttrium and the rare earth elements. Geochim. Cosmochim. Acta 39, 569–584.

Boynton, W. V., and Wark, D. A. (1987) Origin of CAI rims. I. The evidence from the rare earth elements. Lunar Planet. Sci. 18, 117–118.

Boynton, W. V., Starzyk, P. M., and Schmitt, R. A. (1976) Chemical evidence for the genesis of the ureilites, the achondrite chassigny and the nakhlites. Geochim. Cosmochim. Acta 40, 1439–1447.

Bratsch, S. G. (1983) Electron affinities of the lanthanides. Chem. Phys. Lett. 98, 133–117.

Bretscher, M. S. (1985) The molecules of the cell membrane. Sci. Am. 253, 100–109.

Brewer, P. G., Nazaki, Y., Spencer, D. W., and Fleer, A. P. (1980) Sediment trap experiments in the deep North Atlantic: Isotopic and elemental fluxes. J. Mar. Res. 38, 703–728.

Broecker, W. S., and Peng. T. H. (1982) *Tracers in the Sea*. Columbia University Press, New York.

Brown, E. T., Edmond, J. M., Raisbeck, G. M., Bourles, D. L., Yiou, F., and Measures, C. I. (1992) Beryllium isotope geochemistry in tropical river basins. Geochim. Cosmochim. Acta 56, 1607–1624.

Brown, H. (1949) A table of relative abundances of nuclear species. Rev. Mod. Phys. 21, 625–634.

Brownlee, D. E. (1978) Interplanetary dust: possible implications for comets and presolar interstellar grains. In *Protostars and Planets*, ed. Gehrels, T. University of Arizona Press, Tucson, Ariz., pp. 134–150.

———. (1985) Cosmic dust: collection and research. Annu. Rev. Earth Planet. Sci. 13, 147–173.

Brownlee, D. E., Rajan, R. S., and Tomandl, D. A. (1977) A chemical and textural comparison between carbonaceous chondrites and interplanetary dust. In *Comets, Asteroids, Meteorites*, ed. Delsemme, A. Univ. of Toledo Press, Toledo, Ohio, pp. 137–141.

Bruegmann, G. E., Arndt, N. T., Hofmann, A. W., and Tobschall, H. J. (1987) Noble metal abundances in komatiite suites from Alexo, Ontario, and Gorgona Island, Colombia. Geochim. Cosmochim. Acta 51, 2159–2169.

Bruland, K. W. (1980) Oceanographic distributions of cadmium, zinc, nickel and copper in the North Pacific. Earth Planet. Sci. Lett. 47, 176–198.

Bruland, K. W., and Franks, R. P. (1983) Mn, Ni, Zn, and Cd in the western North Atlantic. In *Trace Metals in Seawater*, eds. Wong, C. S. Boyle, E., Bruland, K. W., Burton J. D., and Goldberg, E. D. Plenum, New York, pp. 395–414

Buat-Menard, P., and Arnold, M. (1978) The heavy metal chemistry of atmospheric particulate matter emitted by Mount Etna volcano. Geophys. Res. Lett. 5, 245–248.

Burbidge, E. M., Burbidge, G. R., Fowler, W. A., and Hoyle, F. (1957) Synthesis of the elements in stars. Rev. Mod. Phys. 29, 548–647.

Burnett, W. C., and Froelich, P. N., eds. (1988) The origin of marine phosphorite. The results of the R. V. Robert D. Conrad cruise 23-06 to the Peru shelf. Mar. Geol. 80, Special Issue.

Burns, J. A. (1986) Some background about satellites. In *Satellites*, eds. Burns, J. A., and Mathews, M. S. University of Arizona Press, Tucson, Ariz, pp. 1–38.

Burns, R. G. (1970) *Mineralogical Applications of Crystal Field Theory*. Cambridge University Press, Cambridge, England.

Buseck, P. R. (1977) Pallasite meterorites—mineralogy, petrology and geochemistry. Geochim. Cosmochim. Acta 41, 711–740.

Buseck, P. R., and Goldstein, J. I. (1969) Olivine compositions and cooling rates of pallasitic meteorites. Geol. Soc. Am. Bull. 80, 2141–2158.

Buseck, P. R., and Hua, X. (1993) Matrices of carbonaceous chondrite meteorites. Annu. Rev. Earth Planet. Sci. 21, 255–305.

Cameron, A. G. W. (1957) *Stellar Evolution, Nuclear Astrophysics, and Nucleogenesis*. Atomic Energy of Canada, Ltd. Report CRL-41, Chalk River, Ontario.

———. (1968) A new table of abundances of elements in the solar system. In *Origin and Distribution of the Elements*, ed. Ahrens, L. H. Pergamon, Elmsford, N.Y., pp. 125–143.

———. (1973) Abundances of the elements in the solar system. Space Sci. Rev. 15, 121–146.

———. (1982) Elementary and nuclidic abundances in the solar system. In *Essays in Nuclear Astrophysics*. eds. Barnes, C. H., Clayton, D. D., and Shramm, D. N. Cambridge University Press, Cambridge, England.

———. (1985) Formation and evolution of the primitive solar nebula. In *Protostars and Planets*, eds. Black, D. C., and Matthews, M. S. University of Arizona Press, Tucson, Ariz.

Cartledge, G. H. (1928) Periodic system. Part I, Ionic potential as a periodic function. Part II, Ionic potential and related problems. Am. Chem. Soc. J. 50, 2855–2872.

———. (1930) Periodic system. Part III, Relation between ionizing potentials and ionic potentials. Am. Chem. Soc. J. 52, 3076–3083.

Cassidy, W. A. and Harvey, R. P. (1991) Are there real differences between Antarctic finds and modern falls meteorites? Geochim. Cosmochim. Acta 55, 99–104.

Chan, L. H., Edmond, J. M., Thompson, G. and Gillis, K. (1992) Lithium isotopic composition of submarine basalts: Implications for the lithium cycle in the oceans. Earth Planet. Sci. Lett. 108, 151–160.

Chen, C. Y., Frey, F. A., and Garcia, M. O. (1990) Evolution of alkalic lavas at Haleakala Volcano, east Maui, Hawaii. Contrib. Mineral. Petrol. 105, 197–218.

Chen, C. Y., Frey, F. A., Garcia, M. O., Dalrymple, G. B., and Hart, S.R. (1991) The tholeiite to alkalic basalt transition at Haleakala Volcano, Maui, Hawaii. Contrib. Mineral. Petrol. 106, 183–200.

Chen, J. H., Wasserburg, G. J., von Damm, K. L. and Edmond, J. M. (1986) The U-Th-Pb systematics in hot springs on the East Pacific Rise at 21°N and Guaymas Basin. Geochim. Cosmochim. Acta 50, 2467–2479.

Chou, C. L., Shaw, D. M. and Crocket, J. H. (1983) Siderophile trace elements in the earth's oceanic crust and upper mantle. J. Geophys. Res. Suppl. 88, A507–A518.

Clark, R. N., Fanale, F. P. and Gaffey, M. J. (1986) Surface composition of natural satellites. In *Satellites*, eds. Burns, J. A. and Matthews, M. S. University of Arizona Press, Tucson, Ariz.

Clarke, F. W. (1924) *The Data of Geochemistry*, 5th ed., USGS Bulletin 770. U.S. Government Printing Office, Washington, D.C.

Clayton, D. D. (1968) *Principles of Stellar Evolution and Nucleosynthesis*. McGraw-Hil, New York.

Clayton, R. N., and Mayeda, T. K. (1977a) Carbonaceous chondrites. Lunar Sci. VIII, 193–195.

———. (1977b) Correlated oxygen and magnesium isotope anomalies in Allende inclusions, I: Oxygen. Geophys. Res. Lett. 4, 295–302.

———. (1978a) Genetic relations between iron and stony meteorites. Earth Planet. Sci. Lett. 40, 168–174.

———. (1978b) Multiple parent bodies of polymict brecciated meteorites. Geochim. Cosmochim. Acta 42, 325–325.

———. (1983) Oxygen isotopes in eucrites, shergottites, nakhlites and chassignites. Earth Planet. Sci. Lett. 62, 1–6.

———. (1984a) The oxygen isotope record in Murchison and other carbonaceous chondrites. Earth Planet. Sci. Lett. 67, 151–161.

———. (1984b) Oxygen isotopic compositions of enstatite chondrites and aubrites. J. Geophys. Res. Suppl. 89, C245–C249.

———. (1985) Oxygen isotopes in chondrules from enstatite chondrites: Possible identification of a major nebular reservoir. Lunar Planet. Sci. 16, 142–143.

———. (1988) Formation of ureilites by nebular processes. Geochim. Cosmochim. Acta 52, 1313–1318.

Clayton, R. N., Grossman, L., and Mayeda, T. K. (1973) A component of primitive nuclear composition in carbonaceous meteorites. Science 182, 485–488.

Clayton, R. N., Onuma, N., and Mayeda, T. K. (1976) A classification of meteorites based on oxygen isotopes. Earth Planet. Sci. Lett. 30, 10–18.

Clayton, R. N., Onuma, N., Grossman, L., and Mayeda, T. K. (1977) Distribution of the pre-solar component in Allende and other carbonaceous chondrites. Earth Planet. Sci. Lett. 34, 209–224.

Clayton, R. N., Mayeda, T., Keil, K., and Olsen, E. J. (1981) Redox processes in chondrules and chondrites. Lunar Planet Sci. 12, 154–156.

Clayton, R. N., Onuma, N., Ikeda, Y., Mayeda, T. K., Hutcheon, I. D., Olsen, E. J., and Molini-Velsko, C. (1983a) Oxygen isotopic compositions of chondrules in Allende and Ordinary chondrites. In *Chondrules and their Origins*, ed. King, E. A. Lunar and Planetary Institute, Houston, Tex.

Clayton, R. N., Mayeda, T. K., Olsen, E. J., and Prinz, M. (1983b) Oxygen isotope relationships in iron meteorites. Earth Planet. Sci. Lett. 65, 229–232.

Clayton, R. N., Mayeda, T., and Yanai, K. (1984) Oxygen isotopic compositions of some Yamato meteorites. Mem. Natl. Inst. Polar Res. Spec. Issue 35, 267–271.

Clayton, R. N., Mayeda, T., and Molini-Velsko, C. A. (1985) Isotopic variations in solar system material: Evaporation and condensation of silicates. In *Protostars and Planets II*. eds. Black, D. C., and Matthews, M. S. University of Arizona Press, Tucson, Ariz.

Clayton, R. N., Mayeda, T. K., Prinz, M., Nehru, C. E., and Delaney, J. S. (1986) Oxygen isotope confirmation of a genetic association between achondrites and IIIAB iron meteorites. Lunar Planet. Sci. 17, 141.

Clayton, R. N., Mayeda, T. K., Rubin, A. E., and Wasson, J. T. (1987) Oxygen isotopes in Allende chondrules and coarse grained rims. Lunar Planet. Sci. 18, 187–189.

Clayton, R. N., Mayeda, T. K., Goswami, J. N., and Olsen, E. J. (1991) Oxygen isotope studies of ordinary chondrites. Geochim. Cosmochim. Acta 55, 2317–2337.

Clube, S. V. M. ed. (1989) *Catastrophes and Evolution: Astronomical Foundations*. Cambridge University Press, Cambridge, England.

Cohen, R. S., O'Nions, R. K., and Dawson, J. B. (1984) Isotope geochemistry of xenoliths from East Africa: Implications for development of mantle reservoirs and their interaction. Earth Planet. Sci. Lett. 68, 209–220.

Colman, S. M. (1982) *Chemical Weathering of Basalts and Andesities: Evidence from Weathering Rinds*, U.S.G.S. Prof. Paper 1246. U.S. Government Printing Office, Washington, D.C.

Colodner, D. (1991) *The Marine Geochemistry of Rhenium, Iridium and Platinum*. Ph.D. Thesis. WHOI-MIT.

Condie, K. C. (1991) Another look at rare earth elements in shales. Geochim. Cosmochim. Acta 55, 2527–2531.

———. (1993) Chemical composition and evolution of the upper continental crust: Contrasting results from surface samples and shales. Chem. Geol. 104, 1–37.

Conway, E. J. (1942) Mean geochemical data in relation to oceanic evolution. R. Irish Acad. Proc. 48B, 119–159.

Cotton, F. A., and Wilkinson, G. (1988) *Advanced Inorganic Chemistry*, 5th ed. Wiley, New York.

Cowen, J. A. (1993) *Inorganic Biochemistry: An Introduction*. VCH, New York.

Cox, K. G., Bell, J. D., and Pankhurst, R. J. (1979) *The Interpretation of Igneous Rocks*. George Allen & Unwin, London.

Craig, D. C., and Loughnan, F. C. (1964) Chemical and mineralogical transformations accompanying the weathering of basic volcanic rocks from New South Wales. Australian J. Soil Res. 2, 218–234.

Craig, H. (1961) Standard for reporting concentrations of deuterium and oxygen-18 in natural waters. Science 133, 1833–1835.

Cruikshank, D. P., Brown, R. H., and Clark, R. N. (1984) Nitrogen on Triton. Icarus 58, 293–305.

Cutter, G. A., and Cutter, L. S. (1995) Behavior of dissolved antimony, arsenic, and selenium in the Atlantic Ocean. Mar. Chem. 49, 295–306.

Das, N., Horita, J., and Holland, H. D. (1990) Chemistry of fluid inclusions in halite from the Salina Group of the Michigan Basin: Implications for Late Silurian seawater and the origin of sedimentary brines. Geochim. Cosmochim. Acta 54, 319–327.

Davis, A. M., Tanaka, T., Grossman, L., Lee, T., and Wasserburg, G. J. (1982) Chemical composition of HAL, an isotopically unusual Allende inclusion. Geochim. Cosmochim. Acta 46, 1627–1651.

Davis, J. A., and Kent, D. B. (1990) Surface complexation modeling in aqueous geochemistry. In *Mineral-Water Interface Geochemistry*, eds. Hochella, M. F., Jr., and White, A. F. Reviews in Mineralogy, vol. 23. Mineralogical Society of America, Washington, D.C.

Davis, J. C. (1973) *Statistics and Data Analysis in Geology*. Wiley, New York.

Davy, R. (1983) Part A: a contribution of iron formations based on depositional environments. In *Iron-Formation: Facts and Problems*, eds. Trendall, A. F., and Morris, R. C. Elsevier, New York.

Dean, J. A., ed. (1985) *Lange's Handbook of Chemistry*, 13th ed. McGraw-Hill, New York.

Deer, W. A., Howie, R. A., and Zussman, J. (1992) *An Introduction to the Rock-Forming Minerals*. Longman Sci. Tech., New York.

Delsemme, A. H. (1991) Nature and history of the organic compounds in comets: an astrophysical view. In *Comets in the Post-Halley Era, Volume 1*, eds. Newburn, Jr., Neugebauer, M., and Rahe, J. Kluwer, Dordrecht, pp. 377–428.

Deuser, W. G., and Ross, E. H. (1980) Seasonal change in the flux of organic carbon to the deep Sargasso Sea. Nature 283, 364–365.

Dodd R. T. (1981) *Meteorites: A Petrologic-Chemical Synthesis*. Cambridge University Press, Cambridge, England.

Dott, R. H. (1964) Wacke, graywacke and matrix—what approach to immature sandstone classification. Sed. Petrol. 34, 625–632.

Dreibus, G., and Waenke, H. (1979) On the chemical composition of the Moon and the eucrite parent body and a comparison with the composition of the Earth. Lunar Planet. Sci. X, 315–317.

Drever, J. I. (1974) Geochemical model for the origin of Precambrian banded iron formations. Geol. Soc. Am. Bull. 85, 1099–1106.

———. (1988) *The Geochemistry of Natural Water*. Prentice-Hall, Englewood Cliffs, N.J.

Duarte, C. M. (1992). Nutrient concentration of aquatic plants: patterns across species. Limnol. Oceanogr. 37(4), 882–889.

Dugger, D. L., Stanton, J. H., Irby, B. N., McConnell, B. L., Cummings, W. W., and Maatman, R. W. (1964) The exchange of twenty metal ions with the weakly acidic silanol group of silica gel. J. Phys. Chem. 68, 757–760.

Duncan, M. J., and Levison, H. F. (1997) A disk of scattered icy objects and the origin of Jupiter-family comets. Science 276, 1670–1672.

Duncan, M., Quinn, T., and Tremaine, S. (1988) The origin of short period comets. Astrophys. J. Lett. 328, L69, 74–F9.

Durbin, P. W. (1960) Metabolic characteristics within a chemical family. Heath Physics 2, 225–238.

Dziewonski, A. M., and Anderson, D. L. (1981) Preliminary reference earth model. Phys. Earth Planet. Inter. 25, 297–356.

Easton, A. J. (1985) E-chondrites: Significance of the partition of elements between silicate and sulphide. Meteoritics 20, 89–101.

Edmond, J. M., Measures, C., McDuff, R. E., Chan, L. H., Collier, R., Grant, B., Gordon, L. I., and Corliss, J. B. (1979) Ridge-crest hydrothermal activity and the balances of the major and minor elements in the ocean: the Galapagos data. Earth Planet. Sci. Lett. 46, 1–18

Eisler, R. (1981) *Trace Metal Concentrations in Marine Organisms*. Pergamon, New York.

Eisma, E., and Jurg, J. W. (1969). Fundamental aspects of the generation of petroleum. In *Organic Geochemistry*, eds. Eglinton, G., and Murphy, M. T. Springer-Verlag, New York.

Elderfield, H., and Schultz, A. (1996) Mid-ocean ridge hydrothermal fluxes and the chemical composition of the ocean. Annu. Rev. Earth Planet. Sci. 24, 191–224.

Elderfield, H., Upstill-Goddard, R., and Sholkovitz, E. R. (1990) The rare earth elements in rivers, estuaries, and coastal seas and their significance to the composition of ocean waters. Geochim. Cosmochim. Acta 54, 971–991.

Engel, A. E., Engle, C. G., and Haven, R. G. (1965) Chemical characteristics of oceanic basalts and the upper mantle. Geol. Soc. Am. Bull. 76, 719–734.

Engelmann, J. J., Goret, P., Juliusson, E., Koch-Miramond, L., Masse, P., Petrou, N., Rio, Y., Soutoul, A., Byrnak, B., Jakobsen, H., Lund, N., Peters, B., Rasmussen, M., Rotenberg, M., and Westergaard, N. (1981) The elemental composition of cosmic rays from Be to Zn as measured by the French-Danish instrument on HEAO-3, 17th International Cosmic Ray Conference vol. 9, 97–100.

Erdman, J. A., Shaklette, H. T., and Keith, J. R. (1976) *Elemental Composition of Selected Native Plants and Associated Soils from Major Vegetation Areas in Missouri*, U.S.G.S. Prof. Paper 954-C. U.S. Govt. Printing Office, Washington, D.C., 87 pp.

Erel, Y., and Morgan, J. J. (1991) The effect of surface reactions on the relative abundances of trace metals in deep-ocean water. Geochim. Cosmochim. Acta 55, 1807–1814.

Esser, B. K. (1991) *Osmium Isotope Geochemistry of Terrigenous and Marine Sediments*. Ph.D. Thesis, Yale University.

Fahey, A. J., Goswani, J. N., McKeegan, K. D., and Zinner, E. (1987a) ^{26}Al, ^{244}Pu, ^{50}Ti, REE, and trace element abundances in hibonite grains from CM and CV meteorites. Geochim. Cosmochim. Acta 51, 329–350.

———. (1987b) More isotopic measurements in CM hibonites: Carbon, oxygen and silicon. Lunar Planet. Sci. 18, 297–298.

Falkner, K. K., and Edmond, J. M. (1990) Gold in seawater. Earth Planet. Sci. Lett. 98, 208–221.

Farinelia, P. (1987) Small satellites. In *The Evolution of the Small Bodies of the Solar System*, Proceedings of the International School of Physics "Enrico Fermi"; course 98. North-Holland Physics Publications, Amsterdam, pp. 276–300.

Faure, G. (1986) *Principles of Isotope Geology*, 2nd ed. Wiley, New York.

Fausto da Silva, J. J. R., and Williams, R. J. P. (1991) *The Biological Chemistry of the Elements*. Clarendon Press, Oxford.

Fegley, B., Jr., and Lewis, J. B. (1980) Volatile element chemistry in the solar nebula: Na, K, F, Cl, Br, and P. Icarus 41, 439–455.

Fegley, B., Jr., and Palme, H. (1985) Evidence for oxidizing conditions in the solar nebula from Mo and W depletions in refractory inclusions in carbonaceous chondrites. Earth Planet. Sci. Lett. 72, 311–326.

Filby, R. H. (1975) The nature of metals in petroleum. In *The Role of Trace Metals in Petroleum*, ed. Yen, T. F. Ann Arbor Sci., Michigan.

Finnegan, D. L., Miller, T. L., and Zoller, W. H. (1990) Indium and other trace-metal enrichments from Hawaiian volcanoes. In *Global Catastrophes in Earth History*, eds. Sharpton, V. L., Ward, P. D., and Museum, T. B. Geol. Soc. Am. Special paper 247. Geological Society of America, Boulder, Colo.

Flanagan, F. J., Moore, R., and Arusacvage, P. J. (1982) Mercury in geologic reference samples. Geostandards Newsletter 6, 25–46.

Flegal, A. R., and Patterson, C. C. (1985) Thallium concentrations in seawater. Mar. Chem. 15, 327–331.

Flegel, A. R., Settle, D. M., and Patternson, C. C. (1986) Thallium in marine plankton. Mar. Biol. 90, 501–503.

Flegal, A. R., Sanudo-Wilhelmy, S. A., and Scelfo, G. M. (1995) Silver in the eastern Atlantic Ocean. Mar. Chem. 49, 315–320.

Fleischer, M., tech. ed. (1963–1979) *The Data of Geochemistry*, U.S.G.S. Prof. Paper 440. U.S. Government Printing Office, Washington, D.C.

Floran, R. J., Prinz, M., Hlava, P. F., Keil, K., Nehru, C. E., and Hinthorne, J. R. (1978) The Chassigny meteorite: A cumulate dunite with hydrous amphibole-bearing melt inclusion. Geochim. Cosmochim. Acta 42, 1213–1229.

Fowler, S. W. (1977) Trace elements in zooplankton particulate products. Nature 269, 51–53.

Fowler, W. A. (1984) Experimental and theoretical nuclear astrophysics: the quest for the origin of the elements. Rev. Mod. Phys. 56, 149–179.

Fredriksson, K., Noonan, A., and Brenner, P. (1976) Bulk and major phase composition of eight hypersthene achondrites. Meteoritics 11, 278–280.

Frieden, E. (1984) *Biochemistry of the Essential Ultratrace Elements*. Plenum, New York.

Friedlander, G., Kennedy, J. W., Macias, E. S., and Miller, J. M. (1981) *Nuclear and Radiochemistry*, 3rd ed. Wiley, New York.

Froelich, P. N., and Andreae, M. O. (1980) Germanium in the oceans: Eka-silicon. EOS 61, 987 (abstract).

Fukai, R., and Meinke, W. W. (1962) Activation analysis of V, As, Mo, W, Rh and Au in marine organisms. Limnol. Oceanogr. 7, 186–200.

Fukai, Y., and Suzuki, T. (1986) Iron-water reaction under high pressure and its implication in the evolution of the earth. J. Geophys. Res. 91 (B9), 9222–9230.

Fukuoka, T., Boynton, W. V., Ma, M. S., and Schmitt, R. A. (1977) Genesis of howardites, diogenites, and eucrites. Geochim. Cosmochim. Acta Suppl. 8, 187–210.

Fulton C. R., and Rhodes, J. M. (1984) The chemistry and origin of the ordinary chondrites: implications for refractory-lithophile and siderophile elements. J. Geophys. Res. suppl. 89, B543–B558.

Gaffey, M. J. (1976) Spectral reflectance characteristics of the meteorite classes. J. Geophys. Res. 81, 905–920.

———. (1990) Thermal history of the asteroid belt: implications for accretion of the terrestrial planets. In *Origin of the Earth*, eds. Newsom, H. E., and Jones, J. H. Oxford University Press, Oxford England.

Gaffey, M. J., Bell, J. F., and Cruikshank, D. P. (1989) Reflectance spectroscopy and asteroid surface mineralogy. In *Asteroids II*, eds. Binzel, R. P., Gehrels, T., and Matthews, M. S. University of Arizona Press, Tucson, Ariz.

Garlick, G. D. (1974) The stable isotopes of oxygen, carbon and hydrogen in the marine environment. In *The Sea*, vol. 5, ed. Goldberg, E. D. Wiley, New York.

Garrels, R. M. (1967) Genesis of some ground waters from igneous rocks. In *Researches in Geochemistry*, vol. 2, ed. Abelson, P. H. Wiley, New York, p. 420.

Garrels, R. M., and Mackenzie, F. T. (1971) *Evolution of Sedimentary Rocks*. Norton, New York.

Gast, P. W. (1968) Trace element fractionation and the origin of the tholeiitic and alkaline magma types. Geochim. Cosmochim. Acta 32, 1057–1085.

Gibson, E. K., Jr. (1976) Nature of the carbon and sulfur phases and inorganic gases in the Kenna ureilite. Geochim. Cosmochim. Acta 40, 1459–1464.

Gladney, E. S., O'Malley, B. T., Roelandts, I., and Gills, T. E. (1987) *Standard Reference Materials: Compilation of Elemental Concentration Data for NBS Clinical, Biological, Geological, and Environmental Standard Reference Materials*, NBS Spec. Publ. 260-111. National Bureau of Standards, Gaithersburg, Md.

Glanz, J. (1998) Astronomers see a cosmic antigravity force at work. Science 279, 1298–1299.

Glasby, G. P., ed. (1977) *Marine Manganese Deposits*. Elsevier, Amsterdam.

Gluskoter, H. J., Ruch, R. R., Miller, W. G., Cahill, R. A., Dreher, G. B., and Kuhn, J. K. (1977) *Trace Elements in Coal: Occurrence and Distribution*. Illinois State Geological Survey Circular 499, Urbana, Ill.

Goldberg, E. D., (1954) Marine geochemistry I—chemical scavengers of the sea. J. Geol. 62, 249–265.

Goldberg, E. D., and Arrhenius, G. O. S. (1958) Chemistry of Pacific pelagic sediments. Geochim. Cosmochim. Acta 13, 153–212.

Goldberg, E. D., Broecker, W. S., Gross, M. G., and Turekian, K. K. (1971) Marine chemistry. In *Radioactivity in the Marine Environment*. National Science Foundation, Washington, D.C., pp. 137–272.

Goldberg, E. D., Hodge, V. F., Kay, P., Stallard, M., and Koide, M. (1986) Some comparative marine chemistries of platinum and iridium. Appl. Geochem. 1, 227–232.

Goldberg, S. (1985) Chemical modeling of anion competition on goethite using the constant capacitance model. Soil Sci. Soc. Am. J. 49, 851–856.

———. (1986) Chemical modeling of arsenate adsorption on aluminum and iron oxide minerals. Soil Sci. Soc. Am. J. 50, 1154–1157.

Goldberg, S., and Glaubig, R. A. (1985) Boron adsorption on aluminum and oxide minerals. Soil Sci. Soc. Am. J. 49, 1374–1378.

Goldich, S. S. (1938) A study in rock-weathering. J. Geol. 46, 17–58.

Goldschmidt, V. M. (1937) Goechemische Verteilungsgesetze der Elemente IX. Skr. Nor. Vidensk. Akad. Oslo. I. Mat. Natur. K1, No. 4.

Goldschmidt, V. M. (1954) *Geochemistry*, Oxford University Press, Oxford, England.

Goldstein, S. J., and Jacobsen, S. B. (1987) The Nd and Sr isotopic systematic or riverwater dissolved material: Implication for the sources of Nd and Sr in seawater. Chem. Geol. 66, 245–272.

———. (1988) Rare earth elements in river water. Earth Planet. Sci. Lett. 89, 35–47.

Goodfellow, W. D., and Peter, J. M. (1995) Sulphur isotope composition of the Brunswick No. 12 massive sulphide deposit, Bathurst Mining Camp, New Brunswick: implications for ambient environment, sulphur source, and ore genesis. Can. J. Earth Sci. 33, 231–251.

Goodwin, A. M. (1977) Archean volcanism in Superior Province, Canadian shield. In *Volcanic Regimes in Canada*, eds. Baragar, W. R. A., Coleman L. C., and Hall, J. M. Geol. Ass. Canada Spec. Paper pp. 16, 205–241. Geological Association of Canada, Waterloo, Ont.

Gordon, M., Tracey, J. I., and Ellis, M. W. (1958) *Geology of the Arkansas Bauxite Region*, U.S.G. Prof. Paper 299. U.S. Government Printing Office, Washington, D.C.

Gordy, W., and Thomas W. J. O. (1956) Electronegativities of the elements. J. Chem. Phys. 24, 439–444.

Gorsline, D. S., and Emery, K. O. (1959) Turbidity current deposits in San Pedro and Santa Monica Basins off Southern California. Bull. Geol. Soc. Am. 70, 279–290.

Govindaraju, K. (1989) 1989 compilation of working values and sample description for 272 geostandards. Geostandards Newsletter 13, 1–113.

———. (1994) 1994 compilation of working values and sample description for 383 geostandards. Geostandards Newsletter 18, 1–158.

Graham, A. L., Bevan, A. W. R., and Hutchison R. (1985) *Catalogue of Meteorites*, 4th ed. University of Arizona Press, Tucson, Ariz.

Graham, A. L., Easton, A. J., and Hutchison, R. (1977) Forsterite chondrites; the meteorites Kakangari, Mount Morris (Wisconsin), Pontlyfni, and Winona. Mineral Mag. 41, 201–210.

Gratz, A. J., Nellis, W. J., and Hinsey, N. A. (1993) Observations of high velocity, weakly shocked ejecta from experimental impacts. Nature 363, 522–524.

Greenberg, R. and Brahic, A. eds. (1984) *Planetary Rings*. University of Arizona Press, Tucson, Ariz.

Gross, G. A. (1980) A classification of iron formations based on depositional environments. Can. Min. 18, 215–222.

Grossman, J. N., and Wasson, J. T. (1983) Refractory precursor components of Semarkona chondrules and the fractionation of refractory elements among chondrites. Geochim. Cosmochim. Acta 47, 759–771.

———. (1985) The origin and history of the metal and sulfide components of chondrules. Geochim. Cosmochim. Acta 49, 925–939.

Grossman, J. N., Rubin, A. E., Rambaldi, E. R., Rajan, R. S., and Wasson, J. T. (1985) Chondrules in the Qingzhen type 3 enstatite chondrite: Possible precursor components and comparison to ordinary chondrite chondrules. Geochim. Cosmochim. Acta 49, 1781–1795.

Grossman, J. N., Clayton, R. N., and Mayeda, T. K. (1987) Oxygen isotopes in the matrix of Semarkona (LL3.0) chondrite. Meteoritics 22, 395–396.

Grossman, J. N., Rubin, A. E., Nagahara, H., and King, E. A. (1988) Properties of chondrules. In *Meteorites and the Early Solar System*, eds. Kerridge, J. F., and Matthews, M. S. University of Arizona Press, Tucson, Ariz.

Grossman L. (1975) Petrography and mineral chemistry of Ca-rich inclusions in the Allende meteorite. Geochim. Cosmochim. Acta 39, 433–454.

Grossman, L., and Clark, S. P. Jr. (1973) High temperature condensates in chondrites and the environment in which they formed. Geochim. Cosmochim. Acta 37, 635–649.

Grossman, L., and Ganapathy, R. (1976) Trace elements in the Allende meteorite-I, Coarse grained, Ca-rich inclusion. Geochim. Cosmochim. Acta 40, 331–344.

Grossman, L., and Larimer, J. W. (1974) Early chemical history of the solar system. Rev. Geophys. Space Sci. 12, 71–101.

Grossman, L., Ganapathy, R., and Davis, A. M. (1977) Trace elements in the Allende meteorite III, Coarse grained inclusions revisited. Geochim. Cosmochim. Acta 41, 1647–1664.

Gulbrandsen, R. A. (1966) Chemical compositions of phosphorites of the Phosphoria Formation. Geochim. Cosmochim. Acta 30, 769–778.

Haber, F. (1919) Theory of the heat of reaction. Verhandl Deut. Physik Ges. 21, 750–768.

Hagemann, R., Nief, G., and Roth, E. (1970) Absolute isotopic scale for deuterium analysis of natural waters. Absolute D/H ratio for SMOW. Tellus 23, 712–715.

Haissinsky, M. (1946) Echelle des electronegativites de Pauling et Chaleurs de formation des composes inorganiques. J. Phys. Radium 7, 7–11.

Halbout, J., Robert, F., and Javoy, M. (1986) Oxygen and hydrogen isotope relations in water and acid residues of carbonaceous chondrites. Geochim. Cosmochim. Acta 50, 1599–1609.

Halliday, I., Blackwell, A. T., and Griffin, A. A. (1978) The Innisfree meteorite and the Canadian Camera Network. J. R. Astro. Soc. Canada 72, 15–39.

Hamelin, B., and Allègre, C. J. (1985) Large scale regional units in the depleted upper mantle revealed by an isotope study of the South-West Indian Ridge. Nature 315, 196–199.

Hamilton, E. I. (1979) *The Chemical Elements and Man*. Charles C. Thomas Publ., Springfield, Ill.

Hamilton, E. L. (1976) Variations of density and porosity with depth in deep sea sediments. J. Sediment Petrol. 46, 280–300.

Hamlyn, P. R., Keays, R. R., Cameron, W. E., Crawford, A. J., and Waldron, H. M. (1985) Precious metals in magnesium low-Ti lavas: Implications for metallogenesis and sulfur saturation in primary magma. Geochim. Cosmochim. Acta 49, 1797–1811.

Hart, S. (1984) A large scale isotope anomaly in the southern hemisphere mantle. Nature 309, 753–757.

———. (1988) Heterogeneous mantle domains: Signature, genesis and mixing chronologies. Earth Planet. Sci. Lett. 90, 273–296.

Harte, B. (1987) Metasomatic events recorded in mantle xenoliths: An overview. In *Mantle Xenoliths*, ed. Nixon, P. H. Wiley, New York, pp. 625–640.

Hartman, W. K., Tholen, D. J., and Cruikshank, D. P. (1987) The relationship of active comets, "extinct" comets, and dark asteroids. Icarus 69, 33–50.

Harvey, R. D., and Ruch, R. R. (1986) Mineral matter in Illinois and other U.S. coals. In *Mineral Matter and Ash in Coal*, ACS Symposium Series 301. American Chemical Society, Washington, D.C.

Hashimoto, A., and Grossman T. (1987) Alteration of Al-rich inclusions inside amoeboid olivine aggregates in the Allende meteorite. Geochim. Cosmochim. Acta 51, 1685–1704.

Hayashi, C. (1981) Structure of the solar nebula, growth and decay of magnetic fields and effects of magnetic and turbulent viscosities on the nebula. Prog. Theor. Phys. Suppl. 70, 35–53.

Hayashi, C., Nakagawa, K., and Nakagawa, Y. (1985) Formation of the solar system. In *Protostars and Planets, II*, eds. Black, D. C., and Matthews, M. S. University of Arizona Press, Tucson, Ariz.

Hein, J. R., Shwab, W. C., and Davis, A. S. (1988) Cobalt- and platinum-rich ferromanganese crusts and associated substrate rocks from the Marshall Islands. Mar. Geol. 78, 255–283.

Helin, E., Brown, D., and Rabinowitz, D. (1997) Minor Planet. Electron. Circ. B19.

Hertogen, J., Janssens, M. J., and Palme (1980) Trace elements in ocean ridge basalt glasses: Implications for fractionations during mantle evolution and petrogenesis. Geochim. Cosmochim. Acta 44, 2125–2143.

Hewins, R. H. (1988) Experimental studies of chondrules. In *Meteorites and the Early Solar System*, eds. Kerridge, J. F., and Matthews, M. S. University of Arizona Press, Tucson, Ariz.

———. (1997) Chondrules. Annu. Rev. Earth Planet. Sci. 25, 61–83.

Hewins, R. H., Jones, R. H., and Scott, E. R. D., eds. (1996) *Chondrules and the Protoplanetary Disk.* Cambridge University Press, Cambridge, England.

Hinton, R. W., Davis, A. M., Scatena-Wachel, D. E., Grossman, L., and Draus, R. J. (1988) A chemical and isotopic study of hibonite-rich refractory inclusions in primitive meteorites. Geochim. Cosmochim. Acta 52, 2573–2598.

Hirata, K., and many others (1987) The supernova SN 1987A. Phys. Rev. Lett. 58, 1490–93.

Hitchon, B., Filby, R. H., and Shah, K. R. (1975) Geochemistry of trace elements in crude oils, Alberta, Canada, in *The Role of Trace Metals in Petroleum*, ed. Yen, T. F. Ann Arbor Sci., Michigan.

Hodge, V. F., Stallard, M., Koide, M., and Goldberg, E. D. (1985) Platinum and the platinum anomaly in the marine environment. Earth Planet. Sci. Lett. 72, 158–162

———. (1986) Determination of platinum and iridium in marine waters, sediments and organisms. Anal. Chem. 58, 610–620.

Hofmann, A. W. (1988) Chemical differentiation of the Earth: The relationship between mantle, continental crust, and oceanic crust. Earth Planet. Sci. Lett. 90, 297–314.

———. (1997) Mantle geochemistry: the message from oceanic volcanism. Nature 385, 219–229.

Hofmannn, A. W., and White, W. M. (1983) Ba, Rb and Cs in the Earth's mantle. Z. Naturforsch. 38a, 256–266.

Holland, H. D. (1973) The oceans: a possible source of iron formations. Econ. Geol. 68, 1169–1172.

———. (1978) *The Chemistry of the Atmosphere, and Oceans.* Wiley, New York.

———. (1984) *The Chemical Evolution of the Atmosphere and Ocean.* Princeton University Press, Princeton, N.J.

Holweger, H. (1979) Abundances of the elements in the sun. XXIInd Liege International Astrophysical Symposium, Universitė de Liege, pp. 117–140.

Honjo, S. (1978) Sedimentation of material in the Sargasso Sea at a 5367 m deep station. J. Mar. Res. 36, 469–492.

———. (1980) Material fluxes and modes of sedimentation in the mesopelagic and bathypelagic zones. J. Mar. Res. 38, 53–97.

Horita, J., Friedman, T. J., Lazar, B., and Holland, H. D. (1991) The composition of Permian seawater. Geochim. Cosmochim. Acta 55, 417–432.

Hotop, H., and Lineberger, W. C. (1985) Binding energy in atomic negative ions: II. J. Phys. Chem. Ref. Data 14, 731–750.

Hoyle, F., Fowler, W. A., Burbidge, E. M., and Burbidge, G. R. (1956) Origin of the elements in stars. Science, 124, 611–614.

Hufen, T. H. (1974) *A Geohydrological Investigation of Honolulu's Basal Waters Based on Isotopic and Chemical Analysis of Water Samples*. Ph.D. Thesis, University of Hawaii, Honolulu.

Hughes, M. N. (1981) *The Inorganic Chemistry of Biological Processes*, 2nd ed. Wiley, New York.

Huss, G. R., Keil, K., and Taylor, G. J. (1981) The matrices of unequilibrated ordinary chondrites: Implications for the origin and history of chondrites. Geochim. Cosmochim. Acta 45, 33–51.

Ida, S., Canup, R. M., and Stewart, G. R. (1997) Lunar accretion from an impact-generated disk. Nature 389, 353–360.

Ikeda, Y. (1980) Petrology of Allen Hill-764 chondrites (LL3). Mem. Natl. Inst. Polar Res. Spec. Issue 17, 50–82.

———. (1982) Petrology of the ALH-77003 chondrite (C3). Mem. Natl. Inst. Polar Res. Spec. Issue 25, 34–65.

———. (1983) Major element chemical compositions and chemical types of chondrules in unequilibrated E, O, and C chondrites from Antarctica. Mem. Natl. Inst. Polar Res. Spec. Issue 30, 122–145.

Imamura, K., and Honda, M. (1976) Distribution of tungsten and molybdenum between metal, silicate, and sulphide phases of meteorites. Geochim. Cosmochim. Acta 40, 1073–1080.

Irving, A. J., and Frey, F. A. (1984) Trace element abundances in megacrysts and their host basalts: Constraints on partition coefficients and megacryst genesis. Geochim. Cosmochim. Acta 48, 1201–1221.

Ito, E., and Takahashi, E. (1987) Ultra high-pressure phase transformations and the constitution of the deep mantle. In *High-Pressure Research in Mineral Physics*, Geophysics Monograph 39, eds. Manghnani, M. H., and Syono, Terra Science, Tokyo, pp. 221–230.

Iyengar, G. V., Kollmer, W. E., and Bowen, H. J. M. (1978) *The Elemental Compositions of Human Tissues and Body Fluids*. Verlag Chemie, Weinheim.

Jacobs, L., and Emerson, S. (1982) Trace metal solubility in an anoxic fjord. Earth Planet. Sci. Lett. 60, 237–252.

Jacobsen, S. B., and Pimentel-Klose, M. (1988) A Nd isotopic study of the Hamersley and Michipicoten banded iron formations: the source of REE and F in Archean oceans. Earth Planet. Sci. Lett. 87, 29–44.

Jagoutz, E., Palme, H., Baddenhausen, H., Blum, K., Cendales, M., Dreibus, G., Spettel, B., Lorenz, V., and Waenke, H. (1979) The abundances of major, minor and trace elements in the earth's mantle as derived from primitive ultramafic nodules. Proc. Lunar Planet. Sci. Conf. 10, 2031–2050.

James, H. L. (1983) distribution of banded iron-formation in space and time. In *Iron-Formation: Facts and Problems*, eds. Trendall, A. F., and Morris, R. C. Elsevier, New York.

James, R. D., and Healy, T. W. (1972) Adsorption of hydrolysable metal ions at the oxide-water interface III. A thermodynamic model of adsorption. J. Colloid Interface Sci. 40, 65–81.

Jana, D. and Walker, D. (1997a) the influence of sulfur on partitioning of siderophile elements. Geochim. Cosmochim. Acta 61, 5255–5277.

———. (1997b) The impact of carbon on element distribution during core formation. Geochim. Cosmochim. Acta 61, 2759–2763.

———. (1997c) The influence of silicate melt composition on distribution of siderophile elements among metal and silicate liquids. Earth Planet. Sci. Lett. 150, 463–472.

Janssens, M. J., Hertogen, J., Wolf, R. Ebihara, M., and Anders, E. (1987) Ureilites: Trace element clues to their origin. Geochim. Cosmochim. Acta 51, 2275–2283.

Jaques, A. L., and Green, D. H. (1980) Anhydrous melting of peridotite at 0–15 kb pressure and the genesis of tholeiitic basalts. Contrib. Mineral. Petrol. 73, 287–310.

Jessberger, E. K., Christoforidis, A., and Kissel, J. (1988) Aspects of the major element compositions of Halley's dust. Nature 332, 691–695.

Jewitt, D. C., and Luu, J. X. (1993) Discovery of the candidate Kuiper belt object 1992 QB1. Nature 362, 730–732.

Johnson, J. E., Scrymgour, J., Jarosewich, E., and Mason, B. (1977) Brachina meteorite—a chassignite from South Australia. Records of the South Australian Museum 17, 309–319.

Jones, J. H., and Drake, M. J. (1983) Experimental investigations of trace element fractionation in iron meteorites, II: The influence of sulfur. Geochim. Cosmochim. Acta 47, 1199–1209.

———. (1986) Geochemical constraints on core formation in the Earth. Nature 322, 221–228.

Jones, R. H. (1994) Petrology of Fe-poor, porphyritic pyroxene chondrules in Semarkona chondrite. Geochim. Cosmochim. Acta 58, 5325–5340.

———. (1996) FeO-rich, porphyritic pyroxene chondrules in unequilibrated ordinary chondrites. Geochim. Cosmochim. Acta 60, 3115–3138.

Kabata-Pendias, A., and Pendias, H. (1984) *Trace Elements in Soils and Plants*. CRC Press, Boca Raton, Fla.

Kallemeyn G. W., and Wasson J. T. (1981) The compositional classification of chondrites—I. The carbonaceons chondrite group. Geochim. Cosmochim. Acta 45, 1217–1230.

———. (1985) The compositional classification of chondrites: IV. Ungrouped chondritic meteorites and clasts. Geochim. Cosmochim. Acta 49, 261–270.

———. (1986) Composition of enstatite (EH 3, EH 4, 5 and EL 6) chondrites: implication regarding their formation. Geochim. Cosmochim. Acta 50, 2153–2164.

Kallemeyn, G. W., Boynton, W. V., Willis J., and Wasson, J. T. (1978) Formation of the Bencubbin polymict meteoritic breccia. Geochim. Cosmochim. Acta 42, 507–515.

Karapet'yants, M. Kh., and Karapet'yants, M. L. (1970) *Thermodynamic Constants of Inorganic and Organic Compounds*, trans. Schmorak, J. Ann Arbor Humphrey Sci. Pub., Ann Arbor, Mich.

Kato, T., Ringwood, A. E., and Irifune, T. (1988) Experimental determination of element partitioning between silicate perovskites, garnets and liquids: constraints on early differentiations of the mantle. Earth Planet. Sci. Lett. 89, 123–145.

Kaufmann, W. J. III (1985) *Universe*. Freeman, New York.

Keil, K. (1968) Mineralogical and chemical relationships among enstatite chondrites. J. Geophys. Res. 73, 6945–6976.

Keller, G. V. (1966) Electrical properties of rocks and minerals. In *Handbook of Physical Constants*, ed. Clark, S. P. Geol. Soc. Am. Mem. 97, 553–578.

Kelly, W. R., and Larimer J. W. (1977) Chemical fractionations in meteorites-VIII. Iron meteorites and the cosmochemical history of the metal phase. Geochim. Cosmochim. Acta 41, 93–111.

Kerr, C. C. (1997) *Asteroid Impact and Mass Extinction at the K-T Boundary, an Extinct Red Herring?* Blackwell, Oxford.

Kerridge, J. F., and Matthews, M. S. (1988) *Meteorites and the Early Solar System.* University of Arizona Press, Tucson, Ariz.

Kesson, S. E., FitzGerald, J. D., and Shelley, J. M. G. (1994) Mineral chemistry and density of subducted basaltic crust at lower mantle pressures. Nature 372, 767–769.

Kissel, J., and Krueger, F. R. (1987) The organic component in dust from comet Halley as measured by the PUMA mass spectrometer on board Vega 1. Nature 326, 755–760.

Klein, D. H., Andren, A. W., Carter, J. A., Emery, J. F., Feldman, C., Fulkerson, W., Lyon, W. S., Ogle, J. C., and Talmi, Y. (1975) Pathways of thirty-seven trace elements through coal-fired power plant. Environ. Sci. Technol. 9, 973.

Kloeck, W., and Palme H. (1987) Partitioning of siderophile and chalcophile elements between sulfide, olivine, and glass in a naturally reduced basalt from Disko Island, Greenland. Proc. Lunar Planet. Sci. Conf. 18 471–483.

Knauss, K., and Ku, T. L. (1983) The elemental composition and decay series radionuclide content of plankton from the east Pacific. Chem. Geol. 39, 125–145.

Knittle, E., and Jeanloz, R. (1991) Earth's core-mantle boundary: Results of experiments at high pressures and temperatures. Science 251, 1438–1443.

Koeberl, C., Weinke, H. H., Kluger, F., and Kiesl, W. (1986) Cape York IIIAB iron meteorite: Trace element distribution in mineral and metallic phases. Mem. Natl. Inst. Polar Res. Spec. Issue 41 (Tokyo), 297–313.

Koide, M., Stallard, M., Hodge, V. F., and Goldberg, E. D. (1986) Preliminary studies on the marine chemistry of ruthenium. Netherlands J. Sea Res. 20, 163–166.

Koide, M., Goldberg, E. D., and Walker, R. (1996) The analysis of seawater osmium. Deep Sea Res. II 43, 53–55.

Kornacki, A. S., and Fegley, B., Jr. (1986) The abundance and relative volatility of refractory trace elements in Allende Ca, Al-rich inclusions: implications for chemical and physical processes in the solar nebula. Earth Planet. Sci. Lett. 79, 217–234.

Krankowsky, D., Laemmerzahl, P., Herrwerth, I., Woweries, J., Eberhardt, P., Dolder, U., Herrman, U., Schulte, W., Berthelier, J. J., Illiano, J. M., Hodges, R. R., and Hoffman, J. H. (1986) In situ gas and iron measurements at comet Halley. Nature 321, 326–329.

Krauskopf, K. B. (1956) Factors controlling the concentrations of thirteen rare metals in sea water. Geochim. Cosmochim. Acta 9, 1–32.

Kresak, L. (1967) Relation of meteor orbits to the orbits of comets and asteroids. In *Meteor Orbits and Dust*, Smithsonian Contribution to Astrophysics, vol. 11, ed. Hawkins, G. S. Sci. Tech. Information Division, NASA, Washington D.C., pp. 9–34.

———. (1979) Dynamical interrelations among comets and asteroids. In *Asteroids*, eds. Gehrels, T., and Matthews, M. S. University of Arizona Press, Tucson, Ariz.

———. (1982) Comet discoveries, statistics, and observational selection. In *Comets*, ed. Wilkening, L. L. University of Arizona Press, Tucson, Ariz., pp 56–82.

———. (1987) The systems of interplanetary objects. In *The Evolution of the Small Bodies of the Solar System*, Proceedings of the International School of Physics "Enrico Fermi", course 98. North-Holland Physics Publications, Amsterdam.

Kulkarni, S. R. (1997) Brown Dwarfs: a possible missing link between stars and planets. Science 276, 1350–1354.

Landsberger, S., Vermette, S. J., and Barrie, L. A. (1990) Multielemental composition of the arctic aerosol. J. Geophys. Res. 95, 3509–3515.

Lantzy, R. J., and Mackenzie, F. T. (1979) Atmospheric trace metals: global cycles and assessment of man's impact. Geochim. Cosmochim. Acta 43, 511–425.

Larimer, J. W. (1967) Chemical fractionations in meteorites —I. Condensation of the elements. Geochim. Cosmochim. Acta 31, 1215–1238.

———. (1973) Chemical fractionation in meteorites —VII. Cosmothermometry and cosmobarometry. Geochim. Cosmochim. Acta 37, 1603–1623.

Larimer, J. W., and Bartholomay, M. (1979) The role of carbon and oxygen in cosmic gases: Some applications to the chemistry and mineralogy of enstatite chondrite. Geochim. Cosmochim. Acta 43, 1455–1466.

Larimer, J. W., and Ganapathy, R. (1987) The trace element chemistry of CaS in enstatite chondrites and some implications regarding its origin. Earth Planet. Sci. Lett. 84, 123–134.

Laul, J. C., Keays, R. R., Ganapathy, R., Anders, E., and Morgan, J. W. (1972) Chemical fractionation in meteorites—V. volatile and siderophile elements in achondrites and ocean ridge basalts. Geochim. Cosmochim. Acta 36, 329–345.

Le Bas, M. J., Le Maitre, R. W., Streckeisen, A., and Zanettin, B. (1986) A chemical classification of volcanic rocks based on the total alkali-silica diagram. J. Petrol. 27, 745–750.

Le Maitre, R. W. (1976) The chemical variability of some common igneous rocks. J. Petrol. 17, 589–637.

———. (1989) *A Classification of Igneous Rocks and Glossary of Terms*. Blackwell Science Publications, Oxford, England.

Lee, T. (1979) New isotopic clues to solar system formation. Rev. Geophys. Space Phys. 17, 1591–1613.

———. (1987) Implications of isotopic anomalies on nucleosynthesis. In *Meteorites and the Early Solar System*, ed. Kerridge, J. F., University of Arizona Press, Tucson, Ariz.

Lee, T., Mayeda, T. K., and Clayton, R. N. (1980) Oxygen isotopic anomalies in Allende inclusion HAL. Geophys. Res. Lett. 7, 493–496.

Legrand, M. R., and Delmas, R. J. (1987) A 220 yr continuous record of volcanic H_2SO_4 in the Antarctic ice sheet. Nature 327, 671–676.

Legrand, M. R., and Mayewski, P. (1997) Glaciochemistry of polar ice cores: a review. Rev. Geophys. 35, 219–243.

Lemarchand, F., Villemont, B., and Calas, G. (1987) Trace element distribution coefficients in alkaline series. Geochim. Cosmochim. Acta 51, 1071–1081.

Levi, G. G. (1992) COBE measures anisotropy in cosmic microwave background radiation. Phys. Today 45(6), 17–20.

Levy, E. H., and Lunine, J. I. (1993) *Protostars and Planets, III*. University of Arizona Press, Tucson, Ariz.

Lewis, J. S. (1972) Low temperature condensation from the solar nebula. Icarus 16, 241–252.

———. (1974) The temperature gradient in the solar nebula. Science 186, 440–442.
———. (1997) *Physics and Chemistry of the Solar System.* Academic Press, San Diego.
Lewis, J. S. and Prinn, R. G. (1984) *Planets and Their Atmospheres.* Academic Press, San Diego.
Li, J., and Agee, C. B. (1996) Geochemistry of mantle-core differentiation at high pressure. Nature 381, 686–689.
Li, Y. H. (1972) Geochemical mass balance among lithosphere, hydrosphere, and atmosphere. Am. J. Sci. 272, 119–137.
———. (1981) Ultimate removal mechanisms of elements from the ocean. Geochim. Cosmochim. Acta 45, 1659–1664.
———. (1982) A brief discussion on the mean oceanic residence time of elements. Geochim. Cosmochim. Acta 46, 2671–2675.
———. (1984) Why are the chemical compositions of living organisms so similar? Schweiz. Z. Hydrol. 4612, 176–184.
———. (1988) Denudation rates of the Hawaiian Islands by rivers and groundwaters. Pacific Science 42, 253–266.
———. (1991) Distribution patterns of the elements in the ocean: A synthesis. Geochim. Cosmochim. Acta 55, 3223–3240.
———. (1992) Seasalt and pollution inputs over the continental United States. Water, Air, Soil Pollution 64, 561–573.
Li, Y. H., Burkhardt, L., Buchholtz, M. O'Hara, P., and Santschi, P. H. (1984) Partition of radiotracers between suspended particles and seawater. Geochim. Cosmochim. Acta 48, 2011–2019.
Livingston, D. A. (1963) The sodium cycle and the age of the ocean. Geochim. Cosmochim. Acta 27, 1055–1069.
London, F. (1930) Theory and system of molecular forces. Z. Physik 63, 245–279.
Loss, R. D., Rosman, K. J. R., and De Laeter, J. R. (1983) Ag, Te and Pd in 17 geochemical reference materials by mass spectrometric isotope dilution analysis. Geostandards Newsletter 7, 321–324.
Luu, J., Marsden, B. G., Jewitt, D., Trujillo, C. A., Hergenrother, C. W., Chen, J., and Offutt, W. B. (1997) A new dynamical class of object in the outer solar system. Nature 387, 658–659.
Lyon, W. S. ed. (1977) *Trace Element Measurements at the Coal-Fired Steam Plant.* CRC Press, Boca Raton, Fla.
Ma, M. S., Murali, A. V., and Schmitt, R. A. (1977) Genesis of the Angra dos Reis and other achondritic meteorites. Earth Planet. Sci. Lett. 35, 331–346.
Maas, R., and McCulloch, M. T. (1991) The provenance of Archean clastic metasediments in the Narryer Gneiss Complex, Western Australia: Trace element geochemistry, Nd isotopes, and U-Pb ages for detrital zirons. Geochim. Cosmochim. Acta 55, 1915–1932.
Mackinnon, I. D. R., and Rietmeijer, F. J. M. (1987) Mineralogy of chondritic interplanetary dust particles. Rev. Geophys. 25, 1527–1553.
MacPherson, G. J., and Grossman, L (1984). Fluffy type A Ca-, Al-rich inclusions in the Allende meteorite. Geochim. Cosmochim. Acta 48, 29–46.
MacPherson, G. J., Wark, D. A., and Armstrong, J. T. (1988) Primitive material surviving in chondrites: refractory inclusions. In *Meteorites and the Early Solar System,* eds. Kerridge, J. F., and Matthews, M. S. University of Arizona Press, Tucson, Ariz.

Maenhaut, W., Cornille, P., Pacyna, J. M. and Vitols, V. (1989) Trace element composition and origin of the atmospheric aerosol in the Norwegian Arctic. Atmos. Environ. 23, 2551–2569.

Mahoney, J. J., Natland, J. H., White, W. M., Poreda, R., Bloomer, S. H., Fisher, R. L., and Baxter, A. N. (1989) Isotopic and geochemical provinces of the western Indian Ocean Spreading centers. J. Geophys. Res. 94, 4033–4052.

Mahood, G. and Hildreth W. (1983) Large partition coefficients for trace elements in high-silica rhyolites. Geochim. Cosmochim. Acta 47, 11–30.

Manheim, F. T., Pratt, R. M., and McFarlin, P. F. (1980) Composition and origin of phosphorite deposits of the Black Plateau. In *Marine Phosphorites*, ed. Bentor, Y. K. Society of Economic Paleontologists and Mineralogists, E. Brothers, Inc., Ann Arbor, Mich.

Margolis, S. H., and Black, J. B. (1985) The heaviest cosmic ray nuclei. Astrophys. J. 299, 334–340.

Marowsky, G., and Wedepohl, K. H. (1971) General trends in the behavior of Cd, Hg, Tl and Bi in some major rock forming processes. Geochim. Cosmochim. Acta 35, 1255–1267.

Martin, J. H., and Knauer, G. A. (1973) The elemental composition of plankton. Geochim. Cosmochim. Acta 37, 1639–1653.

Martin, J. M., and Meybeck, M. (1979) Elemental mass-balance of material carried by major world rivers. Mar. Chem. 7, 173–206.

Martin, J. M., and Whitfield, M. (1981) The significance of the river input of chemical elements to the ocean. In *Trace Elements in Seawater*, ed. Wong, C. S. Plenum, New York.

Martin, P. M., and Mason, B. (1974) Major and trace elements in the Allende meteorite. Nature 249, 333–334.

Mason, B. (1966a) *Principles of Geochemistry*, 3rd ed. Wiley, New York.

———. (1966b) The enstatite chondrites. Geochim. Cosmochim. Acta 30, 23–39.

———. (1971) *Handbook of Elemental Abundances in Meteorites.* Gordon and Breach, New York.

———. (1978) Antarctic meteorite data sheet; Sample No. 30081. Antarct. Meteorite Newsl. 1, 19

———. (1979) *Cosmochemistry. Part I. Meteorites.* U.S.G.S. Prof. Paper 440-B-1. U.S. Government Printing Office, Washington, D.C.

Mason, B., and Graham, A. L. (1970) Minor and trace elements in meteoritic minerals. Smithsonian Contr. Earth Sci. 3.

Mason, B., and Jarosewich, E. (1967) The Winona meteorite. Geochim. Cosmochim. Acta 31, 1097–1098.

Mason, B., and Martin, P. M. (1977) Geochemical differences among components of the Allende meteorite. Smithson. Contr. Earth Sci. 19, 84–95.

Mason, B., and Moore, C. B. (1982) *Principles of Geochemistry*, 4th ed. Wiley, New York.

Mason, B. and Taylor, S. R. (1982) Inclusions in the Allende meteorite. Smithson. Contr. Earth Sci. 25, pp. 30.

Mason, B., and Wilk, H. B. (1962) The Renazzo Meteorite, Am. Mus. Novitates 2106, 1–11.

———. (1966) The compositions of the Bath, Frankfort, Kakangari, Rose City and Tadjera meteorites, Am. Mus. Novitates, 2272.

Mason, B., Jarosewich, E., and Nelen, J. A. (1979) The pyroxene-plagioclase achondrites. Smithson. Contr. Earth Sci. 22, 27–45.

Massalski, T. B. (1986) *Binary Alloy Phase Diagrams*, vols. 1 and 2, American Society of Metals, Metals Park Ohio.

Masuzawa, T., Noriki, S., Kurosaki, T., Tsunagai, S., and Koyama, M. (1989) Compositional change of settling particles with water depth in the Japan Sea. Mar. Chem. 27, 61–78.

Matsunami, S. (1984) The chemical compositions and textures of matrices and chondrule rims of eight unequilibrated ordinary chondrites: A preliminary report. Mem. Natl. Inst. Polar Res. Spec. Issue 35, 126–148.

Mayeda, T. K., and Clayton, R. N. (1980) Oxygen isotopic composition of aubrites and some unique meteorites. Proc. Lunar Planet. Sci. Conf. 11, 1145–1151.

Mayeda, T. K., Clayton, R. N., and Olsen, E. J. (1980) Oxygen isotopic anomalies in an ordinary chondrite. Meteoritics 15, 330–331.

Mayeda, T. K., Clayton, R. N., and Nagasawa, H. (1986) Oxygen isotope variations within Allende refractory inclusions. Lunar Planet. Sci. 17, 526–527.

Mayeda, T., Clayton, R. N., and Yanai, K. (1987) Oxygen isotope compositions of several Antarctic meteorites. Mem. Natl. Polar Res. Spec. Issue 46, 144–150.

Mayewsky, P. A., Lyons, W. B., Twickler, M., Dansgaard, W., Koci, Davidson, C. I., and Honrath, R. E. (1986) Sulfate and nitrate concentrations from a South Greenland ice core. Science 232, 975–977.

McCarthy, M., Pratum,T., Hedges, J., and Benner, R. (1997) Chemical composition of dissolved organic nitrogen in the ocean. Nature 390, 150–154.

McCarthy, T. S., Ahrens, L. H., and Erlank, A. J. (1972) Further evidence in support of the mixing model for howardite origin. Earth Planet. Sci. Lett. 15, 86–93.

McCarthy, T. S., Erland, A. J., and Willis, J. P. (1973) On the origin of eucrites and diogenites. Earth Planet. Sci. Lett. 18, 433–442.

McCarthy, T. S., Erland, A. J., Willis, J. P., and Ahrens, L. H. (1974) New chemical analysis of six achondrites and one chondrite. Meteoritics 9, 215–221.

McCrosky, R. E., Posen, A., Schwartz, G., and Shao, C. Y. (1971) Lost City meteorite—its recovery and a comparison with other fireballs. J. Geophys. Res. 76, 4090–4108.

McDonough, W. F., and Sun, S. S. (1995) The composition of the Earth. Chem. Geol. 129, 223–253.

McKay, G. A. (1989) Partitioning of rare earth elements between major silicate minerals and basaltic melts. In *Geochemistry and Mineralogy of Rare Earth Elements*, eds. Lipin, B. R., McKay, G. A. Rev. Mineral. 21, 45–78.

McKeegan, K. D. (1987) Oxygen isotopes in refractory stratospheric dust particles: Proof of extraterrestrial origin. Science 237, 1468–1471.

McLennan, S. M., and Taylor, S. R. (1979) Rare earth element mobility associated with uranium mineralisation. Nature 282, 247–250.

———. (1991) Sedimentary rocks and crustal evolution: Tectonic setting and secular trends. J. Geol. 99, 1–21.

McSween, H. Y. (1977) Chemical and petrogaphic constraints on the origin of chondrules and inclusions in carbonaceous chondrites. Geochim. Cosmochim. Acta 41, 1843–1860.

———. (1989) Achondrites and igneous processes on asteroids. Annu. Rev. Earth Planet. Sci. 17, 119–140.

Mead, W. J. (1914) The average igneous rocks. J. Geol. 22, 772–781.

Measures, C. I., and Burton, J. D. (1980) Behavior and speciation of dissolved selenium in estuarine waters. Nature 273, 293–295.

Measures, C. I., and Edmond, J. M. (1983) The geochemical cycle of ^{9}Be: A reconnaissance. Earth Planet. Sci. Lett. 66, 101–110.

Mendis, D. A., and Marconi, M. L. (1986) A note on the total mass of comets in the solar system. Earth, Moon and Planets 36, 187–191.

Metzler, K., Bischoff, A., and Stöffler, D. (1992) Accretionary dust mantles in CM chondrites: Evidence for solar nebula processes. Geochim. Cosmochim. Acta 56, 2873–2897.

Meybeck, M. (1979) Concentrations des eaux fluviales en élémens majeurs et apports en solution aux océans. Rev. Géol. Dyn. Géogr. Phys. 21(b), 215–246.

———. (1982) Carbon, nitrogen and phosphorus transport by world rivers. Am. J. Sci. 282, 401–450.

Michard, A., and Albarebe, F. (1986) The REE content of some hydrothermal fluids. Chem. Geol. 55, 51–60.

Miknis, F. P., and McKay, J. F., eds. (1983) *Geochemistry and Chemistry of Oil Shales*. ACS Symposium series 230. American Chemical Society, Washington, D.C.

Milliman, J. D., and Meade, R. H. (1983) World-wide delivery of river sediment to the oceans. J. Geol. 91, 1–21.

Mills, K. C. (1974) *Thermodynamic Data for Inorganic S, Se and Te*. Butterworth, London.

Miyashiro, A. (1974) Volcanic rock series in island arcs and active continental margins. Am. J. Sci. 274, 321–355.

Moore, D. G., and Scott, M. R. (1986) Behavior of Ra-226 in the Mississippi River mixing zone. J. Geophys. Res. 91, 14317–14329.

Moore, J. G., and Fabbi, B. P. (1971) An estimate of the juvenile sulfur content of basalts. Contrib. Mineral. Petrol. 33, 118–127.

Morfill, G. E., Tscharnuter, W., and Völk, H. J. (1985) Dynamical and chemical evolution of the protoplanetary nebula. In *Protostars & Planets II*, eds. Black, D. C., and Matthews, M. S. University of Arizona Press, Tucson, Ariz.

Morgan, J. J., and Stumm, W. (1964) Colloid-chemical properties of manganese dioxide. J. Colloid Sci. 19, 347–359.

Morgan, J. W., Higuchi, H., Takahashi, H., and Hertogen, J. (1978) Achondritic eucrite parent body: Inference from trace elements. Geochim. Cosmochim. Acta 42, 27–38.

Morse, J. W., and Mackenzie, F. T. (1990) *Geochemistry of Sedimentary Carbonates*. Elsevier, Amsterdam.

Morton, J. L., and Sleep, N. H. (1985) A mid-ocean ridge thermal model: Constraints on the volume of axial hydrothermal heat flux. J. Geophys. Res. 90, 11345–11353.

Mottl, M. J. (1983) Metabasalts, axial hot springs, and the structure of hydrothermal systems at mid-ocean ridges. Bull. Geol. Soc. Am. 94, 161–180.

Mulliken, R. S. (1934) A new electroaffinity scale: together with data on valence states and on valance ionization potentials and electron affinities. J. Chem. Phys. 2, 782–793.

Murozumi, M., and Nakamura, S. (1980) Isotope dilution mass spectrometry of copper, cadmium, thallium and lead in marine environments. In *Isotope Marine Chemistry*, eds. E. D. Goldberg, E. D., Horibe, Y., and Suruhashi, K. Uchida Rokakuho, Tokyo, pp. 439–471.

Murozumi, M., Chow, T. J., and Patterson, C. C. (1969) Chemical concentrations of pollutant lead aerosols, terrestrial dusts and sea salts in Greenland and Antarctic snow strata. Geochim. Cosmochim. Acta 45, 1247–1294.

Murray, J. W. (1975) The interaction of metal ions at the manganese dioxide-solution interface. Geochim. Cosmochim. Acta 39, 505–519.

Murrell, M. T., and Burnett, D. S. (1982) Actinide microdistributions in the enstatite meteorites. Geochim. Cosmochim. Acta 46, 2453–2460.

NADP/NTN Annual Data Summary. (1987) *Precipitation Chemistry in the United States, 1987.* Natural Resource Ecology Lab., Colorado State University, Fort Collins, Colo., 353 pp.

Nagahara, H. (1983) Texture of chondrules. Mem. Natl. Inst. Polar Res. Spec. Issue 30, 61–83.

———. (1984) Matrices of type 3 ordinary chondrites—primitive nebular records. Geochim. Cosmochim. Acta 48, 2581–2595.

Nagahara, H., and Ozawa, K., (1985) Petrology of Yamato-791493, Lodranite: melting, crystallization, cooling history, and relationship to other meteorite. Mem. Natl. Inst. Polar Res. Special isssue 36, 181–205.

Nance, W. B., and Taylor, S. R. (1976) Rare earth element patterns and crustal evolution—I. Australian post-Archean sedimentary rocks. Geochim. Cosmochim. Acta 40, 1539–1551.

Nash, D. B., Carr, M. H., Gradie, J., Hunten, D. M. and Yoder, C. F. (1986) Io. In *Satellites*, eds. Burns, J. A., and Matthews, M. S. University of Arizona Press, Tucson, Ariz.

Nash, W. P., and Crecraft, H. R. (1985) Partition coefficients for trace elements in silicic magmas. Geochim. Cosmochim. Acta 49, 2309–2322.

National Environmental Monitoring Center of China (1990) *Elemental Background Values in Chinese Soils* (in Chinese). China Environmental Sci. Press, Beijing.

———. (1994) *The Atlas of Soil Environmental Background Values in the People's Republic of China.* China Environmental Sci. Press, Beijing.

Neftel, A., Oeschger, H., Zhrcher, F., and Finkel, R. C. (1985) Sulphate and nitrate concentrations in snow from South Greenland 1895–1978. Nature 314, 611–613.

Nesbitt, H. W. (1979) Mobility and fractionation of rare earth elements during weathering of a granodiorite. Nature 279, 206–210.

Nesbitt, R. W., and Sun, S. S. (1976) Geochemistry of Archean spinifex textured peridotites and magnesian and low magnesian tholeiites. Earth Planet. Sci. Lett. 31, 433–453.

Neumann, H.-J., Paczynska-Lahme, B., and Severin, D. (1981) *Composition and Properties of Petroleum.* Halsted Press, New York.

Newsom, H. E., and Palme, H. (1984) The depletion of siderophile elements in the Earth's mantle: New evidence from molybdenum and tungsten. Earth Planet. Sci. Lett. 69, 354–364.

Newsom, H. E., White, W. M. Jochum, K. P., and Hofmann, A. W. (1986) Siderophile and chalcophile element abundances in oceanic basalts, Pb isotope evolution and growth of the Earth's core. Earth Planet. Sci. Lett. 80, 299–313.

Niederer, F. R., and Papanastassiou, D. A. (1984) Ca isotope in refractory inclusions. Geochim. Cosmochim. Acta 48, 1279–1293.

Nittler, L. R., O'D. Alexander, C. M., Gao, X., Walker, R. M., and Zinner, E. K. (1994) Interstellar oxide grains from the Tieschitz ordinaary chondrite. Nature 370, 443–446.

Nriagu, J. O. (1989) A global assessment of natural sources of atmospheric trace metals. Nature 338, 47–49.

Nyffeler, U. P., Li, Y. H., and Santschi, P. H. (1984) A kinetic approach to describe trace element distribution between particles and solution in natural aquatic systems. Geochim. Cosmochim. Acta 48, 1513–1522.

Ochiai, E. I. (1987) *General Principles of Biochemistry of the Elements*. Plenum, New York.

Ohtani, E., and Ringwood, A. E. (1984) Composition of the core I: Solubility of oxygen in molten iron at high temperature. Earth Planet. Sci. Lett. 71, 85–93.

Ohtani, E., Ringwood, A. E., and Hibberson, W. (1984) Composition of the core II: Effect of high pressure on solubility of FeO in molten iron. Earth Planet. Sci. Lett. 71, 94–103.

Olmez, I., Finnegan, D. L., and Zoller, W. H. (1986) Iridium emissions from Kilauea volcano. J. Geophys. Res. 91, 653–663.

Oort, J. H. (1950) The structure of the cometary cloud surrounding the solar system and a hypothesis concerning its origin. Bull. Astron. Inst. Netherland 11, 91–110.

Orians, K. J., Boyle, E. A., and Bruland, K. W. (1990) Dissolved titanium in the open ocean. Nature 348, 322–325.

Ostdick, V. J., and Bord, D. J. (1991) *Inquiry into Physics*, 2nd ed. West Publ. Co., St. Paul, Minn.

Ostwald, J. (1988) Mineralogy of the Groote Eyland manganese oxides: A review. Ore. Geol. Rev. 4, 3–45.

Othman, D. B., White, W. M., and Patchett (1989) The geochemistry of marine sediments, island arc magma genesis, and crust-mantle recycling. Earth Planet. Sci. Lett. 94, 1–21.

Ott, U. (1993) Interstellar grains in meteorites. Nature 364, 25–33.

O'Dell, C. B., and Beckwith, S. V. W. (1997) Young stars and their surroundings. Science 276, 1355–1359.

Ozima, M., and Podosek, F. A.(1983) *Noble Gas Geochemistry*. Cambridge University Press, Cambridge, England.

Palacz, Z., and Saunders, A. D. (1986) Coupled trace element and isotope enrichment in the Cook-Austral-Samoa islands, southwest Pacific. Earth Planet. Sci. Lett. 79, 270–280.

Palme, H., and Wlotzka, F. (1976) A metal particle from a Ca, Al rich inclusion from the meteoritic Allende, and the condensation of refractory siderophile elements. Earth Planet. Sci. Lett. 33, 45–60.

Palme, H., Schultz, L., Spettel, B., Weber, H. W., Wanke, H., Christophe Michel-Levy, M., and Lorin, J. C. (1981) The Acapulco meteorite; chemistry, mineralogy and irradiation effects. Geochim. Cosmochim. Acta 45, 727–752.

Palmer, M. R., and Edmond, J. M. (1989) The strontium isotope budget of the modern ocean. Earth Planet. Sci. Lett. 92, 11–26.

Parks, G. A. (1967) Aqueous surface chemistry of oxides and complex oxide minerals. Isoelectronic point and zero point of charge. In *Equilibrium Concepts in Natural Water Systems*, Advances in chemistry series 67 ed. Stumm, W.G. American Chemical Society, Washington, D.C., 344 pp.

Patterson, S. H. (1971) *Investigation of Ferruginous Bauxite and Other Mineral Resources on Kauai and a Reconnaissance of Ferruginous Bauxite Deposits on Maui, Hawaii.* U.S.G.S. Prof. Paper 656. U.S. Government Printing Office, Washington, D.C.

Pauling, L. (1927) The theoretical prediction of the physical properties of many electron atoms and ions; mole refraction, diamagnetic susceptibility and extension in space. Proc. R. Soc. London A 114, 181–211.

———. (1932) The nature of the chemical bond IV. The energy of single bonds and the relative electronegativity of atoms. J. Am. Chem. Soc. 54, 3570–3582.

———. (1960) *The Nature of the Chemical Bond*, 3rd ed. Cornell University Press, Ithaca, N.Y.

Pauling, L., and Pauling, P. (1975) *Chemistry*. W. H. Freeman & Co., New York.

Penzias, A. A., and Wilson, R. W. (1965) A measurement of excess antenna temperature at 4080 Mc/s. Astrophys. J. 142, 419–421.

Perlmutter, S., Aldering, G., Della Valle, M., Deustua, S., Ellis, R. S., Fabbro, S., Fruchter, A., Goldhaber, G., Groom, D. E., Hook, I. M., Kim, A. G., Kim, M. Y., Knop, R. A., Lidman, C., McMahon, R. G., Nugent, P., Pain, R., Panagia, N., Pennypacker, C. R., Ruiz- Lapuente, P., Schaefer, B., and Walton, N. (1998) Discovery of a supernova explosion at half the age of the Universe. Nature 391, 51–54.

Pernicka, E., and Wasson, J. T. (1987) Ru, Re, Os, Pt and Au in iron meteorites. Geochim. Cosmochim. Acta 51, 1717–1726.

Perry E. C., Jr., Tan, F. C., and Morey, G. B. (1973) Geology and stable isotope geochemistry of the Biwabik iron formation, northern Minnesota. Econ. Geol. 68, 1110–1125.

Peterson, P. J. (1971) Unusual accumulations of elements by plants and animals. Sci. Prof. Oxf. 59, 505–526.

Pettijohn, F. J. (1957) *Sedimentary Rocks*, 2nd ed. Harper & Brothers, New York.

Pettijohn, F. J., Potter, P. E., and Siever, R. (1987) *Sand and Sandstone*, 2nd ed. Springer-Verlag, New York.

Pieters, C. M., and McFadden, L. A. (1994) Meteorite and asteroid reflectance spectroscopy. Annu. Rev. Earth Planet. Sci. 22, 457–497.

Pollack, J. B. (1985) Formation of the giant planets and their satellite-ring system: An overview. In *Protostars and Planets, II*, ed. Black, D. C., and Matthews, M. S. University of Arizona Press, Tucson, Ariz.

Potts, P. J. (1993) Laboratory methods of analysis. In *Analysis of Geological Materials*. ed. Riddle, C. Marcel Dekker.

Prinz, M., Waggoner, D. G., and Hamilton, P. J. (1985) Winonaites: a primitive achondritic group related to silicate inclusions in IAB irons. Lunar Planet. Sci. 14, 902–904.

Rambaldi, E. (1976) Trace element content of metals from L-group chondrites. Earth Planet. Sci. Lett. 31, 224–238.

———. (1977) The content of Sb, Ge and refractory siderophile elements in metals of L-group chondrites. Earth Planet. Sci. Lett. 33, 407–419.

Rambaldi, E., and Cendales, M. (1980) Siderophile element fractionation in enstatite chondrites. Earth Planet. Sci. Lett. 48, 325–334.

Rammensee, W., Palme, H., and Waenke, H. (1983) Experimental investigation of metal-silicate partitioning of some lithophile elements (Ta, Mn, V, Cr). Proc. Lunar Planet. Sci. Conf. 14, 628–629.

Rankama, K., and Sahama, Th. G. (1950) *Geochemistry*. University of Chicago Press, Chicago.

Rard, J. A. (1985) Chemistry and thermodynamics of ruthenium and some of its inorganic compounds and aqueous species. Chem. Rev. 85, 1–39.

Redfield, A. C., Ketchum, B. H., and Richards, F. A. (1963) The influence of organisms on the composition of sea-water. In *The Sea*, Vol. 2, ed. Hill, M. N. Interscience, New York, pp. 26–77.

Remy, F., and Mignard, F. (1985) Dynamical evolution of the Oort cloud I. A Monte Carlo simulation. Icarus 63, 1–19.

Reyment, R., and Joereskog, K. G. (1993) *Applied Factor Analysis in the Natural Sciences*. Cambridge University Press, Cambridge, England.

Richardson, S. H., Erlank, A. J., Duncan, A. R., and Reid, D. L. (1982) Correlated Nd, Sr, and Pb isotope variation in Walvis Ridge basalts and implications for the evolution of their mantle source. Earth Planet. Sci. Lett. 59, 327–342.

Richter, F. M. (1979) Focal mechanism and seismic energy release of deep and intermediate earthquakes in the Tonga-Kermadec region and their bearing on the depth extent of mantle flow. J. Geophys. Res. 84, 6783–6795.

Rietmeijer, F. J. M. (1988) On a chemical continuum in early solar system dust at > 1.8 AU. Chem. Geol. 70, 33 (abstract).

Ringwood, A. E. (1975) *Composition and Petrology of the Earth's Mantle*. McGraw-Hill, New York.

———. (1989) Significance of the terrestrial Mg/Si ratio. Earth Planet. Sci. Lett. 95, 1–7.

Riordan, M., and Schramm, D. N. (1991) *The Shadows of Creation: Dark Matter and the Structure of the Universe*. Freeman., New York.

Robie, R. A., Hemingway, B. S., and Fisher, J. R. (1978) *Thermodynamic Properties of Minerals and Related Substances at 298.15 K and 1 Bar Pressure and at Higher Temperature*. U.S.G.S. Bulletin 1452. U.S. Government Printing Office, Washington, D.C.

Roine, A. (1994) *HSC Chemistry for Windows*. Outkumpu Research Oy, Finland.

Ronov, A. B. (1982) The Earth's sedimentary shell. Int. Geology Rev. 24, 1313–1388.

Ronov, A. B., and Yaroshevskiy, A. A. (1976) A new model for the chemical structure of the earth's crust. Geochem. Int. 13 (no. 6), 89–121.

Ross, J. E., and Allen, L. H. (1976) The chemical composition of the sun. Science 191, 1223–1229.

Rubey, W. W. (1951) Geologic history of sea water. An attempt to state the problem. Geol. Soc. Am. Bull. 62, 1111–1148.

Rubin, A. E. (1986) Elemental compositions of major silicic phases in chondrules of un-equilibrated chondritic meteorites. Meteoritics 21, 283–293.

Rubin, A. E., and Wasson, J. T. (1987) Chondrules, matrix and coarse-grained chondrule rims in the Allende meteorite: Origin, interrelationships and possible precursor components. Geochim. Cosmochim. Acta 51, 1923–1937.

———. (1988) Chondrules and matrix in the Ornans CO3 meteorite: possible precursor components. Geochim. Cosmochim. Acta 52, 425–432.

Ruch, R. R., Gluskoter, H. J., and Shimp, N. F. (1974) *Occurrence and Distribution of Potentially Volatile Trace Elements in Coal: A Final Report.* Illinos State Geological Survey Environmental Geology Note 72. Urbana, Ill.

Rudnick, R. L., and Goldstein, S. L. (1990) The Pb isotopic compositions of lower crustal xenoliths and the evolution of lower crustal Pb. Earth Planet. Sci. Lett. 98, 192–207.

Safronov, V. S., and Ruzmaikina, T. V. (1985) Formation of the solar nebula and the planets. In *Protostars and Planets II*, eds. Black, D. C., and Matthews, M. S. University of Arizona Press, Tucson, Ariz.

Sagdeev, R. Z., Elyasberg, P. E., and Moroz, V. I. (1988) Is the nucleus of comet Halley a low density body? Nature 331, 240–242.

Samsonov, G. V. (1973) *The Oxide Handbook*, trans. Turton, C. N., and Turton, T. L. IFI/Plenum, New York.

Sandage, A. (1957) Observational approach to evolution, II. A computed luminosity function for KO-K2 stars from Mv = +5 to Mv = −4.5. Astrophys. J. 125, 435–444.

Sasaki, S., and Nakazawa, K. (1986) Metal-silicate fractionation in the growing earth: Energy source for the terrestrial magma ocean. J. Geophys. Res. 91(B9), 9231–9238.

Sato, H., Sacks, I. S., Takahashi, E., and Scarfe, C. M. (1988) Geotherms in the Pacific Ocean from laboratory and seismic attenuation studies. Nature 336, 154–156.

Saxena, S. K. (1977) A new electronegativity scale for geochemists. In *Energetics of Geological Processes*, eds. Saxena, S. K., and Bhattacharji, S. Springer-Verlag, New York.

Saxena, S. K. and Eriksson, G. (1983) High temperature phase equilibria in a solar composition gas. Geochim. Cosmochim. Acta 47, 1865–1874.

Schaule, B. K., and Patterson, C. C. (1981) Lead concentrations in the northeast Pacific: Evidence for global anthropogenic perturbations. Earth Planet. Sci. Lett. 54, 97–116.

———. (1983) Perturbations of the natural lead depth profile in the Sargasso Sea by industrial lead. In *Trace Metals in Sea Water*, ed. Wong, C. S. Plenum, New York, pp. 487–503.

Schilling, J. G., Zajac, M., Evans, R. Johnston, T., White, W., Devine, J. D., and Kingsley, R. (1983) Petrologic and geochemical variations along the mid-Atlantic ridge from 29°N to 73°N, Am. J. Sci. 283, 510–586.

Schindler, P. W. (1975) Removal of trace metals from the oceans: A zero order model. Thalassia Jugosl. 11, 101–111.

Schindler, P. W., and Stumm, W. (1987) The surface chemistry of oxides, hydroxides, and oxide minerals, in *Aquatic Surface Chemistry*, ed. Stumm, W. Wiley, New York, pp. 83–110.

Schindler, P. W., Furst, B., Dick, R., and Wolf, P. U. (1976) Ligand ties of surface silanol groups I. Surface complex formation with Fe^{3+}, Cu^{2+}, Cd^{2+} and Pb^{2+}. J. Colloid Interface Sci. 55, 469–475.

Schmitt, W., Palme, H., and Waenke, H. (1989) Experimental determination of metal/silicate partition coefficients for P, Co, Ni, Cu, Ga, Ge, Mo and W and some implications for the early evolution of the Earth. Geochim. Cosmochim. Acta 53, 173–185.

Scholl, H. (1987) Dynamics of asteroids. In *The Evolution of the Small Bodies of the Solar System*, Proceedings of the International School of Physics "Enrico Fermi", Course 98. North-Holland Physics Publications, Amsterdam.

Schramm, L. S., Brownlee, D. E., and Wheelock, M. M. (1988) The elemental composition of interplanetary dust. Lunar Planet. Sci. Conf. 19, 1033–1034.

Schubert, G., Spohn, T., and Reynolds, R. T. (1986) Thermal histories, compositions and internal structures of the moon of the solar system. In *Satellites*, eds. Burns, J. A., and Matthews, M. S. University of Arizona Press, Tucson, Ariz.

Schultz, L. G., Tourtelot, H. A., Gill, J. R., and Boerngen, J. G. (1980) *Composition and Properties of the Pirre Shale and Equivalent Rocks: Northern Great Plains Region*, U.S.G.S. Prof. Paper 1064-B. U.S. Government Printing Office, Washington D.C.

Schwarzenback, G. (1957) *Complexometric Titrations*. Interscience, New York.

Scott, E. R. D. (1972) Chemical fractionation in iron meteorites and its implication. Geochim. Cosmochim. Acta 36, 1205–1236.

Scott, E. R. D., and Taylor, G. J. (1983) Chondrules and other components in C, O, and E chondrites: Similarities in their properties and origins. J. Geophys. Res. Suppl. 88 B275–B286.

Scott, E. R. D., and Wasson, J. A. (1975) Classification and properties of iron meteorites. Rev. Geophys. Space Phys. 13, 527–546.

Sears, D. W. G., and Dodd, R. T. (1988) Overview and classification of meteorites. In *Meteorites and the Early Solar System*. eds. Kerridge, J. F., and Matthews M. S. University of Arizona Press, Tucson, Ariz.

Seeger, P. A., Fowler, W. A., and Clayton, D. D. (1965) Nucleosynthesis of heavy elements by neutron capture. Astrophys. J. Suppl. 11, 121–166.

Shacklette, H. T., Hamilton, J. G., Boerngen, J. G., and Bowles, J. M. (1971) *Elemental Composition of Surficial Materials in the Conterminous United States*, U.S.G.S. Prof. Paper 574. U.S. Government Printing Office, Washington, D.C.

Shanks III, W. C., Boehlke, J. K., and Seal II, R. R. (1995) Stable isotopes in mid-ocean ridge hydrothermal systems: interactions between fluids, minerals and organisms. In *Seafloor Hydrothermal Systems*. eds. Humphris, S. F., Zierenberg, R. A., Mullineaux, L. S., and Thomson, R. E. American Geophysical Union, Washington, D.C.

Shannon, R. D. (1976) Revised effective ionic radii and systematic studies of interatomic distances in halides and chalcogenides. Acta Cryst. A 32, 751–767.

Shannon, R. D., and Prewitt, C. T. (1969) Effective ionic radii in oxides and fluorides. Acta Cryst. B 25, 925–946.

Shaw, A. L., and Guilbert, J. M. (1990) Geochemistry and metallogeny of Arizona paer-aluminous granitoids with reference to Appalachian and European occurrences. In *Ore-Bearing Granite Systems; Petrogenesis and Mineralizing Processes*, Geol. Soc. Am. Special Paper 246 317–356. Geological Society of America, Boulder, Colo.

Shaw, D. M. (1970) Trace element fractionation during anatexis. Geochim. Cosmochim. Acta 34, 237–243.

Shaw, D. M., Reilly, G. A., Muysson, J. R., Pattenden, G. E., and Campbell, F. E. (1967) An estimate of the chemical composition of the Canadian Precambrian Shield. Can. J. Earth Sci. 4, 829–853.

Shaw, D. M., Dostal, J., and Keays, R. R. (1976) Additional estimates of continental surface Precambrian shield composition in Canada. Geochim. Cosmochim. Acta 40, 77–83.

Sherman, G. D. (1952) The titanium content of Hawaiian soils and its significance. Soil Sci. Proc. 16, 15–18.

Shiller, A. M., and Boyle, E. A. (1985) Dissolved zinc in rivers. Nature 317, 49–52.

———. (1987) Dissolved vanadium in rivers and estuaries. Earth Planet. Sci. Lett. 86, 214–224.

Sholkovitz, E. R. (1988) Rare earth elements in the sediments of north Atlantic Ocean, Amazon Delta, and East China Sea: Reinterpretation of terrigenous input patterns to the oceans. J. Am. Sci. 288, 236–281.

Sibley, D. F., and Wilband, J. T. (1977) Chemical balance of the Earth's crust. Geochim. Cosmochim. Acta 41, 545–554.

Silk, J. (1989) *The Big Bang*. Freeman, New York.

Simpson, A. B., and Ahrens, L. H. (1977) The chemical relationship between howardites and the silicate fraction of mesosiderites. In *Comets, Asteroids and Meteorites*, ed. Delsemme, A. University of Toledo Press, Toledo, Ohio, pp. 445–450.

Simpson, J. A. (1983) Elemental and isotopic composition of the Galactic cosmic rays. Annu. Rev. Nucl. Part. Sci. 33, 327–381.

Smith, B. A., and Terrile, R. J. (1984) A circumstellar disk around Pictoris. Science 226, 1421–1424.

Smith, E. V. P., and Jacobs, K. C. (1973) *Introductory Astronomy and Astrophysics*. Saunders, Philadelphia.

Smith, J. W. (1983) The chemistry that formed Green River Formation oil shale. In *Geochemistry and Chemistry of Oil Shales*, eds. Miknis, F. P., and McKay, J. F. ACS Symposium Series 230. American Chemical Society, Washington D.C.

Smith, M. R., Laul, J. C., Ma, M. S., Huston, T., Verkouteren, R. M., Lipschutz, M. E., and Schmitt, R. A. (1984) Petrogenesis of the SNC meteorites: Implication for their origin from a large dynamic planet, possibly Mars. J. Geophys. Res. Suppl. 89, B612–B630.

Smith, R. M., and Martell, A. E. (1976) *Critical Stability Constants: Inorganic Complexes*, vol. 4. Plenum, New York.

Snow, J. E., and Schmidt, G. (1998) Constraints on Earth accretion deduced from noble metals in the oceanic mantle. Nature 391, 166–169.

Snyder, W. S., Cook, M. I., Nasset, E. S., Karhausen, L. R., Howells, G. P., and Tipton, I. H. (1975) *Report of the Task Group on Reference Man*, International Commission on Radiological Protection No. 23. Pergamon, New York.

Spencer, D. W., Brewer, P. G., Fleer, A., Honjo, S., Krishnaswami, S., and Nazaki, Y. (1978) Chemical fluxes from a sediment trap experiment in the deep Sargasso Sea. J. Mar. Res. 36, 493–523.

Spencer, J. R., and Mitton J. eds. (1995) *The Great Comet Crash: the Impact of Comet Shoemaker-Levy 9 on Jupiter*. Cambridge University Press, Cambridge, England.

Stallard, R. F., and Edmond, J. M. (1981) Geochemistry of the Amazon 1: Precipitation chemistry and the marine contribution to the dissolved load. J. Geophys. Res. 86(C10), 9844–9858.

Statistical Abstract of the United States (1941, 1989) U.S. Dept. of Commerce, Bureau of the Census, Washington, D.C.

Staudigel, H., Plank, T., White, W., and Schmincke, H. U. (1996) Geochemical fluxes during seafloor alteration of the upper oceanic crust: DSDP sites 417–418. In *Subduction Top to Bottom*, eds. Bebout, G. E., Scholl, D. W., Kirby, S. H., and Platt, J. P. American Geophysical Union, Washington, D.C.

Stevenson, D. J. (1981) Model of the Earth's core. Nature 214, 611–619.

———. (1982) Interiors of the giant planets. Ann. Rev. Earth Planet. Sci. 10, 257–295.

Stewart, G. R. (1997) The frontier beyond Neptune. Nature 387, 658–659.

Stolper, E. (1982) Crystallization sequences of Ca-Al-rich inclusions from Allende: An experimental study. Geochim. Cosmochim. Acta 46, 2159–2180.

Stumm, W., and Morgan, J. J. (1981) *Aquatic Chemistry*, 2nd ed. Wiley, New York.

Stumm, W., Huang, C. P., and Jenkins, S. R. (1970) Specific chemical interaction affecting the stability of dispersed system. Croat. Chem. Acta 42, 223–245.

Suess, H. E., and Urey, H. C. (1956) Abundances of the elements. Rev. Mod. Phys. 28, 53–74.

Sugimura, Y., and Suzuki, Y. (1985) A method of chemical speciation of metallic elements dissolved in sea water by using XAD-2 resin. Papers Meteorol. Geophys. 36, 187–207.

Sun, S. S. (1982) Chemical composition and origin of the earth's primitive mantle. Geochim. Cosmochim. Acta 46, 179–192.

Sverdrup, H. U., Johnson, M. W., and Fleming, R. H. (1942) *The Oceans: Their Physics, Chemistry, and General Biology*. Prentice-Hall Englewood, Cliffs, N.J.

Symonds, R. B., Rose, W. I., Reed, M. H., Lichte, F. E., and Finnegan, D. L. (1987) Volatilization, transport and sublimation of metallic and non-metallic elements in high temperature gases at Merapi Volcano, Indonesia. Geochim. Cosmochim. Acta 51, 2083–2102.

Takahashi, E. (1983) Melting of a Yamato L3 chondrite (Y74191) up to 30 Kbar. Mem. Natl. Inst. Polar Res. Spec. issue 30, 168–180.

———. (1986) Melting of a dry peridotite KLB-1 up to 14 GPa: Implications on the origin of peridotite upper mantle. J. Geophys. Res. 91, B9, 9367–9382.

Takahashi, E., and Scarfe, E. M. (1985) Melting of peridotite to 14 GPa and the genesis of komatiite. Nature 351, 566–570.

Takahashi, E., and Ito, E. (1987) Mineralogy of mantle peridotite along a model geotherm up to 700 km depth. In *High-Pressure Research in Mineral Physic*, Geophysical Monograph 39, eds. Manghnani, M. H., and Syono, Y. American Geophysical Union, Washington, D.C., pp. 427–438.

Takeda, H. (1986/87) Mineralogy of Antarctic ureilites and a working hypothesis for their origin and evolution. Earth Planet. Sci. Lett. 81, 358–370.

Taran, Y. A., Hedenquist, J. W., Korzhinsky, M. A., Tkachenko, S. I., and Shmulovich, K. I. (1995) Geochemistry of magmatic gases from Kudryavy volcano, Iturup, Kuril Islands. Geochim. Cosmochim. Acta 59, 1759–1762.

Tardy, Y., and Garrels, R. M. (1976) Prediction of Gibbs free energies of formation, I. Relationships among Gibbs free energies of formation of hydroxides, oxides and aqueous ions. Geochim. Cosmochim. Acta 40, 1051–1056.

Taylor, S. R. (1964) Abundance of elements in the continental crust: A new table. Geochim. Cosmochim. Acta 28, 1273–1285.

Taylor, S. R., and McLennan, S. M. (1985) *The Continental Crust: Its Composition and Evolution*. Blackwell Science Publications, Boston.

Taylor, S. R., McLennan, S. M., and McCulloch, M. T. (1983) Geochemistry of loess, continental crust composition and crustal model ages. Geochim. Cosmochim. Acta 47, 1897–1906.

The National Atlas of the United States of America (1970) U.S. Geological Survey, Washington, D.C.

Thiemens, M. H., and Heidenreich, J. E. (1983) The mass independent fractionation of oxygen: A novel isotope effect and its possible cosmochemical implication. Science 219, 1073–1075.

Tholen, D. J. (1984) *Asteroid Taxonomy from Cluster Analysis of Photometry*. Ph.D. dissertation, University of Arizona.

———. (1989) Results from five years of Pluto-Charon mutual event observations. Eos, Trans. Am. Geophys. Union, 70, 385.

Tomeoka, K., and Buseck, P. R. (1985) Indicators of aqueous alteration in CM carbonaceous chondrites: Microtextures of a layered mineral containing Fe, S, O, and Ni. Geochim. Cosmochim. Acta 49, 2149–2163.

Trimble, V. (1975) The origin and abundances of the chemical elements. Rev. Mod. Phys. 47, 877–968.

———. (1982) Supernovae Part I: the events. Rev. Mod. Phys. 54, 1183–1218.

———. (1983) Supernovae Part II: the aftermath. Rev. Mod. Phys. 55, 511–552.

Trivedi, B. M. P. (1984) Mass loss from the protosun: formation and evolution of the solar nebula. Astrophys. J. 218, 375–380.

———. (1987) Chemical condensation in the outflowing matter from the protosun and its application to meteorites. Astrophys. J. 320, 430–436.

Tromp, J., and Dziewonski, A. M. (1998) Two views of the deep mantle. Science 281, 655–656.

Truran, J. W. (1984) Nucleosynthesis. Annu. Rev. Nucl. Part. Sci. 34, 53–97.

Tsurutani, B. T. (1989) The Voyager 2 Neptune encounter. Eos, Trans. Am. Geophys. Union, 70, 915–921.

Tuncel, G., Aras, N. K., and Zoller, W. H. (1989) Temporal variations and sources of elements in the south pole atmosphere. J. Geophys. Res. 94, 13025–13038.

Turekian, K. K. (1971) Rivers, tributaries, and estuaries. In *Impingement of Man on the Ocean*, ed. Hood, D. W. Wiley-Interscience, New York.

Turekian, K. K., and Wedepohl, K. H. (1961) Distribution of the elements in some major units of the Earth's crust. Geol. Soc. Am. Bull. 72, 175–192.

Urban, N. R. (1994) Retention of sulfur in lake sediments. In *Environmental Chemistry of Lakes and Reservoirs*, ACS Advances in Chemistry Series 237, ed. Baker, L. A. American Chemical Society, Washington, D.C.

Urey, H. C. (1952) Chemical fractionation in the meteorites and the abundance of the elements. Geochim. Cosmochim. Acta 2, 267–282.

Urey, H. C., and Craig, H. (1953) The composition of the stone meteorites and the origin of the meteorites. Geochim. Cosmochim. Acta 4, 36–82.

U.S.G.S. Water Resources Data Louisiana (1987) National Technical Information Services, Springfield, Va.

van der Hilst, R. D., Engdahl, R., Spakman, W., and Nolet, G. (1991) Tomographic imaging of subducted lithosphere below northwest Pacific island arcs. Nature 353, 37–43.

van der Hilst, R. D., Widiyantoro, S., and Engdahl, E. R. (1997) Evidence for deep mantle circulation from global tomography. Nature 386, 578–584.

van der Marel, R. P., Tim de Zeeuw, P., Rix, H.-W., and Quinlan, G. D. (1997) A massive black hole at the centre of the quiescent galaxy M32. Science 385, 610–612.

Van Zeggeren, F., and Storey, S. H. (1970) *The Computation of Chemical Equilibria*. Cambridge University Press, New York.

Vidale, J. E., and Hedlin, M. A. H. (1998) Evidence for partial melt at the core-mantle boundary north of Tonga from the strong scattering of seismic waves. Nature 391, 682–685.

Villemant, B., Jaffrezic, H., Joron, J., and Treuil, M. (1981) Distribution coefficients of major and trace elements; fractional crystallization in the alkali basalt series of Chaine des Puys (Massif Central, France). Geochim. Cosmochim. Acta 45, 1997–2016.

Vinogradov, A. P. (1953) *The Elementary Chemical Composition of Marine Organisms*. Sears Foundation, New York.

———. (1956) Regularities of distribution of chemical elements in the Earth's crust. Geokhimiya 1956, 1–43.

———. (1959) *Geochemistry of Rare and Dispersed Elements in Soils*. Consultants Bureau Enterprises. New York.

Von Damm, K. L. (1990) Sea floor hydrothermal activity: Black smoker chemistry and chimneys. Annu. Rev. Earth Planet. Sci. 18, 173–204.

———. (1995) Controls on the chemistry and temporal variability of seafloor hydrothermal fluids. In *Seafloor Hydrothermal Systems*. eds. Humphris, S. F., Zierenberg, R. A., Mullineaux, L. S., and Thomson, R. E. American Geophysical Union, Washington, D.C.

Von Damm, K. L., Edmond, J. M., Grant, B., Measures, C. I., Walden, B., and Weiss, R. F. (1985) Chemistry of submarine hydrothermal solutions at 21°N, East Pacific Rise. Geochim. Cosmochim. Acta 49, 2197–2220.

Waenke, H., Dreibus, G., and Jagoutz, E. (1984) Mantle chemistry and accretion history of the Earth. In *Archaean Geochemistry*, eds. Kroner, A., Hanson, G. N., and Goodwin, A. M. Springer-Verlag, New York.

Wagman, D. D., Evans, W. H., Parker, V. B., Schumm, R. H., Halow, I., Bailey, S. M., Churney, K. L., and Nuttall, R. L. (1982) The NBS tables of chemical thermodynamics properties. J. Phys. Chem. Ref. Data. 11, Supplement 2.

Wai, C. M., and Wasson, J. T. (1977) Nebular condensation of moderately volatile elements and their abundances in ordinary chondrites. Earth Planet. Sci. Lett. 36, 1–13.

———. (1979) Nebular condensation of Ga, Ge and Sb and the chemical classification of iron meteorites. Nature 282, 790–793.

Wai, C. M., Wasson, J. T., Willis, J., and Kracher, A. (1978) Nebular condensation of moderately volatile element, their abundances in iron meteorites, and the quantization of Ge and Ga abundances. Lunar Planet. Sci. 9, 1193–1195.

Walker, D., Norby, L., and Jones, J. H. (1993) Superheating effects on metal-silicate partitioning of siderophile elements. Science 262, 1858–1861.

Walker, R. J. , Morgan, J. W., and Horan, M. F. (1995) Osmium-187 enrichment in some plumes: evidence for core-mantle interaction. Science 269, 819–822.

Wang, D., and Xie, X. (1977) Preliminary investigation of mineralogy, petrology and chemical composition of Qingzhen enstatite chondrite. Geochimica (China) 277–286.

Wark, D. A. (1981) The pre-alteration compositions of Allende Ca-Al-rich condensates. Lunar Planet. Sci. 12, 1148–1150.

———. (1987) Plagioclase-rich inclusions in carbonaceous chondrite meteorites: Liquid condensates? Geochim. Cosmochim. Acta 51, 221–242.

———. (1996) Evidence for successive episodes of condensation at high temperature in a part of the solar nebula. Earth Planet. Sci. Lett. 77, 129–148.

Wark, D. A., and Lovering, J. F. (1982) The nature and origin of type B1 and B2 Ca-Al-rich inclusions in Allende meteorite. Geochim. Cosmochim. Acta 46, 2581–2594.

Wark, D. A., Boynton, W. V., Keays, R. R., and Palme, H. (1987) Trace element and petrologic clues to the formation of forsterite-bearing Ca-Al-rich inclusions in the Allende meteorite. Geochim. Cosmochim. Acta 51, 607–622.

Warneck, P. (1988) *Chemistry of the Natural Atmosphere*. Academic Press, New York.

Wasserburg, G. J., Lee, T., and Papanastassiou (1977) Correlated O and Mg isotopic anormalies in Allende inclusions: II Mg. Geophys. Res. Lett. 4. 299–302.

Wasserburg, G. J., Papanastassion, D. A., and Lee, T. (1980) Isotopic heterogeneities in the solar system. In *Early Solar System Processes and the Present Solar System*, ed. Lal, D. Italian Physical Society, North-Holland, Amsterdam, pp. 144–191.

Wasson, J. T. (1974) *Meteorites: Classification and Properties*. Springer-Verlag, New York.

———. (1985) *Meteorites—Their Record of Early Solar System History*. Freeman, New York.

Wasson, J. T., and Kallemeyn, G. W. (1988) Compositions of chondrites. Phil. Trans. R. Soc. London A325, 535–544.

Wasson, J. T., and Wetherill, G. W. (1979) Dynamical chemical and isotopic evidence regarding the formation locations of asteroids and meteorites. In *Asteroids*, ed. Gehrels, T. University of Arizona Press, Tucson, Ariz., pp. 926–974.

Wasson, J. T., Chou, C. L., Bild, R., and Baedecker, P. A. (1976) Classification of an elemental fractionation among ureilites. Geochim. Cosmochim. Acta 40, 1449–1458.

Watters, T. R., and Prinz, M. (1979) Aubrites: Their origin and relationship to enstatite chondrites. Proc. Lunar Sci. Conf. 10, 1073–1093.

Weaver, T. A., and Woosley, S. E. (1980) Evolution and explosion of massive stars. In *Ninth Texas Symposium on Relativistic Astrophysics*. Ann. N.Y. Acad. Sci., 336, 335–357.

Weaver, T. A., Woosley, S. E., and Fuller, G. M. (1983) In *Proceedings of the Conference on Numerical Astrophysics*, eds. Bowers, R., Centrella, J., LeBlanc, J., and LeBlanc, M., Science Books International, Boston.

Wedpohl, K. H., executive ed. (1969–1978) *Handbook of Geochemistry*. Springer-Verlag, New York, 5 volumes.

———. (1995) The composition of the continental crust. Geochim. Cosmochim. Acta 59, 1217–1232.

Wedepohl, K. H., and Muramatsu, Y. (1979) The chemical composition of Kimberlites compared with the average composition of three basaltic magma types. In *Kimberlites, Diatremes, and Diamonds: Their Geology, Petrology and Geochemistry*, ed. Boyd, F. R. Proc. Int.

Weidner, D. J. and Ito, E. (1987) Mineral physics constraints on a uniform mantle composition. In *High-Pressure Research in Mineral Physics*, Geophysical Monograph 39, eds. Manghnani, M. H., and Syono, Y. American Geophysical Union, Washington, D.C., pp. 439–446.

Weissman, P. R. (1985) The origin of comets: implications for planetary formation. In *Protostars and Planets II*, eds. Black, D. C., and Matthews, M. S. University of Arizona Press, Tucson, Ariz.

Wells, M. C., Boothe, P. N., and Presley, B. J. (1988) Iridium in marine organisms. Geochim. Cosmochim. Acta 52, 1737–1739.

Wen, X., De Carlo, E. H., and Li, Y. H. (1997) Interelement relationships in ferromanganese crusts from the central Pacific ocean: their implications for crust genesis. Mar. Geol. 136, 277–297.

West, H. B., and Leeman, W. P. (1987) Isotopic evolution of lavas from Haleakala Crater, Hawaii. Earth Planet. Sci. Lett. 84, 211–225.

Whitfield, M. (1979) Activity coefficients in natural water. In *Activity Coefficients in Electrolyte Solution*, vol. 2, ed. Pytkowics, M. CRC Press, Boca Raton, Fla., pp. 153–300.

Whitfield, M., and Turner, D. R. (1987) The role of particles in regulating the composition of seawater. In *Aquatic Surface Chemistry*. ed. Stumm, W. Wiley, New York.

WHO/IAEA (1989) *Minor and Trace Elements in Breast Milk*. Report of a Joint WHO/IAEA Collaborative Study. World Health Organization, Geneva.

Widom, E., and Shirey, S. B. (1996) Os isotope systematics in the Azores: implications for mantle plume sources. Earth Planet. Sci. Lett. 142, 451–465.

Wiebe, P. H., Boyd, S. H., and Winget, C. (1976) Particulate matter sinking to the deep sea floor at 2000 m in the Tongue of the Ocean, Bahamas, with a description of a new sedimentation trap. J. Mar. Res. 34, 341–354.

Wiik, H. B. (1956) The chemical composition of some stony meteorites. Geochim. Cosmochim. Acta 9, 219–289.

Williams, Q., Knittle, E., and Jeanloz, R. (1989) Geophysical and crystal chemical significance of (Mg, Fe)SiO_3 perovskite. In *Perovskite: A Structure of Great Interest to Geophysics and Materials Science*, Geophysical Monograph 45, ed. Navrotsky, A., and Weidner, D. J. American Geophysical Union, Washington, D.C., pp. 1–10.

Wilson, M. (1989) *Igneous Petrogenesis—a Global Tectonic Approach*. Unwin Hyman, London.

Wolf, R., Ebihara, M., Richter, G. R., and Anders, E. (1983) Aubrites and diogenites: Trace element clues to their origin. Geochim. Cosmochim. Acta 47, 2257–2270.

Wolff, E. W., and Peel, D. A. (1985) The record of global pollution in polar snow and ice. Nature 313, 535–540.

Wood, B. (1993) Carbon in the core. Earth Planet. Sci. Lett. 117, 593–607.

Woolum, D. S., Cochrane, R. B., Joyce, D., Goresy, A. E., Benjamin, T. M., Rogers, P. S. Z., Maggiore, C. M., and Duffy, C. J. (1983) Trace element Pixe, studies of Qinzhen (EH3) metal and sulfides. Lunar Planet. Sci. 9, 935–936.

Woosley, S. E., and Weaver, T. A. (1986a) The physics of supernova explosions. Annu. Rev. Astron. Astrophys. 24, 205–253.

———. (1986b) Theoretical models for Type I and Type II supernova. In *Nucleosynthesis and its Implications on Nuclear and Particle Physics*, NATO ASI series C, 169. Reidel, Boston, pp. 145–166.

———. (1989) The great supernova of 1987. Sci. Am. (August), 32–40.

Yamamoto, T. (1972) Chemical studies on the seaweeds (27): The relations between concentration factor in seaweeds and residence time of some elements in seawater. Rec. Oceanogr. Works Japan 11, 65–72.

———. (1983) *Distribution of Trace Elements in Marine Algae—Comparative Biological Data.* Kyoto University of Education, Kyoto, Japan.

Yamamoto, T., Otsuka, Y., Aoyama, K., Tabata, H., and Okamoto, K. (1985) The distribution of chemical elements in selected marine organisms: Comparative biological data. In *Marine and Estuarine Geochemistry*, eds. Sigleo, A. C. and Hattori, A. Lewis Publisher, Ann Arbor, Mich.

Yee, H. S., Measures, C. I., and Edmond, J. M. (1987) Selenium in the tributaries of the Orinoco in Venezuela. Nature 326, 686–689.

Yen, T. F. (1975) Chemical aspects of metals in native petroleum. In *The Role of Trace Metals in Petroleum*, ed. Yen, T. F. Ann Arbor Sci., Ann Arbor, Mich.

Yoder, H. S., Jr. (1976) *Generation of Basaltic Magma.* National Academy of Science, Washington, D.C.

Yoshihara, K., Kudo, H., and Sekine, T. (1985) *Periodic Table with Nuclides and Reference Data.* Springer-Verlag, New York.

Zindler, A. (1982) Nd and Sr isotopic studies of komatiites and related rocks. In *Komatiites*, eds. Arndt, N. T., and Nisbet, E. G. Allen and Unwin, London, pp. 399–420.

Zindler, A., and Hart, S. (1986) Chemical geodynamics. Annu. Rev. Earth Planet. Sci. 14, 493–571.

Zindler, A., Jagoutz, E., and Goldstein, S. (1982) Nd, Sr and Pb isotopic systematics in a three-component mantle: A new perspective. Nature 58, 519–523.

Zolensky, M. E., and McSween, H. Y., Jr. (1988) Aqueous alteration. In *Meteorites and the Early Solar System*, eds. Kerridge, J. F., and Matthews, M. S. University of Arizona Press, Tucson, Ariz.

Zolensky, M. E., Barrett, R., and Browning, L. (1993) Mineralogy and composition of matrix and chondrule rims in carbonaceous chondrites. Geochim. Cosmochim. Acta 57, 3123–3148.

Zreda-Gostynska, G., Kyle, P. R., Finnegan, D., and Prestbo, K. M. (1997) Volcanic gas emissions from Mount Erebus and their impact on the Antarctic environment. J. Geophys. Res. 102, B7, 15039–15055.

INDEX

Achondrites, 91
  classification of, tIV-1, 120; fIV-2, 125; fIV-3, 127
  composition of, tIV-4, 128; tIV-15, 180; fIV-21–fIV-24, 181–184
  eucrite-diogenite-pallasite, tIV-15, 180; fIV-21–fIV-23, 181–183
  oxygen isotopes of, fIV-5, 130; fIV-7, 132
  primitive, fIV-2, 125; fIV-3, 127; tIV-4, 128
  ureilite, tIV-15, 180; fIV-23, 183; fIV-24, 184
Acid dissociation constant, 49; fVI-19, 300
Activity vs. activity phase diagrams, 262; fVI-6, 263
Adsorption model of James and Healy, 299
  chemical bond energy, 299
  coulombic electrostatic energy, 299
  secondary solvation energy, 299
Aerosols, Arctic and Antarctic, 400; tVIII-11, 401; fVIII-20, 403
Affinity of aqueous cation to oxide, fI-21, 50; 52
Albedo of celestial bodies, tIII-1, 87; 92; tIII-4b, 101
Allende chondrite (CV3):
  bulk and chondrules, tIV-11, 159
  Calcium aluminum rich inclusions (CAI), fIV-17, 170; tIV-13, 173; fIV-18, 174
  Fremdlinge, 177; tIV-14, 178
  minerals, tIV-12, 163
  oxygen isotopes, fIV-6, 131; fIV-16, 166
  silicon isotopes, fIV-19, 175
Allende mixing line, 130; fIV-6, 131; fIV-7, 132 (*see also* Oxygen isotopes of meteorites)
Amoeboid olivine rich inclusion (AOI), 91; fIV-17, 170
Amor object, fIII-3, 89; 90
Aphelion, 84; fIII-1, 84
Apollo object, fIII-3, 89; 90

Note: t = Table, ta = Appendix Table, f = Figure

Archean:
  iron formation, 288; fVI-16, 292
  komatiite, 209
  shale, 280; tVI-7, 284
  upper and bulk crusts, tV-5, 227
Area-loading or area density, 104; fIII-8a, 105
Arkose or feldspathic arenite, 266; fVI-8, 266
Asteroids, 86
  albedo of, tIII-1, 87
  Amor and Apollo objects, fIII-3, 89; 90
  distribution of, fIII-2, 88; fIII-3, 89; tIII-2, 95
  Kirkwood gaps, 87; fIII-3, 89
  reflectance spectra of, 90; fIII-4, 92; fIII-5, 93; tIII-2, 95
  three largest, tIII-1, 87
Astronomical unit (AU), 84
Asymptotic giant branch, fII-6b, 69; 81
Atmophile elements, 58; fII-1, 58
Atomic mass units (amu), 13
Atomic nuclei, 13
  isobar, isotone, and isotope, fI-3, 14
  mass number (A), fI-4, 15; fI-5, 17; fII-3, 61; fII-4, 62
  shielding and shielded, fI-3, 14; fII-4, 62; fII-7, 72
Atomic number (Z), even and odd, 3; fI-1, 10; fI-7, 20; fI-9, 26; fI-12, 28; fI-18, 43; fII-2, 59; fII-7, 72
A-type cations and anions, 24; fI-12–fI-18, 28–43
Average binding energy of electron, 24; tI-3, 25; fI-11, 27

Bauxite, tVI-2, 259
Big Bang theory, 63; fII-5, 66
  big Chill, 68
  big Crunch, 68
  cosmic singularity, 63
  era of recombination or decoupling, fII-5, 66; 68
  grand unified theories, 64; fII-5, 66
  Planck time, 64; fII-5, 66
  symmetry breaking, 67
  theory of everything, 63; fII-5, 66

Binding energy of:
electron, 24; tI-3, 25; fI-11, 27; ta-1, 415
average, tI-3, 25
nucleus, 14; fI-4, 15
Biologically mediated reduction, 315
Biological standard reference materials, 355; fVIII-2, 357; tVIII-1, 358
aquatic plant (SRM8030), 358
bovine liver (SRM1577), 358
citrus leave (SRM1572), 358
coal (SRM1632), tVIII-5, 377
kale, 358
oyster (SRM1566), 358
pine needle (SRM1575), 358
spinach (SRM1570), 358
tomato leave (SRM1573), 358
Biophile and biophobe elements, 321; fVII-6, 321; 323; fVII-17, 339; 362; fVIII-3, 363; fVIII-4, 364
Black body radiation, 86
Black hole, 81
Black shale, 386; tVIII-5, 377; fVIII-14, 386
Boiling point of:
element, tI-2, 21; fI-8, 22
oxide, fVIII-16, 393
Box model for ocean, fVII-3, 311
Brown dwarf, 81
β-stability valley, 17; fI-6, 18; fII-7, 72
B-type cations, 24; fI-12–fI-18, 28–43

Calc alkali series of igneous rocks, 193
Calcareous ooze, 329; tVII-5, 329; tVII-6, 331
Calcium aluminum rich inclusions (CAI), 91, 167
classification of, fIV-17, 170
compositions of, tIV-13, 173; fIV-18, 174
oxygen isotopes of, fIV-6, 131
silicon isotopes of, fIV-19, 175; fIV-20, 176
Carbonaceous chondrites, 55, 90
classification of, tIV-1, 120; tIV-3, 124
compositions of, tIV-6, 134
factor analysis, fIV-8, 142; fIV-9, 144
primitive (C1 or CI), tII-1, 56; tIV-6, 134
Carbonate, tVI-4, 269; fVI-10, 270; 283 (*see also* Limestone)
Carbon burning in stars, 73; fII-6, 69; tII-3, 78; ta-2, 418
Carbon reservoirs on Earth's surface, tVIII-2, 365
Chalcophile elements, 144; tIV-8, 145; fIV-10, 149; fIV-12, 153
Chandrasekhar limit, 75
Chemical bond energy, 299
Chemical speciations of elements in ocean, 312; tVII-2, 313; fVII-4, 316
Chemical weathering, 254; fVI-1–fVI-6, 255–263
Chondrites, 119
average compositions of, tIV-6,134
chondrules, tIV-9, 155; fIV-14, 161
classification of, tIV-1, 120; fIV-1, 123; tIV-3, 124; fIV-3, 127
factor analysis of, fIV-8, 142; fIV-9, 144
matrix and rim, 154; tIV-10, 157; fIV-13, 158; tIV-13, 173
mineral components, tIV-8a–tIV-8c, 145–148; tIV-11, 159; tIV-12, 163; tIV-13, 173
oxygen isotopes of, 129; fIV-5, 130; fIV-6, 131
unequilibrated, 122
Chondrules, 91, 158; fIV-14, 161
compositions of, tIV-11, 159; tIV-12, 163; tIV-13, 173
factor analysis of, fIV-15, 165
oxygen isotopes, fIV-16, 166
silicon isotopes, fIV-19, 175; fIV-20, 176
types, tIV-9, 155; fIV-14, 161
Classification scheme for:
asteroids and comets, 86; fIII-3, 89; 94; tIII-2, 95; fIII-6, 96
concentration profiles of elements in oceans, tVII-1, 304; 308; fVII-2, 310; fVII-16a, 337
elements, tI-1, 4; tIV-7, 139; tIV-8, 145; fIV-10, 149; fIV-12, 153
igneous rocks, fV-1–fV-3, 190–192
meteorites, 119; fIV-1–fIV-4, 123–127
sedimentary rocks, tVI-3, 264
sandstone and mudstone, fVI-8, 266
Coal, 376; tVIII-5, 377
different fractions, tVIII-6, 381
factor analysis of, fVIII-12, 382
fly ash, slag, and vapor, tVIII-8, 388; fVIII-15, 391; fVIII-17, 396
thermodynamic model for combustion of, fVIII-16, 393; tVIII-9, 394; fVIII-17, 396; fVIII-18, 398

Comets, 94
  distribution of, fIII-3, 89; fIII-6, 96
  Halley type, 95; fIII-6, 96; tIII-3, 98
  Jupiter family, 95; fIII-3, 89; fIII-6, 96
  Kuiper belt, 97
  Oort cloud comets, 95; fIII-6, 96
Compatible and incompatible elements, 209
Complexation constants (first) of cations with:
  carbonate, chloride, and hydroxyl ions, fVII-4, 316
  EDTA, fVII-9, 324
Composition of:
  achondrites: eucrite (Juvinas), diogenite (Johnstown), ureilite (Kenna), tIV-15, 180
  Allende (CV3), Semarkona (LL3), and Qingzhen(EH3):
    bulk, chondrule, minerals, tIV-11, 159; tIV-12, 163
  Archean bulk and upper crusts, tV-5, 227
  basalt, tV-4a, 222
  bauxite, tVI-2, 259
  biological standard reference materials, tVIII-1, 358
  black shale, tVIII-5, 377
  calcareous ooze, tVII-5, 329; tVII-6, 331
  chondrites, tIV-6, 134
  coal, tVIII-5, 377
    different fractions, tVIII-6, 381
    fly ash and slag, tVIII-8, 388
  continental crust:
    bulk, tV-5, 227
    upper, tV-4a, 222; tVI-5a, 274; fVI-12, 278
  continental sediments, average, tVI-4, 269
  cratonic shales (Archean, Proterozoic, and Phanerozoic), tVI-7, 284
  crude oil, tVIII-5, 377
    different fractions, tVIII-7, 385
  evaporites, tVI-4, 269
  fecal pellets of euphausiid, tVII-4a, 326
  fly ash and slag of burnt coal, tVIII-8, 388
  Fremdlinge from Allende, tIV-14, 178
  global average sediments, tVI-4, 269
  granite, tV-4a, 222
  Halley comet, particles, tIII-3, 98
  hemipelagic mud, tVI-5a, 274
  human blood, brain, kidney, liver, lung, lymph, muscle, etc., tVIII-3b, 367
  human skeleton, average diet, urine, milk, tVIII-4, 371
  hydrothermal vent solution, tVII-7, 349
  interplanetary dust particles (IDP), tIII-3, 98
  iron formation (IF), tVI-9, 290
  island arc andesite, tV-5, 227
  juvenile upper crusts (Archean, Proterozoic, and Phanerozoic), tV-4b, 226
  limestone or carbonate, tVI-4, 269; tVI-8, 287
  loess, tVI-5a, 274
  mangenese nodule and crust, tVII-6, 331
  marine algae, plankton, tVII-3, 318
  marine or oceanic sediments, average, tVII-5, 329
  matrix+rim of carbonaceous chondrites, tIV-10, 157
  metals and sulfides in meteorites, tIV-8a, 145; tIV-8b, 147; tIV-8c, 148
  mica schist, tVI-5a, 274
  mid-ocean ridge basalt (MORB), tV-5, 227
  minerals in Modoc chondrite (L6), tIV-8a, 145
  oceanic crust, tV-5, 227
  oil shale, tVIII-5, 377
  pelagic clay, tVII-5, 329; tVII-6, 331
  petroleum, tVIII-5, 377
  phosphorite, tVII-6, 331
  Pierre Shale Member, tVI-6, 281
  primitive upper mantle, tV-2, 212
  reference man, tVIII-4, 371
  river mud, tVI-5a, 274
  river suspended particles, tVI-10, 294
  river water, tVI-10, 294
  sandstone, tVI-4, 269; tVI-8, 287
  saprolite, tVI-1, 257; tVI-2, 259
  seamount manganese crust, tVII-6, 331
  seawater, tVII-1, 304
  sediment trap materials, tVII-4c, 326
  shale, tVI-5a, 274
  siliceous ooze, tVII-5, 329; tVII-6, 331
  soil profiles, tVI-1, 257; tVI-2, 259
  soils, tVI-5a, 274
  solar atmosphere, tII-1, 56
  solar nebula, ta-2, 418
  type 1 carbonaceous chondrites (C1), tII-1, 56; tIV-6, 134

Concentration profiles of elements in ocean, tVII-1, 304; tVII-2, 313; fVII-2, 310
  conservative type, 304
  mixed type, 304
  nutrient type, 304
  pseudo-conservative type, 304
  scavenged type, 304
Concentration ratios of elements for:
  continental crust/primitive mantle, fV-15, 233
  gas/glass during burning of coal, tVIII-9, 394
  iron formation/crust, fVI-17, 293
  manganese nodules/seawater, fVII-21, 346
  marine algae/seawater, fVII-8, 323; fVII-17, 339
  mid oceanic ridge basalt/hydrothermal vent solution, fVII-24, 351
  pelagic clay/seawater, fVII-15–fVII-18, 336–340
  reference man/urine, fVIII-8, 375
  river particles/river water, tVI-10, 294; fVI-18, 297
  river water/seawater, fVII-1, 308; fVII-15, 336
Condensation temperature, 109; fIII-9, 111; tIII-5, 112; fIII-10, 114; fIII-11, 116
Congeners, 3
Conservative type of elements in ocean, 308; tVII-1, 304; tVII-2, 313; fVII-16b, 337
Continental crust, 221; tV-4, 222; tV-5, 227; fV-15, 233; fV-16a, 234; tVI-5a, 274; fVI-12, 278
Continental sediments, average, tVI-4, 269
Core collapse and bounce in supernova, tII-3, 78
Core formation, Earth, 215; tV-3, 218; fV-14, 220
Core helium burning, 71; fII-6, 69; tII-3, 78
Core hydrogen burning, 69; fII-6, 69
Correlation coefficients matrix, 141 (*see also* Factor analysis)
Correlation plots of elemental concentrations for:
  algae/fungi/angiosperms/mollusca/pisces, fVIII-1, 356
  algae/phytoplankton/zooplankton, fVII-5, 320; fVIII-1, 356
  algae/shale, fVII-6, 321
  C1/solar photosphere, fII-1, 58
  human liver/bovine liver, fVIII-10, 376
  human muscle/other organs, fVIII-5, 369
  pelagic clay/calcareous ooze/siliceous ooze, fVII-12, 328
  reference man/diet/spinach, fVIII-7, 374
  shale/coal, fVIII-11, 380
  shale/crude oil/petroleum, fVIII-13, 384
  shale/limestone/sandstone, fVI-10, 270; fVI-15, 288
  shale/upper crust, fVI-12, 278
  soils/buckbush/hickory/oak/pine, fVIII-4, 364
  spinach/soil, fVIII-3, 363
  upper crust/global average sediments, fVI-11, 271
Cosmic singularity, 63
Cosmochemical classification of elements, tIV-7, 139; tIV-8, 145; fIV-10, 149; fIV-12, 153
Cosmological nucleosynthesis, 63 (*see also* Big Bang Theory)
Coulombic electrostatic energy, 299
Crude oils, tVIII-5, 377; 383
  different fractions, tVIII-7, 385
Crust:
  Archean, bulk and upper, tV-5, 227
  continental, bulk and upper, tV-4, 222; tV-5, 227; tVI-5a, 274; fVI-12, 278
  differentiation from mantle, fV-15–fV-20, 233–243
  formation rate of new oceanic crust, 351
  juvenile upper (Archean, Proterozoic, and Phanerozoic), tV-4b, 226
  oceanic, tV-5, 227
Crystal field splitting parameter, 28, tI-5, 30
Crystal lattice energy, 44; tI-10, 46; fI-20, 47
Cumulative ionization energy, 23

Degeneracy of electrons and neutrons, 75
Depleted mantle (DM1), 232; fV-16–fV-19, 234–241
Diogenite, 179 (*see* Achondrites)
Dissociation of oxyacid, 48, 299; fI-23, 52; fVI-19, 300
Dissolved products of chemical weathering, 260; fVI-4–fVI-6, 261–263
Distribution or partition coefficient of elements, 198, 201
  bulk (igneous rock/melt), 198; fV-8, 203; fV-10, 208; tV-3, 218; fV-17, 235

mineral (mineral/melt), 199; tV-1, 204; fV-9, 206
pelagic clay/seawater, fVII-18, 340
D" region, 193
Dry and wet precipitations, 405; fVIII-22, 406; fVIII-23, 407

Earth:
core formation, 215; tV-3, 218; fV-14, 220
density and velocity, fV-4, 194
heterogeneous accretion model of, 218
isotopic ratios of daughter-parent pairs, tV-6, 238
orbital and physical parameters, tIII-4a, 100
partition of elements between mantle and core, tV-3, 218; fV-14, 220
structure and mineral composition of, fV-4–fV-6, 194–196
two stage evolution model of, 234; fV-18, 240
Eccentricity, 83; fIII-2, 88; fIII-3, 89
Ecliptic plane, 85
Eigenvalues, 141 (*see also* Factor analysis)
Electric polarizability of ions, 35; tI-6, 36; fI-16, 37; relative, tI-7, 38
Electron affinity, 23; ta-1, 415
Electron binding energy, 24; tI-3, 25; fI-11, 27 (*see also* Binding energy of, electron; Ionization energy or electron binding energy)
Electron configuration of atom, tI-1b, 5; 10; fI-2, 11; high- and low-spin, 30; tI-5, 30
Electronegativity, 39
cationic, tI-8, 40; tI-9, 41; 42; fI-17, 42; fI-18, 43; fI-19, 44
nonmetal ions, tI-8, 40
total cationic, 41; fI-21, 50
Enrichment factor of elements in, 133
black shale, fVIII-14a, 386
CAI, fIV-18, 174
chondrites, tIV-7, 139
chondrules, tIV-11, 159
cratonic shales, tVI-7, 284
fecal pellets, fVII-10, 327
iron formations, fVI-16, 292
manganese nodules and crusts, fVII-19, 342
matrix and rim of chondrites, tIV-10, 157
oil shale, fVIII-14b, 386
pelagic clay, fVII-13, 335
phosphorites, fVII-23, 347
polar aerosol, tVIII-11, 401; fVIII-20, 403
primitive upper mantle, tV-2, 212; fV-13, 216
sediment trap materials fVII-10, 327
soil, shale and related materials, tVI-5b, 280
volcanic ash, gas, plum, tVIII-10, 397; fVIII-18, 398; fVIII-19, 399
Enstatite chondrites (EC), 91; tIV-1, 120; tIV-6, 134; tIV-8c, 148; tIV-11, 159; tIV-12, 163
Enstatite meteorite mixing line, 129; fIV-5, 130
Enthalpy of sublimation, 20 (*see also* Sublimation heat of)
Equilibrium condensation models, 108
condensation temperature, fIII-9, 111; fIII-10, 114; fIII-11, 116
50% condensation temperature, tIII-5, 112
Era of recombination or decoupling, fII-5, 66; 68
Escape velocity, 85; tIII-4a, 100
Eucrite, 179; tIV-15, 180; fIV-21–fIV-23, 181–183 (*see also* Achondrites)
Evaporite, tVI-3, 264; 268; tVI-4, 269
Excess volatiles, 273 (*see also* Primary magmatic volatiles)
Explosive nucleosynthesis, 78; fII-10, 80; ta-2, 418
Extent of leaching in hydrothermal vents, tVII-7, 349; 350
Extra-depleted mantle (DM2), 232; fV-16, 234

Factor analysis, 141
correlation coefficient matrix, 141
eigenvalue, 141
factor loading, 141
factor score, 141
principal component, 141
varimax, 141
Factor analysis of
Arctic aerosol, fVIII-21, 404
coal, fVIII-12, 382
chondrites, fIV-8, 142; fIV-9, 144
chondrules, fIV-15, 165
manganese crusts, seamount, fVII-20, 345
marine algae, fVII-7, 322

Factor analysis of (*cont.*)
  matrix and rim of chondrites, fIV-13, 158
  mid Atlantic ridge basalts, fV-20, 243
  oceanic island basalts, fV-23, 247
  sediments and nodules, Pacific, fVII-14, 335
  shale, Pierre Shale Members, fVI-13, 282
  spinel lherzolite xenoliths, fV-12, 215
  two mica granites, fV-25, 250
Fecal pellets, 325; tVII-4a, 326; fVII-10, 327
Fly ash and slag of burnt coal, 387; tVIII-8, 388; fVIII-15, 391
Fossil fuel burning, 376 (*see also* Coal)
  effects on rain and river water, fVIII-22, 406; fVIII-23, 407
  on polar aerosol, fVIII-20, 403
Fractional crystallization models for igneous magma, 201; fV-7, 201; fV-8, 203
Fractionation factor, 19
Fremdlinge, 177; tIV-14, 178 (*see also* Allende chondrite)
FUN (Fractionation and Unknown Nuclear effects) inclusions, 130; fIV-6, 131; fIV-19, 175
Fundamental forces and particles in the Universe, tII-2, 64

Galactic clusters, 70
Geostandards:
  basalt (W1), tV-4a, 222
  black shale (SDO-1), tVIII-5, 377
  calcareous ooze (OOPE-401), tVII-6, 331
  coal (SRM-1632), tVIII-5, 377
  granite (GA), tV-4a, 222
  hemipelagic mud (MAG-1), tVI-5a, 274
  iron formation (IF-G), tVI-9, 290
  limestone (GSR-6), tVI-8, 287
  loess (GSS-8), tVI-5a, 274
  manganese nodule (OOPE-601), tVII-6, 331
  mica schist (SDC-1), tVI-5a, 274
  oil shale (SGR-1), tVIII-5, 377
  pelagic clay (OOPE-501), tVII-6, 331
  river mud (GSD-9), tVI-5a, 274
  sandstone (GSR-4), tVI-8, 287
  shale (SCO-1), tVI-5a, 274
  siliceous silt (OOPE-402), tVII-6, 331
  soils (SO-4), tVI-5a, 274
Glacial till, 268
Global average sediments, tVI-4, 269; fVI-11, 271
Global vent water flux, 351
Globular cluster, 71 (*see also* Stars)
Grand unified theories, 64; fII-5, 66
Granites, fV-1, 190; tV-4a, 222
Graywackes or wackes, fVI-8, 266

Halley-type comets, 95; fIII-6, 96; tIII-3, 98
Harker diagrams for:
  Mid-Atlantic ridge basalts, fV-20, 243; fV-21, 244; fV-22, 246
  oceanic island basalts, fV-24, 248
  two mica granites, fV-26, 251
Helium burning, 71; tII-3, 78; ta-2, 418
Helium core flash, fII-6, 69; 81
Hemipelagic mud, 273; tVI-5a, 274
Hertzsprung-Russell diagram of stars, fII-6, 69
Heterogeneous accretion model of the Earth, 218
Human body, 366; tVIII-4, 371
  diet, fVIII-7, 374
  milk, 370
  organs (blood, brain, kidney, liver, lung, lymph, muscle), tVIII-3, 366; fVIII-5, 369
  reference man, fVIII-7, 374
  residence time of elements in, fVIII-8, 375
  skeleton and total soft tissue, fVIII-6, 373
  urine, 370; fVIII-8, 375
Hydrogen burning in stars, 69; tII-3, 78
Hydrolysis constant of cation, first, 48; tI-11, 49; fI-21, 50; fI-22, 51
Hydrothermal vents of mid ocean ridge, 348; tVII-7, 349; fVII-24, 351; fVII-25, 352

Igneous magmatic differentiation:
  achondrites and iron meteorites, 179
  fractional crystallization models, 201; fV-7, 201; fV-8, 203
  partial melting models, 198; fV-7, 201; fV-16, 234; fV-17, 235
Igneous rocks:
  alkali magma series, fV-3, 192; 193
  basalt, fV-1, 190; fV-3, 192
  calc alkali series, fV-3, 192; 193
  classification of, fV-1–fV-3, 190–192
  granite, fV-1, 190; tV-4a, 222

komatiite, 192
mid ocean ridge basalts (MORB), fV-18, 240; fV-19–fV-22, 241–246
oceanic island basalts (OIB), fV-22, 246; fV-23, 247; fV-24, 248
picrite, 192
spinel lherzolite xenolith, fV-11, 210; fV-12, 215
subalkali magma series, 193
tholeiitic series, 193
two mica granites, fV-25, 250; fV-26, 251
ultramafic rocks, fV-2, 191
Inclination, orbital and spin, 85; fIII-6, 96; tIII-4a, 100
Interplanetary dust particles (IDP), 98; tIII-3, 98; fIV-6, 131
Intrinsic acidity and stability constants, 297; fVI-19, 300
Ionic potential, 32; fI-14, 33
Ionic radii, 28; tI-4, 29; fI-13, 31; fI-15, 34
Ionization energy or electron binding energy, 23; fI-9, 26; fI-10, 27; fI-11, 27; ta-1, 415
Ionization potential (*see* Ionization energy or electron binding energy)
Iron formation (IF), tVI-3, 264; 288; tVI-9, 290; fVI-16, 292; fVI-17, 293
Iron meteorites, 91, tIV-2, 122; tIV-5, 128; fIV-4, 127; 185; fIV-25, 186
Island arc andesite, tV-5, 227
Isotopic heterogeneity of the mantle, 234; fV-19, 241; fV-22, 246

Jupiter family of comets, 95; fIII-6, 96

Kirkwood gaps, 87; fIII-3, 89
Komatiite, 192
Kuiper belt, 97

Lanthanide and actinide contractions, 30; fI-13, 31
Limestone, tVI-4, 269; tVI-8, 287; fVI-15b, 288
Lithophile elements, tIV-7, 139; 144; tIV-8, 145; fIV-10, 149; fIV-12, 153
Loess, tVI-5a, 274; 279

Magic numbers of nuclides, 16
Main sequence stars, 69; fII-6, 69
Major carbon reservoirs, 365; tVIII-3a, 366
Manganese nodule and crust, 330; tVII-6, 331; fVII-20, 345; fVII-21, 346
Mantle, 231
depleted, fV-16–fV-19, 234–241
extra depleted, fV-16, 234
isotopic heterogeneity of, 234; fV-19, 241; fV-22, 246
partition of elements between mantle and core, tV-3, 218
primitive upper, tV-2, 212; fV-13, 216
Marine algae and plankton, 317; tVII-3, 318; fVII-5, 320; fVII-7, 322; fVII-8, 323; fVII-17, 339
Marine sediments, 329
average composition of, tVII-5, 329; tVII-6, 331
calcareous ooze, fVII-12, 328
hemipelagic mud, tVI-5a, 274
manganese nodule and crusts, fVII-14, 335; 341; fVII-19–fVII-22, 342–346
pelagic clay, fVII-12–fVII-18, 328–340
phosphorite, 347; fVII-23, 347
siliceous ooze, fVII-12, 328
Mars- and Earth-crossing orbits, 87; fIII-3, 89
Mass balance, igneous rocks and weathered products, 268
Mass dependent fractionation of isotopes, 129 (*see also* Terrestrial fractionation)
Mass number, even and odd, 13; fI-4, 15; fI-5, 17; fII-3, 61; fII-4, 62
Matrix+rim of carbonaceous chondrites, 154; tIV-10, 157; fIV-13, 158
compositions of, tIV-10, 157; tIV-13, 173
factor analysis, fIV-13, 158; fIV-15, 165
Mean residence time of elements, in:
ocean, 303; fVII-15, 336
human body, tVIII-4, 371; fVIII-8, 375
Metal ion activated enzyme, 365
Metallic bond, 19
Metalloenzyme, 365
Metalloids, tI-1a, 4; 12
Metals, tI-1a, 4; 12
Metals and sulfides in meteorites, tIV-8a, 145; tIV-8b, 147; tIV-8c, 148
Metasomatic fluid, 214
Meteorites, 119
achondrites, tIV-1, 120; tIV-4, 128
chondrites, tIV-1, 120; tIV-3, 124; tIV-4, 128
iron meteorites, tIV-2, 122; fIV-4, 127

Meteorites (*cont.*)
statistics, tIV-5, 128
stony-irons meteorites, tIV-1, 120
Mica schist, tVI-5a, 274
Mid-ocean ridge basalts (MORB), tV-5, 227; 242; fV-20, 243; fVII-24, 351
Mixed type of elements in ocean, tVII-1, 304; 309; fVII-2, 310; tVII-2, 313; fVII-16a, 337
Moderately volatile lithophiles and siderophiles, 138; tIV-7, 139
Molar volumes of elements, fI-1, 10
Mudstone, tVI-3, 264; 266; fVI-8, 266

Neon burning in stars, 73; tII-3, 78
Neutronized core, fII-10, 80
Neutron star, 77
Non-metals, tI-1, 4; 12
Normalized planetary masses, 103
Nuclear statistical equilibrium, 75
Nucleosynthesis:
cosmological, 63 (*see also* Big Bang Theory)
stellar, 68 (*see also* Stellar Nucleosynthesis)
Nutrient type of elements in oceans, 309; tVII-1, 304; tVII-2, 313; fVII-2, 310; fVII-16b, 337

Oceanic crust, 221, 232 (*see also* Mid-ocean ridge basalts)
Oceanic island basalt (OIB), fV-19, 241; 245; fV-23, 247; fV-24, 248
Oceanic pelagic clay, 329 (*see also* Pelagic clay)
Oceanic sediments, average, tVI-4, 269
Oil shale, tVIII-5, 377; 387
Oort cloud comets, 95; fIII-6, 96
Opal, tVII-4b, 326; fVII-11, 328
Orbit of celestial body, 83; fIII-1, 84
aphelion, 84
eccentricity, 83; fIII-2, 88; fIII-3, 89
inclination, 85; fIII-2, 88; fIII-6, 96
perihelion, 84
semimajor and semiminor axes, 83; fIII-3, 89; fIII-6, 96
Ordinary chondrite (OC), 91 (*see also* Chondrites)
Ordinary chondrite mixing line, 129; fIV-5, 130
Organic rich shale, 386 (*see also* Black shale; Oil shale)
Oxidative uptake, 338
Oxygen burning in stars, 74; tII-3, 78
Oxygen isotopes of meteorites, 129
Allende mixing line, fIV-6, 131; fIV-7, 132
chondrules, 164, fIV-16, 166
ordinary chondrite and enstatite meteorite mixing lines, fIV-5, 130
terrestrial fractionation line or mass dependent fractionation line, fIV-5, 130

Partial melting model for igneous rocks, 197; fV-7, 202; fV-16, 235; fV-17, 236
Partition of elements between mantle and core, 215; tV-3, 218
Pelagic clay, fVII-12–fVII-18, 328–340; 329; tVII-5, 329; tVII-6, 331
Perihelion, 84
Periodic table, tI-1, 4; fIV-10, 149
pH, fVI-20, 301
pH of zero point of charge (pH_{zpc}), 297
Phanerozoic upper continental crust and shale, tV-4b, 226; tVI-7, 284
Phosphorites, marine, 330; tVII-6, 331; fVII-23, 347
Physical weathering, 254 (*see also* Weathering)
Picritic rock, 192
Planck time, 64; fII-5, 66
Planets of solar system, 99
area loading, or area-density, fIII-8, 105; 140
condensation temperature, fIII-11, 116
formation temperature, fIII-8, 105
inner and outer planets, 99; tIII-4a, 100
normalized planetary mass, 103
orbital and physical parameters, tIII-4a, 100
satellites and rings, tIII-4b, 101
surface temperature, fIII-8, 105
Polarizing power of ion, 24; fI-12, 28
Population I and II stars, 70
Post-Archean shales and crusts, tVI-7, 284; fVI-14, 286
Primary magmatic volatiles, 253
Primitive mantle (PM) or primitive upper mantle, tV-2, 212; fV-13, 216; 232; fV-15, 233; fV-16, 234; fV-18, 240
p (proton)-process, 80; fII-4, 62; ta-2, 418

Proterozoic upper continental crust and shale, tV-4b, 226; tVI-7, 284
Pseudo-conservative type of elements in ocean, tVII-1, 304; 309; tVII-2, 313; fVII-16b, 337

QAPF triangle diagram, 189; fV-1, 190
Quantum number, 10
  angular momentum, 10
  magnetic, 10
  principal, 10
  spin, 10

Radius and mass relationship of celestial objects, fIII-7, 103
Rain, 405
  distribution of major ions over United States, fVIII-22, 406
  fluxes of major ions, fVIII-23, 407; tVIII-12, 409
r(rapid)-process, 78; fII-4, 62; fII-7, 72; ta-2, 418
Rare earth elements (REE):
  Archean shale and crust/chondrites, fVI-14, 286
  distribution coefficients in igneous rocks, fV-9b, 207
  hydrothermal vents/upper crust, fVII-25, 352
  manganese nodule and crust/seawater, fVII-22, 346
  pelagic clay/seawater, fVII-22, 346
  Phanerozoic shale and crust, fVII-14, 335
  river particles/river water, fVII-22, 346
  river water/upper crust, fVII-25, 352
  seawater/upper crust, fVII-25, 352
Rayleigh evaporation and condensation models for silicon isotopes, 175, fIV-19, 175; fIV-20, 176; 413
Rayleigh partial melting, 187
Red clay, 329 (*see also* Pelagic clay)
Redfield ratios, 320
Red giant, 69; fII-6, 69
Red giant branch, fII-6, 69; 81
Red supergiant, 69, fII-6, 69; 73
Reference man, 366
  composition of, tVIII-4, 371
  dry and wet weight of organs, tVIII-3a, 366
  mean residence time of elements in, tVIII-4, 371; fVIII-8, 375
Reflectance spectra of asteroids and meteorites, 90; fIII-4, 92; fIII-5, 93
Refractory lithophiles and siderophiles, 138; tIV-7, 139
Relative mobility of elements during weathering, 256; fVI-2, 256; fVI-3, 258
Relative polarizability of ions, 35; tI-7, 38
Relative reactivity of elements in ocean, 308 (*see also* Mean residence time of elements, in)
Relative volatility of elements, tVIII-8, 388; fVIII-15, 391; fVIII-16, 393; tVIII-9, 394
Rings of outer planets, 106
River:
  Amazon and Orinoco, fVI-20, 301
  average compositions of suspended particles and water, tVI-10, 294; fVI-18, 297
  Mississippi, fVIII-23, 407; tVIII-12, 409
  riverwater/seawater, fVII-1, 308; fVII-15, 336
  Yangtze river mud, tVI-5a, 274
Roche limit, 107
Rock forming minerals of meteorites, ta-3, 424; of igneous rocks, ta-4, 426

Sandstone, tVI-4, 269; 283; fVI-10, 270; tVI-8, 287
Saprolite, tVI-1, 257; tVI-2, 259; fVI-3, 258
Satellites, tIII-4b, 101; 104
Scavenged type of elements in ocean, tVII-1, 304; 309; fVII-2, 310; tVII-2, 313; fVII-16b, 337
Seamount manganese crust, 330; tVII-6, 331 (*see also* Manganese crust)
Seawater, 303
  composition of, tVII-1, 304
  speciation of, tVII-2, 313; fVII-4, 316; fVII-16b, 337
Secondary minerals, 254; fVI-1, 255
Secondary solvation energy, 299
Sedimentary rocks, 264
  carbonate or limestone, tVI-4, 269; fVI-10, 270; tVI-8, 287
  classification of, tVI-3, 264; fVI-8, 266
  evaporites, tVI-4, 269
  grain size distribution, tVI-3, 264; fVI-7, 265; fVI-9, 267
  iron formations (IF), tVI-9, 290

Sedimentary rocks (*cont.*)
relative abundance of, 268
sandstone, 266; tVI-4, 269; fVI-10, 270; tVI-8, 287
shale, tVI-5a, 274; fVI-12–fVI-15, 278–288
turbidite deposits, 266
Sediment trap materials, 325
composition of, tVII-4c, 326
fecal pellets, tVII-4a, 326
vertical fluxes of, tVII-4b, 326; fVII-11, 328
Separation energy of neutron, fI-5, 17
Shale, 268, 273; tVI-5a, 274; fVI-12–fVI-15, 278–288
Archean, Proterozoic, and Phanerozoic, tVI-7, 284
black shales, fVIII-5, 369
geostandard (SCO-1), tVI-5a, 274
oil shales, tVIII-5, 377
Pierre shale member, tVI-6, 281; fVI-13, 282
Shielding and shielded nuclei, fI-3, 14; 17; fII-4, 62; fII-7, 72
Siderophile elements, tIV-7, 139; 144; tIV-8a, 145; fIV-10, 149; fIV-12, 153
Silica oversaturated and undersaturated, 192
Siliceous ooze, fVII-12, 328; 329; tVII-5, 329; tVII-6, 331
Silicon burning in stars, 75; fII-8, 76; tII-3, 78
Silicon isotopes, Allende, fIV-19, 175
s(slow)-process, 62; fII-4, 72; fII-7, 72; ta-2, 418
Skeletal bone, human, tVIII-3a, 366; tVIII-4, 371
Soil profiles, fVI-1, 255; fVI-2, 256; tVI-1, 257; fVI-3, 258; tVI-2, 259
Soils, tVI-5a, 274; 279
Solar atmosphere or photosphere, 55; tII-1, 56
Solar mass, fII-9, 77; fII-10, 80
Solar nebula, 107; ta-2, 418
abundance of elements and isotopes, tII-1, 56; fII-2, 59; fII-3, 61; fII-4, 62
Solubility product, 52
Spinel lherzolite xenolith, 209; fV-11a, 210; fV-12, 215
Spin inclination, 85; tIII-4a, 100
Stability sequence of minerals during chemical weathering, 254
Stable nuclei, fI-3, 14
Standard mean ocean water (SMOW), 18
Standard reference materials (SRM), 354 (*see also* Biological standard reference materials; Geostandards)
Stars:
asymptotic giant branch, fII-6, 69
black hole, 81
brown dwarf, 81
Chandrasekhar limit, 75
degeneracy state of electrons and neutrons, 75
evolution of, fII-6, 69
explosion of, tII-3, 78; fII-10, 80
Hertzsprung-Russell diagram, fII-6, 69
galactic cluster, 70
globular cluster, 71
main sequence stars, fII-6, 69
neutron star, 77
population I and II stars, 70
red giant, fII-6, 69
red giant branch, fII-6, 69
red supergiant, fII-6, 69; 73
supernova, fII-6, 69; 77; fII-10, 80
white dwarf, fII-6, 69
Stellar nucleosynthesis, 68
carbon burning, shell and core, fII-6, 69; tII-3, 78
CNO cycle, 70
explosive oxygen (neon) and silicon burnings, 79; ta-2, 418
helium burning, shell and core, 69; fII-6, 69; tII-3, 78
hydrogen burning, shell and core, fII-6, 69; tII-3, 78
Ne-Na cycle, 70
neon burning, tII-3, 78
nuclear statistical equilibrium, 75; fII-9, 77
oxygen burning, tII-3, 78
p-process, fII-4, 62
r-process, fII-4, 62; fII-7, 72
silicon burning, 75; fII-8, 76
s-process, fII-4, 62; fII-7, 72
stages of star evolution, tII-3, 78
x-process, 81
Stony iron meteorites, 91
Subalkali magma series, 193
Subgraywacke or lithic arenite, 266; fVI-8, 266

Sublimation heat of:
element, fI-7, 20; tI-2, 21; fI-8, 22
oxide, fVIII-16, 393
Sun:
orbital and physical parameters, tIII-4a, 100
solar atmosphere or photosphere, tII-1, 56
Supernova, 77
Surface complexation model, 296
Symmetry breaking, 67

Terrestrial fractionation line, 129; fIV-5, 130; fIV-6, 131
Theory of Everything, 63; fII-5, 66
Thermal pulses (TP), fII-6, 69; 81
Thermodynamic model for:
chemical speciation of elements in seawater, tVII-2, 313; fVII-4, 316
classification of elements in ordinary chondrites, tIV-8, 145; fIV-12, 153
volatility of elements during coal burning, 392; fVIII-16, 393; tVIII-9, 394
Tholeiitic series of igneous rocks, 193
Threshold temperature of elementary particles, tII-2b, 65; 66
Total alkali silica (TAS) diagram, 192; fV-3, 192
Transition metal and its cations, fI-1, 10; tI-5, 30; fI-13, 31; fI-15, 34; tI-7, 38
T-Tauri wind, 108
Turbidite deposit, 266
Turbidity current, 266
Two mica (biotite-muscovite) granite, 249; fV-25, 250; fV-26, 251
Type 1 carbonaceous chondrite (C1), 55; tII-1, 56; 90; tIV-1, 120; tIV-6, 134 (*see also* Carbonaceous chondrite)
Type II supernova, 77

Ultramafic rocks, 191; fV-2, 191
Unequilibrated chondrites, 122
Ureilite, fIV-3, 127; fIV-5, 130; fIV-7, 132; tIV-15, 180; 183; fIV-23, 183; fIV-24, 184

Volatile elements in chondrites, 138; tIV-7, 139
Volcanic emissions, 396; fVIII-18, 398; fVIII-19, 399
Volcanic rocks, fV-1, 190; fV-3, 192 (*see also* Igneous rocks)
Volcanic tuffs, 268

Weak Van der Waals force, 19
Weathering, 254
chemical and physical, fVI-1–fVI-6, 255–263
dissolved products of, fVI-4–fVI-6, 261–263
relative mobility of elements, 256; fVI-2, 256; fVI-3, 258
secondary minerals, fVI-1, 255; fVI-2, 256
stability sequence of primary minerals, 254
White dwarf, 69; fII-6a, 69

x-process, 81; ta-2, 418

Zooplankton fecal pellets, 325; tVII-4a, 326; fVII-10, 327 (*see also* Fecal pellets)